Springer Undergraduate Texts in Mathematics and Technology

Springer Undergraduate Texts in Mathematics and Technology (SUMAT) publishes textbooks aimed primarily at the undergraduate. Each text is designed principally for students who are considering careers either in the mathematical sciences or in technology-based areas such as engineering, finance, information technology and computer science, bioscience and medicine, optimization or industry. Texts aim to be accessible introductions to a wide range of core mathematical disciplines and their practical, real-world applications; and are fashioned both for course use and for independent study.

More information about this series at https://link.springer.com/bookseries/7438

Heather A. Moon · Thomas J. Asaki ·
Marie A. Snipes

Application-Inspired
Linear Algebra

 Springer

Heather A. Moon
Department of Physics
and Astronomy
Washington State University
Pullman, WA, USA

Thomas J. Asaki
Department of Mathematics
and Statistics
Washington State University
Pullman, WA, USA

Marie A. Snipes
Department of Mathematics
and Statistics
Kenyon College
Gambier, OH, USA

ISSN 1867-5506   ISSN 1867-5514 (electronic)
Springer Undergraduate Texts in Mathematics and Technology
ISBN 978-3-030-86157-5   ISBN 978-3-030-86155-1 (eBook)
https://doi.org/10.1007/978-3-030-86155-1

Mathematics Subject Classification: 15-Axx

This Springer imprint is published by the registered company Springer Nature Switzerland AG
The registered company address is: Gewerbestrasse 11, 6330 Cham, Switzerland

To the curious learner, the creative thinkers, and to the explorers.

–Heather

To the curious student, the motivated reader, the independent person in each of us,
and to the amazing people who inspire them.

–Tom

To Ian

–Marie

Preface

We wrote this book with the aim of inspiring the reader to explore mathematics. Our goal is to provide opportunities for students to discover mathematical ideas *in the context of applications*. Before any formal mathematics, the text starts with two main data applications—radiography/tomography of images and heat diffusion—to inspire the creation and development of Linear Algebra concepts.

The applications are presented primarily through a sequence of explorations. Readers first learn about one aspect of a data application, and then, in an inquiry framework, they develop the mathematics necessary to investigate the application. After each exploration, the reader will see the standard definitions and theorems for a first-year Linear Algebra course, but with the added context of the applications.

A novel feature of this approach is that the applied problem *inspires* the mathematical theory, rather than the applied problem being presented after the relevant mathematics has been learned. Our goal is for students to organically experience the relevance and importance of the abstract ideas of linear algebra to real problems. We also want to give students a taste of research mathematics. Our explorations ask students to make conjectures and answer open-ended questions; we hope they demonstrate for students the living process of mathematical discovery.

Because of the application-inspired nature of the text, we created a path through introductory linear algebra material to naturally arise in the process of investigating two data applications. This led to a couple key content differences from many standard introductory linear algebra texts. First, we introduce vector spaces very early on as the appropriate settings for our problems. Second, we approach eigenvalue computations from an application/computation angle, offering a determinant-free method as well as the typical determinant method for calculating eigenvalues.

Although we have focused on two central applications that inspire the development of the linear algebra ideas in this text, there are a wide array of other applications and mathematical paths, many of which relate to data applications, that can be modeled with linear algebra. We have included "sign posts" for these applications and mathematical paths at moments where the reader has learned the necessary tools for exploring the application or mathematical path. These applications and mathematical paths are separated into three main areas: Data and Image Analysis (including Machine Learning), Dynamical Modeling, and Optimization and Optimal Design.

Outline of Text

In Chapter 1 we outline some of the fundamental ways that linear algebra is used in our world. We then introduce, with more depth, the applications of radiography/tomography, diffusion welding, and heat warping of images, which will frame our discussion of linear algebra concepts throughout the book.

Chapter 2 introduces systems of equations and vector spaces in the context of the applications. The chapter begins with an exploration (Section 2.1) of image data similar to what would be used for radiographs or the reconstructed images of brain slices. Motivated by a question about combining images, Section 2.2 outlines methods for solving systems of equations. (For more advanced courses, this chapter can be skipped.) In Section 2.3, we formalize properties of the set of images (images can be added, multiplied by scalars, etc.) and we use these properties to define a vector space. While Section 2.3 focuses on the vector spaces of images and Euclidean spaces, Section 2.4 introduces readers to a whole host of new vector spaces. Some of these (like polynomial spaces and matrix spaces) are standard, while other examples introduce vector spaces that arise in applications, including heat states, 7-bar LCD digits, and function spaces (including discretized function spaces). We conclude the chapter with a discussion of subspaces (Section 2.5), again motivated by the setting of images.

Chapter 3 delves into the fundamental ideas of linear combinations, span, and linear independence, and concludes with the development of bases and coordinate representations for vector spaces. Although the chapter does not contain any explorations, it is heavily motivated by explorations from the previous chapter. Specifically, the goal of determining if an image is an arithmetic combination of other images (from Section 2.1) drives the definition of linear combinations in Section 3.1, and also adds context to the abstract concepts of the span of a set of vectors (Section 3.2) and linear independence (Section 3.3). In Sections 3.4 and 3.5, we investigate how linearly independent spanning sets (bases) in the familiar spaces of images and heat states are useful for defining coordinates on those spaces. This allows us to match-up images and heat states with vectors in Euclidean spaces of the same dimension. We conclude the chapter with a "sign post" for regression analysis.

Chapter 4 covers linear transformations. In Section 4.1, readers are taken through an exploration of the radiographic transformation beginning with a definition of the transformation. Next, they use coordinates to represent this transformation with a matrix, and in Section 4.2, they investigate transformations more generally. In Section 4.3 readers see how the heat diffusion operator can be represented as a matrix, and in Section 4.4 they explore more generally how to represent arbitrary transformations between vector spaces by matrices of real numbers. In Section 4.6, the reader will return to the radiographic transformation and explore properties of the transformation, considering whether it is possible for two objects to produce the same radiograph, and whether there are any radiographs that are not produced by any objects. This exploration leads to the definitions of one-to-one and onto linear transformations in Section 4.7. This is also where the critical idea of invertibility is introduced; in the radiographic transformation setting, if the transformation is invertible then reconstruction (tomography) is possible.

The goal of Chapter 5 is to understand invertibility so that we can solve inverse problems. In Section 5.1 readers consider what would happen if the radiographic transformation is not invertible. This leads to a study of subspaces related to the transformations (nullspace and range space). The section concludes with the rank-nullity theorem. In Section 5.2, the corresponding ideas for matrix representations of transformations (nullspace, row space, and column space) are discussed along with the introduction of the Invertible Matrix Theorem. In Section 5.3 the reader will reconstruct brain slice images for certain radiographic setups after developing the concept of a left inverse. We conclude this chapter with a "sign post" for linear programming.

Chapter 6 introduces eigenvector and eigenvalue concepts in preparation for simplifying iterative processes. Section 6.1 revisits the heat diffusion application. In this exploration, readers examine a variety of initial heat states, and observe that some heat states have a simple evolution while others do not. Combining this with the linearity of the diffusion operator leads to the idea of creating bases of these simple heat states. Section 6.2 formalizes the ideas of the previous heat diffusion exploration and introduces eigenvectors, eigenvalues, eigenspaces, and diagonalization. Using these constructs, in Section 6.3 readers, again, address the long-term behavior of various heat states, and start to make connections to other applications. We follow the application with Section 6.4 where we present many more applications described by repeated matrix multiplication or matrix/vector sequences. Within this chapter are "sign posts" for Fourier analysis, nonlinear optimization and optimal design, and for dynamical processes.

Chapter 7 includes the discussion on how to find suitable solutions to inverse problems when invertibility is not an option. In Section 7.1, motivated by the idea of determining the "degree of linear independence" of a set of images, readers will be introduced to the concepts of inner product and norm in a vector space. This chapter also develops the theory of orthogonality. Section 7.2 uses orthogonality to define orthogonal projections in Euclidean space along with general projections. The tools built here are then used to construct the Gram-Schmidt Process for producing an orthonormal basis for a vector space. In Section 7.3, motivated by ideas from earlier tomography explorations, we develop orthogonal transformations and related properties of symmetric matrices. Section 7.4 is an exploration in which the reader will learn about the concepts of maximal isomorphisms and pseudo-invertibility. In Section 7.5, readers will combine their knowledge about diagonalizable and symmetric transformations and orthogonality to more efficiently invert a larger class of radiographic transformations using singular value decomposition (SVD). The Final exploration, in Section 7.6, makes use of SVD to perform brain reconstructions. Readers will discover that SVD works well with clean data, but poorly for noisy data. At the end of this section, readers explore ideas for nullspace enhancements to reconstruct brain images from noisy data. This final section is set up so that the reader can extend their knowledge of Linear Algebra in a grand finale exploration reaching into an active area of inverse problem research. Also included throughout Chapter 7 are "sign posts" for data analysis tools, including support vector machines, clustering, and principle component analysis.

Finally, Chapter 8 wraps up the text by describing the exploratory nature of applied mathematics and encourages the reader to continue using similar techniques on other problems.

Using This Text

The text is designed around a semester-long course. For a first course in Linear Algebra, we suggest including Chapters 1-6 with selected topics from Chapter 7 as time allows. Although the heat diffusion application is not fully resolved until Section 6.3 and the tomography application is not fully resolved until Section 7.6, one could reasonably conclude a 1-semester course after Section 6.2. At that point, some (relatively elementary) brain images have been reconstructed from radiographic data, a good exploration of Heat Diffusion has completed a study of Eigenvectors and Diagonalization, and tomography has motivated ideas that will lead to inner product, vector norm, projection, and the Gram-Schmidt Process. An outline from an example of our introductory courses is included on page xi.

Chapter 8 can be a great source of ideas for student projects. We encourage anyone using this text to consider applications discussed there.

This text has also been used for a more advanced second course in Linear Algebra. In this scenario, the instructor can move more rapidly through the first three chapters highlighting connections with the applications. The course could omit Sections 5.3 and 6.3 in order to have adequate time to complete the tomographic explorations in Chapter 7. Such a course could additionally include the derivation of the diffusion equation in Appendix B and/or a deeper understanding of radiographic transformations described in Appendix A.

Exercises

As mathematics is a subject best learned by doing, we have included exercises of a variety of types at many levels: concrete practice/computational, theoretical/proof-based, application-based, application-inspired/inquiry, and open-ended discussion exercises.

Computational Tools

One of the powerful aspects of Linear Algebra is its ability to solve large-scale problems arising in data analysis. We have designed our explorations to highlight this aspect of Linear Algebra. Many explorations include

Chapter	Sections	Title	# of (50-min) Classes
Ch 1		Introduction to Applications	1 Class
Ch 2		Vector Spaces	
	§2.1	Exploration: Digital Images	1 Class
	§2.2	Systems of Equations	2 Classes
	§2.3	Vector Spaces	1 Class
	§2.4	Vector Space Examples	1 Classes
	§2.5	Subspaces	2 Classes
Ch 3		Vector Space Arithmetic and Representations	
	§3.1	Linear Combinations	2 Classes
	§3.2	Span	2 Classes
	§3.3	Linear Independence	2 Classes
	§3.4	Bases	2 Classes
	§3.5	Coordinates	1 Class
Ch 4		Linear Transformations	
	§4.1	Exploration: Computing Radiographs	2 Classes
	§4.2	Linear Transformations	2 Classes
	§4.3	Exploration: Heat Diffusion	1 Class
	§4.4	Matrix Representations of Linear Transformations	
	§4.5	The Determinant of a Matrix	1 Class
	§4.6	Exploration: Tomography	1 Class
	§4.7	Transformation Properties (1-1 and Onto)	3 Classes
Ch 5		Invertibility	
	§5.1	Transformation Spaces	2 Classes
	§5.2	The Invertible Matrix Theorem	1 Class
	§5.3	Exploration: Tomography Without an Inverse	1 Class
Ch 6		Diagonalization	
	§6.1	Exploration: Heat State Evolution	1 Class
	§6.2	Eigenspaces	3 Classes
	§6.3	Exploration: Diffusion Welding	1 Class
	§6.4	Markov Processes	1 Class
Ch 7		Inner Product Spaces	
	§7.1	Inner Products	1 Class
	§7.2	Projections	1 Class

code for students to run in either MATLAB or the free, open-source software Octave. In most cases, the code can be run in online programming environments, eliminating the need for students to install software on their own computers.

Ancillary Materials

Readers using this text are invited to visit our website (www.imagemath.org) to access data sets and code for explorations. Instructors are able to create an account at our website so that they can download ancillary materials.

Materials available to instructors include all code and data sets for the explorations and instructor notes and expected solution paths for the explorations.

Pullman, USA   Heather A. Moon
Pullman, USA   Thomas J. Asaki
Gambier, USA   Marie A. Snipes

The original version of the book has been revised. A correction to this book can be found at https://doi.org/10.1007/978-3-030-86155-1_9

Acknowledgements

This book is an extension of the IMAGEMath collection of application-inspired modules, developed as a collaborative NSF project[1]. We are grateful for the financial support from the NSF that made the IMAGE-Math modules possible, and we are deeply grateful for the encouragement and mentorship of our program officer John Haddock for both the IMA-GEMath and this follow-on book project.

We are grateful to the Park City Mathematics Institute's Undergraduate Faculty Program, where we met and laid the groundwork for our friendship, this work, and many more collaborations.

We extend our deepest gratitude to Dr. Amanda Harsy Ramsay for her enthusiasm for this work, early adoption of IMAGEMath materials, and invaluable feedback on the explorations in this text.

To Chris Camfield, thank you for being part of our first project.

We thank all of our linear algebra students for their willingness to look at mathematics from a different perspective, for insightful discussions through this material, for finding our typos, and for inspiring us to find better ways to ask you questions.

The genesis of this text was the body of notes HM wrote for her 2014 linear algebra class at St. Mary's College of Maryland weaving together IMAGEMath explorations with linear algebra concepts. We particularly thank students in this class for their formative comments and feedback.

We thank Kenyon College for its hospitality and financial support for several visits that HM and TA made to Gambier, Ohio, and we thank the Mathematics and Statistics Department for its support and encouragement for developing and using these materials at Kenyon College.

We thank Washington State University Mathematics Department for its support and encouragement for developing, disseminating, and using these materials.

We also thank Lane and Shirley Hathaway, at Palouse Divide Lodge, for providing a venue with such beautiful views and great walks.

To Ben Levitt, thank you for your interest in this work.

To the reviewers, thank you for your insightful feedback.

To the editors, thank you for all your help in preparing this work.

---

[1]NSF-DUE 1503856, 1503929, 1504029, 1503870, and 1642095.

Contents

About the Authors

Heather A. Moon received her Ph.D. in Mathematics at Washington State University. She spent 8 years as an Assistant/Associate Professor of Mathematics at liberal arts colleges, where she mentored numerous undergraduate research projects in data and image analysis. Currently, she is pursuing her scientific interests, in the doctoral program in Physics at Washington State University, where she is excited to use her expertise in mathematics and data and image analysis to study the world around us.

Thomas J. Asaki received his Ph.D. in Physics at Washington State University. He spent 13 years as a postdoc and technical staff member at Los Alamos National Laboratory exploring inverse problems, image and data analysis methods, nondestructive acoustic testing and related optimization problems. He is currently an Associate Professor in the Department of Mathematics and Statistics at Washington State University where, since 2008, he has enjoyed continuing research and helping students discover richness in applied mathematics.

Marie A. Snipes received her Ph.D. in Mathematics from the University of Michigan. Prior to her doctoral studies, she spent four years in the US Air Force as a data analyst. She is currently an Associate Professor of Mathematics at Kenyon College, where she regularly teaches inquiry-based courses and mentors undergraduate research projects. Her research interests include geometric function theory, analysis in metric spaces, and applications of these areas to data problems.

Introduction To Applications

Welcome to your first course in linear algebra—arguably the most useful mathematical subject you will encounter. Not only are the skills important for solving linear problems, they are a foundation for many advanced topics in mathematics, physics, engineering, computer science, economics, business, and more.

This book focuses on the investigation of specific questions from real-world applications. Your explorations will inspire the development of necessary linear algebra tools. In other words, you will be given interesting open problems and as you create your own solutions, you will also be discovering and developing linear algebra tools. This is similar to the way that original investigators recognized key concepts in linear algebra and which led to the standardization of linear algebra as a branch of mathematics. Rather than introduce linear algebra from a vacuum, we will create linear algebra as the skill set necessary to solve our applications. Along the way we hope that you will discover that linear algebra is, in itself, an exciting and beautiful branch of mathematics. This book aims to integrate the exploratory nature of open-ended application questions with the theoretical rigor of linear algebra as a mathematical subject.

1.1 A Sample of Linear Algebra in Our World

We began this text with a very bold claim. We stated that linear algebra is the most useful subject you will encounter. In this section, we show how linear algebra provides a whole host of important tools that can be applied to exciting real-world problems and as well as to advance your mathematical skills. We also discuss just a few interesting areas of active research for which linear algebra tools are essential. Each of these problems can be expressed using the language of linear algebra.

1.1.1 Modeling Dynamical Processes

The word "dynamic" is just a fancy way of describing something that is changing. All around us, from microscopic to macroscopic scales, things are changing. Scientists measure the spread of disease and how to mitigate such spread. Biologists use population models to show how population dynamics change when new species are introduced to an ecosystem. Astronomers use orbital mechanics to study the interactions of celestial bodies in our universe. Manufacturers are interested in the cooling process after diffusion welding is complete. Chemists and nuclear physicists study atomic and subatomic

© Springer Nature Switzerland AG 2022
H. A. Moon et al., *Application-Inspired Linear Algebra*, Springer Undergraduate Texts
in Mathematics and Technology, https://doi.org/10.1007/978-3-030-86155-1_1

interactions. And, geologic and atmospheric scientists study models that predict the effects of geologic or atmospheric events. These are only a small subset of the larger set of dynamic models being studied today.

Differential equations provide the mathematical interpretation of dynamic scenarios. Solutions to the differential equations provide scientists deeper understanding of their respective problem. Solutions help scientists determine the best orbit for space telescopes like Hubble or the James Webb Space telescope, determine how different mitigation strategies might affect disease spread, and predict the location of a tsunami as it spreads across the Pacific Ocean.

1.1.2 Signals and Data Analysis

In our world, data is collected on just about everything. Researchers are interested in determining ways to glean information from this data. They might want to find patterns in the data that are unique to a particular scenario. How often does the apparent luminosity of a distant star fall because of the transit of an orbiting planet? When is the best time to advertise your product to users on a social media platform? How do we determine climate change patterns? Researchers are also interested in how to determine what the data tells them about grouping like subjects in their study. Which advertisement type matches with which social media behavior? How can we classify animal fossils to link extinct animals to the animals we see around us? Scientists are interested in determining the division between unlike subjects. Will a particular demographic vote for a particular candidate based on key features in an ad campaign? They might want to predict outcomes based on data already collected. What can downstream flow and turbidity data reveal about forest health and snow pack in the upstream mountains? Image analysts are interested in recovering images, determining image similarity, and manipulating images and videos. How can we recover the 3D structure of an object from many 2D pictures? How can we view brain features from CAT scans using fewer scans? How do we create smooth transitions between images in a video?

Machine learning is an active area of data analysis research. The overarching goal of machine learning is to create mathematical tools that instruct computers on classification of research subjects or how to predict outcomes. Some key tools of machine learning are Fourier analysis (for pattern recognition), regression analysis (for predictions), support vector machines (for classification), and principle component analysis (for finding relevant features or structure).

1.1.3 Optimal Design and Decision-Making

Optimization is the mathematical process of making optimal decisions. Every day we are confronted with an endless stream of decisions which we must make. What should I eat for breakfast? Should I check my email? What music will I listen to? Pet the cat or not? Most decisions are made easily and quickly without much attention to making a "best" choice and without significantly affecting the world around us. However, researchers, business owners, engineers, and data analysts cannot often ignore best choices. Mathematical optimization seeks a best choice when such a choice is crucial. For example, an engineer designing a tail fin for a new aircraft cannot produce a design based on a personal desire for it to "look cool." Rather, the relevant criteria should include performance, structural integrity, cost, manufacturing capability, etc.

Optimization is ubiquitous in the sciences and business. It is the tool for engineering design, multi-period production planning, personnel and/or event scheduling, route-planning for visits or deliveries, linear and nonlinear data fitting, manufacturing process design, packaging, designing facility layouts, partitioning data into classes, bid selection (tendering), activities planning, division into teams, space probe trajectory planning, highway improvements for traffic flow, allocation of limited resources, security camera placement, and many others.

1.2 Applications We Use to Build Linear Algebra Tools

This book will consider a variety of pedagogical examples and applications. However, you will primarily find two real-world application tasks woven throughout the book, inspiring and motivating much of the material. Therefore, we begin our path into linear algebra with an introduction to these two questions.

1.2.1 CAT Scans

Consider the common scenario of computerized axial tomography (CAT) scans of a human head. The CAT scan machine does not have a magic window into the body. Instead, it relies on sophisticated mathematical algorithms in order to interpret X-ray data. The X-ray machine produces a series (perhaps hundreds) of radiographs such as those shown in Figure 1.1, where each image is taken at a different orientation. Such a set of 2D images, while visually interesting and suggestive of many head features, does not directly provide a 3D view of the head. A 3D view of the head could be represented as a set of head slices (see Figure 1.2), which when stacked in layers provide a full 3D description.

Task #1: *Produce a 3-dimensional image of a human head from a set of 2-dimensional radiographs.*

Each radiograph, and each head slice, is shown as a grayscale image with a fixed number of pixels. Grayscale values in radiographs are proportional to the intensity of X-rays which reach the radiographic film at each location. Grayscale values in the head images are proportional to the mass at each location. The radiographic process is quite complex, but with several reasonable assumptions can be modeled

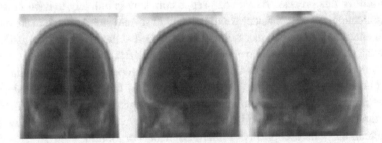

Fig. 1.1 Three examples of radiographs of a human head taken at different orientations.

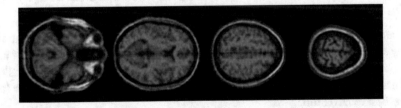

Fig. 1.2 Four examples of density maps of slices of a human head.

Fig. 1.3   A diffusion welding apparatus. Picture Credit: AWS Welding Handbook.

as a process described well within a linear algebra context. In fact, we used linear algebra concepts discussed later in this text to produce the images of brain slices seen in Figure 1.2 from surprisingly few radiographs like those in Figure 1.1. As you can see, the details in the slices are quite remarkable! The mathematical derivation and brief explanations of the physical process are given in Appendix A.

1.2.2   Diffusion Welding

Consider the following scenario involving diffusion welding. Like other types of welding, the goal is to adjoin two or more pieces together. Diffusion welding is used when it is important not to have a visible joint in the final product while not sacrificing strength. A large apparatus like the one found in Figure 1.3 is used for this process. As you can see, a company is unlikely to have many of these on hand. In our application, we will consider welding together several small rods to make one longer rod. The pieces will enter the machine in place. Pressure is applied at the long rod ends as indicated at the bottom of Figure 1.4. The red arrows show where heat is applied at each of the joints. The temperature at each joint is raised to a significant fraction of the melting temperature of the rod material. At these temperatures, material can easily diffuse through the joints. At the end of the diffusion process, the rod temperature is as indicated in the plot and color bar in Figure 1.4. The temperature is measured relative to the temperature at the rod ends, which is fixed at a cool temperature. After the rod cools there are no indications (macroscopically or microscopically) of joint locations.

   Task #2: *Determine the earliest time at which a diffusion-welded rod can be safely removed from the apparatus, without continuous temperature monitoring.*

   The rod can be safely removed from the apparatus when the stress due to temperature gradients is sufficiently small. The technician can determine a suitable thermal test. It is in the interest of the manufacturer to remove the rod as soon as possible, but not to continuously monitor the temperature profile. In later chapters, we discuss the linear algebra techniques that will allow us (and you) to determine the best removal time.

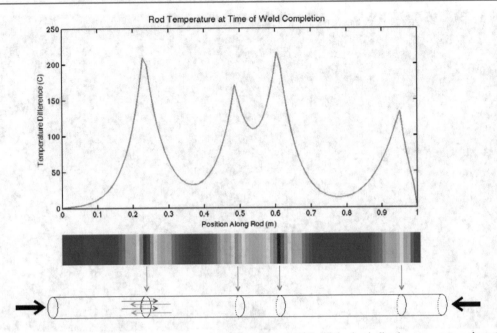

Fig. 1.4   Bottom: Diffusion welding to make a rod with four joints. Middle: Color bar indicating temperature along the rod. Top: Heat difference profile corresponding to temperature difference from the ends of the rod.

1.2.3   Image Warping

Consider the scenario of image warping, where you have two images, and you want to create a video or a sequence of images that flows smoothly from one image to the other. In our application, we will consider techniques similar to those used in the diffusion welding application. This application extends the techniques to 2 dimensions. Linear algebra techniques discussed in later chapters will allow us to transition between two or more images as depicted in Figure 1.5.

1.3   Advice to Students

The successful mastery of higher-level mathematical topics is something quite different from memorizing arithmetic rules and procedures. And, it is quite different from proficiency in calculating numerical answers. Linear algebra may be your first mathematics course that asks you for a deeper understanding of the nature of skills you are learning, knowing "why" as well as "how" you accomplish tasks. This transition can be challenging, yet the analytic abilities that you will acquire will be invaluable in future mathematics courses and elsewhere. We (the authors) offer the following guidelines for success that we feel are worth your serious consideration.

▶ **Keys to Success in Linear Algebra**

- You must be willing to explore, conjecture, guess. You must be willing to fail. Explorations are key to building your understanding.
- You will need to practice learned concepts through exercises, written descriptions, and verbal discussion. Always explain your reasoning.

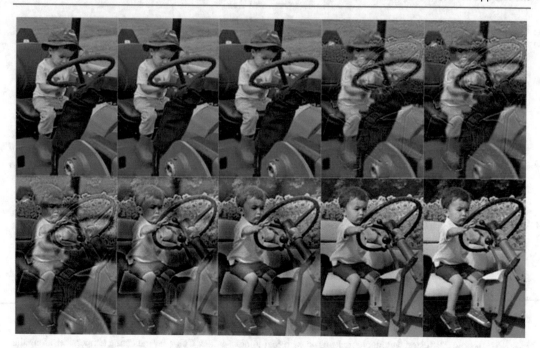

Fig. 1.5 Ten images in the warping sequence that begins with a young boy on a tractor and ends with the same boy a little older.

- You must be willing to ask questions and participate in classroom activities. Never be satisfied with a yes/no response.
- Whether you approach mathematics from a computational, theoretical, or applied view, you must be willing to accept that Linear Algebra is most vibrant when it incorporates each of these aspects.
- You will need to be willing to read actively and with understanding, working through the examples and questions posed in the text.

1.4   The Language of Linear Algebra

In this text, you will learn many new definitions and terminology. In order to fully understand the topics in this text, effectively communicate your understanding, and to have effective conversations with other professionals, you will need to become fluent in the language of linear algebra. In order to help you recognize appropriate uses of this language, we added "Watch Your Language!" indicators such as the one below to exemplify this language.

⋆⋆ **Watch Your Language!**   When communicating about equations and/or expressions, it is important to use the proper language surrounding each.

✓ We **solve equations**.
✓ We **simplify expressions**.

We do not say

✗ We solve an expression.

1.5 Rules of the Game

In this text, we will pose questions for which there isn't always one correct answer. Sometimes answers do not exist, proofs can't be had, and statements are just not true or are open to interpretation. We do this, not to cause frustration, but rather to give you an opportunity to exercise your creative mathematical talent, learn to explore possible strategies to recognize truths (and untruths), and to strengthen your understanding. However, we do abide by the following list:

▶ **Rules of the Game**

- If we ask you to prove something, you can assume the statement is true. On the other hand, we may ask you whether or not a statement is true. In this case, we will expect you to justify your assertion.
- If we ask you to find something, you may or may not be able to find it. We want you to experience the satisfaction of determining the existence as if you were researching a new problem.
- If we ask you to compute something, you can be sure that the task is possible.

1.6 Software Tools

There are a variety of excellent software tools for exploring and solving linear algebra tasks. In this text, we highlight essential MATLAB/OCTAVE commands and functions to accomplish key computational and graphical needs. All of the exploratory tasks are written with these software tools in mind. Several of these tasks make use of application data files which are collected in MATLAB-compatible format. The code and data files can be found on the IMAGEMATH website at http://imagemath.org/AILA.html.

1.7 Exercises

1. Which of the two application tasks discussed in this chapter do you feel will be the most challenging and why?
2. In analyzing a CAT scan, how many radiographs do you feel would be sufficient to obtain accurate results? How many do you believe would be too many?
3. In the tomography application, do you believe that it is possible that a feature in an object will not show up in the set of radiographs?
4. Consider the tomography application. Suppose the data-gathering instrument returns radiographs with missing data for some pixels. Do you believe that an accurate 3D head reconstruction might still be obtained?
5. Consider the diffusion welding application. For what purpose are the ends of the rod held at constant temperature during the cooling process?
6. Consider the diffusion welding application. Do you believe it is possible for a rod to have a temperature profile so that at some location the temperature does not simply decrease or simply increase throughout the diffusion process?
7. Consider the diffusion welding application. If at some location along the rod the initial temperature is the same as the temperature at the ends, do you believe the temperature at that location will remain the same throughout the diffusion process?

Vector Spaces

2

In this chapter, we will begin exploring the Radiography/Tomography example discussed in Section 1.2.1. Recall that our goal is to produce images of slices of the brain similar to those found in Figure 2.1 from radiographic data such as that seen in Figure 2.2.

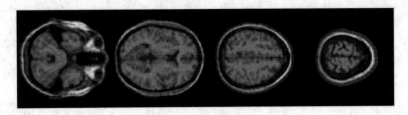

Fig. 2.1 Four examples of density maps of slices of a human head.

In order to meet this goal, we will begin exploring images as mathematical objects. We will discuss how we can perform arithmetic operations on images. Finally, we will explore properties of the set of images. With this exploration, we will begin recognizing many sets with similar properties. Sets with these properties will be the sets on which we focus our study of linear algebra concepts.

2.1 Exploration: Digital Images

In order to understand and solve our tomography task (Section 1.2.1), we must first understand the nature of the radiographs that comprise our data. Each radiograph is actually a digitally stored collection of numerical values. It is convenient for us when they are displayed in a pixel arrangement with colors or grayscale. This section explores the nature of pixelized images and provides exercises and questions to help us understand their place in a linear algebra context.

We begin by formalizing the concept of an image with a definition. We will then consider the most familiar examples of images in this section. In subsequent sections we will revisit this definition and explore other examples.

Definition 2.1.1

An **image** is a finite ordered list of real numbers with an associated geometric arrangement. □

© Springer Nature Switzerland AG 2022, corrected publication 2023
H. A. Moon et al., *Application-Inspired Linear Algebra*, Springer Undergraduate Texts in Mathematics and Technology, https://doi.org/10.1007/978-3-030-86155-1_2

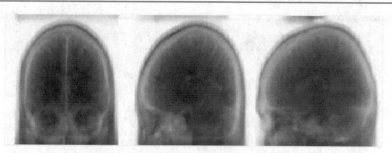

Fig. 2.2 Three examples of radiographs of a human head taken at different orientations.

Fig. 2.3 Digital images are composed of pixels, each with an associated brightness indicated by a numerical value.
Photo Credit: Sharon McGrew.

First, let us look at an image from a camera in grayscale. In Figure 2.3, we see one of the authors
learning to sail. When we zoom in on a small patch, we see squares of uniform color. These are the
pixels in the image. Each square (or pixel) has an associated intensity or brightness. Intensities are
given a corresponding numerical value for storage in computer or camera memory. Brighter pixels are
assigned larger numerical values.

Consider the 4×4 grayscale image in Figure 2.4. This image corresponds to the array of numbers at
right, where a black pixel corresponds to intensity 0 and increasingly lighter shades of gray correspond
to increasing intensity values. A white pixel (not shown) corresponds to an intensity of 16.

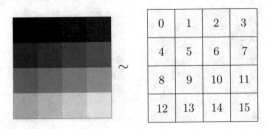

Fig. 2.4 Pixel intensity assignment.

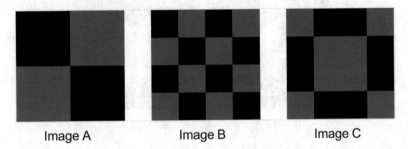

Fig. 2.5 Three 4×4 grayscale images. The pixel intensities in these particular images are all either 0 or 8.

A given image can be displayed on different scales; in Figure 2.3, a pixel value of 0 is displayed as black and 255 is displayed as white, while in Figure 2.4 a pixel value of 0 is displayed as black and 16 is displayed as white. The display scale does not change the underlying pixel values of the image.

Also, the same object may produce different images when imaged with different recording devices, or even when imaged using the same recording device with different calibrations. For example, this is what a smart phone is doing when you touch a portion of the screen to adjust the brightness when you take a picture with it.

Our definition of an image yields a natural way to think about arithmetic operations on images such as multiplication by a scalar and adding two images. For example, suppose we start with the three images A, B, and C in Figure 2.5.

Multiplying Image A by one half results in Image 1 in Figure 2.6. Every intensity value is now half what it previously was, so all pixels have become darker gray (representing their lower intensity). Adding Image 1 to Image C results in Image 2 (also in Figure 2.6); so Image 2 is created by doing arithmetic on Images A and C.

Caution: Digital images and *matrices* are both arrays of numbers. However, not all digital images have rectangular geometric configurations like matrices[1], and even digital images with rectangular configurations are not matrices, since there are operations[2] that can be performed with matrices that do not make sense for digital images.

2.1.1 Exercises

For some of these exercises you will need access to OCTAVE or MATLAB software. The following exercises refer to images found in Figures 2.5 and 2.6.

---

[1] See Page 39 for examples of non-rectangular geometric configurations.

[2] An example of an operation on matrices that is meaningless on images is row reduction.

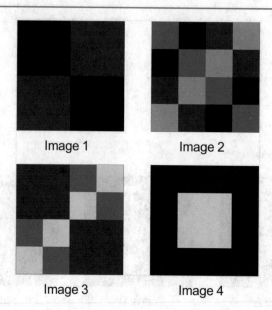

Fig. 2.6 Four additional images. Image 1 is (0.5)(Image A) and Image 2 is (Image 1)+(Image C).

1. Express Image 3 using arithmetic operations on Images A, B, and C. (Note that the pixel intensities in Image 3 are all either 4, 8, or 16.)
2. Express Image 4 using arithmetic operations on Images A, B, and C. (Note that the pixel intensities in Image 4 are all either 0 or 16.)
3. Input the following lines of code into the command window of OCTAVE/MATLAB. Note that ending a line with a semicolon suppresses terminal output. If you want to show the result of a computation, delete the semicolon at the end of its line. Briefly describe what each of these lines of code produces.

```
M_A = [0  0  8  8;  0  0  8  8;  8  8  0  0;  8  8  0  0];
M_B = [0  8  0  8;  8  0  8  0;  0  8  0  8;  8  0  8  0];
M_C = [8  0  0  8;  0  8  8  0;  0  8  8  0;  8  0  0  8];

figure('position', [0,0,1200,360]);
GrayRange=[0 16];

subplot(1,3,1);
imagesc(M_A,GrayRange);
title('Image A');

subplot(1,3,2);
imagesc(M_B,GrayRange);
title('Image B');

subplot(1,3,3);
imagesc(M_C,GrayRange);
title('Image C');

colormap(gray(256));
```

4. Enter the following lines of code one at a time and state what each does.

```
M_A
M_1 = .5*M_A
M_2 = M_1 + M_C
figure('position', [0,0,1200,360]);
GrayRange=[0 16];
subplot(1,2,1);
imagesc(M_1,GrayRange);
title('Image 1');
subplot(1,2,2);
imagesc(M_2,GrayRange);
title('Image 2');
colormap(gray(256));
```

5. Write your own lines of code to check your conjectures for producing Images 3 and/or 4. How close are these to Images 3 and/or 4?

6. We often consider display scales that assign pixels with value 0 to the color black. If a recording device uses such a scale then we do not expect any images it produces to contain pixels with negative values. However, in our definition of an image we do not restrict the pixel values. In this problem you will explore how OCTAVE/MATLAB displays an image with negative pixel values, and you will explore the effects of different gray scale ranges on an image.

Input the image pictured below into OCTAVE/MATLAB. Then display the image using each of the following five grayscale ranges.

-10	0	20
10	5	-20

  (i)   GrayRange= $[0, 20]$,
 (ii)   GrayRange= $[0, 50]$,
(iii)   GrayRange= $[-20, 20]$,
 (iv)   GrayRange= $[-10, 10]$,
  (v)   GrayRange= $[-50, 50]$.

(a) Describe the differences in the display between: setting (i) and setting (ii); setting (i) and setting (iii); setting (iii) and setting (iv); and finally between setting (iii) and setting (v).

(b) Summarize what happens when pixel intensities in an image exceed the display range as input into the imagesc function.

(c) Summarize what happens when the display range becomes much larger than the range of pixel values in an image.

(d) Discuss how the pixels with negative values were displayed in the various gray scale ranges.

7. How should we interpret pixel intensities that lie outside our specified grayscale range?

8. Consider digital radiographic images (see Figure 1.1). How would you interpret intensity values? How would you interpret scalar multiplication?

9. What algebraic properties does the set of all images have in common with the set of real numbers?

10. Research how color digital images are stored as numerical values. How can we modify our concepts of image addition and scalar multiplication to apply to color images?

11. Describe how a heat state on a rod can be represented as a digital image.
12. Think of two different digital image examples for which negative pixel intensities make real-world sense.

2.2 Systems of Equations

In Section 2.1, we considered various 4×4 images (see page 11). We showed that Image 2 could be formed by performing image addition and scalar multiplication on Images A, B, and C. In particular,

$$(\text{Image 2}) = \left(\frac{1}{2}\right)(\text{Image A}) + (0)(\text{Image B}) + (1)(\text{Image C}).$$

We also posed the question about whether or not Images 3 and 4 can be formed using any arithmetic operations of Images A, B, and C. One can definitely determine, by inspection, the answer to these questions. Sometimes, however, trying to answer such questions by inspection can be a very tedious task. In this section, we introduce tools that can be used to answer such questions. In particular, we will discuss the method of elimination, used for solving systems of linear equations. We will also use matrix reduction on an augmented matrix to solve the corresponding system of equations. We will conclude the section with a key connection between the number of solutions to a system of equations and a reduced form of the augmented matrix.

2.2.1 Systems of Equations

In this section we return to one of the tasks from Section 2.1. In that task, we were asked to determine whether a particular image could be expressed using arithmetic operations on Images A, B, and C. Let us consider a similar question. Suppose we are given the images in Figures 2.7 and 2.8. Can Image C be expressed using arithmetic operations on Images A, D, and F?

For this question, we are asking whether we can find real numbers α, β, and γ so that

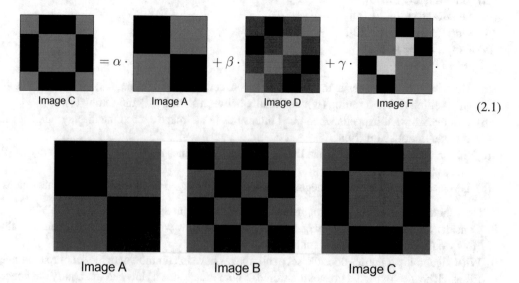

$$\text{Image C} = \alpha \cdot \text{Image A} + \beta \cdot \text{Image D} + \gamma \cdot \text{Image F}. \tag{2.1}$$

Image A   Image B   Image C

Fig. 2.7 Images A, B, and C from Section 2.1.

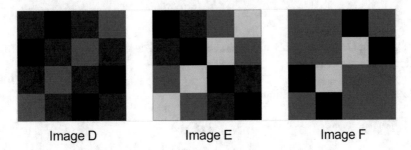

Image D   Image E   Image F

Fig. 2.8 Images D, E, and F are more example 16-pixel images.

First, in order to make sense of this question, we need to define what it means for images to be equal.

Definition 2.2.1

Let I_1 and I_2 be images. We say that $I_1 = I_2$ if each pair of corresponding pixels from I_1 and I_2 has the same intensity. □

The convention in Figure 2.4, Definition 2.2.1, and Equation 2.1 give us a means to write an equation, corresponding to the upper left pixel of Image D,

$$8 = 0\alpha + 4\beta + 8\gamma. \tag{2.2}$$

This equation has a very specific form: it is a linear equation. Such equations are at the heart of the study of linear algebra, so we recall the definition below.

Definition 2.2.2

A **linear equation** is an equation of the form

$$a_1x_1 + a_2x_2 + \cdots + a_nx_n = b,$$

where $b \in \mathbb{R}$, $a_1, \cdots, a_n \in \mathbb{R}$ are called **coefficients** and $x_1, x_2, \ldots, x_n$ are called **variables**. □

In the definition above, we use the symbol "$\in$" and mean "element of" or "in the set." We write, above, that $b \in \mathbb{R}$. This means that b is an element of the set of all real numbers. Typically, we read this as "b is real." We will use this notation throughout the text for many different sets.

This definition considers coefficients which are real numbers. Later, we will encounter some generalizations where coefficients are elements of other fields. See Appendix D for a discussion of fields. Let us consider some examples of linear equations.

Example 2.2.3 The equation

$$3x + 4y - 2z = w$$

is linear with variables x, y, z, and w, and with coefficients 3, 4, -2, and -1 we see this by putting the equation in the form given in Definition 2.2.2. In this form, we have

$$3x + 4y - 2z - w = 0.$$

$\square$

The form given in Definition 2.2.2 will often be referred to as *standard form*. We now follow with some equations that are not linear.

Example 2.2.4 The following equations are not linear:

$$3x + 2yz = 3$$
$$x^2 + 4x = 2$$
$$\cos x - 2y = 3.$$

As we can see, in each equation above, operations applied on the variables that are more complicated than addition and multiplication by a constant coefficient. $\square$

Let us, now, return to the question at hand. In Equation 2.2, the variables are α, β, and γ. We seek values of these variables so that the equation is true, that is, so that both sides are equal. Appropriate values for α, β, and γ constitute a *solution* to Equation 2.2, which we define below.

Definition 2.2.5

Let

$$a_1 x_1 + a_2 x_2 + \cdots + a_n x_n = b$$

be a linear equation in n variables, $x_1, x_2, \cdots, x_n$. Then

$$(v_1, v_2, \cdots, v_n) \in \mathbb{R}^n$$

is a **solution** to the linear equation if

$$a_1 v_1 + a_2 v_2 + \cdots + a_n v_n = b.$$

$\square$

To understand Definition 2.2.5, we recall that when we solve an equation, we can check our solution by substituting it into the original equation to determine whether the equation is true. The definition tells us that if we do this substitution and obtain a true statement, then the ordered n-tuple of the substituted values is a solution. For example, $(3, 0, 1)$ is a solution to Equation 2.2 because

$$8 = 0 \cdot 3 + 4 \cdot 0 + 8 \cdot 1.$$

Notice, also, that $(1, 1, 1)$ is not a solution to Equation 2.2:

$$0 \cdot 1 + 4 \cdot 1 + 8 \cdot 1 = 12$$

is false because $8 \neq 12$.

In order for Equation 2.1 to be true, we need α, β, and γ so that all linear equations (corresponding to all pixel locations) are true. Let us write out these 16 linear equations (starting in the upper left corner and going across the top before moving to the second row of pixels).

$$8 = 0 \cdot \alpha + 4 \cdot \beta + 8 \cdot \gamma$$
$$0 = 0 \cdot \alpha + 0 \cdot \beta + 8 \cdot \gamma$$
$$0 = 8 \cdot \alpha + 4 \cdot \beta + 0 \cdot \gamma$$
$$8 = 8 \cdot \alpha + 8 \cdot \beta + 8 \cdot \gamma$$
$$0 = 0 \cdot \alpha + 0 \cdot \beta + 8 \cdot \gamma$$
$$8 = 0 \cdot \alpha + 4 \cdot \beta + 8 \cdot \gamma$$
$$8 = 8 \cdot \alpha + 8 \cdot \beta + 16 \cdot \gamma \qquad (2.3)$$
$$0 = 8 \cdot \alpha + 4 \cdot \beta + 0 \cdot \gamma$$
$$0 = 8 \cdot \alpha + 4 \cdot \beta + 0 \cdot \gamma$$
$$8 = 8 \cdot \alpha + 8 \cdot \beta + 16 \cdot \gamma$$
$$8 = 0 \cdot \alpha + 4 \cdot \beta + 8 \cdot \gamma$$
$$0 = 0 \cdot \alpha + 0 \cdot \beta + 8 \cdot \gamma$$
$$8 = 8 \cdot \alpha + 8 \cdot \beta + 8 \cdot \gamma$$
$$0 = 8 \cdot \alpha + 4 \cdot \beta + 0 \cdot \gamma$$
$$0 = 0 \cdot \alpha + 0 \cdot \beta + 8 \cdot \gamma$$
$$8 = 0 \cdot \alpha + 4 \cdot \beta + 8 \cdot \gamma$$

Because these equations are all related by their solution (α, β, γ), we call this list of equations a *system* of equations as defined below.

Definition 2.2.6

Let $m, n \in \mathbb{N}$. A set of m linear equations, each with n variables $x_1, x_2, \ldots, x_n$, is called a **system of equations** with m equations and n variables.

We say that $(v_1, v_2, \ldots, v_n) \in \mathbb{R}^n$ is a **solution to the system of equations** if it is simultaneously a solution to all equations in the system.

Finally, the set of all solutions to a system of equations is the **solution set** of the system of equations.   □

Before continuing toward a solution to the system of equations in (2.3), we should observe that several equations are repeated. Therefore, thankfully, we need only write the system without repeated equations.

$$8 = 0 \cdot \alpha + 4 \cdot \beta + 8 \cdot \gamma$$
$$0 = 0 \cdot \alpha + 0 \cdot \beta + 8 \cdot \gamma$$
$$0 = 8 \cdot \alpha + 4 \cdot \beta + 0 \cdot \gamma \qquad (2.4)$$
$$8 = 8 \cdot \alpha + 8 \cdot \beta + 8 \cdot \gamma$$
$$8 = 8 \cdot \alpha + 8 \cdot \beta + 16 \cdot \gamma$$

Though we were able to use symmetry to simplify this particular system of equations, you can, likely, imagine that most important applications using systems of equations require very large numbers (quite possibly thousands or even millions) of equations and variables[3]. In these cases, we use computers to find solutions. Even though technology is used for solving such problems, we find value in understanding and making connections between these techniques and properties of the solutions. So, in the next section we will describe by-hand methods for solving systems of equations.

2.2.2   Techniques for Solving Systems of Linear Equations

In this section, we will describe two techniques for solving systems of equations. We use these two techniques to solve systems like the one presented in the previous section that arose from a question about images.

Method of elimination

In this section, we solve the system of equations in 2.4 using the method of elimination. You may have used this method before, but we include it here to introduce some terminology to which we will refer in later sections. We will also give a parallel method later in this section.

Definition 2.2.7

Two systems of equations are said to be **equivalent** if they have the same solution set.   □

The idea behind the method of elimination is that we seek to manipulate the equations in a system so that the solution is easier to obtain. Specifically, in the new system, one or more of the equations will be of the form $x_i = c$. Since one of the equations tells us directly the value of one of the variables, we can substitute that value into the other equations and the remaining, smaller system has the same solution (together, of course, with $x_i = c$).

Before we solve the system in (2.4), we provide the list of allowed operations for solving a system of equations, using the method of elimination.

▶ **Allowed operations for solving systems of equations**

   (1) Multiply both sides of an equation by a nonzero number.
   (2) Change one equation by adding a nonzero multiple of another equation to it.
   (3) Change the order of equations.

You may find these operations familiar from your earlier experience solving systems of equations; they do not change the solution set of a system. In other words, every time we change a system using one of these operations, we obtain an *equivalent* system of equations.

Fact 2.2.8

Two systems of equations that differ by one or more operations allowed in the method of elimination (as outlined in the list above) are equivalent.

Let us, now, use these operations to solve our example system of equations.

---

[3] In practice, we typically use m to represent the number of equations and n to represent the number of variables.

Example 2.2.9 We want to solve the system of equations in (2.4). We will use annotations to describe which of the allowed operations were used. The convention with our annotations will be that E_k represents the new equation k and e_k represents the previous equation k. For example, $E_3 = 2e_3 + e_1$ means that we will multiply the third equation by 2 and then add the first equation. The result will then replace the third equation.

$$
\begin{array}{l}
8 = 0 \cdot \alpha + 4 \cdot \beta + 8 \cdot \gamma \\
0 = 0 \cdot \alpha + 0 \cdot \beta + 8 \cdot \gamma \\
0 = 8 \cdot \alpha + 4 \cdot \beta + 0 \cdot \gamma \\
8 = 8 \cdot \alpha + 8 \cdot \beta + 8 \cdot \gamma \\
8 = 8 \cdot \alpha + 8 \cdot \beta + 16 \cdot \gamma
\end{array}
\xrightarrow{\text{Simplify}}
\begin{array}{l}
8 = \qquad 4\beta + 8\gamma \\
0 = \qquad\qquad 8\gamma \\
0 = 8\alpha + 4\beta \\
8 = 8\alpha + 8\beta + 8\gamma \\
8 = 8\alpha + 8\beta + 16\gamma
\end{array}
$$

$$
\xrightarrow[\substack{E_3=\frac{1}{4}e_3 \\ E_4=\frac{1}{8}e_4,\ E_5=\frac{1}{8}e_5}]{E_1=\frac{1}{4}e_1,\ E_2=\frac{1}{8}e_2}
\begin{array}{l}
2 = \qquad \beta + 2\gamma \\
0 = \qquad\qquad \gamma \\
0 = 2\alpha + \beta \\
1 = \alpha + \beta + \gamma \\
1 = \alpha + \beta + 2\gamma
\end{array}
$$

$$
\xrightarrow[E_5=-e_4+e_5]{E_3=-2e_4+e_3}
\begin{array}{l}
2 = \qquad \beta + 2\gamma \\
0 = \qquad\qquad \gamma \\
-2 = \qquad -\beta - 2\gamma \\
1 = \alpha + \beta + \gamma \\
0 = \qquad\qquad \gamma
\end{array}
$$

$$
\xrightarrow{E_3=e_3+2e_2}
\begin{array}{l}
2 = \qquad \beta + 2\gamma \\
0 = \qquad\qquad \gamma \\
-2 = \qquad -\beta \\
1 = \alpha + \beta + \gamma \\
0 = \qquad\qquad \gamma.
\end{array}
$$

Notice that the second and fifth equations tell us that $\gamma = 0$ and the third equation says that $\beta = 2$. We can use substitution in the fourth equation to find α as follows:

$$
\begin{aligned}
1 &= \alpha + 2 + 0 \\
\Rightarrow -1 &= \alpha.
\end{aligned}
$$

There are no other possible choices of α, β, and γ that satisfy our system. Hence, the solution set of the system of equations (2.4) is $\{(-1, 2, 0)\}$. You can check that, as Fact 2.2.8 asserts, this solution is a solution to each of the systems in each step of the elimination performed above. $\square$

We have, now, answered our original question: Image C can be written as

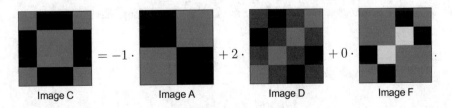

Image C $\qquad$ Image A $\qquad$ Image D $\qquad$ Image F

Method of matrix reduction

In this section, we first describe how to represent a system of equations using matrix notation, and then we give a systematic algorithm for *reducing* the matrix until the solution set is apparent.

As with the previous method, we will use a model example from our image exploration to introduce the process. Can we represent Image A using arithmetic operations on Images D, E, and F? That is, we want to find α, β, and γ so that

| Image A | Image D | Image E | Image F | (2.5) |

As before, we will write a system of equations by matching up corresponding pixels in Equation 2.5, leaving out repeated equations. The system is shown below:

$$\begin{aligned}
4\alpha + 0\beta + 8\gamma &= 0 \\
0\alpha + 4\beta + 8\gamma &= 0 \\
4\alpha + 8\beta + 0\gamma &= 8 \\
8\alpha + 12\beta + 8\gamma &= 8 \\
8\alpha + 12\beta + 16\gamma &= 8.
\end{aligned} \tag{2.6}$$

To use matrix notation, we store the coefficients and the constants of the equations in an *augmented matrix* as follows:

$$\begin{pmatrix}
4 & 0 & 8 & 0 \\
0 & 4 & 8 & 0 \\
4 & 8 & 0 & 8 \\
8 & 12 & 8 & 8 \\
8 & 12 & 16 & 8
\end{pmatrix}. \tag{2.7}$$

Notice that the entries to the left of the vertical line (or augment) in the matrix are the coefficients of the variables in the linear equations. Notice also that the entries to the right of the augment are the right-hand sides (constant coefficients) in the equations.

In general, the system of equations with m equations in n variables

$$\begin{aligned}
a_{1,1}x_1 + a_{1,2}x_2 + \ldots + a_{1,n}x_n &= b_1 \\
a_{2,1}x_1 + a_{2,2}x_2 + \ldots + a_{2,n}x_n &= b_2 \\
&\vdots \\
a_{m,1}x_1 + a_{m,2}x_2 + \ldots + a_{m,n}x_n &= b_m,
\end{aligned}$$

has a corresponding augmented matrix with m rows and $n + 1$ columns, and is written as

$$\begin{pmatrix}
a_{1,1} & a_{1,2} & \ldots & a_{1,n} & b_1 \\
a_{2,1} & a_{2,2} & \ldots & a_{2,n} & b_2 \\
\vdots & & \ddots & & \vdots \\
a_{m,1} & a_{m,2} & \ldots & a_{m,n} & b_m
\end{pmatrix}.$$

We will *reduce* the augmented matrix corresponding to the system in (2.6) using a set of allowed *row operations*. As you will see, the key fact (presented below) about row operations on an augmented matrix is that they do not change the solution set of the corresponding system.

Matrix reduction is nearly the same as the elimination method for systems of equations. In fact, the differences between the two methods are that, instead of working with a system that has equations, you work with a matrix that has rows. Here, we write the allowed steps to standardize this process.

▶ **Row operations to reduce a matrix**   The following are operations allowed to be used when reducing a matrix.

   (1) Multiply a row by a nonzero number.
   (2) Change one row by adding a nonzero multiple of another row to it.
   (3) Change the order of rows.

When solving systems of equations using elimination, we discussed equivalent systems. We have a similar definition for matrices. The following definition gives us terminology for equivalence between matrices that should not be confused with equality.

Definition 2.2.10

Let P and Q be two $m \times (n + 1)$ augmented matrices, corresponding to systems with m equations in n variables. If Q can be obtained by performing a finite sequence of row operations on P, then P is **row equivalent** to Q, which we denote

$$P \sim Q.$$

□

In Definition 2.2.10, we specify that for two matrices to be row equivalent, we require that the matrix reduction process begins at P and terminates at Q after only finitely many row operations. It is important to recognize that row operations can be performed on non-augmented matrices. So, although we then don't have the interpretation as corresponding to a system of equations, we can still make sense of row equivalent non-augmented matrices.

In Exercise 33, you will show that row operations can be reversed, that is, if $P \sim Q$, then $Q \sim P$. Hence, rather than writing "P is row equivalent to Q," it makes sense to write "P and Q are row equivalent."

Now we can restate Fact 2.2.8 in terms of augmented matrices.

Fact 2.2.11

Two augmented matrices that are row equivalent correspond to equivalent systems.

Clearly, any given matrix (augmented or not) is row equivalent to many other matrices. In order to determine whether two matrices are row equivalent, we need to be able to use row operations to transform one matrix into the other. Alternatively, we could use row operations to transform both matrices into the same third matrix.

Lemma 2.2.12

Let P, Q, and R be three $m \times n$ matrices and suppose that $P \sim R$ and $Q \sim R$. Then $P \sim Q$.

The proof is Exercise 34.

We now return to pursuing our overall goal of finding the solution set of a given linear system. We will use row operations to get the corresponding augmented matrix into a format for which we can readily find the solution set; this solution set is also the solution set for the original system.

What simple format might we require for the row equivalent matrix? There are many possible forms we might choose, but we will define and focus on two in this text: echelon form and reduced echelon form. Both forms include conditions on the *leading entries* of the matrix.

Definition 2.2.13

The **leading entries** in a matrix are the first nonzero entries in each row, when reading from left to right. □

For example, the leading entries in the matrix P below are boxed.

$$P = \begin{pmatrix} 0 & \boxed{1} & 2 & 4 & 0 \\ \boxed{2} & 0 & 1 & 5 & 2 \\ 0 & 0 & 0 & \boxed{1} & 1 \end{pmatrix}.$$

Definition 2.2.14

A matrix is in **echelon form** if the following three statements are true:

- All leading entries are 1.
- Each leading entry is to the right of the leading entry in the row above it.
- Any row of zeros is below all rows that are not all zero. □

The matrix P, above, is not in echelon form. Two of the three criteria above are not true. First, 2 is a leading entry so the first rule is not true. Second, the leading entry in the second row is to the left of the leading entry in the first row, violating the second rule.

However, if we start with P, then multiply row two by $1/2$ and then switch the first two rows, we get

$$Q = \begin{pmatrix} 1 & 0 & 1/2 & 5/2 & 1 \\ 0 & 1 & 2 & 4 & 0 \\ 0 & 0 & 0 & 1 & 1 \end{pmatrix},$$

which *is* in echelon form. It is not just a coincidence that we could row reduce P to echelon form; in fact, *every* matrix can be put into echelon form. It is indeed convenient that every matrix is row equivalent to a matrix that is in echelon form, and you should be able to see that an augmented matrix in echelon form corresponds to a simpler system to solve.

However, augmented matrices in echelon form are not the *simplest* to solve, and moreover, echelon form is not unique. For example, both the matrices R and S below are also in echelon form and are row equivalent to Q.

$$R = \begin{pmatrix} 1 & -1 & -3/2 & -1 & | & 3/2 \\ 0 & 1 & 2 & 6 & | & 2 \\ 0 & 0 & 0 & 1 & | & 1 \end{pmatrix},$$

$$S = \begin{pmatrix} 1 & 0 & 1/2 & 0 & | & -3/2 \\ 0 & 1 & 2 & 0 & | & -4 \\ 0 & 0 & 0 & 1 & | & 1 \end{pmatrix}.$$

Since the matrices P, Q, R, and S are all row equivalent, they correspond to equivalent systems. Which would you rather work with? What makes your choice nicer to solve than the others? Likely, what you are discovering is that matrices corresponding to systems that are quite easy to solve are matrices in *reduced echelon form*.

Definition 2.2.15

A matrix is said to be in **reduced echelon form** if the matrix is in echelon form and all entries above each leading entry are zeros. □

The matrix S above is in reduced echelon form, but the matrices Q and R, although in echelon form, do not have only zeros above and below each leading entry. Therefore, Q and R are not in reduced echelon form. The following is an algorithm (list of steps) that will row reduce a matrix to reduced echelon form.

▶ **Row Reduction** Steps to perform Row Reduction on an $m \times n$ matrix.
Start with $r = 1$ and $c = 1$.

1. Find a leading entry in column c.
      If possible Go to step 2.
      If not possible Increase c by 1.
          If $c \leq n$ Go to step 1.
          If $c > n$ Algorithm finished: matrix is in reduced echelon form.
2. Arrange the rows in an order so that there is a leading entry in row r and column c.
3. Use row operations (1) or (2) to make the leading entry in row r a leading 1, without changing any rows above row r.
4. Using the leading 1 and row operation (2) to get a 0 in each of the entries above and below the leading 1.
5. Increase r by 1 and increase c by 1.
6. Go to step 1.

There are many other sequences of row operations that could be used to put a matrix into reduced echelon form, but they all lead to the same result, as the next theorem shows

---

Theorem 2.2.16
Each matrix is row equivalent to exactly one matrix in reduced echelon form.

The proof of Theorem 2.2.16 can easily distract us from our applications. Therefore, we leave the proof out of the text, but encourage the reader to explore how such a proof can be written or to read one of many versions already available.

Example 2.2.17 To illustrate matrix reduction, we will reduce the matrix in (2.7) to echelon form. We will annotate our steps in a way similar to our annotations in the method of elimination. Here, R_k will represent the new row k and r_k will represent the old row k.

$$
\begin{pmatrix}
4 & 0 & 8 & | & 0 \\
0 & 4 & 8 & | & 0 \\
4 & 8 & 0 & | & 8 \\
8 & 12 & 8 & | & 8 \\
8 & 12 & 16 & | & 8
\end{pmatrix}
\xrightarrow[\substack{R_1=(1/4)r_1 \\ R_2=(1/4)r_2}]{}
\begin{pmatrix}
1 & 0 & 2 & | & 0 \\
0 & 1 & 2 & | & 0 \\
4 & 8 & 0 & | & 8 \\
8 & 12 & 8 & | & 8 \\
8 & 12 & 16 & | & 8
\end{pmatrix}
$$

$$
\xrightarrow[\substack{R_3=-4r_1+r_3 \\ R_4=-8r_1+r_4 \\ R_5=-8r_1+r_5}]{}
\begin{pmatrix}
1 & 0 & 2 & | & 0 \\
0 & 1 & 2 & | & 0 \\
0 & 8 & -8 & | & 8 \\
0 & 12 & -8 & | & 8 \\
0 & 12 & 0 & | & 8
\end{pmatrix}
\xrightarrow[\substack{R_3=-8r_2+r_3 \\ R_4=-12r_2+r_4 \\ R_5=-12r_2+r_5}]{}
\begin{pmatrix}
1 & 0 & 2 & | & 0 \\
0 & 1 & 2 & | & 0 \\
0 & 0 & -24 & | & 8 \\
0 & 0 & -32 & | & 8 \\
0 & 0 & -24 & | & 8
\end{pmatrix}
$$

$$
\xrightarrow[\substack{R_3=-(1/24)r_3 \\ R_4=-(1/32)r_4 \\ R_5=-r_3+r_5}]{}
\begin{pmatrix}
1 & 0 & 2 & | & 0 \\
0 & 1 & 2 & | & 0 \\
0 & 0 & 1 & | & -1/3 \\
0 & 0 & 1 & | & -1/4 \\
0 & 0 & 0 & | & 0
\end{pmatrix}
\xrightarrow[R_4=-r_3+r_4]{}
\begin{pmatrix}
1 & 0 & 2 & | & 0 \\
0 & 1 & 2 & | & 0 \\
0 & 0 & 1 & | & -1/3 \\
0 & 0 & 0 & | & 1/12 \\
0 & 0 & 0 & | & 0
\end{pmatrix}.
$$

We have reduced the original augmented matrix to echelon form. What additional row operations[4] would be used to put the matrix into reduced echelon form? □

The system of equations corresponding to the echelon form found in Example 2.2.17 is

$$
\begin{aligned}
\alpha \quad\; + 2\gamma &= 0 \\
\beta + 2\gamma &= 0 \\
\gamma &= -1/3 \\
0 &= 1/12 \\
0 &= 0.
\end{aligned}
$$

Clearly, the fourth equation is false. This means that we cannot find any α, β, and γ so that Equation 2.5 is true. In this case, we would say the system in (2.6) is inconsistent, or has no solution.

Let us consider one more example similar to those above.

Example 2.2.18 Let us, now, ask whether Image A can be written using arithmetic combinations of any of the other 5 images. That is, can we find α, β, γ, τ, and δ so that

---

[4] Reduced echelon form is particularly useful if you want to find a solution to the system. In this case, since the system has no solution, we did not present these additional row operations.

$$= \alpha \cdot \quad + \beta \cdot \quad + \gamma \cdot \quad + \tau \cdot \quad + \delta \cdot \quad ? \tag{2.8}$$

Again, we set up the system of equations corresponding to Equation (2.8). Without repeated equations, the corresponding system of linear equations is

$$
\begin{aligned}
8\beta + 4\gamma \quad\quad\quad + 8\delta &= 0 \\
8\alpha \quad\quad\quad + 4\tau + 8\delta &= 0 \\
4\gamma + 8\tau \quad\quad &= 8 \\
8\alpha + 8\beta + 8\gamma + 12\tau + 8\delta &= 8 \\
8\alpha + 8\beta + 8\gamma + 12\tau + 16\delta &= 8.
\end{aligned}
\tag{2.9}
$$

We will use matrix reduction to find a solution $(\alpha, \beta, \gamma, \tau, \delta)$ to the system in (2.9). The corresponding augmented matrix is

$$
\left(\begin{array}{ccccc|c}
0 & 8 & 4 & 0 & 8 & 0 \\
8 & 0 & 0 & 4 & 8 & 0 \\
0 & 0 & 4 & 8 & 0 & 8 \\
8 & 8 & 8 & 12 & 8 & 8 \\
8 & 8 & 8 & 12 & 16 & 8
\end{array}\right) .
\tag{2.10}
$$

We will reduce this matrix to reduced echelon form. To begin, we will interchange rows so that we begin with the current rows 2, 4, and 5 at the top, without changing the order in any other way. In our second step, we will multiply every row by 1/4.

$$
\left(\begin{array}{ccccc|c}
8 & 0 & 0 & 4 & 8 & 0 \\
8 & 8 & 8 & 12 & 8 & 8 \\
8 & 8 & 8 & 12 & 16 & 8 \\
0 & 8 & 4 & 0 & 8 & 0 \\
0 & 0 & 4 & 8 & 0 & 8
\end{array}\right)
\longrightarrow
\left(\begin{array}{ccccc|c}
2 & 0 & 0 & 1 & 2 & 0 \\
2 & 2 & 2 & 3 & 2 & 2 \\
2 & 2 & 2 & 3 & 4 & 2 \\
0 & 2 & 1 & 0 & 2 & 0 \\
0 & 0 & 1 & 2 & 0 & 2
\end{array}\right)
$$

$$
\begin{array}{c}
R_2 = -r_1 + r_2 \\
\longrightarrow \\
R_3 = -r_1 + r_3
\end{array}
\left(\begin{array}{ccccc|c}
2 & 0 & 0 & 1 & 2 & 0 \\
0 & 2 & 2 & 2 & 0 & 2 \\
0 & 2 & 2 & 2 & 2 & 2 \\
0 & 2 & 1 & 0 & 2 & 0 \\
0 & 0 & 1 & 2 & 0 & 2
\end{array}\right)
\begin{array}{c}
R_2 = (1/2)r_2 \\
\longrightarrow \\
R_3 = (1/2)r_3
\end{array}
\left(\begin{array}{ccccc|c}
2 & 0 & 0 & 1 & 2 & 0 \\
0 & 1 & 1 & 1 & 0 & 1 \\
0 & 1 & 1 & 1 & 1 & 1 \\
0 & 2 & 1 & 0 & 2 & 0 \\
0 & 0 & 1 & 2 & 0 & 2
\end{array}\right)
$$

$$
\begin{array}{c}
R_3 = -r_2 + r_3 \\
\longrightarrow \\
R_4 = 2r_2 - r_4
\end{array}
\left(\begin{array}{ccccc|c}
2 & 0 & 0 & 1 & 2 & 0 \\
0 & 1 & 1 & 1 & 0 & 1 \\
0 & 0 & 0 & 0 & 1 & 0 \\
0 & 0 & 1 & 2 & -2 & 2 \\
0 & 0 & 1 & 2 & 0 & 2
\end{array}\right)
\begin{array}{c}
R_3 = r_5 \\
\longrightarrow \\
R_4 = r_3 \\
R_5 = r_4
\end{array}
\left(\begin{array}{ccccc|c}
2 & 0 & 0 & 1 & 2 & 0 \\
0 & 1 & 1 & 1 & 0 & 1 \\
0 & 0 & 1 & 2 & 0 & 2 \\
0 & 0 & 0 & 0 & 1 & 0 \\
0 & 0 & 1 & 2 & -2 & 2
\end{array}\right)
$$

$$\begin{array}{c} R_2=-r_3+r_2 \\ \xrightarrow{} \\ R_5=r_3-r_5 \end{array} \left(\begin{array}{ccccc|c} 2 & 0 & 0 & 1 & 2 & 0 \\ 0 & 1 & 0 & -1 & 0 & -1 \\ 0 & 0 & 1 & 2 & 0 & 2 \\ 0 & 0 & 0 & 0 & 1 & 0 \\ 0 & 0 & 0 & 0 & 2 & 0 \end{array}\right) \quad \begin{array}{c} R_1=-2r_4+r_1 \\ \xrightarrow{} \\ R_5=-2r_4+r_5 \end{array} \left(\begin{array}{ccccc|c} 2 & 0 & 0 & 1 & 0 & 0 \\ 0 & 1 & 0 & -1 & 0 & -1 \\ 0 & 0 & 1 & 2 & 0 & 2 \\ 0 & 0 & 0 & 0 & 1 & 0 \\ 0 & 0 & 0 & 0 & 0 & 0 \end{array}\right)$$

$$\begin{array}{c} R_1=(1/2)r_1 \\ \xrightarrow{} \end{array} \left(\begin{array}{ccccc|c} 1 & 0 & 0 & 1/2 & 0 & 0 \\ 0 & 1 & 0 & -1 & 0 & -1 \\ 0 & 0 & 1 & 2 & 0 & 2 \\ 0 & 0 & 0 & 0 & 1 & 0 \\ 0 & 0 & 0 & 0 & 0 & 0 \end{array}\right).$$

The corresponding system of equations for the reduced echelon form of the matrix in (2.10) is

$$\begin{array}{rcl} \alpha \quad\quad + (1/2)\tau & = & 0 \\ \beta \quad - \quad \tau & = & -1 \\ \gamma + \quad 2\tau & = & 2 \\ \delta & = & 0 \\ 0 & = & 0 \\ 0 & = & 0. \end{array} \tag{2.11}$$

From this system, we can deduce that $\delta = 0$ and α, β, and γ can be obtained for a given τ. That is, for each real number we choose for τ, we get a solution. Some example solutions are

$$\begin{array}{ll} (0, -1, 2, 0, 0) & \text{(for } \tau = 0) \\ (-1/2, 0, 0, 1, 0) & \text{(for } \tau = 1) \\ (-1, 1, -2, 2, 0) & \text{(for } \tau = 2). \end{array}$$

Since we can choose any real value for τ, we see that there are infinitely many solutions to Equation 2.8.
□

In Example 2.2.18, we had five variables in our system of equations. The corresponding augmented matrix had a column, to the left of the augment, corresponding to each variable. In the reduction process, one column did not have a leading entry. This happened to be the column corresponding to the variable τ. So, the solutions depend on the choice of τ. In cases like this, we say τ is a *free variable*. There are a few items to note about such scenarios. The system of equations in Example 2.2.18 need not be written with the variables in the order we wrote them. Indeed, the system is still the same system if we rewrite it by moving the α term in all equations to be the fourth term instead of the first

$$\begin{array}{rcl} 8\beta + 4\gamma \quad\quad\quad\quad + 8\delta & = & 0 \\ 4\tau + 8\alpha + 8\delta & = & 0 \\ 4\gamma + 8\tau \quad\quad\quad\quad & = & 8 \\ 8\beta + 8\gamma + 12\tau + 8\alpha + 8\delta & = & 8 \\ 8\beta + 8\gamma + 12\tau + 8\alpha + 16\delta & = & 8. \end{array} \tag{2.12}$$

Matrix reduction (we will not show the steps) leads to the reduced echelon form

$$
\begin{pmatrix}
8 & 4 & 0 & 0 & 8 & | & 0 \\
0 & 0 & 4 & 8 & 8 & | & 0 \\
0 & 4 & 8 & 0 & 0 & | & 8 \\
8 & 8 & 12 & 8 & 8 & | & 8 \\
8 & 8 & 12 & 8 & 16 & | & 8
\end{pmatrix}
\sim
\begin{pmatrix}
1 & 0 & 0 & 2 & 0 & | & -1 \\
0 & 1 & 0 & -4 & 0 & | & 2 \\
0 & 0 & 1 & 2 & 0 & | & 0 \\
0 & 0 & 0 & 0 & 1 & | & 0 \\
0 & 0 & 0 & 0 & 0 & | & 0
\end{pmatrix}. \tag{2.13}
$$

We can write system of equations corresponding to the reduced echelon form because we know in which order the variables were in the system. Indeed, the corresponding system is

$$
\begin{aligned}
\beta \qquad + 2\alpha \quad &= -1 \\
\gamma \quad - 4\alpha \quad &= 2 \\
\tau + 2\alpha \quad &= 0 \\
\delta &= 0 \\
0 &= 0.
\end{aligned} \tag{2.14}
$$

In this case, we say that α is a free variable because we can choose any real number for α and obtain a solution.

Not only can we rewrite the system of equations with the variables in a different order, we can use different elimination steps (or row operations in the case of matrix reduction). Indeed, the matrix corresponding to the system of equations in Example 2.2.18 can be reduced as follows:

$$
\begin{pmatrix}
8 & 0 & 0 & 4 & 8 & | & 0 \\
8 & 8 & 8 & 12 & 8 & | & 8 \\
8 & 8 & 8 & 12 & 16 & | & 8 \\
0 & 8 & 4 & 0 & 8 & | & 0 \\
0 & 0 & 4 & 8 & 0 & | & 8
\end{pmatrix}
\begin{array}{c}
R_1=(1/4)r_1 \\
R_2=-r_1+r_2 \\
\longrightarrow \\
R_3=-r_1+r_3
\end{array}
\begin{pmatrix}
2 & 0 & 0 & 1 & 2 & | & 0 \\
0 & 8 & 8 & 8 & 0 & | & 8 \\
0 & 8 & 8 & 8 & 8 & | & 8 \\
0 & 8 & 4 & 0 & 8 & | & 0 \\
0 & 0 & 4 & 8 & 0 & | & 8
\end{pmatrix}
\begin{array}{c}
R_2=-r_4+r_2 \\
\longrightarrow \\
R_3=-r_4+r_3
\end{array}
\begin{pmatrix}
2 & 0 & 0 & 1 & 2 & | & 0 \\
0 & 0 & 4 & 8 & -8 & | & 8 \\
0 & 0 & 4 & 8 & 0 & | & 8 \\
0 & 8 & 4 & 0 & 8 & | & 0 \\
0 & 0 & 4 & 8 & 0 & | & 8
\end{pmatrix}
$$

$$
\begin{array}{c}
R_2=(1/4)r_2 \\
R_3=-r_2+r_3 \\
\longrightarrow \\
R_4=-r_2+r_4 \\
R_5=-r_2+r_5
\end{array}
\begin{pmatrix}
2 & 0 & 0 & 1 & 2 & | & 0 \\
0 & 0 & 1 & 2 & -2 & | & 2 \\
0 & 0 & 0 & 0 & 8 & | & 0 \\
0 & 8 & 0 & -8 & 16 & | & -8 \\
0 & 0 & 0 & 0 & 8 & | & 0
\end{pmatrix}
\begin{array}{c}
R_1=-(1/4)r_3+r_1 \\
R_2=(1/4)r_3+r_2 \\
\longrightarrow \\
R_4=-2r_3+r_4 \\
R_5=-r_3+r_5
\end{array}
\begin{pmatrix}
2 & 0 & 0 & 1 & 0 & | & 0 \\
0 & 0 & 1 & 2 & 0 & | & 2 \\
0 & 0 & 0 & 0 & 8 & | & 0 \\
0 & 8 & 0 & -8 & 0 & | & -8 \\
0 & 0 & 0 & 0 & 0 & | & 0
\end{pmatrix}
$$

$$
\begin{array}{c}
R_2=-2r_1+r_2 \\
R_3=(1/8)r_3 \\
\longrightarrow \\
R_4=2r_1+r_4
\end{array}
\begin{pmatrix}
2 & 0 & 0 & 1 & 0 & | & 0 \\
-4 & 0 & 1 & 0 & 0 & | & 2 \\
0 & 0 & 0 & 0 & 1 & | & 0 \\
16 & 8 & 0 & 0 & 0 & | & -8 \\
0 & 0 & 0 & 0 & 0 & | & 0
\end{pmatrix}
\begin{array}{c}
\longrightarrow \\
R_4=(1/8)r_4
\end{array}
\begin{pmatrix}
2 & 0 & 0 & 1 & 0 & | & 0 \\
-4 & 0 & 1 & 0 & 0 & | & 2 \\
0 & 0 & 0 & 0 & 1 & | & 0 \\
2 & 1 & 0 & 0 & 0 & | & -1 \\
0 & 0 & 0 & 0 & 0 & | & 0
\end{pmatrix}.
$$

The corresponding system of equations is

$$
\begin{aligned}
2\alpha \qquad\quad + \tau &= 0 \\
-4\alpha \quad + \gamma \qquad &= 2 \\
\delta &= 0 \\
2\alpha + \beta \qquad &= -1 \\
0 &= 0.
\end{aligned}
$$

In this case, we want to choose α to be free because α appears in more than one equation.

> **Fact 2.2.19**
>
> The reader should recognize that, in the case of infinitely many solutions, the variables that we say are free depends upon how we choose to eliminate in the system or reduce the matrix.

It should also be noted that one variable in Example 2.2.18 is never free. Do you see which one? (See Exercise 28.)

As mathematicians, we always want to observe patterns in our results, in hopes to make the next similar problem more efficient. So, let us return to our first two examples. First wanted to find a solution (α, β, γ) to the system in (2.4). We found one solution. Let us look at the process using matrix reduction. Consider the augmented matrix corresponding to System 2.4, the augmented matrix corresponding to our last step in elimination, and the reduced echelon form of this matrix,

$$
\begin{pmatrix}
0 & 4 & 8 & 8 \\
0 & 0 & 8 & 0 \\
8 & 4 & 0 & 0 \\
8 & 8 & 8 & 8 \\
8 & 8 & 16 & 8
\end{pmatrix}
\sim
\begin{pmatrix}
0 & 1 & 2 & 2 \\
0 & 0 & 1 & 0 \\
0 & -1 & 0 & -2 \\
1 & 1 & 1 & 1 \\
0 & 0 & 1 & 0
\end{pmatrix}
\sim
\begin{pmatrix}
1 & 0 & 0 & -1 \\
0 & 1 & 0 & 2 \\
0 & 0 & 1 & 0 \\
0 & 0 & 0 & 0 \\
0 & 0 & 0 & 0
\end{pmatrix}. \tag{2.15}
$$

In the reduced echelon form the right column holds the solution because the matrix corresponds to the system

$$
\alpha = -1, \quad \beta = 2, \quad \gamma = 0 \text{ (and two equations that say } 0 = 0\text{)}.
$$

Notice that there were as many leading entries to the left of the augment line as there are variables. This is key! Exercise 29 asks how you know this will always lead to a unique solution.

Second, we sought a solution to Equation 2.5. The reduced form of the corresponding matrix had a leading entry to the right of the augment line. This is key to recognizing that there is no solution. Exercise 30 asks you to explain how this always tells us there is no solution.

Finally, in Example 2.2.18, all leading entries were to the left of the augment line, but there were fewer leading entries than variables. In this scenario, we see there are infinitely many solutions. Exercise 31 asks you to explain this scenario.

It turns out that these are the only types of solution results for systems of linear equations. We put this all together in the next theorem.

Theorem 2.2.20

Consider the system of m equations with n variables

$$
\begin{aligned}
a_{1,1}x_1 + a_{1,2}x_3 + \ldots a_{1,n}x_n &= b_1 \\
a_{2,1}x_1 + a_{2,2}x_3 + \ldots a_{2,n}x_n &= b_2 \\
&\;\;\vdots \\
a_{m,1}x_1 + a_{m,2}x_3 + \ldots a_{m,n}x_n &= b_m.
\end{aligned}
\tag{2.16}
$$

Let $[M|b]$ be the corresponding augmented coefficient matrix with,

$$
M = \begin{pmatrix}
a_{1,1} & a_{1,2} & \cdots & a_{1,n} \\
a_{2,1} & a_{2,2} & \cdots & a_{2,n} \\
\vdots & & \ddots & \vdots \\
a_{m,1} & a_{m,2} & \cdots & a_{m,n}
\end{pmatrix}
\quad \text{and} \quad
b = \begin{pmatrix}
b_1 \\ b_2 \\ \vdots \\ b_m
\end{pmatrix}.
\tag{2.17}
$$

Then the following statements hold.

1. The system of equations in (2.16) has a unique solution if and only if the number of leading entries in the reduced echelon forms of both M and $[M|b]$ is n.
2. The system of equations in (2.16) has no solution if the number of leading entries in the reduced echelon form of M is less than the number of leading entries in the reduced echelon form of $[M|b]$.
3. The system of equations in (2.16) has infinitely many solutions if and only if the number of leading entries in the reduced echelon forms of M and $[M|b]$ is equal, but less than n.

If we consider a system of equations with n equations and n variables then Theorem 2.2.20 tells us that if we reduce M to find that there is a leading entry in every column then the system of equations corresponding to $[M|b]$ (no matter what b is) has a unique solution. So, if $m < n$ then the system cannot have a unique solution.

The phrase "if and only if" is a way of saying that two statements are equivalent. We discuss statements with this phrase more in Appendix C. This language will occur often in this text.

2.2.3 Elementary Matrix

In this section, we will briefly connect matrix reduction to a set of matrix products[5]. This connection will prove useful later. To begin, let us define an elementary matrix. We begin with the identity matrix.

Definition 2.2.21

The $n \times n$ **identity matrix,** I_n is the matrix that satisfies $I_n M = M I_n = M$ for all $n \times n$ matrices M.

□

---

[5] For the definition and examples of matrix products, see Section 3.1.2.

One can show, by inspection, that I_n must be the matrix with n 1's down the diagonal and 0's everywhere else:

$$I = \begin{pmatrix} 1 & 0 & \dots & 0 \\ 0 & 1 & & 0 \\ \vdots & & \ddots & \vdots \\ 0 & & \dots & 1 \end{pmatrix}.$$

Definition 2.2.22

An $n \times n$ **elementary matrix** E is a matrix that can be obtained by performing one row operation on I_n.

$\square$

Let us give a couple examples of elementary matrices before we give some results.

Example 2.2.23 The following are 3×3 elementary matrices:

- $E_1 = \begin{pmatrix} 1 & 0 & 0 \\ 0 & 0 & 1 \\ 0 & 1 & 0 \end{pmatrix}$ is obtained by changing the order of rows 2 and 3 of the identity matrix.

- $E_2 = \begin{pmatrix} 2 & 0 & 0 \\ 0 & 1 & 0 \\ 0 & 0 & 1 \end{pmatrix}$ is obtained by multiplying row 1 of I_3 by 2.

- $E_3 = \begin{pmatrix} 1 & 0 & 0 \\ 3 & 1 & 0 \\ 0 & 0 & 1 \end{pmatrix}$ is obtained by adding 3 times row 1 to row 2.

Since $M = \begin{pmatrix} 2 & 0 & 0 \\ -3 & 1 & 0 \\ 0 & 0 & 1 \end{pmatrix}$ cannot be obtained by performing a single row operation on I_3, so is **not** an elementary matrix.

$\square$

Let us now see how these are related to matrix reduction. Consider the following example:

Example 2.2.24 Let $M = \begin{pmatrix} 2 & 3 & 5 \\ 1 & 2 & 1 \\ 3 & 4 & -1 \end{pmatrix}$. Let us see what happens when we multiply by each of the elementary matrices in Example 2.2.23.

$$E_1 M = \begin{pmatrix} 1 & 0 & 0 \\ 0 & 0 & 1 \\ 0 & 1 & 0 \end{pmatrix} \begin{pmatrix} 2 & 3 & 5 \\ 1 & 2 & 1 \\ 3 & 4 & -1 \end{pmatrix} = \begin{pmatrix} 2 & 3 & 5 \\ 3 & 4 & -1 \\ 1 & 2 & 1 \end{pmatrix}.$$

The matrix multiplication results in M altered by changing rows 2 and 3.

$$E_2 M = \begin{pmatrix} 2 & 0 & 0 \\ 0 & 1 & 0 \\ 0 & 0 & 1 \end{pmatrix} \begin{pmatrix} 2 & 3 & 5 \\ 1 & 2 & 1 \\ 3 & 4 & -1 \end{pmatrix} = \begin{pmatrix} 4 & 6 & 10 \\ 1 & 2 & 1 \\ 3 & 4 & -1 \end{pmatrix}.$$

The matrix multiplication results in M altered by multiplying row 1 by 2.

$$E_3 M = \begin{pmatrix} 1 & 0 & 0 \\ 3 & 1 & 0 \\ 0 & 0 & 1 \end{pmatrix} \begin{pmatrix} 2 & 3 & 5 \\ 1 & 2 & 1 \\ 3 & 4 & -1 \end{pmatrix} \begin{pmatrix} 2 & 3 & 5 \\ 7 & 11 & 16 \\ 3 & 4 & -1 \end{pmatrix}.$$

The matrix multiplication results in M altered by adding 3 times row 1 to row 2.

Indeed, multiplying a matrix M by an elementary matrix has the same result as performing the corresponding row operation! □

In fact, we could reduce a matrix to echelon or reduced echelon form using multiplication by elementary matrices. Indeed, consider the following example.

Example 2.2.25 Consider the problem of reducing the matrix $M = \begin{pmatrix} 1 & 2 & 4 \\ 2 & 4 & 1 \\ 1 & 1 & 1 \end{pmatrix}$ to reduced echelon form. A possible first step is to replace row 2 by multiplying row 1 by -2 and adding it to row 2. This is the same as multiplying M on the left by $E_1 = \begin{pmatrix} 1 & 0 & 0 \\ -2 & 1 & 0 \\ 0 & 0 & 1 \end{pmatrix}$. We get

$$E_1 M = \begin{pmatrix} 1 & 0 & 0 \\ -2 & 1 & 0 \\ 0 & 0 & 1 \end{pmatrix} \begin{pmatrix} 1 & 2 & 4 \\ 2 & 4 & 1 \\ 1 & 1 & 1 \end{pmatrix} = \begin{pmatrix} 1 & 2 & 4 \\ 0 & 0 & -7 \\ 1 & 1 & 1 \end{pmatrix}.$$

The next step might be to replace row 3 by multiplying row 1 by -1 and adding to row 3. This is the same as multiplying $E_1 M$ on the left by $E_2 = \begin{pmatrix} 1 & 0 & 0 \\ 0 & 1 & 0 \\ -1 & 0 & 1 \end{pmatrix}$. We get

$$E_2(E_1 M) = \begin{pmatrix} 1 & 0 & 0 \\ 0 & 1 & 0 \\ -1 & 0 & 1 \end{pmatrix} \begin{pmatrix} 1 & 2 & 4 \\ 0 & 0 & -7 \\ 1 & 1 & 1 \end{pmatrix} = \begin{pmatrix} 1 & 2 & 4 \\ 0 & 0 & -7 \\ 0 & -1 & -3 \end{pmatrix}.$$

The next three steps might be to replace rows 2 and 3 by doing three operations. First, swap rows 2 and 3 and then multiply the new rows 2 and 3 by -1 and $-1/7$, respectively. This is equivalent to left-multiplying by the elementary matrices

$$E_3 = \begin{pmatrix} 1 & 0 & 0 \\ 0 & 0 & 1 \\ 0 & 1 & 0 \end{pmatrix}, \quad E_4 = \begin{pmatrix} 1 & 0 & 0 \\ 0 & -1 & 0 \\ 0 & 0 & 1 \end{pmatrix}, \quad \text{and } E_5 = \begin{pmatrix} 1 & 0 & 0 \\ 0 & 1 & 0 \\ 0 & 0 & -1/7 \end{pmatrix}.$$

This multiplication results in

$$E_5 E_4 E_3(E_2 E_1 M) = \begin{pmatrix} 1 & 0 & 0 \\ 0 & 1 & 0 \\ 0 & 0 & -1/7 \end{pmatrix} \begin{pmatrix} 1 & 0 & 0 \\ 0 & -1 & 0 \\ 0 & 0 & 1 \end{pmatrix} \begin{pmatrix} 1 & 0 & 0 \\ 0 & 0 & 1 \\ 0 & 1 & 0 \end{pmatrix} \begin{pmatrix} 1 & 2 & 4 \\ 0 & 0 & -7 \\ 0 & -1 & -3 \end{pmatrix} = \begin{pmatrix} 1 & 2 & 4 \\ 0 & 1 & 3 \\ 0 & 0 & 1 \end{pmatrix}.$$

Next, we replace row 1 by multiplying row 2 by -2 and adding it to row 1 or equivalently, we left-multiply by $E_6 = \begin{pmatrix} 1 & -2 & 0 \\ 0 & 1 & 0 \\ 0 & 0 & 1 \end{pmatrix}$. This results in

$$E_6(E_5 E_4 E_3 E_2 E_1 M) = \begin{pmatrix} 1 & -2 & 0 \\ 0 & 1 & 0 \\ 0 & 0 & 1 \end{pmatrix} \begin{pmatrix} 1 & 2 & 4 \\ 0 & 1 & 3 \\ 0 & 0 & 1 \end{pmatrix} = \begin{pmatrix} 1 & 0 & 10 \\ 0 & 1 & 3 \\ 0 & 0 & 1 \end{pmatrix}.$$

Finally, the last two steps are completed by left-multiplying by the elementary matrices

$$E_7 = \begin{pmatrix} 1 & 0 & -10 \\ 0 & 1 & 0 \\ 0 & 0 & 1 \end{pmatrix} \quad \text{and } E_8 = \begin{pmatrix} 1 & 0 & 0 \\ 0 & 1 & -3 \\ 0 & 0 & 1 \end{pmatrix}$$

which will replace row 1 by multiplying row 3 by -10 and adding to row 1 and then replacing row 2 by multiplying row 3 by -3 and adding to row 2. This results in

$$E_8 E_7(E_6 E_5 E_4 E_3 E_2 E_1 M) = \begin{pmatrix} 1 & 0 & 0 \\ 0 & 1 & -3 \\ 0 & 0 & 1 \end{pmatrix} \begin{pmatrix} 1 & 0 & -10 \\ 0 & 1 & 0 \\ 0 & 0 & 1 \end{pmatrix} \begin{pmatrix} 1 & 0 & 10 \\ 0 & 1 & 3 \\ 0 & 0 & 1 \end{pmatrix} = \begin{pmatrix} 1 & 0 & 0 \\ 0 & 1 & 0 \\ 0 & 0 & 1 \end{pmatrix} = I.$$

So, we see that we can reduce M, in this case to I, by multiplying by elementary matrices in this way:

$$I = E_8 E_7 E_6 E_5 E_4 E_3 E_2 E_1 M.$$

$\square$

The same process works for matrices that that do not reduce to the identity. Let us see this in a quick example.

Example 2.2.26 Let $M = \begin{pmatrix} 1 & 2 & 3 \\ 1 & 2 & 3 \\ 2 & 4 & 6 \end{pmatrix}$. We can reduce M to reduced echelon form by replacing rows 2 and 3 by two row operations. First, we replace row 2 by multiplying row 1 by -1 and adding to row 2. Second, we replace row 3 by multiplying row 1 by -2 and adding to row 3. These row operations are equivalent to multiplying by the elementary matrices

$$E_1 = \begin{pmatrix} 1 & 0 & 0 \\ -1 & 1 & 0 \\ 0 & 0 & 1 \end{pmatrix} \quad \text{and } E_2 = \begin{pmatrix} 1 & 0 & 0 \\ 0 & 1 & 0 \\ -2 & 0 & 1 \end{pmatrix}.$$

Our final result is

$$E_2 E_1 M = \begin{pmatrix} 1 & 0 & 0 \\ 0 & 1 & 0 \\ -2 & 0 & 1 \end{pmatrix} \begin{pmatrix} 1 & 0 & 0 \\ -1 & 1 & 0 \\ 0 & 0 & 1 \end{pmatrix} \begin{pmatrix} 1 & 2 & 3 \\ 1 & 2 & 3 \\ 2 & 4 & 6 \end{pmatrix} = \begin{pmatrix} 1 & 2 & 3 \\ 0 & 0 & 0 \\ 0 & 0 & 0 \end{pmatrix}.$$

Even though M does not reduce to the identity, multiplying by elementary matrices still works to reduce M to reduced echelon form. $\square$

In this section, we built tools that helped us systematically answer the question of Section 2.1 about Images 3 and 4. More importantly, we have tools to answer many similar questions. To this point, we have used familiar operations (addition and multiplication) on images. In Section 2.3, we will explore the properties of addition and multiplication on the real numbers and how they are related to properties for the same operations, but on images. First, we will review the proper use of terminology presented in the subsection.

★★ **Watch Your Language!** When solving systems of equations or reducing matrices, it is important to use the proper language surrounding each.

    ✓ We solve a system of equations.
    ✓ We reduce a matrix.
    ✓ The system of equations has a free variable.
    ✓ The solution to the system of equations is $(a_1, a_2, ..., a_n)$.

We do not say

    ✗ We solve a matrix.
    ✗ The matrix has a free variable.
    ✗ The solutions to the system of equations are $a_1, a_2, ..., a_n$.

2.2.4 The Geometry of Systems of Equations

It turns out that there is an intimate connection between solutions to systems of equations in two variables and the geometry of lines in $\mathbb{R}^2$. We recall the graphical method to solving systems below. Although you will likely have already done this in previous classes, we include it here so that you can put this knowledge into the context of solution sets to systems of equations as classified in Theorem 2.2.20.

We begin with the following simple example:

Example 2.2.27 Let us consider $u = \begin{pmatrix} 2 \\ -3 \end{pmatrix}$, $v = \begin{pmatrix} 1 \\ 1 \end{pmatrix}$, and $w = \begin{pmatrix} 2 \\ 3 \end{pmatrix} \in \mathbb{R}^2$. Suppose we want to know if we can express u using arithmetic operations on v and w. In other words, we want to know if there are scalars x, y so that

$$\begin{pmatrix} 2 \\ -3 \end{pmatrix} = x \cdot \begin{pmatrix} 1 \\ 1 \end{pmatrix} + y \cdot \begin{pmatrix} 2 \\ 3 \end{pmatrix}.$$

We can rewrite the right-hand side of the vector equation so that we have the equation with two vectors

$$\begin{pmatrix} 2 \\ -3 \end{pmatrix} = \begin{pmatrix} x + 2y \\ x + 3y \end{pmatrix}.$$

The equivalent system of linear equations with 2 equations and 2 variables is

$$x + 2y = 2 \tag{2.18}$$
$$x + 3y = -3. \tag{2.19}$$

Equations (2.18) and (2.19) are equations of lines in $\mathbb{R}^2$, that is, the set of pairs (x, y) that satisfy each equation is the set of points on each respective line. Hence, finding x and y that satisfy both equations amounts to finding all points (x, y) that are on both lines. If we graph these two lines, we can see that they appear to cross at one point, $(12, -5)$, and nowhere else, so we estimate $x = 12$ and $y = -5$ is the only solution of the two equations. (See Figure 2.9.) You can also algebraically verify that $(12, 5)$ is a solution to the system. □

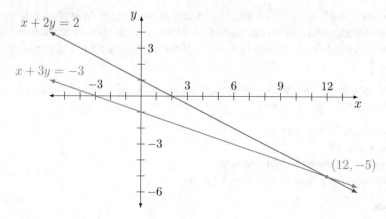

Fig. 2.9 The equations of Example 2.2.27.

The graphical method for solving systems of equations can be very inaccurate when the solution is a point that does not line up very well with integer coordinates. In general, we prefer using algebraic techniques to solve systems.

However, the geometry does give us some intuition about what is going on. Recall that Theorem 2.2.20 classified systems of equations by their solution sets; a system has either a unique solution, no solution, or infinitely many solutions. Example 2.2.27 showed the graphical interpretation of a case where there is a unique solution to the system.

Exercise 35 asks you to explore the other two cases:

- What would a system of two equations in two variables and no solution look like, graphically?
- What would a system of two equations in two variables and infinitely many solutions look like, graphically? (Fun fact: this gives you some intuition about why it is impossible to have a system of equations with exactly two solutions. See Exercise 35.)

For systems with more variables, Theorem 2.2.20 guarantees that we have the same three possible types of solution sets: The empty set, the set containing only one point, and a set containing infinitely many points. To see this geometrically, you need to know that solution sets to linear equations with n variables correspond to $(n-1)$-dimensional hyperplanes in $\mathbb{R}^n$. These hyperplanes can intersect in either a single point, infinitely many points, or not at all.

For ease of visualization, we illustrate the scenario involving a system of three equations in three variables, in which case each equation corresponds to a two-dimensional plane in $\mathbb{R}^3$. Figure 2.10 shows some of the possible ways three planes will cross. There are other ways three planes can intersect. In Exercise 33, you will be asked to describe the other possibilities.

Fig. 2.10 Geometric visualization of possible solution sets (in black) for systems of 3 equations and 3 variables. What are other possible configurations of planes so that there is no solution? (See Exercise 33.)

2.2.5   Exercises

Using the method of elimination, solve each system of equations:

1. $\begin{aligned} x + y &= 2 \\ x - y &= 1 \end{aligned}$

2. $\begin{aligned} 2x + 3y &= -5 \\ 2x - 2y &= 10 \end{aligned}$

3. Which of the following points are solutions to the given system of equations?

$$\begin{aligned} x_1 + 2x_2 - x_3 + x_4 &= 4 \\ x_1 + 2x_2 \quad - x_4 &= 2 \\ -x_1 - 2x_2 - x_3 + 3x_4 &= 0 \end{aligned}$$

   (a) $(2, 0, -2, 0)$
   (b) $(0, 1, -2, 1)$
   (c) $(1, 1, 0, 1)$
   (d) $(2 + a, 0, 2a - 2, a)$, where $a \in \mathbb{R}$

Solve each of the systems of equations below. If the system has infinitely many solutions, choose your free variables, write out 3, and determine a pattern to the solutions.

4.
$$\begin{aligned} x - y - z &= 4 \\ 2x - y + 3z &= 2 \\ -x + y - 2z &= -1 \end{aligned}$$

5.
$$\begin{aligned} 2x - y - 3z &= 1 \\ 3x + y - 3z &= 4 \\ -2x + y + 2z &= -1 \end{aligned}$$

6.
$$\begin{aligned} x - 2y - 3z &= 2 \\ 4x + y - 2z &= 8 \\ 5x - y - 5z &= 10 \end{aligned}$$

7.
$$\begin{aligned} x - 2y - 3z &= 2 \\ 4x + y - 2z &= 8 \\ 5x - y - 5z &= 3 \end{aligned}$$

8.
$$\begin{aligned} x - 2y - 3z &= 2 \\ 2x - 4y - 6z &= 4 \\ -x + 2y + 3z &= -2 \end{aligned}$$

Use an augmented matrix to solve each of the systems of equations below by reducing the matrix to reduced echelon form. If the system has infinitely many solutions, choose your free variables, write out 3, and determine a pattern to the solutions.

9.
$$\begin{aligned} x - y - z &= 4 \\ 2x - y + 3z &= 2 \\ -x + y - 2z &= -1 \end{aligned}$$

10.
$$\begin{aligned} 2x - y - 3z &= 1 \\ 3x + y - 3z &= 4 \\ -2x + y + 2z &= -1 \end{aligned}$$

11.
$$\begin{aligned} x - 2y - 3z &= 2 \\ 4x + y - 2z &= 8 \\ 5x - y - 5z &= 10 \end{aligned}$$

12.
$$\begin{aligned} x - 2y - 3z &= 2 \\ 4x + y - 2z &= 8 \\ 5x - y - 5z &= 3 \end{aligned}$$

13. One needs to be very cautious when combining row operations into a single "step." Start with the matrix A given below:

$$A = \begin{pmatrix} 1 & 1 & 2 & 3 \\ 5 & 8 & 13 & 0 \\ 3 & 2 & 1 & 1 \end{pmatrix}.$$

(a) First perform all of the following row operations on the matrix A in a single step:
  - $R_1 = 2r_1 - r_2$
  - $R_2 = r_2 + 2r_3$
  - $R_3 = r_3 + (1/2)r_2$.

What matrix results?

(b) Next, sequentially perform the row operations on A, showing the intermediate matrix after each step. What matrix results?

(c) Explain what happened. Why does this mean you need to be cautious about combining row operations?

For Exercises 14–20, determine the elementary matrices needed to transform $M = \begin{pmatrix} 2 & 4 & 1 \\ 3 & 2 & 2 \\ 1 & 2 & 1 \end{pmatrix}$ to M'.

Use matrix multiplication to verify your answer.

14. $M' = \begin{pmatrix} 2 & 4 & 1 \\ -6 & -4 & -4 \\ 1 & 2 & 1 \end{pmatrix}.$

15. $M' = \begin{pmatrix} 1 & 2 & 1 \\ 3 & 2 & 2 \\ 2 & 4 & 1 \end{pmatrix}.$

16. $M' = \begin{pmatrix} 0 & 0 & -1 \\ 3 & 2 & 2 \\ 1 & 2 & 1 \end{pmatrix}.$

17. $M' = \begin{pmatrix} -2 & -4 & -1 \\ 6 & 4 & 4 \\ 1 & 2 & 1 \end{pmatrix}$.

18. $M' = \begin{pmatrix} 1 & 2 & 1 \\ 3 & 2 & 2 \\ 2 & 4 & 1 \end{pmatrix}$.

19. $M' = \begin{pmatrix} 1 & 2 & 1 \\ 0 & 0 & -1 \\ 0 & -4 & -1 \end{pmatrix}$.

20. $M' = \begin{pmatrix} 1 & 0 & 0 \\ 0 & 1 & 0 \\ 0 & 0 & 1 \end{pmatrix}$.

Additional Questions.

21. Heather has one week left on her summer job doing landscaping for the city. She can take on three types of jobs which require 5, 6, and 7 hours, respectively. She has agreed to complete seven jobs in the coming week. How many of each type of job should she complete in order to work a full 40-hour week?

22. The water flow rate w at a river gauge at time t is given in the following data table:

time ($hour$)	0	1	2
flow (ft^3/sec)	20	20	40

Find a quadratic function $w = at^2 + bt + c$ that fits the data. Discuss possible benefits of having this function. [Hint: Use the data to find a system of linear equations in the unknowns a, b, and c.]

23. The water flow rate w at a river gauge at time t is given in the following data table:

time ($hour$)	0	1	2	3
flow (ft^3/sec)	20	20	40	60

Show that no quadratic function $w = at^2 + bt + c$ fits the data exactly.

24. Wendy makes three types of wooden puzzles in her shop. The first type of puzzle takes 1 hour to make and sells for $10. The second type of puzzle takes 3 hours to make and sells for $20. The third type of puzzles takes 3 hours to make and sells for $30. What is the maximum revenue that Wendy can realize if she works for 100 hours making puzzles?

25. In Example 2.2.18, one of the variables can never be considered free, no matter the elimination process. Which variable is it? How do you know?

26. If a system of equations has corresponding augmented matrix whose reduced form has the same number of leading entries, on the left of the augment line, as variables, how do we know there is a unique solution?

27. Why do we know that there is no solution to a system of equations whose corresponding augmented matrix has reduced form with a leading entry to the right of the augment line?

28. If a system of equations has corresponding matrix whose reduced form has only leading entries to the left of the augment line, but there are fewer leading entries than variables, how do we know there are infinitely many solutions?

29. Prove that every matrix can be reduced to echelon form.
30. Prove that if a matrix P is row equivalent to a matrix Q (i.e., $P \sim Q$), then Q is row equivalent to P (i.e., $Q \sim P$).
31. Prove Lemma 2.2.12.
32. Find examples of systems of two equations and two variables that exemplify the other possibilities of Theorem 2.2.20.

   (a) What is the graphical interpretation of a system of two equations in two variables that has no solution? Give an example of a system of two equations in two variables that has no solution. Show both graphical and algebraic solutions for your system.
   (b) What is the graphical interpretation of a system of two equations in two variables that has infinitely many solutions? Give an example of a system of two equations in two variables that has infinitely many solutions. Show both graphical and algebraic solutions for your system.

33. Figure 2.10 depicts some of the possible solution sets to a system of 3 equations with 3 variables. Sketch or describe the other possibilities.
34. If $(x_1, x_2, \ldots, x_n)$ and $(y_1, y_2, \ldots, y_n)$ are two solutions to a system of linear equations (with m equations and n variables), what can you say about the n-tuple of averages: $((x_1 + y_1)/2, (x_2 + y_2)/2, \ldots, (x_n + y_n)/2)$? Justify your assertion.
35. Prove that if a system of equations has more than one solution, then it has infinitely many solutions.

2.3 Vector Spaces

In Section 2.1, we saw that the set of images possessed a number of convenient algebraic properties. It turns out that any set that possesses similar convenient properties can be analyzed in a similar way. In linear algebra, we study such sets and develop tools to analyze them. We call these sets *vector spaces*.

2.3.1 Images and Image Arithmetic

In Section 2.1 we saw that if you add two images, you get a new image, and that if you multiply an image by a scalar, you get a new image. We represented a rectangular pixelated image as an array of values, or equivalently, as a rectangular array of grayscale patches. This is a very natural idea in the context of digital photography.

Recall the definition of an image given in Section 2.1. We repeat it here, and follow the definition by some examples of images with different geometric arrangements.

Definition 2.3.1

An **image** is a finite ordered list of real values with an associated geometric arrangement. □

Four examples of arrays along with an index system specifying the order of patches can be seen in Figure 2.11. As an image, each patch would also have a numerical value indicating the brightness of the patch (not shown in the figure). The first is a regular pixel array commonly used for digital photography. The second is a hexagonal pattern which also nicely tiles a plane. The third is a map of the African continent and Madagascar subdivided by country. The fourth is a square pixel set with enhanced resolution toward the center of the field of interest. It should be clear from the definition that images are not matrices. Only the first example might be confused with a matrix.

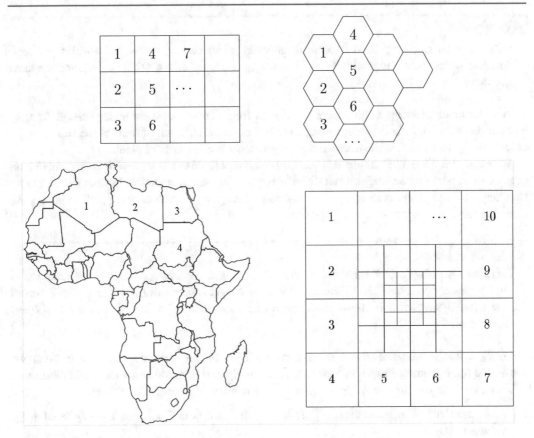

Fig. 2.11 Examples of image arrays. Numbers indicate example pixel ordering.

We first fix a particular geometric arrangement of pixels (and let n denote the number of pixels in the arrangement). Then an image is precisely described by its (ordered) intensity values. With this determined, we formalize the notions of scalar multiplication and addition on images that were developed in the previous section.

Definition 2.3.2

Given two images x and y with (ordered) intensity values $(x_1, x_2, \cdots, x_n)$ and $(y_1, y_2, \cdots, y_n)$, respectively, and the same geometry, the **image sum** written $z = x + y$ is the image with intensity values $z_i = x_i + y_i$ for all $i \in \{1, 2, \cdots, n\}$, and the same geometry. $\square$

Hence, the sum of two images is the image that results from pixel-wise addition of intensity values. Put another way, the sum of two images is the image that results from adding corresponding values of their ordered lists, while maintaining the geometric arrangement of pixels.

Definition 2.3.3

Given scalar α and image x with (ordered) intensity values $(x_1, x_2, \cdots, x_n)$, the **scalar product**, written $z = \alpha x$ is the image with intensity values $z_i = \alpha x_i$ for all $i \in \{1, 2, \cdots, n\}$, and the same geometry. $\square$

A scalar times an image is the image that results from pixel-wise scalar multiplication. That is, a scalar times an image is the image which results from multiplication of each of the values in the ordered list by that scalar, while maintaining the geometric arrangement of pixels.

We found that these basic arithmetic operations on images lead to a key property: any combination of arithmetic operations on images results in an image of the same configuration. In other words, adding two images *always* yields an image, and multiplying an image by a scalar *always* yields an image. We formalize this notion with the concept of closure:

Definition 2.3.4

Consider a set of objects X with scalars taken from $\mathbb{R}$, and operations of addition $(+)$ and scalar multiplication $(\cdot)$ defined on X. We say that X is **closed under addition** if $x + y \in X$ for all $x, y \in X$. We say that X is **closed under scalar multiplication** if $\alpha \cdot x \in X$ for each $x \in X$ and each $\alpha \in \mathbb{R}$. $\square$

Let $\mathcal{I}_{m \times n}$ denote the set of all $m \times n$ rectangular images. We see that the set $\mathcal{I}_{4 \times 4}$, from which we considered some example images in Section 2.1, is closed under addition and scalar multiplication.

This arithmetic with images in $\mathcal{I}_{4 \times 4}$ also satisfies a number of other natural properties:

- (Commutativity of image addition.) If I_A and I_B are images in $\mathcal{I}_{4 \times 4}$, then $I_A + I_B = I_B + I_A$. For example,

- (Associativity of image addition.) If I_A, I_B, and I_C are images in $\mathcal{I}_{4 \times 4}$, then $(I_A + I_B) + I_C = I_A + (I_B + I_C)$. For example,

- (Associativity of scalar multiplication.) If $\alpha, \beta \in \mathbb{R}$, and $I \in \mathcal{I}_{4 \times 4}$, then $\alpha(\beta I) = (\alpha\beta)I$, e.g.,

- (Distributivity of Scalar Multiplication over Image Addition.) If $\alpha \in \mathbb{R}$ and $I_A, I_B \in \mathcal{I}_{4\times 4}$, then $\alpha(I_A + I_B) = \alpha I_A + \alpha I_B$, e.g.,

$$7 \cdot \left(\underset{\text{Image A}}{\blacksquare} + \underset{\text{Image B}}{\blacksquare} \right) = 7 \cdot \underset{\text{Image A}}{\blacksquare} + 7 \cdot \underset{\text{Image B}}{\blacksquare} \, .$$

- (Additive identity image.) There is a zero image in $\mathcal{I}_{4\times 4}$—the image that has every pixel intensity equal to zero. The sum of the zero image and any other image I is just I.
- (Additive inverses.) For every image $I \in \mathcal{I}_{4\times 4}$, there is an image J so that the sum $I + J$ is just the zero image. (Recall that the set of images include those that can be captured by your camera, but there are many more, some with negative pixel intensities as well.)
- (Multiplicative identity.) For any image $I \in \mathcal{I}_{4\times 4}$ the scalar product $1 \cdot I = I$.

The fact that the space $\mathcal{I}_{4\times 4}$ of 4×4 has these properties will enable us to develop tools for working with images. In fact, we will be able to develop tools for *any* set (and field of scalars) that satisfies these properties. We will call such sets *vector spaces*.

2.3.2 Vectors and Vector Spaces

In the last section, we saw that the set of 4×4 images, together with real scalars, satisfies several natural properties. There are in fact many other sets of objects that also have these properties.

One class of objects with these properties are the vectors that you may have seen in a course in multivariable calculus or physics. In those courses, vectors are objects with a fixed number, say n, of values put together into an ordered tuple. That is, the word vector may bring to mind something that looks like $\langle a, b \rangle$, $\langle a, b, c \rangle$, or $\langle a_1, a_2, \ldots, a_n \rangle$. Maybe you have even seen notation like any of the following:

$$(a, b), \quad (a, b, c), \quad (a_1, a_2, \ldots, a_n), \quad \begin{pmatrix} a \\ b \\ c \end{pmatrix}, \quad \begin{bmatrix} a \\ b \\ c \end{bmatrix}, \quad \begin{pmatrix} a_1 \\ a_2 \\ \vdots \\ a_n \end{pmatrix}, \quad \begin{bmatrix} a_1 \\ a_2 \\ \vdots \\ a_n \end{bmatrix}$$

called vectors as well.

In this section, we generalize the notion of a vector. In particular, we will understand that images and other classes of objects can be vectors in an appropriate context. When we consider objects like brain images, radiographs, or heat state signatures, it is often useful to understand them as collections having certain natural mathematical properties. Indeed, we will develop mathematical tools that can be used on all such sets, and these tools will be instrumental in accomplishing our application tasks.

We haven't yet made the definition of a vector space (or even a vector) rigorous. We still have some more setup to do. In this text, we will primarily use two scalar fields[6]: $\mathbb{R}$ and Z_2. The field Z_2 is the two element (or binary) set $\{0, 1\}$ with addition and multiplication defined modulo 2. That is, addition defined modulo 2, means that:

---

[6] The definition of a field can be found in Appendix D. The important thing to remember about fields (for the material in this book) is that there are two operations (called addition and multiplication) that satisfy properties we usually see with real numbers.

$$0 + 0 = 0, \quad 0 + 1 = 1 + 0 = 1, \quad \text{and } 1 + 1 = 0.$$

And, multiplication defined modulo 2 means

$$0 \cdot 0 = 0, \quad 0 \cdot 1 = 1 \cdot 0 = 0, \quad \text{and } 1 \cdot 1 = 1.$$

We can think of the two elements as "on" and "off" and the operations as binary operations. If we add 1, we flip the switch and if we add 0, we do nothing. We know that Z_2 is closed under scalar multiplication and vector addition.

Definition 2.3.5

Consider a set V over a field $\mathbb{F}$ with given definitions for addition ($+$) and scalar multiplication ($\cdot$). V with $+$ and $\cdot$ is called a **vector space over** $\mathbb{F}$ if for all $u, v, w \in V$ and for all $\alpha, \beta \in \mathbb{F}$, the following ten properties hold:

(P1) **Closure Property for Addition** $u + v \in V$.
(P2) **Closure Property for Scalar Multiplication** $\alpha \cdot v \in V$.
(P3) **Commutative Property for Addition** $u + v = v + u$.
(P4) **Associative Property for Addition** $(u + v) + w = u + (v + w)$.
(P5) **Associative Property for Scalar Multiplication**
  $\alpha \cdot (\beta \cdot v) = (\alpha\beta) \cdot v$.
(P6) **Distributive Property of Scalar Multiplication Over Vector Addition** $\alpha \cdot (u + v) = \alpha \cdot u + \alpha \cdot v$.
(P7) **Distributive Property of Scalar Multiplication Over Scalar Addition** $(\alpha + \beta) \cdot v = \alpha \cdot v + \beta \cdot v$.
(P8) **Additive Identity Property** V contains the additive identity, denoted 0 so that $0 + v = v + 0 = v$ for every $v \in V$.
(P9) **Additive Inverse Property** V contains additive inverses z so that for every $v \in V$ there is a $z \in V$ satisfying $v + z = 0$.
(P10) **Multiplicative Identity Property for Scalars** The scalar set $\mathbb{F}$ has an identity element, denoted 1, for scalar multiplication that has the property $1 \cdot v = v$ for every $v \in V$. □

Notation. We will use the notation $(V, +, \cdot)$ to indicate the vector space corresponding to the set V with operations $+$ and $\cdot$.

In Definition 2.3.5, we label the properties as (P1)–(P10). It should be noted that though we use this labeling in other places in this text, these labels are not standard. This means that you should focus more on the property description and name rather than the labeling.

Definition 2.3.6

Given a vector space $(V, +, \cdot)$ over $\mathbb{F}$. We say that $v \in V$ is a **vector**. That is, elements of a vector space are called **vectors**. □

In this text, we will use context to indicate vectors with a letter, such as v or x. In some courses and textbooks, vectors are denoted with an arrow over the name, $\vec{v}$, or with bold type, $\mathbf{v}$. Also, we usually write scalar multiplication using juxtaposition, i.e., we write αx rather than $\alpha \cdot x$ to distinguish from the dot product that we will encounter in Section 7.1.

We will discuss many vector spaces for which there are operations and fields that are typically used. For example, the set of real numbers $\mathbb{R}$ is a set for which we have a common understanding about what it means to add and multiply. For these vector spaces, we call the operations the "standard operations" and the field is called the "standard field." So, we might typically say that $\mathbb{R}$, with the standard operations, is a vector space over itself.

⋆⋆ **Watch Your Language!** The language used to specify a vector space requires that we state the set V, the two operations $+$ and $\cdot$, and the field $\mathbb{F}$. We make this clear with notation and/or in words. Two ways to communicate this are

   ✓ $(V, +, \cdot)$ is a vector space over $\mathbb{F}$.

   ✓ V with the operations $+$ and $\cdot$ is a vector space over the field $\mathbb{F}$.

   We should not say (unless ambiguity has been removed)

   ✗ V is a vector space.

Definition 2.3.5 is so important and has so many pieces that we will take the time to present many examples. As we do so, consider the following. The identity element for scalar multiplication need not be the number 1. The zero vector need not be (and in general is not) the number 0. The elements of a vector space are called **vectors** but need not look like the vectors presented above.

Example 2.3.7 The set of 4×4 images $\mathcal{I}_{4\times4}$ satisfies properties (P1)–(P10) of a vector space.   □

Example 2.3.8 Let us consider the set

$$\mathbb{R}^3 = \left\{ \begin{pmatrix} a_1 \\ a_2 \\ a_3 \end{pmatrix} \;\middle|\; a_1, a_2, a_3 \in \mathbb{R} \right\}.$$

This means that $\mathbb{R}^3$ is the set of all ordered triples where each entry in the triple is a real number. We can show that $\mathbb{R}^3$ is also a vector space over $\mathbb{R}$ with addition and scalar multiplication defined component-wise. This means that, for $a, b, c, d, f, g, \alpha \in \mathbb{R}$,

$$\begin{pmatrix} a \\ b \\ c \end{pmatrix} + \begin{pmatrix} d \\ f \\ g \end{pmatrix} = \begin{pmatrix} a + d \\ b + f \\ c + g \end{pmatrix} \text{ and } \alpha \begin{pmatrix} a \\ b \\ c \end{pmatrix} = \begin{pmatrix} \alpha a \\ \alpha b \\ \alpha c \end{pmatrix}.$$

We show this by verifying that each of the ten properties of a vector space are true for $\mathbb{R}^3$ with addition and scalar multiplication defined this way.   □

Proof. Let $u, v, w \in \mathbb{R}^3$ and let $\alpha, \beta \in \mathbb{R}$. Then, there are real numbers a, b, c, d, f, g, h, k, and ℓ so that

$$u = \begin{pmatrix} a \\ b \\ c \end{pmatrix}, \quad v = \begin{pmatrix} d \\ f \\ g \end{pmatrix}, \text{ and } w = \begin{pmatrix} h \\ k \\ \ell \end{pmatrix}.$$

We will show that properties (P1)–(P10) from Definition 2.3.5 hold. We first show the two closure properties.

(P1)   Now since $\mathbb{R}$ is closed under addition, we know that $a + d$, $b + f$, and $c + g$ are real numbers. So,

$$u + v = \begin{pmatrix} a \\ b \\ c \end{pmatrix} + \begin{pmatrix} d \\ f \\ g \end{pmatrix} = \begin{pmatrix} a + d \\ b + f \\ c + g \end{pmatrix} \in \mathbb{R}^3.$$

Thus, $\mathbb{R}^3$ is closed under addition.

(P2)   Since $\mathbb{R}$ is closed under scalar multiplication, we know that αa, αb, and αc are real numbers. So

$$\alpha v = \alpha \cdot \begin{pmatrix} a \\ b \\ c \end{pmatrix} = \begin{pmatrix} \alpha \cdot a \\ \alpha \cdot b \\ \alpha \cdot c \end{pmatrix} \in \mathbb{R}^3.$$

Thus, $\mathbb{R}^3$ is closed under scalar multiplication.

(P3)   Now, we show that the commutative property of addition holds in $\mathbb{R}^3$. Using the fact that addition on $\mathbb{R}$ is commutative, we see that

$$u + v = \begin{pmatrix} a \\ b \\ c \end{pmatrix} + \begin{pmatrix} d \\ f \\ g \end{pmatrix} = \begin{pmatrix} a + d \\ b + f \\ c + g \end{pmatrix} = \begin{pmatrix} d + a \\ f + b \\ g + c \end{pmatrix} = v + u.$$

Therefore, addition on $\mathbb{R}^3$ is commutative.

(P4)   Using the associative property of addition on $\mathbb{R}$, we show that addition on $\mathbb{R}^3$ is also associative. Indeed, we have

$$(u + v) + w = \begin{pmatrix} (a + d) + h \\ (b + f) + k \\ (c + g) + \ell \end{pmatrix} = \begin{pmatrix} a + (d + h) \\ b + (f + k) \\ c + (g + \ell) \end{pmatrix} = u + (v + w).$$

So, addition on $\mathbb{R}^3$ is associative.

(P5)   Since scalar multiplication on $\mathbb{R}$ is associative, we see that

$$\alpha \cdot (\beta \cdot v) = \begin{pmatrix} \alpha \cdot (\beta \cdot a) \\ \alpha \cdot (\beta \cdot b) \\ \alpha \cdot (\beta \cdot c) \end{pmatrix} = \begin{pmatrix} (\alpha\beta) \cdot a \\ (\alpha\beta) \cdot b \\ (\alpha\beta) \cdot c \end{pmatrix} = (\alpha \cdot \beta) \cdot v.$$

Thus, scalar multiplication on $\mathbb{R}^3$ is associative.

(P6)   Since property (P6) holds for $\mathbb{R}$, we have that

$$\alpha \cdot (u + v) = \begin{pmatrix} \alpha \cdot (a + d) \\ \alpha \cdot (b + f) \\ \alpha \cdot (c + g) \end{pmatrix} = \begin{pmatrix} \alpha \cdot a + \alpha \cdot d \\ \alpha \cdot b + \alpha \cdot f \\ \alpha \cdot c + \alpha \cdot g \end{pmatrix} = \alpha \cdot u + \alpha \cdot v.$$

Thus, scalar multiplication distributes over vector addition for $\mathbb{R}^3$.

(P7)   Next, using the fact that (P7) is true on $\mathbb{R}$, we get

$$(\alpha + \beta) \cdot v = \begin{pmatrix} (\alpha + \beta) \cdot a \\ (\alpha + \beta) \cdot b \\ (\alpha + \beta) \cdot c \end{pmatrix} = \begin{pmatrix} \alpha \cdot a + \beta \cdot a \\ \alpha \cdot b + \beta \cdot b \\ \alpha \cdot c + \beta \cdot c \end{pmatrix} = \alpha \cdot v + \beta \cdot v.$$

This means that, scalar multiplication distributes over scalar addition for $\mathbb{R}^3$ as well.

(P8)   Using the additive identity $z \in \mathbb{R}$, we form the vector

$$z = \begin{pmatrix} 0 \\ 0 \\ 0 \end{pmatrix} \in \mathbb{R}^3.$$

This element is the additive identity in $\mathbb{R}^3$. Indeed,

$$z + v = \begin{pmatrix} 0 + a \\ 0 + b \\ 0 + c \end{pmatrix} = \begin{pmatrix} a \\ b \\ c \end{pmatrix}.$$

Therefore property (P8) holds for $\mathbb{R}^3$.

(P9)   We know that $-a$, $-b$, and $-c$ are the additive inverses in $\mathbb{R}$ of a, b, and c, respectively. Using these, we can form the vector

$$w = \begin{pmatrix} -a \\ -b \\ -c \end{pmatrix} \in \mathbb{R}^3.$$

We see that w is the additive inverse of v as

$$v + w = \begin{pmatrix} a + (-a) \\ b + (-b) \\ c + (-c) \end{pmatrix} = \begin{pmatrix} 0 \\ 0 \\ 0 \end{pmatrix} = 0.$$

Thus, property (P9) is true for $\mathbb{R}^3$.

(P10)   Finally, we use the multiplicative identity, $1 \in \mathbb{R}$. Indeed,

$$1(v) = \begin{pmatrix} 1 \cdot a \\ 1 \cdot b \\ 1 \cdot d \end{pmatrix} = v.$$

Now, because all ten properties from Definition 2.3.5 hold true for $\mathbb{R}^3$, we know that $\mathbb{R}^3$ with component-wise addition and scalar multiplication is a vector space over $\mathbb{R}$. $\square$

In the above proof, many of the properties easily followed from the properties on $\mathbb{R}$ and did not depend on the requirement that vectors in $\mathbb{R}^3$ were made of ordered **triples**. In most cases a person would not go through the excruciating detail that we did in this proof. Because the operations are component-wise defined and the components are elements of a vector space, we can shorten this proof to the following proof.

Alternate Proof (for Example 2.3.8*).* Suppose $\mathbb{R}^3$ is defined as above with addition and scalar multiplication defined component-wise. By definition of the operations on $\mathbb{R}^3$, the closure properties (P1) and (P2) hold true. Notice that all operations occur in the components of each vector and are the standard operations on $\mathbb{R}$. Therefore, since $\mathbb{R}$ with the standard operations is a vector space over itself, properties (P3)–(P10) are all inherited from $\mathbb{R}$. Thus $\mathbb{R}^3$ with component-wise addition and scalar multiplication is a vector space over $\mathbb{R}$.   □

In this proof, we said, "...properties (P3)–(P10) are all inherited from $\mathbb{R}$" to indicate that the justification for each vector space property for $\mathbb{R}^3$ is just repeated use (in each component) of the corresponding property for $\mathbb{R}$.

Example 2.3.9 Because neither proof relied on the requirement that elements of $\mathbb{R}^3$ are ordered **triples**, we see that a very similar proof would show that for any $n \in \mathbb{N}$,

$$\mathbb{R}^n \text{ is a vector space over the scalar field } \mathbb{R}.$$

Here $\mathbb{R}^n$ is the set of ordered n-tuples.   □

Caution: There are instances where a vector space has components that are elements of a vector space, but not all elements of this vector space are allowed as a component and the alternate proof does not work. So, for example, if we had an image set where we only allowed numbers between 0 and 1 in the pixels, we would not have a vector space, because [0, 1] is not a field (we need the rest of the real numbers in order for the set to be closed under addition).

Example 2.3.10 Let $T = \{a \in \mathbb{R} \mid a \neq 0\}$. Now, consider the set

$$T^3 = \left\{ \begin{pmatrix} a_1 \\ a_2 \\ a_3 \end{pmatrix} \; \middle| \; a_1, a_2, a_3 \in T \right\}$$

with addition and scalar multiplication defined component-wise. Notice that all components of T^3 are real numbers because all elements of T are real numbers. But T does not include the real number 0 and this means T^3 is not a vector space over the field $\mathbb{R}$. Which property fails? Exercise 4 asks you to answer this question.   □

Example 2.3.10 does not simplify the story either. Can you think of a vector space over $\mathbb{R}$, made of ordered n-tuples, with addition and scalar multiplication defined component-wise, whose components are in $\mathbb{R}$, *and* there are restrictions on how the components can be chosen? Exercise 8 asks you to explore this and determine whether or not it is possible.

The operations are themselves very important in the definition of a vector space. If we define different operations on a set, the structure of the vector space, including identity and inverse elements, can be different.

Example 2.3.11 Let us consider again the set of real numbers, $\mathbb{R}$, but with different operations. Define the operation $(\oplus)$ to be multiplication ($u \oplus v = uv$) and $\odot$ defined to be exponentiation ($\alpha \odot u = u^{\alpha}$). Notice that $\oplus$ is commutative but $\odot$ is not. We show that $(\mathbb{R}, \oplus, \odot)$ is not a vector space over $\mathbb{R}$.

(P1) We know that when we multiply two real numbers, we get a real number, thus $\mathbb{R}$ is closed under $\oplus$ and this property holds.

(P2) However, $\mathbb{R}$ is not closed under $\odot$. For example, $(1/2) \odot (-1) = (-1)^{1/2} = \sqrt{-1}$ is not a real number.

Since property (P2) does not hold, we do not need to continue checking the remaining eight properties.

□

To emphasize how the definition of the operations can change a vector space, we offer more examples.

Example 2.3.12 Let $V = \mathbb{R}^2$. Let $\oplus$ and $\odot$ be defined as binary operations on V in the following way. If $u = \begin{pmatrix} a \\ b \end{pmatrix}, v = \begin{pmatrix} c \\ d \end{pmatrix} \in V$ and $\alpha \in \mathbb{R}$ (the scalar set), then

$$u \oplus v = \begin{pmatrix} a + c - 1 \\ b + d - 1 \end{pmatrix} \quad \text{and} \quad \alpha \odot u = \begin{pmatrix} \alpha a + 1 - \alpha \\ \alpha b + 1 - \alpha \end{pmatrix}.$$

Then, $(V, +, \cdot)$ is a vector space over $\mathbb{R}$. $\square$

Proof. Let $u = \begin{pmatrix} a \\ b \end{pmatrix}, \; v = \begin{pmatrix} c \\ d \end{pmatrix}, \; w = \begin{pmatrix} k \\ \ell \end{pmatrix} \in V$ and $\alpha, \beta \in \mathbb{R}$ and define $\oplus$ and $\odot$ as above. We will show how three of the 10 properties given in Definition 2.3.5 are affected by the definition of the operations. It is left to the reader to show the other 7 properties are true (Exercise 14). Here, we will only prove the existence of inverses and identities.

(P8) We will find the element $z \in V$ so that $u \oplus z = u$ for every $u \in V$. Suppose we let $z = \begin{pmatrix} 1 \\ 1 \end{pmatrix} \in V$.
We will show that z is the additive identity. Indeed,

$$u \oplus z = \begin{pmatrix} a \\ b \end{pmatrix} \oplus \begin{pmatrix} 1 \\ 1 \end{pmatrix} = \begin{pmatrix} a \\ b \end{pmatrix}.$$

Thus the additive identity is, in fact, z. Therefore, property (P8) holds true.

(P9) Now, since $u \in V$, we will find the additive inverse, call it $\tilde{u}$ so that $u \oplus \tilde{u} = z$. Let $\tilde{u} = \begin{pmatrix} 2 - a \\ 2 - b \end{pmatrix}$.
We will show that $\tilde{u}$ is the additive inverse of u. Notice that

$$u \oplus \tilde{u} = \begin{pmatrix} a \\ b \end{pmatrix} \oplus \begin{pmatrix} 2 - a \\ 2 - b \end{pmatrix} = \begin{pmatrix} a + 2 - a - 1 \\ b + 2 - b - 1 \end{pmatrix} = \begin{pmatrix} 1 \\ 1 \end{pmatrix}.$$

Therefore, $u \oplus \tilde{u} = z$. Thus, every element of V has an additive inverse.

(P10) Here, we will show that $1 \in \mathbb{R}$ is the multiplicative identity. Indeed,

$$1 \odot u = 1 \odot \begin{pmatrix} a \\ b \end{pmatrix} = \begin{pmatrix} 1 \cdot a - 1 + 1 \\ 1 \cdot b - 1 + 1 \end{pmatrix} = \begin{pmatrix} a \\ b \end{pmatrix}.$$

Therefore, property (P10) holds for $(V, \oplus, \odot)$.

Because all 10 properties from Definition 2.3.5 hold true for $(V, \oplus, \odot)$, we know that $(V, \oplus, \odot)$ is a vector space over $\mathbb{R}$. $\square$

Notice that the "additive" identity was not a vector of zeros in this case.

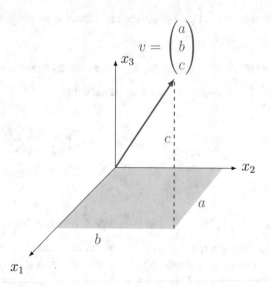

Fig. 2.12 Visualizing a vector in $\mathbb{R}^3$.

2.3.3   The Geometry of the Vector Space $\mathbb{R}^3$

We can visualize the vector space $\mathbb{R}^3$, with the standard operations and field, in 3D space. We typically represent a vector in $\mathbb{R}^3$ with an arrow. For example, the vector $v = \begin{pmatrix} a \\ b \\ c \end{pmatrix}$ can be represented by the arrow pointing from the origin (the 0 vector) to the point (a, b, c) as in Figure 2.12. It can also be represented as any translation of this arrow, that is, an arrow starting from any other vector in $\mathbb{R}^3$ and pointing toward a point a units further in the "x"-direction, b units further in the "y"-direction, and c units further in the "z"-direction from the starting point. The natural question arises: What do the vector space properties mean in the geometric context? In this section, we will discuss the geometry of some of the vector space properties. The rest, we leave for the exercises.

The geometry we discuss here translates nicely to the vector space $\mathbb{R}^n$ (for any $n \in \mathbb{N}$) with standard operations and field.

(P1) The Geometry of Closure under Addition This property says that if we add any two vectors in $\mathbb{R}^3$ together, we never leave $\mathbb{R}^3$.

We briefly recall here the geometric interpretation of addition in $\mathbb{R}^3$. Using the definition of addition, we have that if

$$v = \begin{pmatrix} a \\ b \\ c \end{pmatrix} \text{ and } u = \begin{pmatrix} d \\ f \\ g \end{pmatrix}$$

then

$$v + u = \begin{pmatrix} a + d \\ f + b \\ g + c \end{pmatrix}.$$

That is, $v + u$ is the vector that can be represented by an arrow starting at 0 and pointing toward the point $(d + a, f + b, g + c)$. We can see in Figure 2.13 that the sum is the vector that starts at 0 (the

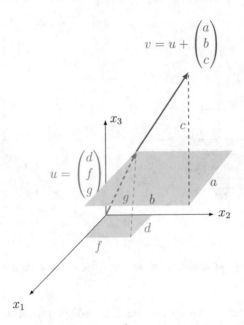

Fig. 2.13 Geometric representation of vector addition in $\mathbb{R}^3$.

start of u) and points to the end of v. Some describe this as drawing "tip to tail" because the tip of u is touching the tail of v.

We see geometrically that if we translate a vector v along the vector u, we can form a new vector that ends at the point to where the tip of v translated.

Of course, this new vector is still in $\mathbb{R}^3$, as indicated by the closure property.

(P2) The Geometry of Closure under Scalar Multiplication Let α be a scalar, then

$$\alpha \cdot v = \begin{pmatrix} \alpha \cdot a \\ \alpha \cdot b \\ \alpha \cdot c \end{pmatrix}.$$

That is, αv can be represented by an arrow starting at 0 and ending at the point $(\alpha a, \alpha b, \alpha c)$. Now, if $\alpha > 1$, this vector points in the same direction as v, but is longer. If $0 < \alpha < 1$, then $\alpha \cdot v$ still points in the same direction as v, but is now shorter. Finally, if $\alpha = 1$, it is the multiplicative identity and $\alpha \cdot v = v$. These can be seen in Figure 2.14.

We will see later that, if $\alpha < 0$ is a negative number, then αv will point in the opposite direction of v but will lie along the same line, and the length of the arrow will correspond to $|\alpha|$.

Any scalar multiple of v is represented by an arrow that points along the line passing through the arrow representing v. Property (P2) says that $\mathbb{R}^3$ contains the entire line that passes through the origin and parallel to v.

(P3) The Geometry of the Commutative Property If we translate v along the vector u (see Figure 2.13) or we translate u along the vector v (we encourage the reader to make this drawing), the vector formed will point toward the same point in $\mathbb{R}^3$. Thus, the commutative property shows us that geometrically, it doesn't matter in which order we traverse the two vectors, we will still end at the same terminal point.

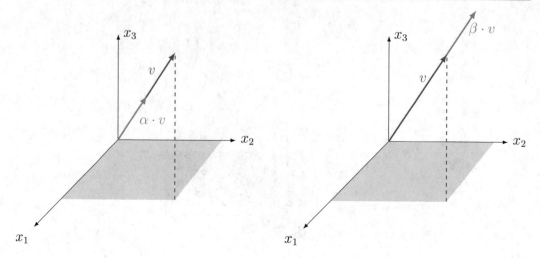

Fig. 2.14 Geometric representation of vector multiplication in $\mathbb{R}^3$ ($0 < \alpha < 1$ and $\beta > 1$).

The remaining seven vector space properties can also be displayed through similar figures. We leave these remaining interpretations to Exercise 7.

2.3.4 Properties of Vector Spaces

A lot of this material can feel tedious, or abstract, or both, and it can be easy to lose sight of why we care about such a long list of properties. Overall, remember that we have defined vector spaces as abstract objects that have certain properties. We can prove theorems about vector spaces, and know that those theorems remain valid in the case of any specific vector space. In much of this book we build machinery that can be used in *any* vector space. This is what makes linear algebra so powerful for so many applications. It is also often useful to recognize when we have a set for which a vector space property does not hold, then we know we cannot use our tools on this set.

We can now deduce some fun facts that hold for all vector spaces. As you read through these properties, think about what they mean in the context of, say, the vector space of images, or any of the vector spaces from the previous section.

> **Theorem 2.3.13**
> If x, y, and z are vectors in a vector space $(V, +, \cdot)$ and $x + z = y + z$, then $x = y$.

Proof. Let x, y, and z be vectors in a vector space. Assume $x + z = y + z$. We know that there exists an additive inverse of z, call it $-z$. We will show that the properties given in Definition 2.3.5 imply that $x = y$. Indeed,

$$
\begin{aligned}
x &= x + 0 & \text{(P8)} \\
&= x + (z + (-z)) & \text{(P9)} \\
&= (x + z) + (-z) & \text{(P4)} \\
&= (y + z) + (-z) & \text{(assumption)} \\
&= y + (z + (-z)) & \text{(P4)} \\
&= y + 0 & \text{(P9)} \\
&= y. & \text{(P8)}
\end{aligned}
$$

$\square$

Each step in the preceding proof is justified by either the use of our initial assumption or a known property of a vector space. The theorem also leads us to the following corollary[7].

Corollary 2.3.14
The zero vector in a vector space is unique. Also, every vector in a vector space has a unique additive inverse.

Proof. We show that the zero vector is unique and leave the remainder as Exercise 11. We consider two arbitrary zero vectors, 0 and $0'$, and show that $0 = 0'$. We know that $0 + x = x = 0' + x$. By Theorem 2.3.13, we now conclude that $0 = 0'$. Therefore, the zero vector is unique. $\square$

The next two theorems state that whenever we multiply by zero (either the scalar zero or the zero vector), the result is always the zero vector.

Theorem 2.3.15
Let $(V, +, \cdot)$ be a vector space over $\mathbb{R}$. Then $0 \cdot x = 0$ for each vector $x \in V$.

The proof follows from Exercise 12, which shows that the vector $0 \cdot x$ (here $0 \in \mathbb{R}$ and $x \in V$) is the additive identity in V.

Theorem 2.3.16
Let $(V, +, \cdot)$ be a vector space over $\mathbb{R}$. Then $\alpha \cdot 0 = 0$ for each scalar $\alpha \in \mathbb{R}$.

---

[7] A *corollary* is a result whose proof follows from a preceding theorem.

The proof follows from Exercise 13, which shows that the vector $\alpha \cdot 0$ (here $\alpha \in \mathbb{R}$ and $0 \in V$) is the additive identity.

Theorem 2.3.17

Let $(V, +, \cdot)$ be a vector space over $\mathbb{R}$. Then $(-\alpha) \cdot x = -(\alpha \cdot x) = \alpha \cdot (-x)$ for each $\alpha \in \mathbb{R}$ and each $x \in V$.

In this theorem it is important to note that "$-$" indicates an additive inverse, not to be confused with a negative sign. Over the set of real (or complex) numbers, the additive inverse is the negative value, while for vector spaces over other fields (including Z_2) this is not necessarily the case. The theorem states the equivalence of three vectors:

1. A vector x multiplied by the additive inverse of a scalar α, $(-\alpha) \cdot x$.
2. The additive inverse of the product of a scalar α and a vector x, $-(\alpha \cdot x)$.
3. The additive inverse of a vector x multiplied by a scalar α, $\alpha \cdot (-x)$.

While these equivalences may seem "obvious" in the context of real numbers, we must be careful to verify these properties using only the established properties of vector spaces.

Proof. Let $(V, +, \cdot)$ be a vector space over $\mathbb{R}$. Let $\alpha \in \mathbb{R}$ and $x \in V$. We begin by showing $(-\alpha) \cdot x = -(\alpha \cdot x)$. Notice that

$$
\begin{aligned}
(-\alpha) \cdot x &= (-\alpha) \cdot x + 0 &&\text{(P8)} \\
&= (-\alpha) \cdot x + (\alpha \cdot x) + (-(\alpha \cdot x)) &&\text{(P9)} \\
&= (-\alpha + \alpha) \cdot x + (-(\alpha \cdot x)) &&\text{(P7)} \\
&= 0 \cdot x + (-(\alpha \cdot x)) &&\text{scalar addition} \\
&= 0 + (-(\alpha \cdot x)) &&\text{(Theorem 2.3.15)} \\
&= -(\alpha \cdot x) &&\text{(P8).}
\end{aligned}
$$

Therefore, $(-\alpha) \cdot x = -(\alpha \cdot x)$. Exercise 15 gives us that $-(\alpha \cdot x) = \alpha \cdot (-x)$. Using transitivity of equality, we know that $(-\alpha) \cdot x = -(\alpha \cdot x) = \alpha \cdot (-x)$. $\qquad\square$

Of particular interest is the special case of $\alpha = -1$. We see that $(-1)x = -x$. That is, the additive inverse of any vector x is obtained by multiplying the additive inverse of the multiplicative identity scalar by the vector x.

2.3.5 Exercises

1. In Example 2.2.27, we considered a system of two equations in two variables, and we visualized the solution in $\mathbb{R}^2$. To find the solution we sketched two lines and found the intersection. Sketch another geometric representation of the solution as the linear combination of the two vectors $u = (1, 1)$ and $v = (2, 3)$ that yields $(2, -3)$. Your sketch should include the vectors u and v.
2. Let $\oplus$ and $\odot$ be defined on $\mathbb{R}$ so that if $a, b \in \mathbb{R}$ then $a \oplus b = a + b + 1$ and $a \odot b = ab - a + 1$. Is $(\mathbb{R}, \oplus, \odot)$ a vector space over $\mathbb{R}$? Justify.

3. Let $V = \mathbb{R}^n$ and z a fixed element of V. For arbitrary elements x and y in V and arbitrary scalar α in $\mathbb{R}$, define vector addition ($\oplus$) and scalar multiplication ($\odot$) as follows:

$$x \oplus y = x + y - z, \text{ and } \alpha \odot x = \alpha(x - z) + z.$$

Show that $(V, \oplus, \odot)$ is a vector space over $\mathbb{R}$.

4. In Example 2.3.10, we stated that T^3 is not a vector space. List all properties that fail to be true. Justify your assertions.

5. Define $\oplus$ and $\odot$ on $\mathbb{R}$ so that if $a, b \in \mathbb{R}$, $a \oplus b = ab$ and $a \odot b = a + b$. Is $(\mathbb{R}, \oplus, \odot)$ a vector space over $\mathbb{R}$? Justify.

6. Consider the set $V = \mathbb{R}^2$. Let (a_1, a_2) and (b_1, b_2) be in V and α in $\mathbb{R}$. Define vector addition and scalar multiplication as follows:

$$(a_1, a_2) + (b_1, b_2) = (a_1 + b_1, 0), \text{ and } \alpha \cdot (a_1, a_2) = (\alpha a_1, \alpha a_2).$$

Show that the set V is not a vector space.

7. Draw similar geometric interpretations for the remaining seven vector space properties not discussed in Section 2.3.3.

8. Find a vector space over $\mathbb{R}$, made of ordered n-tuples (you choose $n > 1$), with addition and scalar multiplication defined component-wise, whose components are in $\mathbb{R}$, but there are restrictions on how the components can be chosen.

9. Consider the set $\mathbb{R}$. Define vector addition and scalar multiplication so that vector space property (P3) is true, but property (P6) is false.

10. Consider the set $\mathbb{R}$. Define vector addition and scalar multiplication so that vector space property (P6) is true, but property (P7) is false.

11. Prove that every vector in a vector space has a unique additive inverse to complete the proof of Corollary 2.3.14.

12. Given that $(V, +, \cdot)$ is a vector space over $\mathbb{F}$, show that if $x \in V$ then $0 \cdot x$ is the additive identity in V. Hint: Strategically choose a vector to add to $0 \cdot x$ and then use one of the distributive properties.

13. Complete the proof of Theorem 2.3.16 by proving the following statement and then employing a theorem from Section 2.3.

   Let $(V, +, \cdot)$ be a vector space over $\mathbb{R}$ and let 0 denote the additive identity in V, then $\alpha \cdot 0$ is the additive identity in V.

14. Complete the proof in Example 2.3.12.

15. Complete the proof of Theorem 2.3.17 by writing a proof of the needed result.

16. Consider the set of grayscale images on the map of Africa in Figure 2.11. Create a plausible scenario describing the meaning of pixel intensity. Image addition and scalar multiplication should have a reasonable interpretation in your scenario. Describe these interpretations.

17. (**Image Warping**) In Chapter 1, we discussed the application of image warping. In order to set up a vector space of images for this task, we will need to add and scale the images as defined in Definitions 2.3.2 and 2.3.3. In order to add two images, what must be required of all images?

2.4 Vector Space Examples

In the previous section, we defined vector spaces as sets that have some of the properties of the real numbers. We looked at two examples in that section: the space of images (given a fixed pixel arrangement) and the Euclidean spaces $\mathbb{R}^n$. We now present a collection of other useful vector spaces to give you a flavor of how broad the vector space framework is. We will use these vector spaces in exercises and examples throughout the book.

2.4.1   Diffusion Welding and Heat States

In this section, we begin a deeper look into the mathematics for the diffusion welding application discussed in Chapter 1. Recall that diffusion welding can be used to adjoin several smaller rods into a single longer rod, leaving the final rod just after welding with varying temperature along the rod but with the ends having the same temperature. Recall that we measure the temperature along the rod and obtain a heat signature like the one seen in Figure 1.4 of Chapter 1. Recall also, that the heat signature shows the temperature *difference* from the temperature at the ends of the rod. Thus, the initial signature (along with any subsequent signature) will show values of 0 at the ends.

The heat signature along the rod can be described by a function $f : [0, L] \to \mathbb{R}$, where L is the length of the rod and $f(0) = f(L) = 0$. The quantity $f(x)$ is the temperature difference on the rod at a position x in the interval $[0, L]$. Because we are detecting and storing heat measurements along the rod, we are only able to collect finitely many such measurements. Thus, we *discretize* the heat signature f by sampling at only m locations along the bar. If we space the m sampling locations equally, then for $\Delta x = \frac{L}{m+1}$, we can choose the sampling locations to be $\Delta x, 2\Delta x, \ldots, m\Delta x$. Since the heat measurement is zero (and fixed) at the endpoints we do not need to sample there. The set of discrete heat measurements at a given time is called a *heat state*, as opposed to a heat signature, which, as discussed earlier, is defined at every point along the rod. We can record the heat state as the vector

$$u = [u_0, u_1, u_2, ..., u_m, u_{m+1}] = [0, f(\Delta x), f(2\Delta x), \ldots, f(m\Delta x), 0].$$

Here, if $u_j = f(x)$ for some $x \in [0, L]$ then $u_{j+1} = f(x + \Delta x)$ and $u_{j-1} = f(x - \Delta x)$. Figure 2.15 shows a (continuous) heat signature as a solid blue curve and the corresponding measured (discretized) heat state indicated by the regularly sampled points marked as circles.

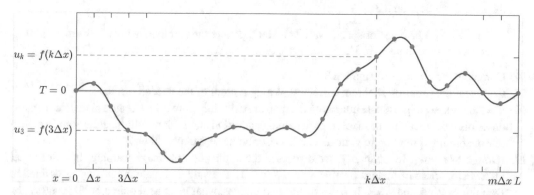

Fig. 2.15 A 1D heat signature, f, is shown as a curve. The corresponding heat state is the discrete collection of $m + 2$ regularly sampled temperatures, $\{u_0, u_1, \cdots, u_{m+2}\}$, shown as dots. Both heat signature and heat state have zero temperature at the end points $x = 0$ and $x = L$.

As the heat diffuses through the rod, the new heat signatures will also be described by functions $f_t : [0, L] \to \mathbb{R}$, where t is the time measured since the welding was completed. The discretized heat states corresponding to these signatures will form vectors as well.

We define scalar multiplication and vector addition of heat states component-wise (in the same way we define the operations on vectors in $\mathbb{R}^{m+2}$). Denote the set of all heat states with $m + 2$ entries (assumed to have zero temperature at the endpoints) by $H_m(\mathbb{R})$. With the operations of addition and scalar multiplication, the set $H_m(\mathbb{R})$ is a vector space (see Exercise 3).

2.4.2 Function Spaces

We have seen that the set of discretized heat states of the preceding example forms a vector space. These discretized heat states can be viewed as real-valued functions on the set of $m + 2$ points that are the sampling locations along the rod. In fact, *function spaces* such as $H_m(\mathbb{R})$ are very common and useful constructs for solving many physical problems. The following are some such function spaces.

Example 2.4.1 Let $\mathcal{F} = \{f : \mathbb{R} \to \mathbb{R}\}$, the set of all functions whose domain is $\mathbb{R}$ and whose range consists of only real numbers.

We define addition and scalar multiplication (on functions) pointwise. That is, given two functions f and g and a real scalar α, we define the sum $f + g$ by $(f + g)(x) := f(x) + g(x)$ and the scalar product αf by $(\alpha f)(x) := \alpha \cdot (f(x))$. $(\mathcal{F}, +, \cdot)$ is a vector space with scalars taken from $\mathbb{R}$. $\quad\square$

Proof. Let $f, g, h \in \mathcal{F}$ and $\alpha, \beta \in \mathbb{R}$. We verify the 10 properties of Definition 2.3.5.

- Since $f : \mathbb{R} \to \mathbb{R}$ and $g : \mathbb{R} \to \mathbb{R}$ and based on the definition of addition, $f + g : \mathbb{R} \to \mathbb{R}$. So $f + g \in \mathcal{F}$ and $\mathcal{F}$ is closed over addition.
- Similarly, $\mathcal{F}$ is closed under scalar multiplication.
- Addition is commutative since

$$\begin{aligned}(f + g)(x) &= f(x) + g(x) \\ &= g(x) + f(x) \\ &= (g + f)(x).\end{aligned}$$

So, $f + g = g + f$.
- And, addition is associative because

$$\begin{aligned}((f + g) + h)(x) &= (f + g)(x) + h(x) \\ &= (f(x) + g(x)) + h(x) \\ &= f(x) + (g(x) + h(x)) \\ &= f(x) + (g + h)(x) \\ &= (f + (g + h))(x).\end{aligned}$$

So $(f + g) + h = f + (g + h)$.
- We see, also, that scalar multiplication is associative. Indeed,

$$(\alpha \cdot (\beta \cdot f))(x) = (\alpha \cdot (\beta \cdot f(x))) = (\alpha\beta)f(x) = ((\alpha\beta) \cdot f)(x).$$

So $\alpha \cdot (\beta \cdot f) = (\alpha\beta) \cdot f$.

- Using the distributive law for real numbers, we see that scalar multiplication distributes over vector addition here as well since

$$
\begin{aligned}
(\alpha \cdot (f+g))(x) &= \alpha \cdot (f+g)(x) \\
&= \alpha \cdot (f(x) + g(x)) \\
&= \alpha \cdot f(x) + \alpha \cdot g(x) \\
&= (\alpha \cdot f + \alpha \cdot g)(x).
\end{aligned}
$$

So $\alpha(f+g) = \alpha f + \alpha g$.

- Using the definition of scalar multiplication and the distributive law for real numbers, we see that scalar multiplication distributes over scalar addition by the following computation:

$$
\begin{aligned}
((\alpha + \beta) \cdot f)(x) &= (\alpha + \beta) \cdot f(x) \\
&= \alpha \cdot f(x) + \beta \cdot f(x) \\
&= (\alpha \cdot f + \beta \cdot f)(x).
\end{aligned}
$$

So, $(\alpha + \beta) \cdot f = \alpha \cdot f + \beta \cdot f$.

- Now, consider the constant function $z(x) = 0$ for every $x \in \mathbb{R}$. We see that

$$
\begin{aligned}
(z+f)(x) &= z(x) + f(x) \\
&= 0 + f(x) = f(x).
\end{aligned}
$$

That is, $z + f = f$. Thus, the function z is the zero vector in $\mathcal{F}$.

- Now, let g be the function defined by $g(x) = -f(x)$ for every $x \in \mathbb{R}$. We will show that g is the additive inverse of f. Observe, $(f+g)(x) = f(x) + g(x) = f(x) + (-f(x)) = 0 = z(x)$, where z is defined above. So, $f + g = z$. That is, g is the additive inverse of f. As f is an arbitrary function in $\mathcal{F}$, we know that all functions in $\mathcal{F}$ have additive inverses.

- Since, $(1 \cdot f)(x) = f(x)$, $1 \cdot f = f$. And, we see that the scalar identity is the real number 1.

Therefore, since all properties from Definition 2.3.5 are true, we know that $(\mathcal{F}, +, \cdot)$ is a vector space over $\mathbb{R}$. □

In the vector space $\mathcal{F}$, vectors are functions. Example vectors in $\mathcal{F}$ include $\sin(x)$, $x^2 + 3x - 5$, and $e^x + 2$. Not all functions are vectors in $\mathcal{F}$ (see Exercise 8).

Example 2.4.2 Let $\mathcal{P}_n(\mathbb{R})$ be the set of all polynomials of degree less than or equal to n with coefficients from $\mathbb{R}$. That is,

$$
\mathcal{P}_n(\mathbb{R}) = \left\{ a_0 + a_1 x + \cdots + a_n x^n \mid a_k \in \mathbb{R}, \ k = 0, 1, \cdots, n \right\}.
$$

Let $f(x) = a_0 + a_1 x + \cdots + a_n x^n$ and $g(x) = b_0 + b_1 x + \cdots + b_n x^n$ be polynomials in $\mathcal{P}_n(\mathbb{R})$ and $\alpha \in \mathbb{R}$. Define addition and scalar multiplication component-wise:

$$
(f+g)(x) = (a_0 + b_0) + (a_1 + b_1)x + \cdots + (a_n + b_n)x^n,
$$

$$
(\alpha \cdot f)(x) = (\alpha a_0) + (\alpha a_1)x + \cdots + (\alpha a_n)x^n.
$$

With these definitions, $\mathcal{P}_n(\mathbb{R})$ is a vector space (see Exercise 11). □

Example 2.4.3 A function $f : \mathbb{R} \to \mathbb{R}$ is called an **even function** if $f(-x) = f(x)$ for all $x \in \mathbb{R}$. The set of even functions is a vector space over $\mathbb{R}$ with the definitions of vector addition and scalar multiplication given in Example 2.4.1 (see Exercise 9). □

Example 2.4.4 Fix real numbers a and b with $a < b$. Let $\mathcal{C}([a, b])$ be the set of all continuous functions $f : [a, b] \to \mathbb{R}$. Then, using limits from Calculus, we can show that $\mathcal{C}([a, b])$ is a vector space with the operations of function addition and scalar multiplication defined analogously to the corresponding operations on $\mathcal{F}$. □

Proof. Suppose $a, b \in \mathbb{R}$ with $a < b$. Let $f, g \in \mathcal{C}([a, b])$ and let $\alpha, \beta \in \mathbb{R}$. We know that $f, g \in \mathcal{C}([a, b])$ implies that, for any $c \in [a, b]$, we have

$$\lim_{x \to c} f(x) = f(c) \quad \text{and} \quad \lim_{x \to c} g(x) = g(c).$$

We will show that the 10 properties from Definition 2.3.5 hold true.

- Using properties of limits, we know that

$$\begin{aligned}
\lim_{x \to c} (f + g)(x) &= \lim_{x \to c} (f(x) + g(x)) \\
&= \lim_{x \to c} f(x) + \lim_{x \to c} g(x) \\
&= f(c) + g(c) \\
&= (f + g)(c).
\end{aligned}$$

Therefore $f + g$ is continuous on $[a, b]$ and $\mathcal{C}([a, b])$ is closed under addition.

- Now,

$$\begin{aligned}
\lim_{x \to c} (\alpha \cdot f)(x) &= \lim_{x \to c} \alpha f(x) \\
&= \alpha \lim_{x \to c} f(x) \\
&= \alpha f(c) \\
&= (\alpha \cdot f)(c).
\end{aligned}$$

Thus, $\alpha f \in \mathcal{C}([a, b])$. So, $\mathcal{C}([a, b])$ is closed under scalar multiplication.

- Since $f, g \in \mathcal{F}$ and no other property from Definition 2.3.5 requires the use of limits, we can say that the last 8 properties are inherited from $\mathcal{F}$.

Therefore, $(\mathcal{C}([a, b]), +, \cdot)$ is a vector space over $\mathbb{R}$. □

Example 2.4.5 Consider the set $\mathcal{C}_0([a, b]) \subset \mathcal{C}([a, b])$ where $\mathcal{C}_0([a, b]) := \{f \in \mathcal{C}([a, b]) : f(a) = f(b) = 0\}$. Then $\mathcal{C}_0([a, b])$ is the set of all continuous functions $f : [a, b] \to \mathbb{R}$ that are zero at the endpoints. The set $\mathcal{C}_0([a, b])$ is a vector space with the operations of function addition and scalar multiplication inherited from $\mathcal{C}([a, b])$. (See Exercise 23.) □

In fact, the space of heat signatures on a 1D rod of length L is modeled by the vector space $\mathcal{C}_0([a, b])$ in Example 2.4.5 with $a = 0$ and $b = L$.

2.4.3 Matrix Spaces

A matrix is an array of real numbers arranged in a rectangular grid, for example, let

$$A = \begin{pmatrix} 1 & 2 & 3 \\ 5 & 7 & 9 \end{pmatrix}.$$

The matrix A has 2 rows (horizontal) and 3 columns (vertical), so we say it is a 2×3 matrix. In general, a matrix B with m rows and n columns is called an $m \times n$ matrix. We say the dimensions of the matrix are m and n.

Any two matrices with the same dimensions are added together by adding their entries entry-wise. A matrix is multiplied by a scalar by multiplying all of its entries by that scalar (that is, multiplication of a matrix by a scalar is also an entry-wise operation, as in Example 2.3.8).

Example 2.4.6 Let

$$A = \begin{pmatrix} 1 & 2 & 3 \\ 5 & 7 & 9 \end{pmatrix}, \ B = \begin{pmatrix} 1 & 0 & 1 \\ -2 & 1 & 0 \end{pmatrix}, \ \text{and } C = \begin{pmatrix} 1 & 2 \\ 3 & 5 \end{pmatrix}.$$

Then

$$A + B = \begin{pmatrix} 2 & 2 & 4 \\ 3 & 8 & 9 \end{pmatrix},$$

but since $A \in \mathcal{M}_{2 \times 3}$ and $C \in \mathcal{M}_{2 \times 2}$, the definition of matrix addition does not work to compute $A + C$. That is, $A + C$ is undefined.

Using the definition of scalar multiplication, we get

$$3 \cdot A = \begin{pmatrix} 3(1) & 3(2) & 3(3) \\ 3(5) & 3(7) & 3(9) \end{pmatrix} = \begin{pmatrix} 3 & 6 & 9 \\ 15 & 21 & 27 \end{pmatrix}.$$

$\square$

With this understanding of operations on matrices, we can now discuss $(\mathcal{M}_{m \times n}, +, \cdot)$ as a vector space over $\mathbb{R}$.

Theorem 2.4.7

Let m and n be in $\mathbb{N}$, the set of natural numbers. The set $\mathcal{M}_{m \times n}$ of all $m \times n$ matrices with real entries together with the operations of addition and scalar multiplication is a vector space over $\mathbb{R}$.

Proof. Consider $\mathcal{M}_{m \times n}$ of real-valued $m \times n$ matrices. Because matrix addition and scalar multiplication are defined entry-wise, neither operation results in a matrix of a different size. Thus, $\mathcal{M}_{m \times n}$ is closed under addition and scalar multiplication. Now, since the entries of every matrix in $\mathcal{M}_{m \times n}$ are all real, and addition and scalar are defined entry-wise, we can say that $\mathcal{M}_{m \times n}$ inherits the remaining vector space properties from $\mathbb{R}$. Thus, $(\mathcal{M}_{m \times n}, +, \cdot)$ is a vector space over $\mathbb{R}$. $\square$

To make the notation less tedious, it is common to write $A = (a_{i,j})_{m \times n}$ or $A = (a_{i,j})$ to denote an $m \times n$ matrix whose entries are represented by $a_{i,j}$, where i, j indicates the position of the entry. That is, $a_{i,j}$ is the entry in matrix A that is located in the ith row and j^{th} column. For example, the number in the first row and second column of the matrix A above is denoted by $a_{1,2}$ and $a_{1,2} = 2$. With this notation, we can formally define matrix addition and scalar multiplication.

Definition 2.4.8

Let $A = (a_{i,j})$ and $B = (b_{i,j})$ be $m \times n$ matrices. We define **matrix addition** by

$$A + B = (a_{i,j} + b_{i,j}).$$

Let $\alpha \in \mathbb{R}$. We define **scalar multiplication** by $\alpha A = (\alpha a_{i,j})$.   □

Caution: The spaces $\mathcal{M}_{m \times n}$ and $\mathcal{M}_{n \times m}$ are not the same. The matrix

$$M_1 = \begin{pmatrix} 1 & -2 & 3 \\ 0 & 4 & 10 \\ 7 & 7 & 8 \\ -1 & 0 & 0 \end{pmatrix}$$

is a 4×3 matrix, whereas

$$M_2 = \begin{pmatrix} 1 & 0 & 7 & -1 \\ -2 & 4 & 7 & 0 \\ 3 & 10 & 8 & 0 \end{pmatrix}$$

is a 3×4 matrix. As we saw above, these two matrices cannot be added together.

2.4.4 Solution Spaces

In this section, we consider solution sets of linear equations. If an equation has n variables, its solution set is a subset of $\mathbb{R}^n$. When is this set a vector space?

Example 2.4.9 Let $V \subseteq \mathbb{R}^3$ be the set of all solutions to the equation $x_1 + 3x_2 - x_3 = 0$. That is,

$$V = \left\{ \begin{pmatrix} x_1 \\ x_2 \\ x_3 \end{pmatrix} \in \mathbb{R}^3 \ \middle|\ x_1 + 3x_2 - x_3 = 0 \right\}.$$

The set V together with the operations $+$ and $\cdot$ inherited from $\mathbb{R}^3$ forms a vector space.   □

Proof. Let V be the set defined above and let

$$u = \begin{pmatrix} u_1 \\ u_2 \\ u_3 \end{pmatrix}, v = \begin{pmatrix} v_1 \\ v_2 \\ v_3 \end{pmatrix} \in V.$$

Let $\alpha \in \mathbb{R}$. Notice that Properties (P3)–(P7) and (P10) of Definition 2.3.5 only depend on the definition of addition and scalar multiplication on $\mathbb{R}^3$ and, therefore, are inherited properties from $\mathbb{R}^3$. Hence we need only check properties (P1), (P2), (P8), and (P9). Since $u, v \in V$, we know that

$$u_1 + 3u_2 - u_3 = 0 \quad \text{and} \quad v_1 + 3v_2 - v_3 = 0.$$

We will use this result to show the closure properties.

First, notice that $u + v = \begin{pmatrix} u_1 + v_1 \\ u_2 + v_2 \\ u_3 + v_3 \end{pmatrix}$. Notice also that

$$(u_1 + v_1) + 3(u_2 + v_2) - (u_3 + v_3) = (u_1 + 3u_2 - u_3) + (v_1 + 3v_2 - v_3)$$
$$= 0 + 0 = 0.$$

Thus, $u + v \in V$. Since u and v are arbitrary vectors in V it follows that V is closed under addition.

Next, notice that $\alpha \cdot u = \begin{pmatrix} \alpha u_1 \\ \alpha u_2 \\ \alpha u_3 \end{pmatrix}$ and

$$\alpha u_1 + 3\alpha u_2 - \alpha u_3 = \alpha(u_1 + 3u_2 - u_3)$$
$$= \alpha(0) = 0.$$

Therefore, $\alpha u \in V$. Hence V is closed under scalar multiplication.

The zero vector $z = \begin{pmatrix} 0 \\ 0 \\ 0 \end{pmatrix} \in \mathbb{R}^3$ is also the zero vector in V because $0 + 3 \cdot 0 - 0 = 0$ and $z + u = u$.

Finally, notice that

$$(-u_1) + 3(-u_2) - (-u_3) = -(u_1 + 3u_2 - u_3) = 0.$$

Therefore, $-u = \begin{pmatrix} -u_1 \\ -u_2 \\ -u_3 \end{pmatrix} \in V$. Since $-u$ is the additive inverse of u in $\mathbb{R}^3$, we know that V contains additive inverses as well.

Therefore, since all properties in Definition 2.3.5 hold, $(V, +, \cdot)$ is a vector space over $\mathbb{R}$. $\square$

We can also visualize this space as a plane in $\mathbb{R}^3$ through the origin. In this case the space V is a plane in $\mathbb{R}^3$ that passes through the points $(0, 0, 0)$, $(1, 0, 1)$, and $(0, 1, 3)$ (three arbitrarily chosen solutions of $x_1 + 3x_2 - x_3 = 0$).

Here is a very similar equation, however, whose solution set does *not* form a vector space.

Example 2.4.10 Let $V \subseteq \mathbb{R}^3$ be the set of all solutions to the equation $x_1 + 3x_2 - x_3 = 5$. In this case, V is a plane in $\mathbb{R}^3$ that passes through the points $(0, 0, -5)$, $(1, 0, -4)$, and $(0, 1, -2)$, in other words, it is the translation of the plane in Example 2.4.9 by 5 units down (in the negative x_3-direction). The set V is not a vector space. The proof is Exercise 13. $\square$

In fact, we can generalize the preceding examples by considering the set of solutions to linear equations that are *homogeneous* or *inhomogeneous*.

Definition 2.4.11

We say that a linear equation, $a_1x_1 + a_2x_2 + \cdots + a_nx_n = b$, is **homogeneous** if $b = 0$ and is **inhomogeneous** if $b \neq 0$. □

Example 2.4.12 Let $\{a_1, a_2, \ldots, a_n\} \subset \mathbb{R}$ and consider the set $V \in \mathbb{R}^n$ of all solutions to the homogeneous linear equation $a_1x_1 + a_2x_2 + \cdots + a_nx_n = 0$. The set V together with the operations $+$ and $\cdot$ inherited from $\mathbb{R}^n$ forms a vector space. The proof is Exercise 14. □

Example 2.4.13 Is the set of all solutions to an *inhomogeneous* linear equation a vector space? In Example 2.4.10, we considered a specific case where the solution set is not a vector space. In Exercise 20, the reader is asked to show that the solution set to any inhomogeneous linear equation is not a vector space. The key steps are to show that neither closure property holds. □

2.4.5 Other Vector Spaces

In many areas of mathematics, we learn about concepts that relate to vector spaces, though the details of vector space properties may be simply assumed or not well established. In this section, we look at some of these concepts and recognize how vector space properties are present.

Sequence Spaces

Here, we explore sequences such as those discussed in Calculus. We consider the set of all sequences in the context of vector space properties. First, we give a formal definition of sequences.

Definition 2.4.14

A **sequence** of real numbers is a function $s : \mathbb{N} \to \mathbb{R}$. That is, $s(n) = a_n$ for $n = 1, 2, \cdots$ where $a_n \in \mathbb{R}$. A sequence is denoted $\{a_n\}$. Let $\mathcal{S}(\mathbb{R})$ be the set of all sequences. Let $\{a_n\}$ and $\{b_n\}$ be sequences in $\mathcal{S}(\mathbb{R})$ and α in $\mathbb{R}$. Define sequence addition and scalar multiplication with a sequence by

$$\{a_n\} + \{b_n\} = \{a_n + b_n\} \text{ and } \alpha \cdot \{a_n\} = \{\alpha a_n\}.$$

□

In Exercise 15, we show that $\mathcal{S}(\mathbb{R})$, with these (element-wise) operations, forms a vector space over $\mathbb{R}$.

Example 2.4.15 (Eventually Zero Sequences) Let $\mathcal{S}_{\text{fin}}(\mathbb{R})$ be the set of all sequences that have a finite number of nonzero terms. Then $\mathcal{S}_{\text{fin}}(\mathbb{R})$ is a vector space with operations as defined in Definition 2.4.14. (See Exercise 16.) □

We find vector space properties for sequences to be very useful in the development of calculus concepts such as limits. For example, if we want to apply a limit to the sum of sequences, we need to know that the sum of two sequences is indeed a sequence. More of these concepts will be discussed later, after developing more linear algebra ideas.

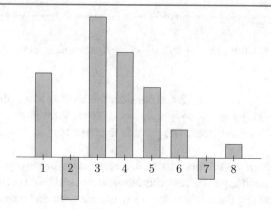

Fig. 2.16 A graphical representation of a vector in $\mathcal{J}_8$.

Bar Graph Spaces

Now we consider the set of bar graphs, one in which visualization is necessary for understanding the elements of the set.

Definition 2.4.16

Let $k \in \mathbb{R}$ and denote by $\mathcal{J}_k$ the set of all **bar graphs** with k bins. Here, we consider a bar graph to be a function from the set $\{1, \ldots, k\}$ to $\mathbb{R}$, and we visualize such an object in the familiar graphical way, as shown in Figure 2.16. Define addition and scalar multiplication on such bar graphs as follows. Let J_1 and J_2 be two bar graphs in $\mathcal{J}_k$ and let α be a scalar in $\mathbb{R}$.

- We define the 0 bar graph to be the bar graph where each of the k bars has height 0.
- $J_1 + J_2$ is defined to be the bar graph obtained by summing the height of the bars in corresponding bins of J_1 and J_2.
- $\alpha \cdot J_1$ is defined to be the bar graph obtained by multiplying each bar height of J_1 by α.

□

With this definition, we can verify that $\mathcal{J}_k$ is a vector space. Indeed, because addition and scalar multiplication are bar-wise defined, and because the heights of the bars are real numbers, the properties of Definition 2.3.5 are inherited from $\mathbb{R}$.

The space of bar graphs is actually a discrete function space with the added understanding that the domain of the functions has a geometric representation (the "bins" for the bar graphs are all lined up in order from 1 to k).

Digital Characters

The following is an example of a vector space over the field Z_2.

Example 2.4.17 Consider the image set of 7-bar LCD characters, $\mathcal{D}(Z_2)$, where Z_2 is the field that includes only the scalar values 0 and 1. Figure 2.17 shows ten example characters along with the image geometry.

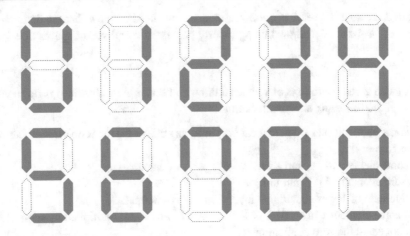

Fig. 2.17 The ten digits of a standard 7-bar LCD display.

For these images, white corresponds to the value zero and green corresponds to the value one. With element-wise definitions (given on page 41) of addition and scalar multiplication as defined for the field Z_2, $\mathcal{D}(Z_2)$ is a vector space. (See Exercise 19.) Here are two examples of vector addition in $\mathcal{D}(Z_2)$:

What happens with scalar multiplication? What are the digital characters that can be obtained just from scalar multiplication (without addition) using the characters in Figure 2.17?

2.4.6 Is My Set a Vector Space?

Given a set and operations of addition and scalar multiplication, we would like to determine if the set is, or is not, a vector space.

If we want to prove that the set is a vector space, we just show that it satisfies the definition.

▶ A set V (with given operations of vector addition and scalar multiplication) is a vector space if it satisfies each of the ten properties of Definition 2.3.5. One must show that these properties hold for arbitrary elements of V and arbitrary scalars in the field (usually $\mathbb{R}$).

In order to determine if a set S is *not* a vector space, we need to show that the definition does not hold for this set, or we need to show that a property possessed by all vector spaces does not hold for the set.

▶ A set V (with given operations of vector addition and scalar multiplication) is **not** a vector space if any one of the following statements is true.

1. For **some** element(s) in V and/or scalar(s) in $\mathbb{R}$, any one of the ten properties of Definition 2.3.5 is **not** true.
2. For some elements x, y, and z in V with $x \neq y$, we have $x + z = y + z$.
3. The zero element of V is not unique.
4. Any element of V has an additive inverse that is not unique.
5. If for some element x in V, $0 \cdot x \neq 0$. That is, the zero scalar multiplied by some element of V does not equal the zero element of V.

2.4.7 Exercises

1. In the properties of a vector space, (P5) and (P6) refer to three different operations, what are they?
2. Suppose $V = \{v \in \mathbb{R} \mid v \neq 0\}$. Let $+$ and $\cdot$ be the standard operations in $\mathbb{R}$. Show why, when considering whether $(V, +, \cdot)$ is a vector space over $\mathbb{R}$, we can say that V is neither closed under scalar multiplication nor is it closed under vector addition.
3. Show that $H_m(\mathbb{R})$, the set of all heat states with $m + 2$ real entries, is a vector space.
4. Plot a possible heat state u for a rod with $m = 12$. In the same graph, plot a second heat state that corresponds to $2u$. Describe the similarities and differences between u and $2u$.
5. Plot a possible heat state u for a rod with $m = 12$. In the same graph, plot the heat state that corresponds to $u + v$, where v is another heat state that is not a scalar multiple of u. Describe the similarities and differences between u and $u + v$.
6. A 12-megapixel phone camera creates images with 12,192,768 pixels, arranged in 3024 rows and 4032 columns, such as the one in Figure 3.9. The color in each pixel is determined by the relative brightness of red, green, and blue light specified by three scalars. So rather than just a single scalar determining grayscale intensity, each pixel is assigned 3 numbers that represent the intensities of red, green, and blue light for the pixel. Verify that the space V of such images is a vector space.
7. Let $\mathcal{P}_2 = \{ax^2 + bx + c \mid a, b, c \in \mathbb{R}\}$. Show that $\mathcal{P}_2$ is a vector space with scalars taken from $\mathbb{R}$ and addition and scalar multiplication defined in the standard way for polynomials.
8. Determine whether or not the following functions are vectors in $\mathcal{F}$ as defined in Example 2.4.1. $f(x) = \tan(x)$, $g(x) = x^5 - 5$, $h(x) = \ln(x)$.
9. Show that the set of even functions (see Example 2.4.3) is a vector space.
10. Let $C^1([a, b])$ be the set of continuous functions whose first derivative is also continuous on the interval $[a, b]$. Is $C^1([a, b])$ a vector space with the definitions of vector addition and scalar multiplication given in Example 2.4.1?
11. Show that the set $\mathcal{P}_n(\mathbb{R})$ of Example 2.4.2 is a vector space.
12. Define the set $\mathcal{P}(\mathbb{R})$ to be the union $\bigcup_{n \in \mathbb{N}} \mathcal{P}_n(\mathbb{R})$. Is this set a vector space, with the operations of addition of polynomials and multiplication by constants? Justify your reasoning.
13. Let $V \subseteq \mathbb{R}^3$ denote the set of all solutions to the linear equation $x_1 + 3x_2 - x_3 = 5$.

    (a) Use the algebraic definition of a vector space to show that V is not a vector space.
    (b) Give a geometric argument for the fact that V is not a vector space.

14. Let $\{a_1, a_2, \ldots, a_n\} \subset \mathbb{R}$ and consider the set $V \in \mathbb{R}^n$ of all solutions to the homogeneous linear equation $a_1 x_1 + a_2 x_2 + \cdots + a_n x_n = 0$. Show that the set V together with the operations $+$ and $\cdot$ inherited from $\mathbb{R}^n$ forms a vector space.

15. Show that the set $S(\mathbb{R})$ of Definition 2.4.14 is a vector space.

16. Show that the set $S_{\text{fin}}(\mathbb{R})$ of Example 2.4.15 is a vector space.

17. For each of the following sets (and operations), determine if the set is a vector space. If the set is a vector space, verify the axioms in the definition. If the set is not a vector space, find an example to show that the set is not a vector space.

   (a) The set of vectors in the upper half plane of $\mathbb{R}^2$, that is, the set of vectors in $\mathbb{R}^2$ whose y-component is greater than or equal to zero. (Addition and scalar multiplication are inherited from $\mathbb{R}^2$.)

   (b) The set of images of the form below, with a, b, and c real numbers.

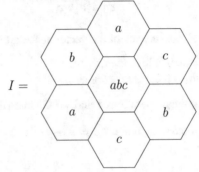

   (c) The set of upper triangular 3×3 matrices. A matrix A is upper triangular if $a_{i,j} = 0$ whenever $i > j$. (Addition and scalar multiplication are inherited from the space of 3×3 matrices.)

   (d) The set of polynomials of any degree that go through the point $(2, 2)$. (Addition and scalar multiplication are inherited from the space of polynomials.)

   (e) The set of solutions to a homogeneous equation with 4 variables, considered as a subset of $\mathbb{R}^4$. (Addition and scalar multiplication are inherited from $\mathbb{R}^4$.)

   (f) The set of solutions to the equation $xy - x^2 = 0$, considered as a subset of $\mathbb{R}^2$. (Addition and scalar multiplication are inherited from $\mathbb{R}^2$.)

   (g) The set of real-valued integrable functions on the interval $[a, b]$. (Addition and scalar multiplication are defined pointwise.)

   (h) The set of (continuous) heat signatures on a rod of length L with at most 3 peaks (local maxima). (Addition and scalar multiplication are inherited from the vector space of heat signatures.)

18. We say that W is a **subset** of V if every element of W is an element of V. In the case where W is a subset of V, we write $W \subset V$. If $W \subset V$ and W does not contain all of the elements of V, we say W is a **proper** subset of V. Now, consider a vector space $(V, +, \cdot)$. Which of the 10 vector space properties are not necessarily true for a proper subset $W \subset V$?

19. Prove that $\mathcal{D}(Z_2)$ from Example 2.4.17 is a vector space.

20. Show that neither closure property holds for the set

$$V\{(x_1, x_2, \ldots, x_n) \in \mathbb{R} \mid a_1 x_1 + a_2 x_2 + \ldots + a_n x_n = b, b \neq 0\}.$$

21. Let V be the finite vector space given by

$$V = \left\{ a = \bigcirc , \ b = \bigcirc , \ c = \bigcirc , \ d = \bigcirc \right\}.$$

Define $\boxplus$, vector addition, and $\boxtimes$, scalar multiplication, according to the following tables:

$\boxplus$	a	b	c	d
a	a	b	c	d
b	b	a	d	c
c	c	d	a	b
d	d	c	b	a

$\boxtimes$	a	b	c	d
0	a	a	a	a
1	a	b	c	d

One can verify by inspection each of the vector space properties hold for V.

(a) What element is the additive identity for V?

(b) What is the additive inverse of a? of c?

22. Consider the set $\mathbb{R}$ and the operations $\dagger$ and $\star$ defined on $\mathbb{R}$ by

Let $u, v \in \mathbb{R}$ be vectors. Define $u \dagger v = u + v - 3$.

and

Let $\alpha \in \mathbb{R}$ be a scalar and $u \in \mathbb{R}$ be a vector. Define $\alpha \star u = \alpha u/2$.

Is $(\mathbb{R}, \dagger, \star)$ a vector space over $\mathbb{R}$? Justify your response with a proof or a counterexample.

23. Show that the closure properties of Definition 2.3.5 hold true for the set of functions $(\mathcal{C}_0([a, b]), +, \cdot)$. Use this result to state how you know that $(\mathcal{C}_0([a, b]), +, \cdot)$ is a vector space over $\mathbb{R}$.

24. Let V be the set of vectors defined below:

$$V = \left\{ \bigcirc , \ \bigcirc \right\}.$$

(a) Define $\boxplus$, vector addition, and $\boxtimes$, scalar multiplication, so that $(V, \boxplus, \boxtimes)$ is a vector space over Z_2. Prove that $(V, \boxplus, \boxtimes)$ is indeed a vector space with the definitions you make.

(b) Give an example where V might be of interest in the real world. (You can be imaginative enough to think of this set as a very simplified version of something bigger.)

(c) Suppose we add another vector to V to get $\tilde{V}$. Is it possible to define $\boxplus$ and $\boxtimes$ so that $(\tilde{V}, \boxplus, \boxtimes)$ is a vector space over Z_2? Justify.

(d) Is it possible to define $\boxplus$ and $\boxtimes$ so that $(V, \boxplus, \boxtimes)$ is a vector space over $\mathbb{R}$? Justify.

2.5 Subspaces

PetPics, a pet photography company specializing in portraits, wants to post photos for clients to review, but to protect their artistic work, they only post electronic versions that have copyright text. The text is

Fig. 2.18 Example Proof photo from PetPics, with copyright text.

added to all images, produced by the company, by overwriting, with zeros, in the appropriate pixels, as shown in Figure 2.18. Only pictures that have zeros in these pixels are considered legitimate images.

The company also wants to allow clients to make some adjustments to the pictures: the adjustments include brightening/darkening, and adding background or little figures like hearts, flowers, or squirrels. It turns out that these operations can all be accomplished by adding other legitimate images and multiplying by scalars, as defined in Section 2.3.

It is certainly true that the set of all legitimate images of the company's standard $(m \times n)$-pixel size is contained in the vector space $\mathcal{I}_{m \times n}$ of all $m \times n$ images, so we could mathematically work in this larger space. But, astute employees of the company who enjoy thinking about linear algebra notice that actually the set of legitimate images itself satisfies the 10 properties of a vector space. Specifically, adding any two images with the copyright text (for example, adding a squirrel to a portrait of a golden retriever) produces another image with the same copyright text, and multiplying an image with the copyright text by a scalar (say, to brighten it) still results in an image with the copyright text. Hence, it suffices to work with the smaller set of legitimate images.

In fact, very often the sets of objects that we want to focus on are actually only *subsets* of larger vector spaces, and it is useful to know when such a set forms a vector space separately from the larger vector space.

Here are some examples of subsets of vector spaces that we have encountered so far.

1. Solution sets of homogeneous linear equations, with n variables, are subsets of $\mathbb{R}^n$.
2. Radiographs are images with nonnegative values and represent a subset of the larger vector space of images with the given geometry.
3. The set of even functions on $\mathbb{R}$ is a subset of the vector space of functions on $\mathbb{R}$.
4. Polynomials of order 3 form a subset of the vector space $\mathcal{P}_5(\mathbb{R})$.
5. Heat states on a rod in a diffusion welding process (the collection of which is $H_m(\mathbb{R})$) form a subset of all possible heat states because the temperature is fixed at the ends of the rod.
6. The set of sequences with exactly 10 nonzero terms is a subset of the set of sequences with a finite number of terms.

Even though operations like vector addition and scalar multiplication on the subset are typically the same as the operations on the larger parent spaces, we still often wish to work in the smaller more relevant subset rather than thinking about the larger ambient space. When does the subset behave like a vector space in its own right? In general, when is a subset of a vector space also a vector space?

2.5.1 Subsets and Subspaces

Let $(V, +, \cdot)$ be a vector space. In this section we discuss conditions on a subset of a vector space that will guarantee the subset is also a vector space. Recall that a subset of V is a set that contains some of the elements of V. We define subset more precisely here.

Definition 2.5.1

Let V and W be sets. We say that W is a **subset** of V if every element of W is an element of V and we write $W \subset V$ or $W \subseteq V$. In the case where $W \neq V$ (there are elements of V that are not in W), we say that W is a **proper subset** of V and we write $W \subsetneq V$. $\square$

In a vector space context, we always assume the same operations on W as we have defined on V.

Let W be a subset of V. We are interested in subsets that also satisfy the vector space properties (recall Definition 2.3.5).

Definition 2.5.2

Let $(V, +, \cdot)$ be a vector space over a field $\mathbb{F}$. If $W \subseteq V$, then we say that W is a **subspace** of $(V, +, \cdot)$ whenever $(W, +, \cdot)$ is also a vector space. $\square$

Now consider which vector space properties of $(V, +, \cdot)$ must also be true of the subset W. Which properties are not necessarily true? The commutative, associative, and distributive properties still hold because the operations are the same, the scalars come from the same scalar field, and elements of W come from the set V. Therefore, since these properties are true in V, they are true in W. We say that these properties are *inherited* from V since V is like a parent set to W. Also, since, we do not change the scalar set when considering a subset, the scalar 1 is still an element of the scalar set. This tells us that we can determine whether a subset of a vector space is, itself, a vector space, by checking those properties that depend on how the subset differs from the parent vector space. The properties we need to check are the following

(P1) W is closed under addition.
(P2) W is closed under scalar multiplication.
(P8) W contains the additive identity, denoted 0.
(P9) W contains additive inverses.

With careful consideration, we see that, because V contains additive inverses, then if (P1), (P2), and (P8) are true for W, it follows that W must also contain additive inverses (see Exercise 14). Hence, as the following theorem states, we need only test for properties (P1), (P2), and (P8) in order to determine whether a subset is a subspace.

Theorem 2.5.3

Let $(V, +, \cdot)$ be a vector space over a field $\mathbb{F}$ and let W be a subset of V. Then $(W, +, \cdot)$ is a subspace of V over $\mathbb{F}$ if and only if $0 \in W$ and W is closed under vector addition and scalar multiplication.

The following corollary is based on the fact that as long as W is nonempty and satisfies (P1) and (P2), then W automatically satisfies (P8).

Corollary 2.5.4

Let $(V, +, \cdot)$ be a vector space over a field $\mathbb{F}$ and let W be a nonempty subset of V. Then $(W, +, \cdot)$ is a subspace of V over $\mathbb{F}$ if and only if W is closed under vector addition and scalar multiplication.

The proof is the subject of Exercise 15.

While Corollary 2.5.4 presents a somewhat simplified test for determining if a subset is a subspace, the additional condition in Theorem 2.5.3 presents a very simple test for possibly determining if a subset is not a subspace. Namely, if $0 \notin W$ then W is not a subspace. When discussing subspaces, we tend to leave off using the notation $(W, +, \cdot)$ and write W instead because the operations are understood in the context of $(V, +, \cdot)$.

The following two results give convenient ways to check for closure under addition and scalar multiplication.

Theorem 2.5.5

Let $(V, +, \cdot)$ be a vector space and let X be a nonempty subset of V. Then X is closed under both addition and scalar multiplication if and only if for all pairs of vectors $x, y \in X$ and for all scalars $\alpha, \beta \in \mathbb{F}$, $\alpha x + \beta y \in X$.

This theorem makes use of an "if and only if" statement. The proof relies on showing both

- ($\Rightarrow$) If X is closed under both addition and scalar multiplication, then for all $x, y \in X$ and for all $\alpha, \beta \in \mathbb{F}$, $\alpha x + \beta y \in X$ and
- ($\Leftarrow$) If for all $x, y \in X$ and for all $\alpha, \beta \in \mathbb{R}$, $\alpha x + \beta y \in X$, then X is closed under both addition and scalar multiplication.

Proof. Let $(V, +, \cdot)$ be a vector space and let X be a nonempty subset of V.

($\Rightarrow$) Suppose X is closed under addition and scalar multiplication, and that x and y are in V and α and β are in $\mathbb{F}$. Since X is closed under scalar multiplication, we know that αx and βy are in X. But then since X is also closed under addition, we conclude that $\alpha x + \beta y$ is in X. Since this is true for all $\alpha, \beta \in \mathbb{F}$ and $x, y \in X$, we conclude that X has the property that for all $x, y \in X$ and for all $\alpha, \beta \in \mathbb{F}$, $\alpha x + \beta y \in X$.

($\Leftarrow$) Suppose X has the property that for all $x, y \in X$ and for all $\alpha, \beta \in \mathbb{F}$, $\alpha x + \beta y \in X$. Now let $x \in X$ and $\alpha \in \mathbb{F}$, and fix $\beta = 0 \in \mathbb{F}$ and $y = 0$, the zero vector. Then $\alpha x + \beta y = \alpha x + 0 = \alpha x$ is in X. Hence X is closed under scalar multiplication. Finally, let $x, y \in X$ and $\alpha = \beta = 1$, the multiplicative identity of $\mathbb{F}$. Then we can also conclude that $\alpha x + \beta y = 1x + 1y = x + y$ is in X, and so X is closed under addition. $\qquad \square$

It is often even simpler to check the conditions for the following similar result.

Corollary 2.5.6
Consider a set of objects X, scalars in $\mathbb{R}$, and operation $+$ (addition) and $\cdot$ (scalar multiplication) defined on X. Then X is closed under both addition and scalar multiplication if and only if $\alpha \cdot x + y \in X$ for all $x, y \in X$ and each $\alpha \in \mathbb{R}$.

The proof is the subject of Exercise 16.

2.5.2 Examples of Subspaces

Every vector space $(V, +, \cdot)$ has at least the following two subspaces.

Theorem 2.5.7
Let $(V, +, \cdot)$ be a vector space. Then V is itself a subspace of $(V, +, \cdot)$.

Proof. Since every set is a subset of itself, the result follows from Definition 2.5.2. □

Theorem 2.5.8
Let $(V, +, \cdot)$ be a vector space. Then the set $\{0\}$ is a subspace of $(V, +, \cdot)$.

The proof is Exercise 19.

Example 2.5.9 Recall Example 2.4.9 from the last section. Let $V \subset \mathbb{R}^3$ be the set of all solutions to the equation $x_1 + 3x_2 - x_3 = 0$. Then V is a subspace of $\mathbb{R}^3$, with the standard operations. □

More generally, as we saw in the last section, the set of solutions to *any* homogeneous linear equation with n variables is a subspace of $(\mathbb{R}^n, +, \cdot)$.

Example 2.5.10 Consider the coordinate axes as a subset of the vector space $\mathbb{R}^2$. That is, let $T \subset \mathbb{R}^2$ be defined by
$$T = \{x = (x_1, x_2) \in \mathbb{R}^2 \mid x_1 = 0 \text{ or } x_2 = 0\}.$$

T is not a subspace of $(\mathbb{R}^2, +, \cdot)$, because although 0 is in T, making $T \neq \emptyset$, T does not have the property that for all $x, y \in T$ and for all $\alpha, \beta \in \mathbb{R}$, $\alpha x + \beta y \in T$. To verify this, we need only produce one example of vectors $x, y \in T$ and scalars $\alpha, \beta \in \mathbb{R}$ so that $\alpha x + \beta y$ is not in T. Notice that $x = (0, 1)$, $y = (1, 0)$ are elements of T and $\alpha = \beta = 1$ are in $\mathbb{R}$. Since $1 \cdot x + 1 \cdot y = (1, 1)$ which is not in T, T does not satisfy the subspace property. □

Example 2.5.11 Consider $W = \{(a, b, c) \in \mathbb{R}^3 \mid c = 0\}$. W is a subspace of $\mathbb{R}^3$, with the standard operations of addition and scalar multiplication. See Exercise 9. □

Example 2.5.12 Consider the set of images

$$V = \{I \mid I \text{ is of the form shown in Figure 2.19 and } a, b, c \in \mathbb{R}\}.$$

V is a subspace of the space of images with the same geometric configuration. $\qquad \Box$

Proof. Let V be defined as above. We will show that V satisfies the hypotheses of Theorem 2.5.3. We can see that the set of images V is a subset of the vector space of all images with the same geometric configuration.

Next, notice that V is nonempty since the image in V with $a = b = c = 0$ is the zero image. Now, we need to show that the set is closed under scalar multiplication and addition. Let $\alpha \in \mathbb{R}$ be a scalar and let I_1 and I_2 be images in V. For the image I_1, let $a = a_1$, $b = b_1$, and $c = c_1$, and for the image I_2, let $a = a_2$, $b = b_2$, and $c = c_2$ (Figure 2.20).

We can see that $\alpha I_1 + I_2$ is also in V with $a = \alpha a_1 + a_2$, $b = \alpha b_1 + b_2$, and $c = \alpha c_1 + c_2$. For example, the pixel intensity in the pixel on the bottom left of $\alpha I_1 + I_2$ is

$$\alpha(a_1 - c_1) + a_2 - c_2 = \alpha a_1 + a_2 - (\alpha c_1 - c_2) = a - c.$$

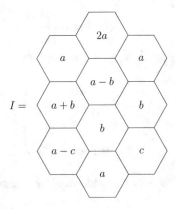

Fig. 2.19 The form of images in the set V of Example 2.5.12.

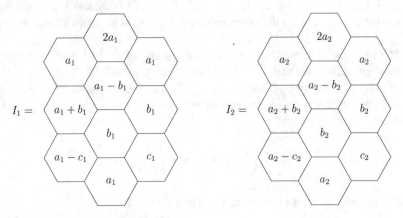

Fig. 2.20 Arbitrary images I_1 and I_2 in the set V of Example 2.5.12.

Similarly, we can see that each pixel intensity satisfies the form shown in Figure 2.19 for the chosen values of a, b, and c. (Try a few yourself to be sure you agree.) Thus V is a subspace of the space of images with this geometric arrangement of ten pixels.   □

This result also means that V is a vector space.

Example 2.5.13 Is the set $V = \{ax^2 + bx + c \mid a > 0\}$ a vector space? Clearly, V is a subset of the vector space $(\mathcal{P}_2, +, \cdot)$. However, the zero vector of $\mathcal{P}_2$, $\mathbf{0}(x) = 0x^2 + 0x + 0$, is not in V. So, by Theorem 2.5.3, V is not a vector space.   □

For the next example, let us first define the transpose of a matrix.

Definition 2.5.14

Let A be the $r \times c$ matrix given by

$$A = \begin{pmatrix} a_{11} & a_{12} & \cdots & a_{1c} \\ a_{21} & a_{22} & \cdots & a_{2c} \\ \vdots & & \ddots & \vdots \\ a_{r1} & a_{r2} & \cdots & a_{rc} \end{pmatrix}.$$

Then, the **transpose** of A, denoted A^T, is the $c \times r$ matrix that is given by

$$A^\mathsf{T} = \begin{pmatrix} a_{11} & a_{21} & \cdots & a_{r1} \\ a_{12} & a_{22} & \cdots & a_{r2} \\ \vdots & & \ddots & \vdots \\ a_{1c} & a_{2c} & \cdots & a_{rc} \end{pmatrix}.$$

In words, A^T is the matrix whose rows are made from the columns of A and whose columns are made from the rows of A.   □

Example 2.5.15 Let $M = \{u \in \mathcal{M}_{2\times 2}(\mathbb{R}) \mid u = u^\mathsf{T}\}$. That is, M is the set of all 2×2 matrices with real-valued entries that are equal to their own transpose. The set M is a subspace of $\mathcal{M}_{2\times 2}(\mathbb{R})$.   □

Proof. Let M be defined as above. First, we note that M is a subset of the vector space $\mathcal{M}_{2\times 2}(\mathbb{R})$, with the standard operations. We will show that M satisfies Theorem 2.5.3. Notice that M contains the zero vector of $\mathcal{M}_{2\times 2}(\mathbb{R})$, the two by two matrix of all zeros. Next, consider two arbitrary vectors u and v in M and arbitrary scalar α in $\mathbb{R}$. We can show that $\alpha u + v = (\alpha u + v)^\mathsf{T}$ so that $\alpha u + v$ is also in M. Indeed, let

$$u = \begin{pmatrix} a & b \\ c & d \end{pmatrix} \quad \text{and} \quad v = \begin{pmatrix} e & f \\ g & h \end{pmatrix}.$$

Then since $u = u^\mathsf{T}$ and $v = v^\mathsf{T}$, we know that $b = c$ and $f = g$. Thus

$$\alpha u + v = \alpha \begin{pmatrix} a & b \\ b & d \end{pmatrix} + \begin{pmatrix} e & f \\ f & h \end{pmatrix}$$

$$= \begin{pmatrix} \alpha a + e & \alpha b + f \\ \alpha b + f & \alpha d + h \end{pmatrix}$$

$$= \begin{pmatrix} \alpha a + e & \alpha b + f \\ \alpha b + f & \alpha d + h \end{pmatrix}^\mathsf{T}$$

$$= (\alpha u + v)^\mathsf{T}.$$

Thus, M is a subspace of $\mathcal{M}_{2\times 2}(\mathbb{R})$. $\qquad\qquad\square$

2.5.3 Subspaces of $\mathbb{R}^n$

In Section 2.3, we introduced the n-dimensional Euclidean spaces $\mathbb{R}^n$. One nice aspect of these spaces is that they lend themselves to visualization: the space $\mathbb{R} = \mathbb{R}^1$ just looks like a number line, $\mathbb{R}^2$ is visualized as the xy-plane, and $\mathbb{R}^3$ can be represented with x-, y-, and z-axes as pictured in Figure 2.21. For larger n, the space is harder to visualize, because we would have to add more axes and the world we live in is very much like $\mathbb{R}^3$. As we continue our study of linear algebra, you will start to gain more intuition about how to think about these higher dimensional spaces.

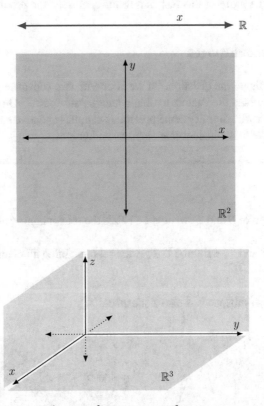

Fig. 2.21 Geometric representations of $\mathbb{R}^1$ (Top), $\mathbb{R}^2$ (Middle), and $\mathbb{R}^3$ (Bottom).

In the last section we looked at a few examples of subspaces of $\mathbb{R}^n$. In this section we explore the geometry of such subspaces.

Theorem 2.5.16
Let L be a line through the origin in $\mathbb{R}^n$. Then L is a subspace of $(\mathbb{R}^n, +, \cdot)$.

Proof. A line through the origin in $\mathbb{R}^n$ is represented as the set $L = \{\beta v \mid \beta \in \mathbb{R}\}$. We show L is a subspace using the property given in Corollary 2.5.6. Let $w_1 = \beta_1 v$ and $w_2 = \beta_2 v$ be two vectors in L and let $\alpha \in \mathbb{R}$ be a scalar. Then

$$\alpha w_1 + w_2 = \alpha \beta_1 v + \beta_2 v = (\alpha \beta_1 + \beta_2)v,$$

and so L is closed under addition and scalar multiplication.

Also, notice that L is nonempty. Thus, by Theorem 2.5.3, L is a subspace of $(\mathbb{R}^n, +, \cdot)$. □

As we explore higher dimensions, we find that planes through the origin (and in fact any geometric space described by the solutions of a homogeneous system of linear equations) are also vector spaces. We define a plane as the set of all linear combinations of two (not colinear) vectors in $\mathbb{R}^3$. We leave it as Exercise 10 to write the proof of this fact as it follows, closely, the proof of Theorem 2.5.16.

2.5.4   Building New Subspaces

In this section, we investigate the question, "If we start with two subspaces of the same vector space, in what ways can we combine the vectors to obtain another subspace?" Our investigation will lead us to some observations that will simplify some previous examples and give us new tools for proving that subsets are subspaces. We first consider intersections and unions.

Definition 2.5.17

Let S and T be sets.

- The **intersection** of S and T, written $S \cap T$, is the set containing all elements that are in both S and T.
- The **union** of S and T, written $S \cup T$, is the set containing all elements that are in either S or T (or both). □

The intersection of two subspaces is also a subspace.

Theorem 2.5.18
Let W_1 and W_2 be any two subspaces of a vector space $(V, +, \cdot)$. Then $W_1 \cap W_2$ is a subspace of $(V, +, \cdot)$.

Proof. Let W_1 and W_2 be subspaces of $(V, +, \cdot)$. We will show that the intersection of W_1 and W_2 is nonempty and closed under scalar multiplication and vector addition. To show that $W_1 \cap W_2$ is nonempty, we notice that since both W_1 and W_2 contain the zero vector, so does $W_1 \cap W_2$.

Now, let u and v be elements of $W_1 \cap W_2$ and let α be a scalar. Since W_1 and W_2 are closed under addition and scalar multiplication, we know that $\alpha \cdot u + v$ is also in both W_1 and W_2. That is, $\alpha \cdot u + v$ is in $W_1 \cap W_2$, so by Corollary 2.5.6 $W_1 \cap W_2$ is closed under addition and scalar multiplication.

Thus, by Corollary 2.5.4, $W_1 \cap W_2$ is a subspace of $(V, +, \cdot)$. $\qquad\square$

An important example involves solutions to homogeneous equations, which we first considered in Example 2.4.12.

Example 2.5.19 The solution set of a single homogeneous equation in n variables is a subspace of $\mathbb{R}^n$ (see Example 2.4.12). By Theorem 2.5.18, the intersection of the solution sets of any k homogeneous equations in n variables is also subspace of $\mathbb{R}^n$. $\qquad\square$

In other words, if a system of linear equations consists only of homogeneous equations, then the set of solutions forms a subspace of $\mathbb{R}^n$. This is such an important result that we promote it from example to theorem.

Theorem 2.5.20

The solution set of a homogeneous system of equations in n variables is a subspace of $\mathbb{R}^n$.

The proof of this theorem closely follows the argument of Example 2.5.19 and is left to the reader. Theorem 2.5.18 also provides a new method for determining if a subset is a subspace. If the subset can be expressed as the intersection of known subspaces, then it also is a subspace.

Example 2.5.21 Consider the vector space of functions defined on $\mathbb{R}$, denoted $\mathcal{F}$. We have shown that the set of even functions on $\mathbb{R}$ and the set of continuous functions on $\mathbb{R}$ are both subspaces. Thus, by Theorem 2.5.18, the set of continuous even functions on $\mathbb{R}$ is also a subspace of $\mathcal{F}$. $\qquad\square$

The union of subspaces, on the other hand, need not be a subspace. Recall Example 2.5.10, where we showed that the coordinate axes in $\mathbb{R}^2$ are not a subspace. Because the coordinate axes can be written as the union of the x-axis and the y-axis, both of which are subspaces of $\mathbb{R}^2$, it is clear that unions of subspaces need not be subspaces.

In fact, the union of subspaces is generally not a subspace. The next theorem tells us that the union of subspaces is a subspace only when one subspace is a subset of the other.

Theorem 2.5.22

Let W_1 and W_2 be any two subspaces of a vector space V. Then $W_1 \cup W_2$ is a subspace of V if and only if $W_1 \subseteq W_2$ or $W_2 \subseteq W_1$.

Proof. Suppose W_1 and W_2 are subspaces of a vector space V.

($\Leftarrow$) First, suppose that $W_1 \subseteq W_2$. Then, we know that $W_1 \cup W_2 = W_2$ which is a subspace. Similarly, if $W_2 \subseteq W_1$ then $W_1 \cup W_2 = W_1$ which is a subspace.

($\Rightarrow$) We will show the contrapositive. That is, we assume that neither subspace (W_1 nor W_2) contains the other. We will show that the union, $W_1 \cup W_2$, is not a subspace. Since $W_1 \nsubseteq W_2$ and $W_2 \nsubseteq W_1$, we know that there is a vector $u_1 \in W_1$, with $u_1 \notin W_2$ and there is another vector $u_2 \in W_2$, with $u_2 \notin W_1$. Since W_1 and W_2 both contain additive inverses and both are closed, $u_1 + u_2 \notin W_1$ and $u_1 + u_2 \notin W_2$. So, $u_1, u_2 \in W_1 \cup W_2$, but $u_1 + u_2 \notin W_1 \cup W_2$. Thus, $W_1 \cup W_2$ is not closed under addition and hence is not a subspace of V. $\qquad\square$

Another way to combine sets is to take the so-called *sum* of the sets. In words, the sum of two sets is the set formed by adding all combinations of pairs of elements, one from each set. We define this more rigorously in the following definition.

Definition 2.5.23

Let U_1 and U_2 be sets. We define the **sum** of U_1 and U_2, denoted $U_1 + U_2$, to be the set $\{u_1 + u_2 \mid u_1 \in U_1, u_2 \in U_2\}$. $\qquad\square$

In order to add two sets, they need not be the same size and the sum of two sets is a set that can be (but is not necessarily) strictly larger than each of the two summand sets. This last comment can be seen in Example 2.5.24.

Example 2.5.24 Let $U_1 = \{3, 4, 5\}$, and $U_2 = \{1, 3\}$. Then

$$\begin{aligned}
U_1 + U_2 &= \{u_1 + u_2 \mid u_1 \in U_1, u_2 \in U_2\} \\
&= \{3 + 1, 3 + 3, 4 + 1, 4 + 3, 5 + 1, 5 + 3\} \\
&= \{4, 5, 6, 7, 8\}.
\end{aligned}$$

$\qquad\square$

Example 2.5.25 Let U_1 be the set of all scalar multiples of image A (see page 11). Similarly, let U_2 be the set of all scalar multiples of image B. We can show that $U_1 + U_2$ is a subspace of the vector space of 4×4 grayscale images. (See Exercise 24.) $\qquad\square$

In both of the previous two examples we see that the sum of two sets can contain elements that are not elements of either set. This means that the sum of two sets is not necessarily equal to the union of the two sets. In Example 2.5.25 the sets U_1 and U_2 are both subspaces and their sum $U_1 + U_2$ is also a subspace. This leads us to consider whether this is always true. We see, in the next theorem, that the answer is yes.

Theorem 2.5.26

Let W_1 and W_2 be subspaces of a vector space $(V, +, \cdot)$. Then $W_1 + W_2$ is a subspace of $(V, +, \cdot)$.

Proof. Let $S = W_1 + W_2$. We will show that S contains the zero vector and is closed under addition and scalar multiplication.

Since W_1 and W_2 are a subspaces of $(V, +, \cdot)$, both contain the zero vector. Now, 0 being the additive identity, gives us $0 = 0 + 0 \in S$.

Let $x = x_1 + x_2$ and $y = y_1 + y_2$ be arbitrary elements of S with $x_1, y_1 \in W_1$ and $x_2, y_2 \in W_2$, and let α be an arbitrary scalar. Notice that, since $\alpha x_1 + y_1 \in W_1$ and $\alpha x_2 + y_2 \in W_2$,

$$\alpha x + y = (\alpha x_1 + y_1) + (\alpha x_2 + y_2) \in S.$$

Thus, by Corollary 2.5.6, S is also closed under scalar multiplication and vector addition. By Theorem 2.5.3, S is a subspace of V. □

In words, Theorem 2.5.26 tells us that the sum of subspaces always results in a subspace. Of special interest are subspaces of the type considered in Example 2.5.25. That is, we are interested in the sum $U_1 + U_2$ for which $U_1 \cap U_2 = \{0\}$.

Definition 2.5.27

Let U_1 and U_2 be subspaces of vector space $(V, +, \cdot)$ such that $U_1 \cap U_2 = \{0\}$ and $V = U_1 + U_2$. We say that V is the **direct sum** of U_1 and U_2 and denote this by $V = U_1 \oplus U_2$. □

Because the direct sum is a special case of the sum of two sets, we know that the direct sum of two subspaces is also a subspace.

Example 2.5.28 Let $U_1 = \{(a, 0) \in \mathbb{R}^2 \mid a \in \mathbb{R}\}$ and $U_2 = \{(0, b) \in \mathbb{R}^2 \mid b \in \mathbb{R}\}$. U_1 can be represented by the set of all vectors along the x-axis and U_2 can be represented by the set of all vectors along the y-axis. Both U_1 and U_2 are subspaces of $\mathbb{R}^2$, with the standard operations. And, $U_1 \oplus U_2 = \mathbb{R}^2$. □

Example 2.5.29 Consider the sets

$$U_1 = \{ax^2 \in \mathcal{P}_2(\mathbb{R}) \mid a \in \mathbb{R}\},$$
$$U_2 = \{ax \in \mathcal{P}_2(\mathbb{R}) \mid a \in \mathbb{R}\},$$
$$U_3 = \{a \in \mathcal{P}_2(\mathbb{R}) \mid a \in \mathbb{R}\}.$$

Notice that each subset is a subspace of $(\mathcal{P}_2(\mathbb{R}), +, \cdot)$. Notice also that $\mathcal{P}_2(\mathbb{R}) = U_1 + U_2 + U_3$. Furthermore, each pair of subsets has the trivial intersection $\{0\}$. Thus, we have $\mathcal{P}_2(\mathbb{R}) = (U_1 \oplus U_2) \oplus U_3$. We typically write $\mathcal{P}_2(\mathbb{R}) = U_1 \oplus U_2 \oplus U_3$ with the same understanding. □

Examples 2.5.28 and 2.5.29 suggest that there might be a way to break apart a vector space into the direct sum of subspaces. This fact will become very useful in Chapter 7.

2.5.5 Exercises

In the following exercises, whenever discussing a vector space $(V, +, \cdot)$ where $+$ and $\cdot$ are the standard operations or clear from the context, we will simplify notation and write only V.

1. Describe or sketch each of the following subsets of $\mathbb{R}^k$ (for the appropriate values of k). Which are subspaces? Show that all conditions for a subspace are satisfied, or if the set is not a subspace, give an example demonstrating that one of the conditions fails.

(a) $S = \left\{ \begin{pmatrix} r \\ s \\ r-s \end{pmatrix} \;\middle|\; r, s \in \mathbb{R} \right\} \subset \mathbb{R}^3$

(b) $S = \left\{ \begin{pmatrix} 1 \\ a \\ b \end{pmatrix} \;\middle|\; a, b \in \mathbb{R} \right\} \subset \mathbb{R}^3$

(c) $S = \left\{ \begin{pmatrix} r \\ \sqrt{1-r^2} \end{pmatrix} \;\middle|\; r \in \mathbb{R} \right\} \subset \mathbb{R}^2$

(d) $S = \left\{ \begin{pmatrix} |a| \\ a \\ -a \end{pmatrix} \;\middle|\; a \in \mathbb{R} \right\} \subset \mathbb{R}^3$

(e) $S = \left\{ c_1 \begin{pmatrix} 0 \\ 1 \\ 1/2 \end{pmatrix} + c_2 \begin{pmatrix} \ln(2) \\ 0 \\ 1 \end{pmatrix} \;\middle|\; c_1, c_2 \in \mathbb{R} \right\} \subset \mathbb{R}^3.$

2. Which of these subsets are subspaces of $\mathcal{M}_{2\times2}(\mathbb{R})$? For each that is not, show the condition that fails.

(a) $\left\{ \begin{pmatrix} a & 0 \\ 0 & b \end{pmatrix} \;\middle|\; a, b \in \mathbb{R} \right\}$

(b) $\left\{ \begin{pmatrix} a & 0 \\ 0 & b \end{pmatrix} \;\middle|\; a + b = 0 \right\}$

(c) $\left\{ \begin{pmatrix} a & 0 \\ 0 & b \end{pmatrix} \;\middle|\; a + b = 5 \right\}$

(d) $\left\{ \begin{pmatrix} a & c \\ 0 & b \end{pmatrix} \;\middle|\; a + b = 0, c \in \mathbb{R} \right\}.$

3. Consider the vector space $\mathbb{R}^2$ and the following subsets W. For each, determine if (i) W is closed under vector addition, (ii) W is closed under scalar multiplication, (iii) W is a subspace of $\mathbb{R}^2$, and (iv.) sketch W in the xy-coordinate plane and illustrate your findings.

(a) $W = \{(1, 2), (2, 1)\}$.
(b) $W = \left\{(3a, 2a) \in \mathbb{R}^2 \mid a \in \mathbb{R}\right\}$.
(c) $W = \left\{(3a - 1, 2a + 1) \in \mathbb{R}^2 \mid a \in \mathbb{R}\right\}$.
(d) $W = \left\{(a, b) \in \mathbb{R}^2 \mid ab \geq 0\right\}$.

4. Recall the space $\mathcal{P}_2$ of degree two polynomials (see Example 2.4.2). For what scalar values of b is $W = \{a_0 + a_1 x + a_2 x^2 \mid a_0 + 2a_1 + a_2 = b\}$ a subspace of $\mathcal{P}_2$?

5. Recall the space $\mathcal{J}_{11}$ of all bar graphs with 11 bins (see Definition 2.4.16). We are interested in the bar graphs that have at most one bar with a value higher than both, or lower than both, of two nearest neighbors. Notice that end bars don't have two nearest neighbors (see Figure 2.22). Call this subset G.

(a) Is G a subspace of $\mathcal{J}_{11}$?
(b) We typically expect grade distributions to follow a bell curve (thus having a histogram that is a bar graph in G). What does your conclusion in part 5a say about a *course* grade distribution if homework, lab, and exam grade distributions are elements of G?

6. What is the smallest subspace of $\mathbb{R}^4$ containing the following vectors:

$$\begin{pmatrix} 1 \\ 0 \\ 1 \\ 0 \end{pmatrix}, \begin{pmatrix} 0 \\ 1 \\ 0 \\ 0 \end{pmatrix}, \text{ and } \begin{pmatrix} 2 \\ -3 \\ 2 \\ 0 \end{pmatrix}?$$

Explain your answer using ideas from this section.

7. Use the discussion about subspaces described by linear equations with 2 variables to discuss the different types of solution sets of a homogeneous system of equations with 2 equations and 2 variables. Include sketches of the lines corresponding to the equations.

8. Use the discussion about subspaces described by linear equations with 3 variables to discuss the different types of solution sets of a homogeneous system of equations with 3 equations and 3 variables. Include sketches of the planes corresponding to the equations.

9. Is $\mathbb{R}^2$ a subspace of $\mathbb{R}^3$? Explain.

10. Show that if a plane in $\mathbb{R}^3$ contains the origin, then the set of all points on the plane form a vector space.

11. Prove: Let $v \in \mathbb{R}^n$, and suppose that V is a subspace of $(\mathbb{R}^n, +, \cdot)$ containing v. Then the line containing the origin and v is contained in V.

12. Show that the set of all arithmetic combinations of images A, B, and C (page 11) is a subspace of the vector space of 4×4 images.

13. A manufacturing company uses a process called diffusion welding to adjoin several smaller rods into a single longer rod. The diffusion welding process leaves the final rod heated to various temperatures along the rod with the ends of the rod having the same temperature. Every a cm along the rod, a machine records the temperature difference from the temperature at the ends to get an array of temperatures called a *heat state*. (See Section 2.4.1.)

    (a) Plot the heat state given below (let the horizontal axis represent distance from the left end of the rod and the vertical axis represent the temperature difference from the ends)

    $$u = (0, 1, 13, 14, 12, 5, -2, -11, -3, 1, 10, 11, 9, 7, 0).$$

    (b) How long is the rod represented by u, the above heat state, if $a = 1$ cm?

    (c) Give another example of a heat state for the same rod, sampled in the same locations.

    (d) Show that the set of all heat states, for this rod, is a vector space.

    (e) What are the distinguishing properties of vectors in this vector space?

14. Show that a subset of a vector space satisfies (P9) as long as (P2) and (P8) are satisfied. [Hint: Theorem 2.3.17 may be useful.]

15. Prove Corollary 2.5.4.

16. Prove Corollary 2.5.6.

17. Prove that the set of all differentiable real-valued functions on $\mathbb{R}$ is a subspace of $\mathcal{F}$. See Example 2.4.1.

18. Prove that the empty set, $\emptyset$, is not a subspace of any vector space.

19. Prove that the set containing only the zero vector of a vector space V is a subspace of V. How many different ways can you prove this?

20. The smallest subspace in $\mathbb{R}^n$ containing two vectors u and v that do not lie on the same line through the origin is the plane containing v, w, and 0. What if a subspace V of $(\mathbb{R}^3, +, \cdot)$ contains a plane through the origin and another line L not in the plane. What can you say about the subspace?

21. Give an example that shows that the union of subspaces is not necessarily a subspace.

22. Consider the sets

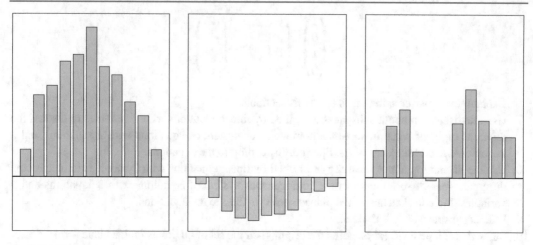

Fig. 2.22 Examples of two histograms in G (left and middle) and an example of a histogram not in G (right).

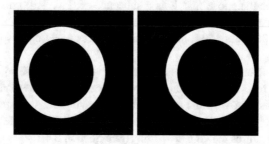

Fig. 2.23 Left image: C_1, Right image: C_2. Here black has a pixel intensity of 0 and white has a pixel intensity of 1.

$$W_1 = \left\{ ax^2 + bx + c \in \mathcal{P}_2(\mathbb{R}) \mid a = c \right\},$$

$$W_2 = \left\{ ax^2 + bx + c \in \mathcal{P}_2(\mathbb{R}) \mid a = 2c \right\}.$$

Show that both are subspaces of $\mathcal{P}_2$. Are $W_1 \cap W_2$ and $W_1 \cup W_2$ subspaces of $\mathcal{P}_2$?

23. Prove that if U_1 and U_2 are subspaces of V such that $V = U_1 \oplus U_2$, then for each $v \in V$ there exist unique vectors $u_1 \in U_1$ and $u_2 \in U_2$ such that $v = u_1 + u_2$.

For exercises 24 through 26, let V be the set of 4×4 grayscale images and consider the example images on page 11. Let U_Q be the set of all scalar multiples of image Q.

24. Complete Example 2.5.25 by showing, without Theorem 2.5.26, that $U_1 + U_2$ is a subspace of the vector space of 4×4 grayscale images.
25. Which of images 1, 2, 3, and 4 are in $U_A + U_B + U_C$?
26. Show that $V \neq U_A \oplus U_B \oplus U_C$.

For exercises 27 through 31, let $V = \mathcal{I}_{256 \times 256}$, the set of 256×256 grayscale images and consider the two images in Figure 2.23. Let W_1 be the set of all scalar multiples of image C_1 and W_2 be the set of all scalar multiples of image C_2. Recall that although the display range (or brightness in the displayed images) may be limited, the actual pixel intensities are not.

27. Describe the elements in W_1. Using the definition of scalar multiplication on images and vector addition on images, describe how you know that W_1 is a subspace of V.

28. Describe the elements in $W_1 + W_2$. Is $W_1 + W_2$ a subspace of V? Explain.
29. Describe the elements in $W_1 \cup W_2$. Is $W_1 \cup W_2$ a subspace of V? Explain.
30. Describe the elements in $W_1 \cap W_2$. Is $W_1 \cap W_2$ a subspace of V? Explain.
31. Describe how you know that $V \neq W_1 \oplus W_2$.

For Exercises 32 to 35, let V denote the space of images with hexagonal grid configuration shown here:

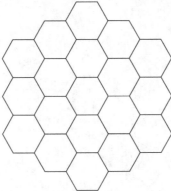

32. Let H be the subset of V consisting of all images whose outer ring of pixels all have value zero. Is H a subspace of V?
33. Let K be the subset of V consisting of all images whose center pixel has value zero. Is K a subspace of V?
34. What images are included in the set $H \cap K$? Is $H \cap K$ a subspace of V?
35. What images are included in the set $H \cup K$? Is $H \cup K$ a subspace of V?

Vector Space Arithmetic and Representations

3

We have seen that it is often quite useful in practice to be able to write a vector as a combination of scalar products and sums of other specified vectors. For example, recall again the task from Section 2.1, where we considered the 4×4 images on page 11. We showed that Image 2 could be formed by performing image addition and scalar multiplication on Images A, B, and C. In particular, we found that

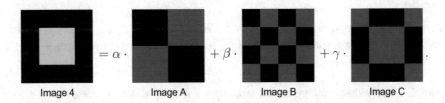

On the other hand, we found that Image 4 could not be formed using any combination of image addition and scalar multiplication with Images A, B, and C. That is, there are no scalars α, β, and γ so that

 = α · + β · + γ · .

Image 4   Image A   Image B   Image C

In this chapter, we will explore vector spaces and subspaces of vector spaces that can be described using linear combinations of particular vectors. We will also discuss optimal sets of vectors that can be used to describe a vector space. Finally, we will discuss a more concrete way to represent vectors in abstract vector spaces.

3.1 Linear Combinations

Suppose we are working within a subspace for which all the radiographs satisfy a particular (significant) property, call it property S. This means that the subspace is defined as the set of all radiographs with property S. Because the subspace is not trivial (that is, it contains more than just the zero radiograph) it consists of an infinite number of radiographs. Suppose also that we have a handful of radiographs

© Springer Nature Switzerland AG 2022, corrected publication 2023
H. A. Moon et al., *Application-Inspired Linear Algebra*, Springer Undergraduate Texts
in Mathematics and Technology, https://doi.org/10.1007/978-3-030-86155-1_3

that we know are in this subspace, but then a colleague brings us a new radiograph, r, one with which we have no experience and the colleague needs to know whether r has property S. Since the set of radiographs defined by property S is a subspace, we can perform a quick check to see if r can be formed from those radiographs with which we are familiar, using arithmetic operations. If we find the answer to this question is "yes," then we know r has property S. We know this because subspaces are closed under scalar multiplication and vector addition. If we find the answer to be "no," we still have more work to do. We cannot yet conclude whether or not r has property S because there may be radiographs with property S that are still unknown to us.

We have also been exploring one-dimensional heat states on a finite interval. We have seen that the subset of heat states with fixed (zero) endpoint temperature differential is a subspace of the vector space of heat states. The collection of vectors in this subspace is relatively easy to identify: finite-valued and zero at the ends. However, if a particular heat state on a rod could cause issues with future functioning of a diffusion welder, an engineer might be interested in whether the subspace of possible heat states might contain this detrimental heat state. We may wish to determine if one such heat state is an arithmetic combination of several others.

In Section 3.1.1, we introduce the terminology of linear combinations for describing when a vector can be formed from a *finite number of* arithmetic operations on a specified set of vectors. In Sections 3.1.3 and 3.1.4 we consider linear combinations of vectors in Euclidean space ($\mathbb{R}^n$) and connect such linear combinations to the inhomogeneous and homogeneous matrix equations $Ax = b$ and $Ax = 0$, respectively. Finally, in Section 3.1.5, we discuss the connection between inhomogeneous and homogeneous systems.

3.1.1   Linear Combinations

We now assign terminology to describe vectors that have been created from (a finite number of) arithmetic operations with a specified set of vectors.

Definition 3.1.1

Let $(V, +, \cdot)$ be a vector space over $\mathbb{F}$. Given a finite set of vectors $v_1, v_2, \cdots, v_k \in V$, we say that the vector $w \in V$ is a **linear combination** of $v_1, v_2, \cdots, v_k$ if $w = a_1 v_1 + a_2 v_2 + \cdots + a_k v_k$ for some scalar coefficients $a_1, a_2, \cdots, a_k \in \mathbb{F}$.   □

Corollary 2.5.4 says that a subspace is a nonempty subset of a vector space that is closed under scalar multiplication and vector addition. Using this new terminology, we can say that a subspace is *closed under linear combinations*.

Following is an example in the vector space $\mathcal{I}_{4 \times 4}$ of 4×4 images.

Example 3.1.2 Consider the 4×4 grayscale images from page 11. Image 2 is a linear combination of Images A,B and C with scalar coefficients $\frac{1}{2}$, 0, and 1, respectively, because

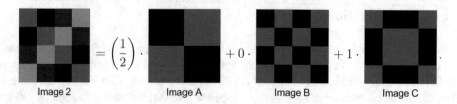

Image 2   Image A   Image B   Image C

□

** **Watch Your Language!** When communicating whether or not a vector can be written as a linear combination of other vectors, you should recognize that the term "linear combination" is a property applied to vectors, not sets. So, we make statements such as

✓ w is a linear combination of v_1, v_2, v_3
✓ w is not a linear combination of $u_1, u_2, u_3, \ldots u_n$.
✓ w is a linear combination of vectors in $U = \{v_1, v_2, v_3\}$.

We do not say

✗ w is a linear combination of U.

In the remainder of this section, we focus on the question: Can a given vector be written as a linear combination of vectors from some specified set of vectors?

In Section 2.2, we set up systems of linear equations to answer such questions. For example, on page 14, we considered whether Image C is a linear combination of Images A, D, and F. To determine the answer to our inquiry, we solved corresponding system of equations and found that, indeed, it is. Also, in Section 2.2, we asked whether it is possible to find scalars α, β, and γ so that

| Image A | Image D | Image E | Image F |

In this case, we used systems of equations to determine that no such scalars exist and so Image A is not a linear combination of Images D, E, and F.

Example 3.1.3 This example illustrates a subtle, yet important, property of linear combinations, the reason we require them to be finite sums. Consider the set $T = \{1, x, x^2, x^3, \ldots\}$ whose vectors are in the vector space of all polynomials. We know, from Calculus, that the Taylor series for cosine is

$$\cos x = \sum_{k=0}^{\infty} \frac{(-1)^k x^{2k}}{(2k)!} = 1 - \frac{x^2}{2!} + \frac{x^4}{4!} - \frac{x^6}{6!} + \cdots .$$

According to the definition of linear combination, this series is not a linear combination of the vectors in T because it does not consist of only finitely many (nonzero) terms. We also know by definition that every polynomial can be written as a linear combination of the vectors in T. Since $\cos x$ cannot be written in this way[1], we have no trouble saying that $\cos x$ is not a polynomial. □

If we can write a vector, v, as an arithmetic combination of infinitely many vectors, $u_1, u_2, \ldots$ like

$$v = \sum_{k=0}^{\infty} a_k u_k, \text{ for some scalars } a_k,$$

---

[1] Do you see why? Hint: how many extrema does a polynomial have?

we may still be able to write v as the arithmetic combination of only finitely many vectors, like

$$v = \sum_{k=0}^{N} b_k u_k, \text{ for some scalars } b_k.$$

In this case, v is a linear combination of $u_0, u_1, ..., u_N$.

⋆⋆ **Watch Your Language!** Remember that linear combinations can only have finitely many terms. Let S be the infinite set $\{u_1, u_2, u_3, ...\}$. We can say

✓ Since $v = 3u_1 + 5u_{12} - 72u_{35}$, v is a linear combination of u_1, u_{12}, u_{35},

✓ $v = \sum_{k=1}^{\infty} a_k u_k$, with $a_k \neq 0$ for $k = 1, 12, 35$, then v is a linear combination of u_1, u_{12}, u_{35}.

It is incorrect to say

✗ If $v = \sum_{k=1}^{\infty} a_k u_k$, then v is a linear combination of the elements of S.

We finish this section with additional examples where we reformulate and answer the linear combination question using systems of linear equations.

Example 3.1.4 Consider the vector space $\mathbb{R}^3$ with the standard operations and field. Can the vector w be written as a linear combination of the vectors v_1, v_2, v_3, where

$$w = \begin{pmatrix} 2 \\ 5 \\ 3 \end{pmatrix}, \quad v_1 = \begin{pmatrix} 1 \\ 0 \\ 1 \end{pmatrix}, \quad v_2 = \begin{pmatrix} 2 \\ 2 \\ 1 \end{pmatrix}, \quad v_3 = \begin{pmatrix} 3 \\ 1 \\ 4 \end{pmatrix}?$$

Equivalently, do coefficients x_1, x_2, x_3 exist so that $w = x_1 v_2 + x_2 v_2 + x_3 v_3$? More explicitly, we seek coefficients such that

$$\begin{pmatrix} 2 \\ 5 \\ 3 \end{pmatrix} = x_1 \begin{pmatrix} 1 \\ 0 \\ 1 \end{pmatrix} + x_2 \begin{pmatrix} 2 \\ 2 \\ 1 \end{pmatrix} + x_3 \begin{pmatrix} 3 \\ 1 \\ 4 \end{pmatrix}. \tag{3.1}$$

Because scalar multiplication and vector addition are defined element-wise, we have the equivalent set of conditions

$$\begin{aligned} 2 &= x_1 + 2x_2 + 3x_3 \\ 5 &= 2x_2 + x_3 \\ 3 &= x_1 + x_2 + 4x_3. \end{aligned} \tag{3.2}$$

So, our original question: Is it possible to write w as a linear combination of the vectors v_1, v_2, v_3? can be reformulated as the question of whether or not the system of equations (3.2) has a solution. We write and row reduce the associated augmented matrix:

$$\begin{pmatrix} 1 & 2 & 3 & | & 2 \\ 0 & 2 & 1 & | & 5 \\ 1 & 1 & 4 & | & 3 \end{pmatrix} \sim \begin{pmatrix} 1 & 0 & 0 & | & -23/3 \\ 0 & 1 & 0 & | & 4/3 \\ 0 & 0 & 1 & | & 7/3 \end{pmatrix}, \tag{3.3}$$

and see that unique coefficients for the desired linear combination are $x_1 = -23/3$, $x_2 = 4/3$, and $x_3 = 7/3$. □

We now consider a similar example in the vector space of degree 2 polynomials $\mathcal{P}_2$.

Example 3.1.5 Consider the following vectors in $\mathcal{P}_2$:

$$v_1 = 3x + 4,\ v_2 = 2x + 1,\ v_3 = x^2 + 2,\ \text{and}\ v_4 = x^2.$$

The polynomial v_1 can be written as a linear combination of the polynomials v_2, v_3, v_4 if there exist scalars α, β, and γ so that

$$v_1 = \alpha v_2 + \beta v_3 + \gamma v_4.$$

If such scalars exist, then

$$3x + 4 = \alpha(2x + 1) + \beta(x^2 + 2) + \gamma(x^2).$$

We match up like terms and obtain the following system of equations:

$$
\begin{array}{ll}
(x^2 \text{ term}) & 0 = \quad\quad \beta + \gamma \\
(x \text{ term}) & 3 = 2\alpha \\
(\text{constant term}) & 4 = \quad \alpha + 2\beta.
\end{array}
$$

Again, we can turn this system into an augmented matrix and row reduce:

$$
\begin{pmatrix}
0 & 1 & 1 & | & 0 \\
2 & 0 & 0 & | & 3 \\
1 & 2 & 0 & | & 4
\end{pmatrix}
\sim
\begin{pmatrix}
1 & 0 & 0 & | & 3/2 \\
0 & 1 & 0 & | & 5/4 \\
0 & 0 & 1 & | & -5/4
\end{pmatrix}. \tag{3.4}
$$

The solution to this system, then, is $\alpha = \frac{3}{2}$, $\beta = \frac{5}{4}$, and $\gamma = -\frac{5}{4}$. In other words,

$$v_1 = \frac{3}{2}v_2 + \frac{5}{4}v_3 - \frac{5}{4}v_4,$$

and so v_1 can be written as a linear combination of v_2, v_3, and v_4. □

We give a final example, this time in the vector space $\mathcal{D}(\mathbb{Z}_2)$ of 7-bar LCD images.

Example 3.1.6 Recall Example 2.4.17, in which we saw the following two examples of linear combinations:

In words, these say that the LCD character is a linear combination of and , and is a linear combination of , , and . The interesting part of this example is that the scalar set for $\mathcal{D}(\mathbb{Z}_2)$ has only two elements. So, if we want to determine whether the "9" digit, , is a linear combination of the other 9 digits, we ask whether there are scalars $\alpha_0, \alpha_1, \ldots, \alpha_8 \in \{0, 1\}$ so that

$$\alpha_0 \; + \alpha_1 \; + \alpha_2 \; + \alpha_3 \; + \alpha_4 \; + \alpha_5 \; + \alpha_6 \; + \alpha_7 \; + \alpha_8 \; = \; .$$

We can then set up a system of equations so that there's an equation for each bar. For example, the equation representing the top bar is

$$\alpha_2 + \alpha_3 + \alpha_5 + \alpha_6 + \alpha_7 + \alpha_8 = 1.$$

Exercise 21 continues this investigation.　　□

3.1.2　Matrix Products

In this section, we state definitions and give examples of matrix products. We expect that most linear algebra students have already learned to multiply matrices, but realize that some students may need a reminder. As with $\mathbb{R}$, $\mathcal{M}_{m \times n}$ can be equipped with other operations (besides addition and scalar multiplication) with which it is associated, namely matrix multiplication. But unlike matrix addition the matrix product is not defined component-wise.

Definition 3.1.7

Given an $n \times m$ matrix $A = (a_{i,j})$ and an $m \times \ell$ matrix $B = (b_{i,j})$, we define the **matrix product** of A and B to be the $n \times \ell$ matrix $AB = (c_{i,j})$ where

$$c_{i,j} = \sum_{k=1}^{m} a_{i,k} b_{k,j}.$$

We call this operation **matrix multiplication**.   □

Notation. There is no "·" between the matrices, rather they are written in juxtaposition to show a difference between the notation of a scalar product and the notation of a matrix product. The definition requires that the number of columns of A is the same as the number of rows of B for the product AB.

Example 3.1.8 Let

$$P = \begin{pmatrix} 1 & 2 \\ 3 & 4 \\ 5 & 6 \end{pmatrix}, \quad Q = \begin{pmatrix} 2 & 1 \\ 1 & -1 \\ 2 & 1 \end{pmatrix}, \text{ and } R = \begin{pmatrix} 2 & 0 \\ 1 & -2 \end{pmatrix}.$$

Since both P and Q are 3×2 matrices, we see that the number of columns of P is not the same as the number of rows of Q. Thus, PQ is not defined. But, since P has 2 columns and R has 2 rows, PR is defined. Let's compute the matrix product PR. We can compute each entry as in Definition 3.1.7.

Position Computation

(i, j)	$p_{i,1}r_{1,j} + p_{i,2}r_{2,j}$
$(1, 1)$	$1 \cdot 2 + 2 \cdot 1$
$(1, 2)$	$1 \cdot 0 + 2 \cdot (-2)$
$(2, 1)$	$3 \cdot 2 + 4 \cdot 1$
$(2, 2)$	$3 \cdot 0 + 4 \cdot (-2)$
$(3, 1)$	$5 \cdot 2 + 6 \cdot 1$
$(3, 2)$	$5 \cdot 0 + 6 \cdot (-2)$

Typically, when writing this out, we write it as

$$PR = \begin{pmatrix} 1 & 2 \\ 3 & 4 \\ 5 & 6 \end{pmatrix} \begin{pmatrix} 2 & 0 \\ 1 & -2 \end{pmatrix}$$

$$= \begin{pmatrix} 1 \cdot 2 + 2 \cdot 1 & 1 \cdot 0 + 2 \cdot (-2) \\ 3 \cdot 2 + 4 \cdot 1 & 3 \cdot 0 + 4 \cdot (-2) \\ 5 \cdot 2 + 6 \cdot 1 & 5 \cdot 0 + 6 \cdot (-2) \end{pmatrix}$$

$$= \begin{pmatrix} 4 & -4 \\ 10 & -8 \\ 16 & -12 \end{pmatrix}.$$

□

In the above example, the result of the matrix product was a matrix of the same size as P. Let's do another example to show that this is not always the case.

Example 3.1.9 Let

$$A = \begin{pmatrix} 1 & 0 & 2 \\ 2 & -1 & 1 \end{pmatrix} \quad \text{and} \quad B = \begin{pmatrix} 2 & 1 & -1 & 0 \\ 1 & -1 & 1 & 1 \\ 2 & 1 & 1 & 2 \end{pmatrix}.$$

We see that A is 2×3 and B is 3×4. So the product AB is defined and is a 2×4 matrix. Following the definition of matrix multiplication, we find AB below.

$$AB = \begin{pmatrix} 1 & 0 & 2 \\ 2 & -1 & 1 \end{pmatrix} \begin{pmatrix} 2 & 1 & -1 & 0 \\ 1 & -1 & 1 & 1 \\ 2 & 1 & 1 & 2 \end{pmatrix}$$

$$= \begin{pmatrix} 6 & 3 & 1 & 4 \\ 5 & 4 & -2 & 1 \end{pmatrix}.$$

$\square$

In Example 3.1.8, RP is not defined, so $RP \neq PR$. Similarly, in Example 3.1.9, $BA \neq AB$. But, one might wonder whether matrix multiplication is commutative when the matrix product is defined both ways. That is, we might wonder if $AB = BA$ when both AB and BA are defined, i.e., when both A and B are $n \times n$ (square) matrices for some n. Let us explore this in the next example.

Example 3.1.10 Let

$$A = \begin{pmatrix} 1 & 2 \\ 3 & -1 \end{pmatrix} \text{ and } B = \begin{pmatrix} 2 & 0 \\ -2 & 3 \end{pmatrix}.$$

Let us compute both AB and BA.

$$AB = \begin{pmatrix} 1 & 2 \\ 3 & -1 \end{pmatrix} \begin{pmatrix} 2 & 0 \\ -2 & 3 \end{pmatrix}$$

$$= \begin{pmatrix} 1 \cdot 2 + 2 \cdot (-2) & 1 \cdot 0 + 2 \cdot 3 \\ 3 \cdot 2 + (-1) \cdot (-2) & 3 \cdot 0 + (-1) \cdot 3 \end{pmatrix}$$

$$= \begin{pmatrix} -2 & 6 \\ 8 & -3 \end{pmatrix}$$

$$BA = \begin{pmatrix} 2 & 0 \\ -2 & 3 \end{pmatrix} \begin{pmatrix} 1 & 2 \\ 3 & -1 \end{pmatrix}$$

$$= \begin{pmatrix} 2 \cdot 1 + 0 \cdot 3 & 2 \cdot 2 + 0 \cdot (-1) \\ -2 \cdot 1 + 3 \cdot 3 & -2 \cdot 2 + 3 \cdot (-1) \end{pmatrix}$$

$$= \begin{pmatrix} 2 & 4 \\ 7 & -7 \end{pmatrix}.$$

Even in the simple case of 2×2 matrices, matrix multiplication is not commutative. $\square$

Because matrix multiplication is not generally defined between two matrices in $\mathcal{M}_{m \times n}$, we can see that $\mathcal{M}_{m \times n}$ is not, in general, closed under matrix multiplication. Since the matrix product (for brevity, let's call it $\otimes$) is not commutative, we can also see that $(\mathcal{M}_{n \times n}, \otimes, \cdot)$ is not a vector space over $\mathbb{R}$.

A special case of matrix multiplication is the computation of a matrix-vector product Mv, where M is in $\mathcal{M}_{m \times n}$ and v is a vector in $\mathbb{R}^n$. Here we show an example of such a product.

Example 3.1.11 Let $M = \begin{pmatrix} -1 & 0 & 3 \\ -2 & 1 & 3 \end{pmatrix}$ and $v = \begin{pmatrix} -1 \\ 1 \\ 2 \end{pmatrix}$. The product is

$$Mv = \begin{pmatrix} -1 & 0 & 3 \\ -2 & 1 & 3 \end{pmatrix} \begin{pmatrix} -1 \\ 1 \\ 2 \end{pmatrix}$$

$$= \begin{pmatrix} -1 \cdot -1 + 0 \cdot 1 + 3 \cdot 2 \\ -2 \cdot -1 + 1 \cdot 1 + 3 \cdot 2 \end{pmatrix}$$

$$= \begin{pmatrix} 7 \\ 9 \end{pmatrix}.$$

□

Let us consider one more example that will connect matrix-vector products to systems of equations.

Example 3.1.12 Let

$$A = \begin{pmatrix} 2 & 3 & 0 \\ 1 & -1 & -1 \\ 0 & 4 & 2 \end{pmatrix} \text{ and } X = \begin{pmatrix} x \\ y \\ z \end{pmatrix}.$$

Since A is 3×3 and X is 3×1, we can multiply to get AX. Remember that the number of columns of A must match the number of rows of X in order for AX to make sense. The product AX must be a 3×1 matrix. Multiplying gives

$$AX = \begin{pmatrix} 2 & 3 & 0 \\ 1 & -1 & -1 \\ 0 & 4 & 2 \end{pmatrix} \begin{pmatrix} x \\ y \\ z \end{pmatrix}$$

$$= \begin{pmatrix} 2 \cdot x + 3 \cdot y + 0 \cdot z \\ 1 \cdot x + (-1) \cdot y + (-1) \cdot z \\ 0 \cdot x + 4 \cdot y + 2 \cdot z \end{pmatrix}.$$

First, notice that the resulting vector looks very much like the variable side of a system of equations. In fact, if we consider the question, "Are there real numbers x, y, and z so that

$$\begin{aligned} 2x + 3y &= 1 \\ x - y - z &= 3 \text{ ?"} \\ 4y + 2z &= -4 \end{aligned}$$

We see that this identical to asking the question, "Given the matrix

$$A = \begin{pmatrix} 2 & 3 & 0 \\ 1 & -1 & -1 \\ 0 & 4 & 2 \end{pmatrix},$$

and vector $b = \begin{pmatrix} 1 \\ 3 \\ -4 \end{pmatrix}$ is there a vector $X = \begin{pmatrix} x \\ y \\ z \end{pmatrix}$ so that $AX = b$?   □

Although matrix product cannot play the role of vector space addition, this operation does play a major role in important linear algebra concepts. In fact, we will begin weaving matrix products into linear algebra in the next section.

3.1.3 The Matrix Equation $Ax = b$

In Section 3.1.1, we used systems of equations and augmented matrices to answer questions about linear combinations. In this section, we will consider the link between linear combinations and matrix-vector products.

We can use the ideas from Section 3.1.2 about matrix-vector products to rewrite a system of equations as a matrix equation. To formalize, we define, in a way similar to images, what it means for two matrices to be equal.

Definition 3.1.13

Given two matrices $A = (a_{i,j})$ and $B = (b_{i,j})$, we say that A and B are **equal**, and write $A = B$, if and only if A and B have the same dimensions and $a_{i,j} = b_{i,j}$ for every pair i, j. That is, corresponding entries of A and B are equal.   □

We now use this definition to rewrite a system of equations.

Example 3.1.14 Consider the system of linear equations given by

$$\begin{aligned}
2a + 3b - \ c &= 2 \\
a - 2b + \ c &= 1 \\
a + 5b - 2c &= 1.
\end{aligned}$$

This system is equivalent to the following vector equation

$$\begin{pmatrix} 2a + 3b - c \\ a \ - 2b + c \\ a \ + 5b - 2c \end{pmatrix} = \begin{pmatrix} 2 \\ 1 \\ 1 \end{pmatrix}.$$

Finally, we can rewrite this vector equation using the matrix-vector product:

$$\begin{pmatrix} 2 & 3 & -1 \\ 1 & -2 & 1 \\ 1 & 5 & -2 \end{pmatrix} \begin{pmatrix} a \\ b \\ c \end{pmatrix} = \begin{pmatrix} 2 \\ 1 \\ 1 \end{pmatrix}.$$

Therefore, we represent the system given in this example as the matrix equation

$$Au = v,$$

where

$$u = \begin{pmatrix} a \\ b \\ c \end{pmatrix}, \quad v = \begin{pmatrix} 2 \\ 1 \\ 1 \end{pmatrix},$$

and A is the corresponding coefficient matrix

$$\begin{pmatrix} 2 & 3 & -1 \\ 1 & -2 & 1 \\ 1 & 5 & -2 \end{pmatrix}.$$

□

These observations allow us to connect solutions to systems of equations with solutions to matrix equations. The following theorem makes this connection precise.

Theorem 3.1.15
Let A be an $m \times n$ matrix and let $b \in \mathbb{R}^m$. The matrix equation $Ax = b$ has the same solution set as the system of equations with augmented coefficient matrix $[A \mid b]$.

In this theorem, we are not distinguishing between the point $(\alpha_1, \alpha_2, \ldots, \alpha_n)$ and the vector $\begin{pmatrix} \alpha_1 \\ \alpha_2 \\ \vdots \\ \alpha_n \end{pmatrix}$

for the solution to a system of equations as the difference is only notational.

Proof. In this proof, we will show that if u is a solution to $Ax = b$ then u is also a solution to the system of equations corresponding to $[A|b]$. Then, we will show that if u is a solution to the system of equations corresponding to $[A|b]$, then u is also a solution to $Ax = b$. Let $A = \begin{pmatrix} a_{1,1} & a_{1,2} & \cdots & a_{1,n} \\ a_{2,1} & a_{2,2} & \cdots & a_{2,n} \\ \vdots & & \ddots & \vdots \\ a_{m,1} & a_{m,2} & \cdots & a_{m,n} \end{pmatrix}$

be an $m \times n$ matrix and let

$$b = \begin{pmatrix} b_1 \\ b_2 \\ \vdots \\ b_m \end{pmatrix} \in \mathbb{R}^m.$$

Suppose, also, that $u = \begin{pmatrix} \alpha_1 \\ \alpha_2 \\ \vdots \\ \alpha_n \end{pmatrix} \in \mathbb{R}^n$ is a solution to the matrix equation $Ax = b$. Then,

$$\begin{pmatrix} a_{1,1} & a_{1,2} & \cdots & a_{1,n} \\ a_{2,1} & a_{2,2} & \cdots & a_{2,n} \\ \vdots & & \ddots & \vdots \\ a_{m,1} & a_{m,2} & \cdots & a_{m,n} \end{pmatrix} \begin{pmatrix} \alpha_1 \\ \alpha_2 \\ \vdots \\ \alpha_n \end{pmatrix} = \begin{pmatrix} b_1 \\ b_2 \\ \vdots \\ b_m \end{pmatrix}.$$

Multiplying the left-hand side gives the vector equation

$$\begin{pmatrix} a_{1,1}\alpha_1 + a_{1,2}\alpha_2 + \cdots + a_{1,n}\alpha_n \\ a_{2,1}\alpha_1 + a_{2,2}\alpha_2 + \cdots + a_{2,n}\alpha_n \\ \vdots & & \ddots & \vdots \\ a_{m,1}\alpha_1 + a_{m,2}\alpha_2 + \cdots + a_{m,n}\alpha_n \end{pmatrix} = \begin{pmatrix} b_1 \\ b_2 \\ \vdots \\ b_m \end{pmatrix}.$$

Thus,

$$\begin{aligned}
a_{1,1}\alpha_1 + a_{1,2}\alpha_2 + \cdots + a_{1,n}\alpha_n &= b_1 \\
a_{2,1}\alpha_1 + a_{2,2}\alpha_2 + \cdots + a_{2,n}\alpha_n &= b_2 \\
&\vdots \\
a_{m,1}\alpha_1 + a_{m,2}\alpha_2 + \cdots + a_{m,n}\alpha_n &= b_m.
\end{aligned}$$

Therefore, $(\alpha_1, \alpha_2, \ldots, \alpha_n)$ is a solution to the system of equations

$$\begin{aligned}
a_{1,1}x_1 + a_{1,2}x_2 + \cdots + a_{1,n}x_n &= b_1 \\
a_{2,1}x_1 + a_{2,2}x_2 + \cdots + a_{2,n}x_n &= b_2 \\
&\vdots \\
a_{m,1}x_1 + a_{m,2}x_2 + \cdots + a_{m,n}x_n &= b_m.
\end{aligned}$$

Now, assume that the system of equations, with coefficient matrix $[A|b]$ has a solution. Then there are values $\alpha_1, \alpha_2, \ldots, \alpha_n$ so that $u = \begin{pmatrix} \alpha_1 \\ \alpha_2 \\ \vdots \\ \alpha_n \end{pmatrix}$ is a solution to the system of equations

$$\begin{aligned}
a_{1,1}x_1 + a_{1,2}x_2 + \ldots + a_{1,n}x_n &= b_1 \\
a_{2,1}x_1 + a_{2,2}x_2 + \ldots + a_{2,n}x_n &= b_2 \\
&\vdots \\
a_{m,1}x_1 + a_{m,2}x_2 + \ldots + a_{m,n}x_n &= b_m.
\end{aligned}$$

Then putting in u, we have

$$\begin{aligned}
a_{1,1}\alpha_1 + a_{1,2}\alpha_2 + \cdots + a_{1,n}\alpha_n &= b_1 \\
a_{2,1}\alpha_1 + a_{2,2}\alpha_2 + \cdots + a_{2,n}\alpha_n &= b_2 \\
&\vdots \\
a_{m,1}\alpha_1 + a_{m,2}\alpha_2 + \cdots + a_{m,n}\alpha_n &= b_m.
\end{aligned}$$

This gives us that

$$\begin{pmatrix} a_{1,1} & a_{1,2} & \cdots & a_{1,n} \\ a_{2,1} & a_{2,2} & \cdots & a_{2,n} \\ \vdots & & & \vdots \\ a_{m,1} & a_{m,2} & \cdots & a_{m,n} \end{pmatrix} \begin{pmatrix} \alpha_1 \\ \alpha_2 \\ \vdots \\ \alpha_n \end{pmatrix} = \begin{pmatrix} b_1 \\ b_2 \\ \vdots \\ b_m \end{pmatrix}.$$

Therefore $Au = b$. Thus, u is a solution to $Ax = b$. □

An important consequence of Theorem 3.1.15 is the following corollary.

Corollary 3.1.16

Let A be an $m \times n$ matrix and let $b \in \mathbb{R}^m$. The matrix equation $Ax = b$ has a unique solution if and only if the corresponding system of equations has a unique solution.

In the proof of Theorem 3.1.15 we see that every solution to the system is also a solution to the matrix equation and every solution of the matrix equation is a solution to the corresponding system. So, the result in Corollary 3.1.16 follows directly from the proof of Theorem 3.1.15. We only state it separately because it is a very important result.

What does all this have to do with linear combinations? Consider again Example 3.1.4.

Example 3.1.17 In Example 3.1.4, we asked whether $w = \begin{pmatrix} 2 \\ 5 \\ 3 \end{pmatrix}$ can be written as a linear combination

of

$$v_1 = \begin{pmatrix} 1 \\ 0 \\ 1 \end{pmatrix}, v_2 = \begin{pmatrix} 2 \\ 2 \\ 1 \end{pmatrix}, \text{ and } v_3 = \begin{pmatrix} 3 \\ 1 \\ 4 \end{pmatrix}.$$

We reformulated this question as the system of equations

$$2 = x_1 + 2x_2 + 3x_3 \tag{3.5}$$
$$5 = 2x_2 + x_3 \tag{3.6}$$
$$3 = x_1 + x_2 + 4x_3. \tag{3.7}$$

Based on discussion in this section, we can rewrite this system of equations as a matrix equation $Ax = b$, where

$$A = \begin{pmatrix} 1 & 2 & 3 \\ 0 & 2 & 1 \\ 1 & 1 & 4 \end{pmatrix} \text{ and } b = \begin{pmatrix} 2 \\ 5 \\ 3 \end{pmatrix}.$$

If we look closely at A, we see that the columns of A are v_1, v_2, and v_3. Moreover, if x is a solution to the equation $Ax = b$, then

$$\begin{pmatrix} 1 & 2 & 3 \\ 0 & 2 & 1 \\ 1 & 1 & 4 \end{pmatrix} \begin{pmatrix} x_1 \\ x_2 \\ x_3 \end{pmatrix} = x_1 \begin{pmatrix} 1 \\ 0 \\ 1 \end{pmatrix} + x_2 \begin{pmatrix} 2 \\ 2 \\ 1 \end{pmatrix} + x_3 \begin{pmatrix} 3 \\ 1 \\ 4 \end{pmatrix} = \begin{pmatrix} 2 \\ 5 \\ 3 \end{pmatrix}.$$

In other words, if x is a solution to the equation $Ax = b$, then b is a linear combination of v_1, v_2, and v_3 with coefficients equal to x_1, x_2, and x_3. □

There was nothing special about the vectors v_1, v_2, v_3 in Example 3.1.4 to make the matrix product relate to a linear combination. In fact, another way to represent the product of a matrix with a vector is as follows.

Theorem 3.1.18
Let A and x be the matrix and vector given by

$$A = \begin{pmatrix} a_{1,1} & a_{1,2} & \cdots & a_{1,n} \\ a_{2,1} & a_{2,2} & \cdots & a_{2,n} \\ \vdots & & \ddots & \vdots \\ a_{m,1} & a_{m,2} & \cdots & a_{m,n} \end{pmatrix} \text{ and } x = \begin{pmatrix} \alpha_1 \\ \alpha_2 \\ \vdots \\ \alpha_n \end{pmatrix}.$$

Then, Ax can be written as a linear combination of the columns of A. In particular,

$$Ax = \alpha_1 c_1 + \alpha_2 c_2 + \ldots + \alpha_n c_n,$$

where $c_1 = \begin{pmatrix} a_{1,1} \\ a_{2,1} \\ \vdots \\ a_{m,1} \end{pmatrix}, c_2 = \begin{pmatrix} a_{1,2} \\ a_{2,2} \\ \vdots \\ a_{m,2} \end{pmatrix}, \ldots, c_n = \begin{pmatrix} a_{1,n} \\ a_{2,n} \\ \vdots \\ a_{m,n} \end{pmatrix}.$

Proof. Let A and x be defined above. Also, let $c_1, c_2, \ldots, c_n$ be the columns of A. Using the definition of a matrix-vector product, we have

$$Ax = \begin{pmatrix} a_{1,1} & a_{1,2} & \ldots & a_{1,n} \\ a_{2,1} & a_{2,2} & \ldots & a_{2,n} \\ \vdots & & \ddots & \vdots \\ a_{m,1} & a_{m,2} & \ldots & a_{m,n} \end{pmatrix} \begin{pmatrix} \alpha_1 \\ \alpha_2 \\ \vdots \\ \alpha_n \end{pmatrix}$$

$$= \begin{pmatrix} a_{1,1}\alpha_1 + a_{1,2}\alpha_2 + \ldots + a_{1,n}\alpha_n \\ a_{2,1}\alpha_1 + a_{2,2}\alpha_2 + \ldots + a_{2,n}\alpha_n \\ \vdots \\ a_{m,1}\alpha_1 + a_{m,2}\alpha_2 + \ldots + a_{m,n}\alpha_n \end{pmatrix}$$

$$= \begin{pmatrix} a_{1,1}\alpha_1 \\ a_{2,1}\alpha_1 \\ \vdots \\ a_{m,1}\alpha_1 \end{pmatrix} + \begin{pmatrix} a_{1,2}\alpha_1 \\ a_{2,2}\alpha_1 \\ \vdots \\ a_{m,2}\alpha_2 \end{pmatrix} + \ldots + \begin{pmatrix} a_{1,n}\alpha_n \\ a_{2,n}\alpha_n \\ \vdots \\ a_{m,n}\alpha_n \end{pmatrix}$$

$$= \alpha_1 \begin{pmatrix} a_{1,1} \\ a_{2,1} \\ \vdots \\ a_{m,1} \end{pmatrix} + \alpha_2 \begin{pmatrix} a_{1,2} \\ a_{2,2} \\ \vdots \\ a_{m,2} \end{pmatrix} + \ldots + \alpha_n \begin{pmatrix} a_{1,n} \\ a_{2,n} \\ \vdots \\ a_{m,n} \end{pmatrix}$$

$$= \alpha_1 c_1 + \alpha_2 c_2 + \ldots + \alpha_n c_n.$$

$\square$

Now that we can connect systems of equations, matrix equations, and linear combinations, we can also connect the solutions to related questions. Below, we present these connections as a theorem.

Theorem 3.1.19
Let A be an $m \times n$ matrix and let $b \in \mathbb{R}^m$. Then the matrix equation $Ax = b$ has a solution if and only if the vector b can be written as a linear combination of the columns of A.

Proof. Let $c_1, c_2, \ldots, c_n$ be the columns of $A = (a_{i,j})$ and let $b = \begin{pmatrix} b_1 \\ b_2 \\ \vdots \\ b_m \end{pmatrix}$. To prove these statements

are equivalent, we will show the following two statements. ($\Rightarrow$) If $Ax = b$ has a solution then there are scalars $\alpha_1, \alpha_2, \ldots, \alpha_n$ so that

$$b = \alpha_1 c_1 + \alpha_2 c_2 + \ldots + \alpha_n c_n.$$

($\Leftarrow$) If there are scalars $\alpha_1, \alpha_2, \ldots, \alpha_n$ so that

$$b = \alpha_1 c_1 + \alpha_2 c_2 + \ldots + \alpha_n c_n$$

then $Ax = b$ has a solution.

($\Rightarrow$) Suppose $Ax = b$ has a solution. Then, there is a vector $u = \begin{pmatrix} \alpha_1 \\ \alpha_2 \\ \vdots \\ \alpha_n \end{pmatrix}$ so that $Au = b$. Then, using the

connection to matrix-vector products and linear combinations, we know that $b = Au = u_1 c_1 + u_2 c_2 + \ldots + u_n c_n$. Therefore, b is a linear combination of $c_1, c_2, \ldots, c_n$.

($\Leftarrow$) Now, suppose that b is a linear combination of the columns of A. Then there are scalars $\alpha_1, \alpha_2, \ldots, \alpha_n$ so that

$$b = \alpha_1 c_1 + \alpha_2 c_2 + \ldots + \alpha_n c_n$$

$$= A \begin{pmatrix} \alpha_1 \\ \alpha_2 \\ \vdots \\ \alpha_n \end{pmatrix}.$$

Therefore $u = \begin{pmatrix} \alpha_1 \\ \alpha_2 \\ \vdots \\ \alpha_n \end{pmatrix}$ is a solution to $Ax = b$. $\qquad \square$

In the proof of Theorem 3.1.19 we see that a solution to $Ax = b$ is given by the vector in $\mathbb{R}^n$ made up of the scalars in the linear combination.

Let us, now, return to another example with vectors in $\mathbb{R}^3$.

Example 3.1.20 Here, we want to determine whether $w = \begin{pmatrix} 1 \\ 2 \\ 2 \end{pmatrix} \in \mathbb{R}^3$ can be written as a linear

combination of $v_1 = \begin{pmatrix} 1 \\ 1 \\ 1 \end{pmatrix}$, $v_2 = \begin{pmatrix} 1 \\ 3 \\ 1 \end{pmatrix}$, and $v_3 = \begin{pmatrix} 2 \\ -2 \\ 2 \end{pmatrix}$. We can write a vector equation to determine if there are scalars α, β, and γ so that

$$w = \alpha v_1 + \beta v_2 + \gamma v_3.$$

This leads to

$$\begin{pmatrix} 1 \\ 2 \\ 2 \end{pmatrix} = \alpha \begin{pmatrix} 1 \\ 1 \\ 1 \end{pmatrix} + \beta \begin{pmatrix} 1 \\ 3 \\ 1 \end{pmatrix} + \gamma \begin{pmatrix} 2 \\ -2 \\ 2 \end{pmatrix}.$$

Matching up entries, we know that α, β and γ must satisfy the following system of equations

$$\begin{aligned} 1 &= \alpha + \beta + 2\gamma \\ 2 &= \alpha + 3\beta - 2\gamma. \\ 2 &= \alpha + \beta + 2\gamma \end{aligned}$$

Using the method of elimination, we get

$$\begin{aligned} 1 &= \alpha + \beta + 2\gamma \\ 2 &= \alpha + 3\beta - 2\gamma \longrightarrow \\ 2 &= \alpha + \beta + 2\gamma \end{aligned} \qquad \begin{aligned} 1 &= \alpha + \beta + 2\gamma \\ 1 &= \quad\quad + 2\beta - 4\gamma. \\ 1 &= 0 \end{aligned}$$

We see that since the last equation is false, this system has no solution and therefore, no scalars α, β, and γ exist so that $w = \alpha v_1 + \beta v_2 + \gamma v_3$. That is, w cannot be written as a linear combination of v_1, v_2 and v_3.   □

Notice that every time we ask whether a vector w is a linear combination of other vectors $v_1, v_2, \ldots, v_n$, we seek to solve an equation like

$$\alpha_1 v_1 + \alpha_2 v_2 + \ldots + \alpha_n v_n = w.$$

We can ask the very similar question:

Are there scalars $\alpha_1, \alpha_2, \ldots, \alpha_n$, β so that $\alpha_1 v_1 + \alpha_2 v_2 + \ldots + \alpha_n v_n + \beta w = 0$?

If we find a solution where $\beta \neq 0$, we know that we can solve for w. So, w can be written as a linear combination of $v_1, v_2, \ldots, v_n$. If β must be zero, then we know that w cannot be written as a linear combination of the v_i's. In the next section, we study the homogeneous equation, $Ax = 0$.

3.1.4 The Matrix Equation $Ax = 0$

Recall that in Definition 2.4.11, we defined a linear equation to be homogeneous if it has the form

$$a_1 x_1 + a_2 x_2 + \cdots + a_n x_n = 0.$$

We now turn our attention to systems of homogeneous equations.

A system of equations made of only homogeneous equations is, then, called a *homogeneous system of equations*.

Example 3.1.21 The system of equations

$$2x_1 + 3x_2 - 2x_3 = 0$$
$$3x_1 - x_2 + x_3 = 0$$
$$x_1 - 2x_2 + 2x_3 = 0$$

is a homogeneous system of equations. On the other hand,

$$2x_1 + 3x_2 - 2x_3 = 0$$
$$3x_1 - x_2 + x_3 = 1$$
$$x_1 - 2x_2 + 2x_3 = 0$$

has two homogeneous equations (the first and third), but is not a homogeneous system because the second equation is not homogeneous. □

When considering a homogeneous system of equations with n variables and m equations, the corresponding matrix equation is $Ax = 0$, for some $m \times n$ matrix A and where 0 represents the vector in $\mathbb{R}^m$ with all zero entries. Notice that there is always a solution to a homogeneous system of equations (see Exercise 24). This also means that if $A = (a_{i,j})$ is the coefficient matrix for the system of equations, then $Ax = 0$ always has a solution, we call this common solution the *trivial solution*.

When working with homogeneous systems of equations, we are then concerned with how many solutions (one or infinitely many) the system has. In this section, we will look at tools to determine the number of solutions.

Recall that we find no solution to the system of equations

$$a_{1,1}x_1 + a_{1,2}x_2 + \ldots + a_{1,n}x_n = b_1$$
$$a_{2,1}x_1 + a_{2,2}x_2 + \ldots + a_{2,n}x_n = b_2$$
$$\vdots \qquad \qquad \ddots \qquad \qquad \vdots$$
$$a_{m,1}x_1 + a_{m,2}x_2 + \ldots + a_{m,n}x_n = b_m$$

after using elimination steps that lead to an equation that is never true. That is, we reduce the system until we arrive at a point where one equation looks like $0 = c$, where c is not 0. This situation cannot possibly occur with a homogeneous system of equations. (See Exercise 26.) Notice that if a system has exactly one solution, then when we reduce the system, we arrive at a system of equations of the form

$$x_1 \qquad \qquad = c_1$$
$$x_2 \qquad \quad = c_2$$
$$\ddots \qquad \vdots$$
$$x_n = c_n$$

whose matrix equation is $I_n x = c$, where I_n is the $n \times n$ matrix given by

$$I_n = \begin{pmatrix} 1 & 0 & \ldots & 0 \\ 0 & 1 & \ldots & 0 \\ \vdots & & \ddots & \\ 0 & \ldots & 0 & 1 \end{pmatrix} \quad \text{and} \quad c = \begin{pmatrix} c_1 \\ c_2 \\ \vdots \\ c_n \end{pmatrix}.$$

If the system of equations corresponding to $Ax = b$ reduces to the system of equations $I_n x = c$, then we will see exactly one solution to the system no matter what values were on the right-hand side of the system.

A system of equations with fewer equations than variables has infinitely many solutions or no solution. In the case of homogeneous systems of equations with more equations than variable, we can immediately say that there are infinitely many solutions to the system without doing any work.

We will use these last two ideas to consider solutions to the equations $Ax = b$ and $Ax = 0$.

Theorem 3.1.22

Let A be an $n \times m$ matrix. The matrix equation $Ax = 0$ has only the trivial solution if and only if the corresponding homogeneous system with coefficient matrix A has only the trivial solution.

Proof. Let A be an $n \times m$ matrix. We know, by Exercise 24, that the trivial solution is always a solution to $Ax = 0$. We know that if $Ax = 0$ has a unique solution, it must be the trivial solution. Applying Corollary 3.1.16, we know that $Ax = 0$ has only the trivial solution if and only if the corresponding system has the trivial solution.   □

Theorem 3.1.22 characterizes homogeneous systems of equations with a unique (trivial) solution. We now consider the set of solutions to a homogeneous matrix equation with infinitely many solutions.

Example 3.1.23 Let A be the matrix given by

$$A = \begin{pmatrix} 1 & 1 & -2 \\ 2 & 1 & -1 \\ 4 & 3 & -5 \end{pmatrix}.$$

We can find the set of all solutions to $Ax = 0$ by solving the corresponding system of equations

$$\begin{aligned} x + y - 2z &= 0 \\ 2x + y - z &= 0. \\ 4x + 3y - 5z &= 0 \end{aligned}$$

We can use the method of elimination (or matrix reduction) to find the set of solutions. We will leave off the intermediate steps and give the reduced echelon form of the corresponding augmented matrix:

$$\left(\begin{array}{ccc|c} 1 & 1 & -2 & 0 \\ 2 & 1 & -1 & 0 \\ 4 & 3 & -5 & 0 \end{array} \right) \sim \left(\begin{array}{ccc|c} 1 & 0 & 1 & 0 \\ 0 & 1 & -3 & 0 \\ 0 & 0 & 0 & 0 \end{array} \right).$$

The corresponding system of equations is

$$\begin{aligned} x + z &= 0 \\ y - 3z &= 0. \\ 0 &= 0 \end{aligned}$$

This means that we $x = -z$ and $y = 3z$ for any $z \in \mathbb{R}$. So, the solution set can be written as

$$S = \left\{ \begin{pmatrix} x \\ y \\ z \end{pmatrix} \in \mathbb{R}^3 \mid x = -z, y = 3z \right\} = \left\{ \begin{pmatrix} -z \\ 3z \\ z \end{pmatrix} \mid z \in \mathbb{R} \right\}.$$

We claim that S is actually a vector space. Indeed, it is a subspace of $\mathbb{R}^3$. This follows either by appealing to Theorem 2.5.20 or the direct proof below. $\square$

Proof. Let $u, v \in S$ and let $\alpha, \beta \in \mathbb{R}$ be scalars. Then, $u = \begin{pmatrix} -z_1 \\ 3z_1 \\ z_1 \end{pmatrix}$ and $v = \begin{pmatrix} -z_2 \\ 3z_2 \\ z_2 \end{pmatrix}$ for some $z_1, z_2 \in \mathbb{R}$. Then

$$\begin{aligned} \alpha u + \beta v &= \alpha \begin{pmatrix} -z_1 \\ 3z_1 \\ z_1 \end{pmatrix} + \beta \begin{pmatrix} -z_2 \\ 3z_2 \\ z_2 \end{pmatrix} \\ &= \begin{pmatrix} -\alpha z_1 - \beta z_2 \\ 3\alpha z_1 + 3\beta z_2 \\ \alpha z_1 + \beta z_2 \end{pmatrix} \\ &= \begin{pmatrix} -(\alpha z_1 + \beta z_2) \\ 3(\alpha z_1 + \beta z_2) \\ (\alpha z_1 + \beta z_2) \end{pmatrix} \in S. \end{aligned}$$

Therefore, by Theorems 2.5.3 and 2.5.5, S is a subspace of $\mathbb{R}^3$. $\square$

This example leads to the next definition.

Definition 3.1.24

The set of solutions to a system of equations (or the corresponding matrix equation) is called a **solution space** whenever it is also a vector space. $\square$

The previous example leads us to ask, "Under what conditions will a solution set be a solution space?" We answer the question immediately with the following theorem.

Theorem 3.1.25

Let A be an $m \times n$ matrix and $b \in \mathbb{R}^m$. The set of solutions to the matrix equation $Ax = b$ (or the corresponding system of equations) is a solution space if and only if $b = 0$.

Proof. Let A be an $m \times n$ matrix and let $b \in \mathbb{R}^m$. Define the set of solutions of $Ax = b$ as

$$S = \{v \in \mathbb{R}^n \mid Av = b\}.$$

We will show two statements:

($\Rightarrow$) If S is a subspace then $b = 0$.

($\Leftarrow$) If $b = 0$ then S is a subspace of $\mathbb{R}^3$.

($\Rightarrow$) We will prove this statement by proving the contrapositive (See Appendix C). That is, we will prove that if $b \in \mathbb{R}^m$ is not the zero vector then S is not a vector space. Notice that for any $m \times n$ matrix

$$A \cdot 0 = 0 \neq b. \tag{3.8}$$

Therefore, $0 \notin S$. Thus, S is not a vector space. So, if S is a solution space to $Ax = b$ then $b = 0$.

($\Leftarrow$) Now, assume $b \in \mathbb{R}^m$ is the zero vector. Then the set S is the solution set of a system of homogeneous equations, and hence by Theorem 2.5.20, it is a subspace of $\mathbb{R}^n$ and therefore a solution space. $\square$

In Equation 3.8 above, we use the symbols "0" in two ways. The 0 on the left side of the equation is a vector in $\mathbb{R}^n$, but the 0 on the right side of the equation is a vector in $\mathbb{R}^m$.

In the proof of Theorem 3.1.25, we use the fact that there is a distributive property for matrix multiplication over vector addition. We also use the property that for any $m \times n$ matrix and vector $x \in \mathbb{R}^n$ we can write $M(\alpha x) = \alpha M x$. These two properties are important properties of matrix-vector products that we will explore later.

We can visualize 3-dimensional vectors, so let us consider solution sets in $\mathbb{R}^3$. The next examples show, visually, the difference between a solution set that is not a solution space and one that is.

We first compare the solution set to a single inhomogeneous equation in three variables, and the solution set to the associated homogeneous equation.

Example 3.1.26 Consider the two equations

$$x + 2y - z = 1$$

and

$$x + 2y - z = 0.$$

We can solve for z in both equations to get

$$z = x + 2y - 1$$

and

$$z = x + 2y.$$

It is apparent that these two solution sets are parallel planes, and that the solution to the homogeneous system goes through the origin, while the solution to the inhomogeneous system is shifted down (in the negative z-direction) by a distance of 1. $\square$

Let's look at a slightly more complicated example to get a sense of the patterns that result.

Example 3.1.27 By Theorem 3.1.25, we know that the solution set to $Ax = b$, where

$$A = \begin{pmatrix} 1 & 2 & -1 \\ 2 & 4 & 2 \\ 3 & 6 & 1 \end{pmatrix} \quad \text{and} \quad b = \begin{pmatrix} 1 \\ 2 \\ 3 \end{pmatrix},$$

is not a solution space. Indeed, the solution set is

$$S = \{x \mid Ax = b\}$$

$$= \left\{ \begin{pmatrix} x \\ y \\ z \end{pmatrix} \middle| \begin{pmatrix} 1 & 2 & -1 \\ 2 & 4 & 2 \\ 3 & 6 & 1 \end{pmatrix} \begin{pmatrix} x \\ y \\ z \end{pmatrix} = \begin{pmatrix} 1 \\ 2 \\ 3 \end{pmatrix} \right\}$$

$$= \left\{ \begin{pmatrix} x \\ y \\ z \end{pmatrix} \middle| x + 2y - z = 1, 2x + 4y + 2z = 2, \text{ and } 3x + 6y + z = 3 \right\}$$

$$= \left\{ \begin{pmatrix} x \\ y \\ z \end{pmatrix} \middle| x + 2y = 1 \text{ and } z = 0 \right\}.$$

Notice that $0 \in \mathbb{R}^3$ is not an element of S. Geometrically, the solution set is the line in $\mathbb{R}^3$ where the planes $\left\{ \begin{pmatrix} x \\ y \\ z \end{pmatrix} \middle| x + 2y - z = 1 \right\}$ and $\left\{ \begin{pmatrix} x \\ y \\ z \end{pmatrix} \middle| 2x + 4y + 2z = 2 \right\}$ intersect. This line does not pass through the origin. □

Example 3.1.28 Also, from Theorem 3.1.25, we know that the set of solutions to $Ax = b$, where

$$A = \begin{pmatrix} 1 & 2 & -1 \\ 2 & 4 & 2 \\ 3 & 6 & 1 \end{pmatrix} \text{ and } b = \begin{pmatrix} 0 \\ 0 \\ 0 \end{pmatrix},$$

is a vector space. In this case, the solution set is

$$S = \left\{ x \middle| Ax = b \right\}$$

$$= \left\{ \begin{pmatrix} x \\ y \\ z \end{pmatrix} \middle| \begin{pmatrix} 1 & 2 & -1 \\ 2 & 4 & 2 \\ 3 & 6 & 1 \end{pmatrix} \begin{pmatrix} x \\ y \\ z \end{pmatrix} = \begin{pmatrix} 0 \\ 0 \\ 0 \end{pmatrix} \right\}$$

$$= \left\{ \begin{pmatrix} x \\ y \\ z \end{pmatrix} \middle| x + 2y - z = 0, 2x + 4y + 2z = 0, \text{ and } 3x + 6y + z = 0 \right\}$$

$$= \left\{ \begin{pmatrix} x \\ y \\ z \end{pmatrix} \middle| x + 2y = 0 \text{ and } z = 0 \right\}.$$

Geometrically, the solution space for this example is the line where the planes $\left\{ \begin{pmatrix} x \\ y \\ z \end{pmatrix} \middle| x+2y-z = 0 \right\}$ and $\left\{ \begin{pmatrix} x \\ y \\ z \end{pmatrix} \middle| 2x + 4y + 2z = 0 \right\}$ intersect. This line passes through the origin and is parallel to the line that describes the solution set found in Example 3.1.27. □

In both cases, the solution sets are parallel to each other, and the homogeneous solution set passes through the origin. It turns out that this is always the case.

3.1.5   The Principle of Superposition

We saw in Examples 3.1.26, 3.1.27, and 3.1.28 that the solution set of a matrix equation (or its corresponding system of equations) is parallel to the solution set of the homogeneous equation (or system) with the same coefficient matrix. This is called the principle of superposition. We state it more clearly in the next theorem.

Theorem 3.1.29 (Superposition)

Let A be an $m \times n$ matrix and $b \in \mathbb{R}^m$. Let $x \in \mathbb{R}^n$ satisfy $Ax = b$. Then $Az = b$ if and only if $z = x + y$ for some $y \in \mathbb{R}^n$ satisfying $Ay = 0$.

Proof. Let $x, z \in \mathbb{R}^n$. Suppose also that $Ax = b$ and $Az = b$. We will show that $z - x$ is a solution to $Ay = 0$. Indeed,

$$
\begin{aligned}
A(x - z) &= Ax - Az \\
&= b - b = 0.
\end{aligned}
$$

Let $y = z - x$. Then $y \in \mathbb{R}^n$, $Ay = 0$, and $z = x + y$. Now, suppose $x, y \in \mathbb{R}^n$ satisfy $Ax = b$ and $Ay = 0$. Suppose also that $z = x + y$. Then

$$
\begin{aligned}
Az &= A(x + y) \\
&= Ax + Ay \\
&= 0 + b.
\end{aligned}
$$

Therefore $Az = b$. $\square$

The principle of superposition tells us that if you know one solution to $Ax = b$, you can find other solutions to the inhomogeneous equation by adding solutions to the homogeneous equation.

Theorem 3.1.29 can be restated in terms of the solution set to the inhomogeneous matrix equation.

Theorem 3.1.30 (Superposition–Restated)

Suppose A is an $m \times n$ matrix, $b \in \mathbb{R}^m$, and $z \in \mathbb{R}^n$ satisfies $Az = b$. Then the solution set to $Ax = b$ is

$$
S = \{x = z + y \mid Ay = 0\}.
$$

To show Theorems 3.1.29 and 3.1.30 are equivalent, we must show

$$\text{Theorem } 3.1.29 \implies \text{Theorem } 3.1.30$$
$$\text{and}$$
$$\text{Theorem } 3.1.30 \implies \text{Theorem } 3.1.29.$$

We leave this proof to Exercise 25.

What does this have to do with parallel solution sets? The solution set to the homogeneous equation can be translated by any vector that solves $Ax = b$ to produce the entire solution set to the equation $Ax = b$. In Exercises 13 and 14, we ask you to draw some examples of solution sets to see how they relate to the Principle of Superposition.

We can also get a solution to the homogeneous equation using two solutions to the inhomogeneous equation $Ax = b$.

Corollary 3.1.31

Let A be an $m \times n$ matrix and $b \in \mathbb{R}^m$. Suppose $u, v \in \mathbb{R}^n$ are solutions to $Ax = b$. Then $u - v$ is a solution to $Ax = 0$.

We leave the proof of this corollary as Exercise 27.

Suppose we are given an $m \times n$ matrix A and vector $b \in \mathbb{R}^m$. Using the Principle of Superposition, we can make a connection between the number of solutions to $Ax = 0$ and $Ax = b$. In fact, every solution to $Ax = b$ is linked to at least one solution of $Ax = 0$. The following corollary states this connection.

Corollary 3.1.32

Let A be an $m \times n$ matrix and $b \in \mathbb{R}^m$. Suppose also that $v \in \mathbb{R}^n$ is a solution to $Ax = b$. Then $Ax = b$ has a unique solution if and only if $Ax = 0$ has a only the trivial solution.

Proof. Let A be an $m \times n$ matrix and $b \in \mathbb{R}^m$. Let $\mathcal{S}$ be the solution set of $Ax = b$.
($\Rightarrow$) Assume $\mathcal{S} = \{v\}$. We will show that $Ax = 0$ has exactly one solution. Suppose, to the contrary, that $x_1, x_2 \in \mathbb{R}^n$ are solutions to $Ax = 0$ with $x_1 \neq x_2$. Then by Theorem 3.1.30, $v + x_1, v + x_2 \in \mathcal{S}$, but $v + x_1 \neq v + x_2$, a contradiction $\rightarrow\leftarrow$. Therefore, the only solution to $Ax = 0$ is the trivial solution.
($\Leftarrow$) Now suppose that $Ax = 0$ has only the trivial solution. Then, by Theorem 3.1.30, $\mathcal{S} = \{v + y \mid Ay = 0\} = \{v + 0\} = \{v\}$. Therefore $Ax = b$ has a unique solution. $\qquad\square$

In Corollary 3.1.32, we assume that $Ax = b$ has at least one solution. We need this assumption because $Ax = 0$ always has a solution (the trivial solution) even in the case where $Ax = b$ does not. So, the connection between solution sets only exists if $Ax = b$ has a solution.

Geometrically, we can see the connection in $\mathbb{R}^3$ in Figure 3.1. Here, we see that the superposition principle tells us that we can find solutions to $Ax = b$ through translations from solutions to $Ax = 0$ by a vector v that satisfies $Av = b$.

Theorem 3.1.29 is very useful in situations similar to the following example.

Example 3.1.33 Suppose we are given some radiographic data $b \in \mathbb{R}^m$ that we know was obtained from some object data $u \in \mathbb{R}^n$. Suppose, also, that we know that the data came from multiplying a

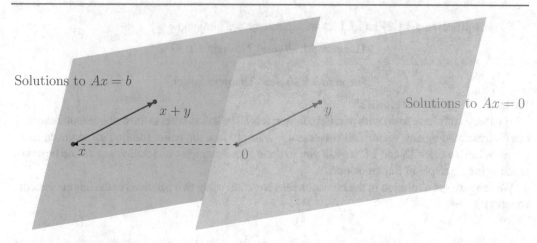

Fig. 3.1 The Principle of Superposition in $\mathbb{R}^3$ showing a translation from the plane describe by $Ax = 0$ to the plane described by $Ax = b$.

matrix A by u. That is, we know

$$Au = b.$$

As is typical in tomographic problems (See Chapter 1), we want to find u so that we know what makes up the object that was radiographed. If $Ax = b$ has infinitely many solutions, we may struggle to know which solution is the correct solution. Because there are infinitely many solutions, it could take a very long time to go through them.

In situations like this, we may want to apply Theorem 3.1.29 and some *a priori* (previously known) information. Exercise 23 asks you to explore a solution path to this problem.   □

★★ **Watch Your Language!**
Since we have drawn many connections between systems of equations and matrices, we close this section with a note about the language we use for solving systems of equations and reducing matrices.
We say

✓   We reduced the augmented matrix to solve the system of equations.
✓   We solved the system of equations with coefficient matrix A.
✓   We solve the matrix equation $Ax = b$.

We **do not** say

✗   We solved the augmented matrix.
✗   We solved the matrix A.

3.1.6   Exercises

1. Write the solution set to the following system as a linear combination of one or more solutions.

$$\begin{aligned} x - y - \ z &= 0 \\ 2x - y + 3z &= 0. \\ -x \ \ \ \ \ \ - 2z &= 0 \end{aligned}$$

2. Write the solution set to the following system as a linear combination of one or more solutions.

$$2x - y - 3z = 0$$
$$4x - 2y - 6z = 0.$$
$$6x - 3y - 9z = 0$$

3. Write the solution set to the following system as a linear combination of one or more solutions.

$$x - 2y - 2z = 0$$
$$-x + 2y + 2z = 0.$$
$$3x - 3y - 3z = 0$$

For Exercises 4 to 8, consider arbitrary vectors u, v, w from a vector space $(V, +, \cdot)$.

4. Write a linear combination of the vectors u, v, and w.
5. Is $3u - 4w + 5$ a linear combination of u and w? Justify your response.
6. Is $5u - 2v$ a linear combination of u, v, and w? Justify your response.
7. Show that any linear combination of u, v, and $3u + 2v$ is a linear combination u and v as well.
8. Is it possible to write a linear combination (with nonzero coefficients) of u, v, and w as a linear combination of just two of these vectors? Justify your response.

In Exercises 9 to 12, determine whether w can be written as a linear combination of u and v. If so, write the linear combination. If not, justify.

9.

$$u = \begin{pmatrix} 1 \\ 3 \\ 5 \end{pmatrix}, v = \begin{pmatrix} -1 \\ -2 \\ 3 \end{pmatrix}, \text{ and } w = \begin{pmatrix} 1 \\ 1 \\ 1 \end{pmatrix}.$$

10. $u = 3x^2 + 4x + 2$, $v = x - 5$, and $w = x^2 + 2x - 1$.

11.

$$u = \begin{pmatrix} 1 \\ 2 \\ 0 \end{pmatrix}, v = \begin{pmatrix} -1 \\ 0 \\ 3 \end{pmatrix}, \text{ and } w = \begin{pmatrix} 1 \\ 1 \\ 4 \end{pmatrix}.$$

12. $u = 3x^2 + x + 2$, $v = x^2 - 2x + 3$, and $w = -x^2 - 1$.

13. Consider the 2×2 matrix M and vector $b \in \mathbb{R}^2$ given by

$$M = \begin{pmatrix} 1 & 2 \\ 2 & 4 \end{pmatrix} \text{ and } b = \begin{pmatrix} 3 \\ 6 \end{pmatrix}.$$

Show that $x = \begin{pmatrix} 1 \\ 1 \end{pmatrix}$ is a solution to $Mx = b$. Find all solutions to $Mx = 0$ and plot the solution set in $\mathbb{R}^2$. Use Theorem 3.1.29 or 3.1.30 to draw all solutions to $Mx = b$.

14. Consider the 3×3 matrix M and vector $b \in \mathbb{R}^3$ given by

$$M = \begin{pmatrix} 1 & 2 & 1 \\ 2 & 1 & 1 \\ 3 & 3 & 2 \end{pmatrix} \text{ and } b = \begin{pmatrix} 3 \\ -1 \\ 2 \end{pmatrix}.$$

Show that $x = \begin{pmatrix} -3 \\ 1 \\ 4 \end{pmatrix}$ is a solution to $Mx = b$. Find all solutions to $Mx = 0$ and plot the solution set in $\mathbb{R}^3$. Use Theorem 3.1.29 or 3.1.30 to draw all solutions to $Mx = b$.

For each of Exercises 15 to 17, do the following. (a) Describe, geometrically, the set X of all linear combinations of the given vector or vectors, and (b) for the value of n specified, either prove or provide a counter example to the statement: "X is a subspace of $\mathbb{R}^n$."

15. $n = 2, u = \begin{pmatrix} 1 \\ 1 \end{pmatrix}$.

16. $n = 3, u = \begin{pmatrix} 1 \\ 0 \\ 2 \end{pmatrix}$ and $v = \begin{pmatrix} 1 \\ 3 \\ -1 \end{pmatrix}$.

17. $n = 3, u = \begin{pmatrix} 1 \\ -2 \\ 1 \end{pmatrix}$ and $v = \begin{pmatrix} -1 \\ 2 \\ -1 \end{pmatrix}$.

18. Solve the system of equations

$$x - 2y + 3z = 6$$
$$x + 2y + 4z = 3$$
$$x + y + z = 1$$

and at each stage, sketch the corresponding planes.

19. A street vendor is selling three types of fruit baskets with the following contents.

	Apples	Pears	Plums
Basket A	4	2	1
Basket B	3	1	2
Basket C	1	2	2

Carlos would like to purchase some combination of baskets that will give him equal numbers of each fruit. Either show why this is not possible, or provide a solution for Carlos.

20. Consider the images in Example 3.1.2. Determine whether Image 3 is a linear combination of Image 2 and Image B.

21. Consider the space of 7-bar LCD characters defined in Example 2.4.17. Define D to be the set of 10 digits seen in Figure 3.2, where d_0 is the character that displays a "0", d_1 is the character that displays a "1", etc. Write the set of all linear combinations of the vectors d_0 and d_9. Can any other elements of D be written as a linear combination of d_0 and d_9? If so, which? Can any other vectors of $\mathcal{D}(Z_2)$ be written as a linear combination of d_0 and d_9? If so, which?

22. Consider the given heat states in $H_4(\mathbb{R})$. (See Figure 3.3)
Draw the result of the linear combination $3h_1 + 2h_4$, where h_1 is the top left heat state and h_4 is the second heat state in the right column.

23. In Example 3.1.33, we discussed using a priori information alongside Theorem 3.1.29 to determine a solution to $Ax = b$ that makes sense for a radiography/tomography problem. Suppose that we expect the data for an object to represent density measurements at specific points throughout the object. Discuss characteristics for a solution u that likely make sense when representing object data. That is, what type of values would you expect to find in the entries of u? How can you use solutions to $Ax = 0$ to find another solution that has values closer to what you expect?

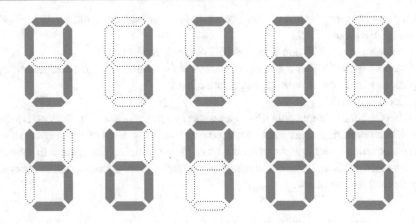

Fig. 3.2 The set $D = \{d_0, d_1, d_2, d_3, d_4, d_5, d_6, d_7, d_8, d_9\}$. (The ten digits of a standard 7-bar LCD display.)

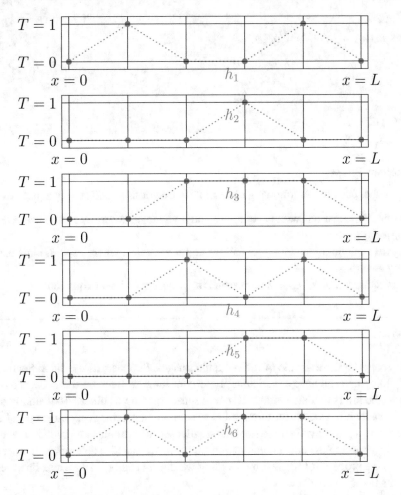

Fig. 3.3 Heat states in $H_4(\mathbb{R})$.

24. Show that every homogeneous system of equations has at least one solution. Why do we call this solution the trivial solution?
25. Prove that Theorem 3.1.29 and Theorem 3.1.30 are equivalent.
26. Describe how we know that a homogeneous system of equations will never reduce to a system of equations where one or more equations are false.
27. Prove Corollary 3.1.31.
28. Recall that the phrase "if and only if" is used when two statements are equivalent. Up to this point, we have been exploring connections between solutions to systems of linear equations, solutions to matrix equations, and linear combinations of vectors. In Section 2.2 and in this chapter, we have presented theorems and corollaries making these connections. Combine these ideas[2] to complete the theorem below.

> **Theorem 3.1.34**
> Let A be an $n \times n$ matrix and $b \in \mathbb{R}^n$. The following statements are equivalent.
>
> - $Ax = 0$ has a unique solution.
> -
> -
> -
> -

Prove Theorem 3.1.34.

29. State whether each of the following statements is true or false and supply a justification.

   (a) A system of equations with m equations and n variables, where $m > n$, must have at least one solution.
   (b) A system of equations with m equations and n variables, where $m < n$, must have infinitely many solutions.
   (c) Every subspace of $\mathbb{R}^n$ can be represented by a system of linear equations.

3.2   Span

Let us, again, consider the example of radiographic images. Suppose, as we did in Section 3.1, that a subspace of radiographs all have a particular property of interest. Because this subspace is not a finite set, and due to the limits on data storage, it makes sense to know whether a (potentially small) set of radiographs is enough to reproduce, through linear combinations, this (very large) set of radiographs holding an important property. If so, we store the smaller set and are able to reproduce any radiograph in this subspace by choosing the linear combination we want. We can also determine whether a particular radiograph has the property of interest by deciding whether it can be written as a linear combination of the elements in the smaller set.

We considered a similar, yet smaller, example in Section 2.1, using Images A, B, and C (also found in Figure 3.4). Our task was to determine whether certain other images could be represented as linear

---

[2] Hint: one bullet should be about the reduced echelon form of A, one should be about a system of equations, one should be about linear combinations of the columns of A, and one should be about the matrix equation $Ax = b$.

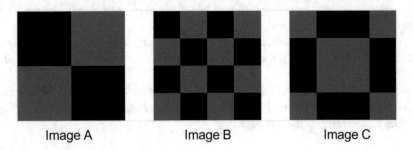

Fig. 3.4 Images A, B, and C of Example 3.1.2.

combinations of these three. Let us define the set S of 4×4 grayscale images that can be obtained through linear combinations of Images A, B, and C. That is, we define

$$S = \{I \in \mathcal{I}_{4 \times 4} \mid I \text{ is a linear combination of Images A, B, and C}\}$$
$$= \{\alpha A + \beta B + \gamma C \in \mathcal{I}_{4 \times 4} \mid \alpha, \beta, \gamma \in \mathbb{R}\}.$$

In Section 2.1, we saw that Image 3 (See Figure 3.5) is in the set S, but Image 4 is not in S. If Image 4 is important to our work, we need to know that it is not in the set S. Or we may simply be interested in exploring the set S because we know that Images A, B, and C represent an important subset of images.

In this section we will study the set of all linear combinations of a set of vectors. We will define *span* and clarify the various ways the word *span* is used in linear algebra. We will discuss this word as a noun, a verb, and as an adjective. Indeed, the term *span* in linear algebra is used in several contexts. The following are all accurate and meaningful uses of the word span that we will discuss in this section.

The set X is the span of vectors u, v, and w.

The vectors x and y span the subspace W.

T is a spanning set for a vector space V.

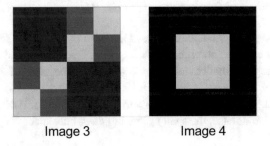

Fig. 3.5 Image 3 is in S, but Image 4 is not.

3.2.1   The Span of a Set of Vectors

We begin by considering a set, X of vectors. As discussed above, we want to know more about the set of all linear combinations of X. We call this set the *span* of X:

Definition 3.2.1

(n.) Let $(V, +, \cdot)$ be a vector space and let $X \subseteq V$. The **span** of the set X, denoted Span X, is the set of all linear combinations of the elements of X. In addition, we define Span $\emptyset := \{0\}$.   □

It is important to note that the object, Span X, is a set of vectors. The set X may have finitely many or infinitely many elements. If $X = \{v_1, v_2, \cdots, v_n\}$, we may also write Span $\{v_1, v_2, \cdots, v_n\}$ to indicate Span X. We can express this set as

$$\text{Span } X = \{\alpha_1 v_1 + \alpha_2 v_2 + \cdots + \alpha_n v_n \mid \alpha_1, \alpha_2, \cdots, \alpha_n \in \mathbb{F}\}.$$

Let us now consider some examples.

Example 3.2.2 In the example above, we defined S to be the set of all linear combinations of Images A, B, and C. We can now write

$$S = \text{Span} \left\{ \text{Image A}, \text{Image B}, \text{Image C} \right\}.$$

And, we know that

$$\text{Image 3} \in S, \text{ but } \text{Image 4} \notin S.$$

□

Definition 3.2.1 includes a separate definition of the span of the empty set. This is a consequence of defining span in terms of linear combinations of vectors. One could equivalently define Span X as the intersection of all subspaces that contain X as is proved in Theorem 3.2.20.

Example 3.2.3 Consider the two polynomials x and 1, both in $\mathcal{P}_1(\mathbb{R})$. We have Span$\{x, 1\} = \{ax + b \mid a, b \in \mathbb{R}\}$. In this particular case, the span is equal to the whole space $\mathcal{P}_1(\mathbb{R})$.   □

Example 3.2.4 Now consider the vectors

$$v_1 = \begin{pmatrix} -1 \\ 0 \\ 1 \end{pmatrix}, \ v_2 = \begin{pmatrix} 1 \\ 0 \\ 0 \end{pmatrix}, \text{ and } v_3 = \begin{pmatrix} 0 \\ 0 \\ 1 \end{pmatrix}$$

in $\mathbb{R}^3$. We can find $W = \text{Span}\{v_1, v_2, v_3\}$ by considering all linear combinations of v_1, v_2, and v_3.

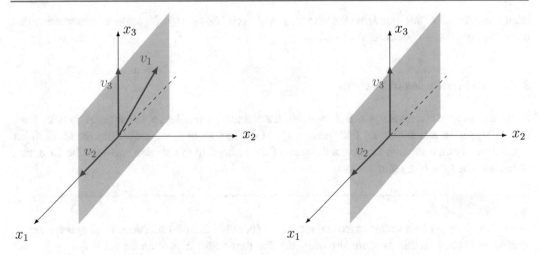

Fig. 3.6 In Example 3.2.4, the same set (the x_1x_3-plane) is represented as the span of two different sets. Left: Span$\{v_1, v_2, v_3\}$, Right: Span$\{v_2, v_3\}$.

$$\text{Span}\{v_1, v_2, v_3\} = \left\{ \alpha \begin{pmatrix} -1 \\ 0 \\ 1 \end{pmatrix} + \beta \begin{pmatrix} 1 \\ 0 \\ 0 \end{pmatrix} + \gamma \begin{pmatrix} 0 \\ 0 \\ 1 \end{pmatrix} \middle| \alpha, \beta, \gamma \in \mathbb{R} \right\}$$

$$= \left\{ \begin{pmatrix} -\alpha + \beta \\ 0 \\ \alpha + \gamma \end{pmatrix} \middle| \alpha, \beta, \gamma \in \mathbb{R} \right\}$$

$$= \left\{ \begin{pmatrix} a \\ 0 \\ b \end{pmatrix} \middle| a, b \in \mathbb{R} \right\}$$

$$= \text{Span} \left\{ \begin{pmatrix} 1 \\ 0 \\ 0 \end{pmatrix}, \begin{pmatrix} 0 \\ 0 \\ 1 \end{pmatrix} \right\}$$

$$= \text{Span}\{v_2, v_3\}.$$

$\square$

This example is interesting because it shows two different ways to write the same set as a span. Another way to say this is that all linear combinations of v_1, v_2, and v_3 is the same set as all linear combinations of the vectors v_2 and v_3. Now, you likely wonder if there are other ways to write this set as a span of other vectors. We leave this exercise to the reader. (See Exercise 8.)

When considering the span of vectors in $\mathbb{R}^n$, as we did in Example 3.2.4, it is helpful to consider the geometry of linear combinations. Figure 3.6 shows that the span of v_1, v_2, and v_3 is the same as the span of v_2 and v_3.

Example 3.2.5 Consider the vectors

$$v_1 = 3x + 4, v_2 = 2x + 1, v_3 = x^2 + 2, \text{ and } v_4 = x^2$$

from Example 3.1.5. We found that the vector v_1 is in Span $\{v_2, v_3, v_4\}$. We know this because v_1 can be written as a linear combination of vectors v_2, v_3, and v_4.   □

3.2.2 To Span a Set of Vectors

In the beginning of this section, we discussed the idea of finding a smaller set whose linear combinations form a bigger subspace that we find interesting. In the last section, we called the set of all linear combinations of a collection of vectors the *span* of the vectors. In this section, we describe the action of creating the span of a set of vectors.

Definition 3.2.6

(v.) Let $(V, +, \cdot)$ be a vector space and let X be a (possibly infinite) subset of V. X **spans** a set W if $W = \text{Span } X$. In this case, we also often say that the vectors in X **span** the set W.   □

In general, it is important to distinguish between a set S and the elements of the set. However, in Definition 3.2.6 above, notice that we do not make a distinction: we can refer to both a set of vectors and the vectors themselves as spanning W.

We continue by considering a few examples to illustrate this terminology.

Example 3.2.7 In Example 3.2.2, we discussed

In this example, we make clear that Images A, B, and C do not span the set of $\mathcal{I}_{4\times4}$ of 4×4 digital images because

Therefore, $\{Image\ A, Image\ B, Image\ C\}$ does not span $\mathcal{I}_{4\times4}$.   □

Example 3.2.8 In Example 3.2.3, we found that $\{x, 1\}$ spans $\mathcal{P}_1(\mathbb{R})$. We can also say that the vectors x and 1 span the set $\mathcal{P}_1(\mathbb{R})$.   □

Example 3.2.9 Let us now consider the solution space V of the system of equations

$$3x + 4y + 2z = 0$$
$$x + y + z = 0$$
$$4x + 5y + 3z = 0.$$

We can write the solution set in different ways. The first just says that V is the solution space of the system of equations.

$$V = \left\{ (x, y, z)^\mathsf{T} \mid 3x + 4y + 2z = 0,\ x + y + z = 0,\ \text{and } 4x + 5y + 3z = 0 \right\}.$$

Next, we can solve the system and write more clearly a description of the solutions.

$$V = \left\{ (-2z, z, z)^\mathsf{T} \mid z \in \mathbb{R} \right\}.$$

We recognize this as a parametrization of a line through the origin and the point $(-2, 1, 1)$ in $\mathbb{R}^3$. Now, we can also write V as a span.

$$V = \mathrm{Span} \left\{ (-2, 1, 1)^\mathsf{T} \right\}.$$

This means that we can say that $(-2, 1, 1)^\mathsf{T}$ *spans* the set V. Or we can say that V is *spanned* by the set $\{(-2, 1, 1)^\mathsf{T}\}$. □

Let us now return to the matrix equation $Ax = b$. Consider the following example.

Example 3.2.10 Consider, again, the question posed in Example 3.1.4. We asked whether it was possible to write $w = \begin{pmatrix} 2 \\ 5 \\ 3 \end{pmatrix}$ as a linear combination of

$$v_1 = \begin{pmatrix} 1 \\ 0 \\ 1 \end{pmatrix},\ v_2 = \begin{pmatrix} 2 \\ 2 \\ 1 \end{pmatrix},\ \text{and } v_3 = \begin{pmatrix} 3 \\ 1 \\ 4 \end{pmatrix}.$$

We found that the answer to this question can be found by solving the matrix equation $Ax = b$, where

$$A = \begin{pmatrix} 1 & 2 & 3 \\ 0 & 2 & 1 \\ 1 & 1 & 4 \end{pmatrix} \text{ and } b = \begin{pmatrix} 2 \\ 5 \\ 3 \end{pmatrix}.$$

We found that w could, indeed, be written as a linear combination of v_1, v_2, and v_3. We also found that $Ax = b$ had a solution. This led to Theorem 3.1.18. Using the vocabulary in this section, we can now say that w is in the span of v_1, v_2, and v_3. □

The result in Example 3.2.10 can be stated for more general situations as the following theorem.

Theorem 3.2.11
Let A be an $m \times n$ matrix and let $b \in \mathbb{R}^m$. The matrix equation $Ax = b$ has a solution if and only if b is in the span of the columns of A.

Proof. Let A and b be as defined above. In particular, let

$$A = \begin{pmatrix} | & | & & | \\ c_1 & c_2 & \ldots & c_n \\ | & | & & | \end{pmatrix},$$

where $c_1, c_2, \ldots, c_n \in \mathbb{R}^m$ are the columns of A.

($\Rightarrow$) Suppose that $Ax = b$ has a solution,

$$x = \begin{pmatrix} \alpha_1 \\ \alpha_2 \\ \vdots \\ \alpha_n \end{pmatrix}.$$

Then, using the understanding of matrix multiplication from Section 3.1, we can write

$$
\begin{aligned}
b &= Ax \\
&= \begin{pmatrix} | & | & & | \\ c_1 & c_2 & \ldots & c_n \\ | & | & & | \end{pmatrix} \begin{pmatrix} \alpha_1 \\ \alpha_2 \\ \vdots \\ \alpha_n \end{pmatrix} \\
&= \alpha_1 c_1 + \alpha_2 c_2 + \ldots + \alpha_n c_n.
\end{aligned}
$$

Thus, $b \in \text{Span}\{c_1, c_2, \ldots, c_n\}$.

($\Leftarrow$) Now, suppose that b is in the span of the columns of A. That is, there are scalars $\alpha_1, \alpha_2, \ldots, \alpha_n$ so that

$$b = \alpha_1 c_1 + \alpha_2 c_2 + \ldots + \alpha_n c_n.$$

Again, using the understanding of matrix-vector products from Section 3.1, we know that

$$b = \begin{pmatrix} | & | & & | \\ c_1 & c_2 & \ldots & c_n \\ | & | & & | \end{pmatrix} \begin{pmatrix} \alpha_1 \\ \alpha_2 \\ \vdots \\ \alpha_n \end{pmatrix}.$$

Therefore, $\begin{pmatrix} \alpha_1 \\ \alpha_2 \\ \vdots \\ \alpha_n \end{pmatrix}$ is a solution to $Ax = b$. $\qquad\square$

We now move to different uses of the word span.

A Spanning Set

A set that spans another set can also be described using the word *span*. Here, we introduce the adjective form to describe a set that spans another set.

Definition 3.2.12

We say that a (possibly infinite) set S is a **spanning set** for W if $W = \text{Span}\, S$. $\qquad\square$

Let us consider how this use of the word can be applied to our previous example.

Example 3.2.13 In Example 3.2.9, we found that $\{(-2, 1, 1)^\mathsf{T}\}$ spans the solution space V. This means that $\{(-2, 1, 1)^\mathsf{T}\}$ is a *spanning* set for V. □

In Definition 3.1.1 we made explicit that a linear combination has finitely many terms, but in Definition 3.2.12, we allow the spanning set to have infinitely many elements. In the next example, we clarify this.

Example 3.2.14 Let $S = \{1, x, x^2, x^3, \ldots\}$. Since every polynomial is a linear combination of (finitely many[3]) elements of S, we see that $\mathcal{P}_\infty(\mathbb{R}) \subseteq \mathrm{Span}\, S$. Also, every element of $\mathrm{Span}\, S$ is a polynomial, so $\mathrm{Span}\, S \subseteq \mathcal{P}_\infty(\mathbb{R})$. Hence, $\mathrm{Span}\, S = \mathcal{P}_\infty(\mathbb{R})$, and so S is a spanning set for $\mathcal{P}_\infty(\mathbb{R})$. □

Example 3.2.15 Going back to Example 3.2.3, again, we can now say that $\{x, 1\}$ is a spanning set for $\mathcal{P}_1(\mathbb{R})$. □

As we saw in earlier sections, spanning sets are not unique. Here is another spanning set for $\mathcal{P}_1(\mathbb{R})$.

Example 3.2.16 Show that $\{x + 1, x - 2, 4\}$ spans $\mathcal{P}_1(\mathbb{R})$.

1. First show that $\mathrm{Span}\{x + 1, x - 2, 4\} \subseteq \mathcal{P}_1(\mathbb{R})$ by showing that an arbitrary vector in the span is also a polynomial of degree 1 or less.
   Indeed, we know that if $p \in \mathrm{Span}\{x + 1, x - 2, 4\}$ then there exists scalars α, β and γ such that

   $$p = \alpha(x + 1) + \beta(x - 2) + \gamma(4).$$

   Now, $p = (\alpha + \beta)x + (\alpha - 2\beta + 4\gamma)$ which is a vector in $\mathcal{P}_1(\mathbb{R})$. Thus, $\mathrm{Span}\{x + 1, x - 2, 4\} \subseteq \mathcal{P}_1(\mathbb{R})$.
2. Next show that $\mathcal{P}_1(\mathbb{R}) \subseteq \mathrm{Span}\{x + 1, x - 2, 4\}$ by showing that an arbitrary polynomial of degree one or less can be expressed as a linear combination of vectors in $\{x + 1, x - 2, 4\}$.
   If $p \in \mathcal{P}_1(\mathbb{R})$, then $p = ax + b$ for some $a, b \in \mathbb{R}$. We want to show that $p \in \mathrm{Span}\{x + 1, x - 2, 4\}$. That is, we want to show that there exist $\alpha, \beta, \gamma \in \mathbb{R}$ so that

   $$p = \alpha(x + 1) + \beta(x - 2) + \gamma(4).$$

If such scalars exist, then as before, we can match up like terms to get the system of equations:

$$\begin{aligned}(x \text{ term:}) \qquad a &= \alpha + \beta, \\ (\text{constant term:}) \; b &= \alpha - 2\beta + 4\gamma.\end{aligned}$$

Thus, if $\alpha = \frac{2a+b}{3}$, $\beta = \frac{a-b}{3}$, and $\gamma = 0$, then

$$p = \alpha(x + 1) + \beta(x - 2) + \gamma(4).$$

(There are infinitely many solutions. The above solution is the particular solution in which $\gamma = 0$.)
So, for any $p \in \mathcal{P}_1(\mathbb{R})$, we can find such scalars. Thus, $\mathcal{P}_1(\mathbb{R}) \subseteq \mathrm{Span}\{x + 1, x - 2, 4\}$.

---

[3] As we discussed in Example 3.1.3, functions like $\cos x$ that can only be written using a Taylor series (that is, using an infinite number of the terms of S) are not in $\mathcal{P}_\infty(\mathbb{R})$; nor are they in $\mathrm{Span}\, S$.

We now have $\mathcal{P}_1(\mathbb{R}) = \text{Span}\{x + 1, x - 2, 4\}$ and, therefore, $\{x + 1, x - 2, 4\}$ spans $\mathcal{P}_1(\mathbb{R})$. □

In the next example, we show that $\{x^2, 1\}$ is not a spanning set of $\mathcal{P}_2(\mathbb{R})$.

Example 3.2.17 Every linear combination of x^2 and 1 is a polynomial in $\mathcal{P}_2(\mathbb{R})$, so $\text{Span}\{x^2, 1\} \subseteq \mathcal{P}_2(\mathbb{R})$. However, not every polynomial in $\mathcal{P}_2(\mathbb{R})$ can be written as a linear combination of x^2 and 1; for example, $x^2 + x + 1$. Thus, $\mathcal{P}_2(\mathbb{R}) \not\subset \text{Span}\{x^2, 1\}$. That is $\text{Span}\{x^2, 1\} \neq \mathcal{P}_2(\mathbb{R})$. □

Example 3.2.18 Recall the vector space $\mathcal{D}(Z_2)$ of 7-bar LCD characters from Example 2.4.17, where white corresponds to the value zero and green corresponds to the value one. With element-wise definitions of addition and scalar multiplication as defined for the field Z_2, $\mathcal{D}(Z_2)$ is a vector space. Here are two examples of vector addition in $\mathcal{D}(Z_2)$:

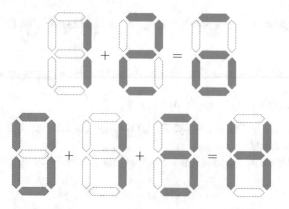

One can show that the set $D = \{d_0, d_1, \cdots, d_9\}$ of the ten digits shown in Figure 2.17 is a spanning set for $\mathcal{D}(Z_2)$. Thus, every character in $\mathcal{D}(Z_2)$ can be written as a linear combination of the vectors in D. □

Because a spanning set will help us describe vector spaces with fewer vectors, it may be useful to find spanning sets for vector spaces of interest. In the next example, we illustrate a possible strategy for finding such a set.

Example 3.2.19 Consider the vector space $\mathcal{P}_1(\mathbb{R})$. Suppose we haven't yet found a spanning set for this vector space, but we want one. We know at least the following three things about a spanning set:

1. The spanning set contains elements of the set it spans.
2. Every element of the set is a linear combination of finitely many elements of the spanning set.
3. A set does not have a unique spanning set. (Many times, there are actually infinitely many spanning sets.)

We can choose an element of $\mathcal{P}_1(\mathbb{R})$ to start. Let's choose

$$v_1 = x + 1.$$

We know that not every element of $\mathcal{P}_1(\mathbb{R})$ is a scalar multiple of v_1. So, $\text{Span}\{v_1\} \neq \mathcal{P}_1(\mathbb{R})$. To find a second vector, we can choose any element of $\mathcal{P}_1(\mathbb{R})$. Let's choose $v_2 = 2x + 2$. Since $v_2 = 2v_1$, we know that any vector not a scalar multiple of v_1 (or of v_2) is not in $\text{Span}\{v_1, v_2\}$. In order to span the whole set, we need to find an element of $\mathcal{P}_1(\mathbb{R})$ that is not a scalar multiple of v_1. We can choose

$$v_3 = 2x + 1.$$

Since there is no scalar α so that $v_3 = \alpha v_1$, we know that $v_3 \notin \text{Span}\{v_1, v_2\}$. Now, it may or may not be clear whether we have $\mathcal{P}_1(\mathbb{R}) = \text{Span}\{v_1, v_2, v_3\}$. We can keep adding vectors until all vectors of $\mathcal{P}_1(\mathbb{R})$ are in the spanning set, but that would not be helpful. After each vector is added, we can check whether there are any elements of $\mathcal{P}_1(\mathbb{R})$ that are still not in the span of the vectors we have chosen. That is, we want to know if for some $a, b \in \mathbb{R}$, we can find scalars α, β, and γ so that

$$ax + b = \alpha v_1 + \beta v_2 + \gamma v_3.$$

That is, we want to know whether or not the system of equations below has a solution for every choice of a and b.
$$\alpha + 2\beta + 2\gamma = a$$
$$\alpha + 2\beta + \gamma = b.$$

Reducing this system, we find that there are infinitely many solutions for α, β, and γ. For example, one solution is $\alpha = 0$, $\beta = \frac{1}{6}a + \frac{1}{3}b$ and $\gamma = \frac{1}{3}(a - b)$. This tells us that $ax + b = 0 \cdot v_1 + (\frac{1}{6}a + \frac{1}{3}b) \cdot v_2 + \frac{1}{3}(a - b) \cdot v_3$. In other words, $ax + b \in \text{Span}\{v_{1,2}, v_3\}$ for any $a, b \in \mathbb{R}$. Since $\text{Span}\{v_1, v_2, v_3\} \subseteq \mathcal{P}_1(\mathbb{R})$ and $\mathcal{P}_1(\mathbb{R}) \subseteq \text{Span}\{v_1, v_2, v_3\}$ (we just showed this part), $\mathcal{P}_1(\mathbb{R}) = \text{Span}\{v_1, v_2, v_3\}$. Thus, one spanning set of $\mathcal{P}_1(\mathbb{R})$ is $\{v_1, v_2, v_3\}$. $\qquad\square$

3.2.3 Span X is a Vector Space

Because of the close connection between the definitions of span and the closure properties of a vector space, it is not surprising that the span of a set is also a vector space. The next theorem formalizes this idea.

Theorem 3.2.19
Let X be a subset of vector space $(V, +, \cdot)$. Then Span X is a subspace of V.

Proof. Suppose X is a subset of vector space $(V, +, \cdot)$. We show that $0 \in \text{Span } X$ and Span X is closed under addition and scalar multiplication. Then, by Theorem 2.5.2, Span X is a subspace of V.

First, $0 \in \text{Span } X$ because either $X = \emptyset$ and $\text{Span } \emptyset = \{0\}$ or 0 is the trivial linear combination (all zero coefficients) of any finite collection of vectors in X.

Let $u, v \in \text{Span } X$ and $\alpha \in \mathbb{F}$. Then u and v are linear combinations of vectors in X, and hence for any scalar α, $\alpha u + v$ is also a linear combination of vectors in X. Thus, Span X is closed under addition and scalar multiplication and is hence a subspace of V. $\qquad\square$

Using Theorem 3.2.19, we can now give an alternate characterization of the span of a set of vectors.

Theorem 3.2.20
Let $(V, +, \cdot)$ be a vector space and let $X \subseteq V$ be a set of vectors. Then, Span X is the intersection of all subspaces of V that contain X.

Proof. Let $\mathcal{S}$ be the set of all subspaces containing X. That is, if $S \in \mathcal{S}$, then S is a subspace of V and $X \subseteq S$. We use the notation $\bigcap_{S \in \mathcal{S}} S$ to denote the intersection of all subspaces[4] in $\mathcal{S}$.

We need to show that Span $X \subseteq \bigcap_{S \in \mathcal{S}} S$ and $\bigcap_{S \in \mathcal{S}} S \subseteq$ Span X.

First, suppose that $v \in$ Span X. Then v is a linear combination of the elements of X. That is,

$$v = \sum_{k=1}^{m} \alpha_k x_k,$$

for some scalars α_k and for $x_k \in X$. Since, for each $k = 1, 2, ..., m$, $x_k \in S$ for each S containing X, and since S is a subspace of V, we know that S is closed under linear combinations. Thus, $v \in S$. Therefore, $v \in \bigcap_{S \in \mathcal{S}} S$. Thus, Span $X \subseteq \bigcap_{S \in \mathcal{S}} S$.

Now assume $w \in \bigcap_{S \in \mathcal{S}} S$. Then $w \in S$ for every $S \in \mathcal{S}$. Notice that

$$\text{Span } X \in \mathcal{S}.$$

Thus, $w \in$ Span X. Therefore Span $X = \bigcap_{S \in \mathcal{S}} S$. $\square$

Theorem 3.2.19 might lead you to consider whether or not every vector space can be written as the span of a set. The answer is "yes." Consider a vector space V. Because V is closed under addition and scalar multiplication, we know that every linear combination of vectors from V are, again, in V. Thus, $V =$ Span V.

Of course this is not very satisfying; a main motivation for considering span to begin with was the desire to have an efficient (that is, *small*) set of vectors that would generate (via linear combinations) a subspace of interest within our vector space.

We will return to the important question of finding efficient spanning sets in upcoming sections. For now, here is a more satisfying example.

Example 3.2.21 Let $V = \{(x, y, z) \in \mathbb{R}^3 \mid x + y + z = 0, 3x + 3y + 3z = 0\}$. We can write V as a span. Notice that V is the solution set of the system of equations

$$\begin{aligned} x + y + z &= 0 \\ 3x + 3y + 3z &= 0 \end{aligned}.$$

We see that after elimination, we get the system

$$\begin{aligned} x + y + z &= 0 \\ 0 &= 0 \end{aligned}.$$

Thus y and z can be chosen to be free variables and we get $(x, y, z)^\mathsf{T} = (-y - z, y, z)^\mathsf{T}$. That is,

---

[4] This notation allows the possibility that there could be an uncountable number of subspaces in $\mathcal{S}$.

$$V = \{(-y - z, y, z)^\mathsf{T} \mid y, z \in \mathbb{R}\}$$
$$= \{y(-1, 1, 0)^\mathsf{T} + z(-1, 0, 1)^\mathsf{T} \mid y, z \in \mathbb{R}\}$$
$$= \mathrm{Span}\{(-1, 1, 0)^\mathsf{T}, (-1, 0, 1)^\mathsf{T}\}.$$

$\square$

This example demonstrated that the parameterized form of a solution set is very useful in finding a spanning set. It also looks forward to the useful idea that any subspace of a vector space can be written as the span of some (usually finite) subset of the vector space. We will develop this idea more fully in the next two sections.

Finally, we introduce the following theorem capturing relationships among subspaces that can be written as the span of some subset.

Theorem 3.2.22

Let X and Y be subsets of a vector space $(V, +, \cdot)$. Then the following statements hold.

(a) $\mathrm{Span}(X \cap Y) \subseteq (\mathrm{Span}\, X) \cap (\mathrm{Span}\, Y)$.
(b) $\mathrm{Span}(X \cup Y) \supseteq (\mathrm{Span}\, X) \cup (\mathrm{Span}\, Y)$.
(c) If $X \subseteq Y$, then $\mathrm{Span}\, X \subseteq \mathrm{Span}\, Y$.
(d) $\mathrm{Span}(X \cup Y) = (\mathrm{Span}\, X) + (\mathrm{Span}\, Y)$.

Proof. We prove (a) and leave the remainder as exercises.

Suppose X and Y are subsets of a vector space $(V, +, \cdot)$. First we consider the case where $X \cap Y = \emptyset$. We have $\mathrm{Span}(X \cap Y) = \{0\}$. Note that $\mathrm{Span}\, X$ and $\mathrm{Span}\, Y$ are subspaces (Theorem 3.2.19), and the intersection of subspaces is also a subspace (Theorem 2.5.18). As every subspace contains the zero vector, $\mathrm{Span}(X \cap Y) \subseteq (\mathrm{Span}\, X) \cap (\mathrm{Span}\, Y)$.

Next, consider the case $X \cap Y \neq \emptyset$. Let u be an arbitrary vector in $X \cap Y$. Then $u \in X, u \in Y$, so $u \in \mathrm{Span}\, X, u \in \mathrm{Span}\, Y$, and thus, $u \in (\mathrm{Span}\, X) \cap (\mathrm{Span}\, Y)$. That is, $\mathrm{Span}(X \cap Y) \subseteq (\mathrm{Span}\, X) \cap (\mathrm{Span}\, Y)$. $\square$

Example 3.2.23 Consider the subsets of $\mathbb{R}^2$: $X = \{(1, 0)\}$ and $Y = \{(0, 1)\}$. We have $X \cap Y = \emptyset$ and $\mathrm{Span}(X \cap Y) = \{0\}$. Also, $\mathrm{Span}\, X$ is the x-axis in $\mathbb{R}^2$ and $\mathrm{Span}\, Y$ is the y-axis in $\mathbb{R}^2$. So, $(\mathrm{Span}\, X) \cap (\mathrm{Span}\, Y) = \{0\}$ and statement (a) in Theorem 3.2.22 holds.

Also, $X \cup Y = \{(1, 0), (0, 1)\}$ and $\mathrm{Span}(X \cup Y) = \mathbb{R}$. But $(\mathrm{Span}\, X) \cup (\mathrm{Span}\, Y)$ is the set of vectors along the x-axis *or* along the y-axis. So statement (b) in Theorem 3.2.22 holds. $\square$

We close this section with a recap of the use of the word *span*.

$\star\star$ **Watch Your Language!** Suppose $X = \{x_1, x_2, \ldots, x_n\}$ is a subset of vector space $(V, +, \cdot)$ such that $\mathrm{Span}\, X = W \subseteq V$. Then the following statements are equivalent in meaning.

✓ X spans the set W.
✓ $x_1, x_2, \ldots, x_n$ span W.
✓ X is a spanning set for W.
✓ W is spanned by X.
✓ The span of $x_1, x_2, \ldots, x_n$ is W.

✓ Span X is a vector space.

It is mathematically (and/or grammatically) incorrect to say the following.

✗ X spans the vectors w_1, w_2, w_3, ...
✗ $x_1, x_2, \ldots, x_n$ spans W.
✗ $x_1, x_2, \ldots, x_n$ is a spanning set for W.
✗ The spanning set of V is a vector space.

3.2.4 Exercises

1. True or False: A set X is always a subset of Span X.
2. True or False: Span X is always a subset of X.
3. In Example 3.2.4, why can we say that

$$\left\{ \begin{pmatrix} -\alpha + \beta \\ 0 \\ \alpha + \gamma \end{pmatrix} \middle| \alpha, \beta, \gamma \in \mathbb{R} \right\} = \left\{ \begin{pmatrix} a \\ 0 \\ b \end{pmatrix} \middle| a, b \in \mathbb{R} \right\} ?$$

4. In Exercise 2 of Section 2.3, you found that some of the sets were subspaces. For each that was a subspace, write it as a span.
5. Use an example to show that the statement $\mathcal{P}_2(\mathbb{R}) \not\subset \mathrm{Span}\{x^2, 1\}$ (in Example 3.2.17) is true.
6. Decide if the given vector lies in the span of the given set. If it does, find a linear combination that makes the vector. If it does not, show that no linear combination exists.

   (a) $\begin{pmatrix} 2 \\ 0 \\ 1 \end{pmatrix}$, $\left\{ \begin{pmatrix} 1 \\ 0 \\ 0 \end{pmatrix}, \begin{pmatrix} 0 \\ 0 \\ 1 \end{pmatrix} \right\}$ in $\mathbb{R}^3$.

   (b) $x - x^3$, $\{x^2, 2x + x^2, x + x^3\}$, in $\mathcal{P}_3(\mathbb{R})$.

   (c) $\begin{pmatrix} 0 & 1 \\ 4 & 2 \end{pmatrix}$, $\left\{ \begin{pmatrix} 1 & 0 \\ 1 & 1 \end{pmatrix}, \begin{pmatrix} 2 & 0 \\ 2 & 3 \end{pmatrix} \right\}$, in $\mathcal{M}_{2 \times 2}(\mathbb{R})$.

7. Determine if the given set spans $\mathbb{R}^3$.

   (a) $\left\{ \begin{pmatrix} 1 \\ 0 \\ 0 \end{pmatrix}, \begin{pmatrix} 0 \\ 2 \\ 0 \end{pmatrix}, \begin{pmatrix} 0 \\ 0 \\ 3 \end{pmatrix} \right\}$

   (b) $\left\{ \begin{pmatrix} 2 \\ 0 \\ 1 \end{pmatrix}, \begin{pmatrix} 1 \\ 1 \\ 0 \end{pmatrix}, \begin{pmatrix} 0 \\ 0 \\ 1 \end{pmatrix} \right\}$

   (c) $\left\{ \begin{pmatrix} 1 \\ 1 \\ 0 \end{pmatrix}, \begin{pmatrix} 3 \\ 0 \\ 0 \end{pmatrix} \right\}$

   (d) $\left\{ \begin{pmatrix} 1 \\ 0 \\ 1 \end{pmatrix}, \begin{pmatrix} 3 \\ 1 \\ 0 \end{pmatrix}, \begin{pmatrix} -1 \\ 0 \\ 0 \end{pmatrix}, \begin{pmatrix} 2 \\ 1 \\ 5 \end{pmatrix} \right\}$

   (e) $\left\{ \begin{pmatrix} 2 \\ 1 \\ 1 \end{pmatrix}, \begin{pmatrix} 3 \\ 0 \\ 1 \end{pmatrix}, \begin{pmatrix} 5 \\ 1 \\ 2 \end{pmatrix}, \begin{pmatrix} 6 \\ 0 \\ 2 \end{pmatrix} \right\}$

8. Is it possible to write the span given in Example 3.2.4 as a span of other vectors? If so, give another example. If not, justify.

9. Find a spanning set for the given subspace.

   (a) The xz-plane in $\mathbb{R}^3$.

   (b) $\left\{ \begin{pmatrix} x \\ y \\ z \end{pmatrix} \middle| \; 3x + 2y + z = 0 \right\}$ in $\mathbb{R}^3$.

   (c) $\left\{ \begin{pmatrix} x \\ y \\ z \\ w \end{pmatrix} \middle| \; 2x + y + w = 0 \text{ and } y + 2z = 0 \right\}$ in $\mathbb{R}^4$.

   (d) $\{a_0 + a_1 x + a_2 x^2 + a_3 x^3 \mid a_0 + a_1 = 0 \text{ and } a_2 - a_3 = 0\}$ in $\mathcal{P}_3(\mathbb{R})$

   (e) The set $\mathcal{P}_4(\mathbb{R})$ in the space $\mathcal{P}_4(\mathbb{R})$.

   (f) $\mathcal{M}_{2 \times 2}(\mathbb{R})$ in $\mathcal{M}_{2 \times 2}(\mathbb{R})$.

10. Briefly explain why the incorrect statements in the "Watch Your Language!" box on page 121 are indeed incorrect.

11. Let u, v be vectors in V.

    (a) Show that $\text{Span}\{u\} = \{au \in V \mid a \in \mathbb{R}\}$.

    (b) Prove that $\text{Span}\{u, v\} = \text{Span}\{u\}$ if and only if $v = au$ for some scalar a.

12. Determine whether or not $\{1, x, x^2\}$ is a spanning set for $\mathcal{P}_1(\mathbb{R})$. Justify your answer using the definitions in this section.

13. Complete the proof of Theorem 3.2.22.

14. Show, with justification, that $\text{Span}(X \cup Y) = (\text{Span } X) \cup (\text{Span } Y)$ (see Theorem 3.2.22) is, in general, false.

15. Show that $(\text{Span } X) \cup (\text{Span } Y)$ is not necessarily a subspace.

16. Show, with justification, that $\text{Span}(X \cap Y) = (\text{Span } X) \cap (\text{Span } Y)$ (see Theorem 3.2.22) is, in general, false.

17. Suppose X is a subset of vector space $(V, +, \cdot)$. Let X^C denote the complement set of X (that is, the set of vectors that are not in X). Compose and prove relationships between $\text{Span } X$ and $\text{Span } X^C$ in the spirit of Theorem 3.2.22.

18. In Exercise 9 Section 3.1 you found that there were infinitely many solutions to the system. Is any equation in the span of the other two?

19. In Exercise 12 Section 3.1 is any equation in the span of the other two?

20. Use Exercises 18 and 19 above to make a similar statement about the rows of the coefficient matrix corresponding to a system of equations.

21. Show (using the allowed operations) that any equation, formed in the elimination process for a system of equations, is in the span of the original equations.

22. Find two different spanning sets (having different number of elements than each other) for each of the following vector spaces.

    (a) $\mathcal{P}_2(\mathbb{R})$.

    (b) $\mathcal{M}_{2 \times 2}(\mathbb{R})$.

    Opinion: Which spanning set in each of the above is likely to be more useful?

23. Consider the space of 7-bar LCD images, $\mathcal{D}(Z_2)$ as defined in Example 2.4.17. Let D be the set of digits of $\mathcal{D}(Z_2)$.

    (a) Sketch the zero vector.

(b) Find the additive inverse element of each of the vectors in D.

(c) How many vectors are in Span$\{d_1\}$?

(d) How many vectors are in Span$\{d_2, d_3\}$?

24. Consider, again, the space of 7-bar LCD images, $\mathcal{D}(Z_2)$ as defined in Example 2.4.17. Let D be the set of digits of $\mathcal{D}(Z_2)$.

    (a) Let d_k be defined as in Exercise 21. Sketch every element of Span$\{d_0, d_9\}$.

    (b) Find one element of D which is in the span of the other elements of D.

    (c) Discuss how you would go about showing that the set of digit images D is a spanning set for $\mathcal{D}(Z_2)$.

25. Consider the 4×4 images of Example 3.1.2. Which of Images 1, 2, 3, and 4 are in the span of Images A, B, and C?

26. Consider the finite vector space V, defined in Section 2.4 Exercise 21, defined by

$$V = \left\{ a = \text{⊖} \; , \; b = \text{⊖} \; , \; c = \text{⊖} \; , \; d = \text{⊖} \right\}.$$

Define $\oplus$, vector addition, and $\odot$, scalar multiplication, according to the following tables:

$\oplus$	a	b	c	d
a	a	b	c	d
b	b	a	d	c
c	c	d	a	b
d	d	c	b	a

$\odot$	a	b	c	d
0	a	a	a	a
1	a	b	c	d

    (a) For each vector space property, state a feature (or features) in the tables above that tells you the property holds.

    (b) Is any one of the vectors of V in the span of the others? Justify.

27. Let H be the set of heat states sampled in 4 places along the rod ($m = 4$). Find a spanning set for H.

28. How many different brain images, $u_1, u_2, \cdots, u_k$, do you think might be needed so that Span$\{u_1, u_2, \cdots, u_k\}$ includes all possible brain images of interest to a physician?

3.3   Linear Dependence and Independence

The concepts of span and spanning sets are powerful tools for describing subspaces. We have seen that even a few vectors may contain all the relevant information for describing an infinite collection of vectors.

Recall, we proved in Section 3.2 that the span of a set of vectors is a vector space. This leads one to consider the question, "Can every vector space be expressed as Span X for some set of vectors X?" The quick answer is "Yes." Indeed, if V is a vector space, $V = $ Span V. But, that is not an impressive answer. As stated above, we have seen that significantly fewer vectors can be used to describe the vector space. So, a better question is "Can every vector space be expressed as a span of a set containing significantly fewer vectors?" The intention is that if we have a vector space V, we hope to find X so that $V = $ Span X is a much simpler way of expressing V.

Consider the problem in which an object is described by intensity values on a 4×4 rectangular grid. That is, objects can be expressed as images in $\mathcal{I}_{4\times4}$. Suppose that an image of interest must be a linear combination of the seven images introduced on page 11. That is, we must choose a solution in the set

$$W = \text{Span} \left\{ \quad , \quad , \quad , \quad , \quad , \quad , \quad \right\}.$$

Image A   Image B   Image C   Image 1   Image 2   Image 3   Image 4

The natural question arises: Is this the simplest description of our set of interest? We discovered that Images 1, 2, and 3 could all be written as linear combinations of Images A, B, and C. So, Images 1, 2, and 3, in some sense, do not add any additional information to the set. In fact, we now understand that the exact same set can be described as

$$W = \text{Span} \left\{ \quad , \quad , \quad , \quad \right\}.$$

Image A   Image B   Image C   Image 4

Is it possible to reduce the description of W further? This is, in general, the key question that we address in this section; the answer in this specific situation is Problem 2 in the exercises.

Consider also the example of 7-bar LCD characters. We know that the set of 10 digit-images is a spanning set for the vector space $\mathcal{D}(Z_2)$ (see Exercise 23 of Section 3.2). That is,

$$\mathcal{D}(Z_2) = \text{Span} \left\{ \quad , \quad , \quad , \quad , \quad , \quad , \quad , \quad , \quad , \quad \right\}.$$

In other words, any possible character can be written as a linear combination of these ten characters. The question arises: Is this the smallest possible set of characters for which this is true? Can we describe $\mathcal{D}(Z_2)$ with a smaller set? For example, is it true that

$$\mathcal{D}(Z_2) = \text{Span} \left\{ \quad , \quad , \quad , \quad , \quad \right\} ?$$

If not, does a smaller spanning set exist or is the 10 digit-images set the smallest possible spanning set? What is the minimum number of vectors necessary to form a spanning set of a vector space? These are important questions that we are now poised to explore and answer.

3.3.1   Linear Dependence and Independence

We have seen that a key component to understanding these questions is whether or not some vectors can be written as linear combinations of others. If so, then our spanning sets are "too large" in the sense that they contain redundant information. Sets that are too large are said to be linearly dependent. Sets that are not too large are said to be linearly independent.

Definition 3.3.1

Let $(V, +, \cdot)$ be a vector space over $\mathbb{F}$ and $W \subseteq V$. We say that W is **linearly dependent** if some vector in W is in the span of the remaining vectors of W. Any set that is not linearly dependent is said to be **linearly independent**.   □

In the preceding definition, we often consider finite subsets W of the vector space V, so $W = \{v_1, v_2, \cdots, v_n\}$. Then W is linearly dependent if for some k between 1 and n, $v_k \in \text{Span} \{v_1, v_2, \cdots, v_{k-1}, v_{k+1}, \cdots, v_n\}$.

Putting the definition in the context of the 4×4 image setting, we see the following:

Example 3.3.2 Consider the 4×4 image example from the beginning of the section. We can say that the set of seven images is linearly dependent because, for example,

We know this is true because Image 2 can be written as a linear combination of the other images:

□

In this case, we do not lose any vectors from the span if we remove Image 2: the remaining 6 images span the same subspace of images. We could then test the remaining vectors (images) to see if this set is still linearly dependent, if it is, we can remove one of the remaining 6 vectors and still span the same subspace of images. Once we end up with a linearly independent set we have an efficient spanning set: removing any additional vector will change the spanned subspace.

If the set $W = \{v_1, v_2, \cdots, v_n\}$ is linearly dependent, the choice of vectors that could be removed is, in general, not unique. (See Exercise 7.) (The exception is the zero vector.)

Example 3.3.3 Suppose $W = \left\{ \begin{pmatrix} 4 \\ 5 \end{pmatrix}, \begin{pmatrix} 1 \\ 2 \end{pmatrix}, \begin{pmatrix} 0 \\ 1 \end{pmatrix}, \begin{pmatrix} 2 \\ 3 \end{pmatrix} \right\} \subseteq \mathbb{R}^2$. W is linearly dependent because $\begin{pmatrix} 4 \\ 5 \end{pmatrix} = 4 \begin{pmatrix} 1 \\ 2 \end{pmatrix} - 3 \begin{pmatrix} 0 \\ 1 \end{pmatrix}$. That is, the first vector in W is a linear combination of the remaining vectors in W. We have

$$\text{Span}\left\{\begin{pmatrix}4\\5\end{pmatrix}, \begin{pmatrix}1\\2\end{pmatrix}, \begin{pmatrix}0\\1\end{pmatrix}, \begin{pmatrix}2\\3\end{pmatrix}\right\} = \text{Span}\left\{\begin{pmatrix}1\\2\end{pmatrix}, \begin{pmatrix}0\\1\end{pmatrix}, \begin{pmatrix}2\\3\end{pmatrix}\right\}.$$

$\square$

Example 3.3.4 Suppose $W = \left\{x, x^2 - x, x^2 + x\right\} \subseteq \mathcal{P}_2(\mathbb{R})$. W is linearly dependent because $(x^2 + x) = (x^2 - x) + 2(x)$. We have

$$\text{Span}\left\{x, x^2 - x, x^2 + x\right\} = \text{Span}\left\{x^2 - x, x\right\}.$$

$\square$

Example 3.3.5 Suppose $W = \left\{\begin{pmatrix}2\\0\end{pmatrix}, \begin{pmatrix}0\\1\end{pmatrix}\right\} \subseteq \mathbb{R}^2$. Neither vector in W can be written as a linear combination of the other. That is $\begin{pmatrix}2\\0\end{pmatrix} \notin \text{Span}\left\{\begin{pmatrix}0\\1\end{pmatrix}\right\}$ and vice versa. Since W is not linearly dependent, W is linearly independent.

$\square$

Let's also revisit our LCD example.

Example 3.3.6 Consider the vector space $\mathcal{D}(Z_2)$. We can say that the set of ten LCD character images is linearly dependent because, for example,

We know that this is true because image d_9 can be written as a linear combination of d_5, d_6, and d_8:

$\square$

Example 3.3.7 In any vector space, any set containing the zero vector is linearly dependent. Consider two cases. First suppose $W = \{0, v_1, v_2, \cdots, v_n\}$. Clearly, 0 is in $\text{Span}\{v_1, v_2, \cdots, v_n\}$ since 0 is the trivial linear combination of vectors given by

$$0 = 0 \cdot v_1 + 0 \cdot v_2 + \ldots + 0 \cdot v_n.$$

Next, suppose $W = \{0\}$. Since, by definition $\text{Span}\,\emptyset = \{0\}$, we know that $\{0\}$ is linearly dependent. In both cases, we conclude that W is linearly dependent.

$\square$

3.3.2 Determining Linear (In)dependence

Determining whether a set is linearly dependent or independent might seem like a tedious process in which we must test enumerable linear combinations in hopes of finding one that indicates linear dependence. Our examples thus far have been with very small sets of a few vectors. Fortunately, we can develop a general test for general sets with a finite number of vectors.

Let $W = \{v_1, v_2, \cdots, v_n\}$ be a subset of a vector space $(V, +, \cdot)$. Suppose, for the sake of argument, that $v_1 = a_2 v_2 + a_3 v_3 + \cdots + a_n v_n$ for some scalars. Then $0 = -v_1 + a_2 v_2 + a_3 v_3 + \cdots + a_n v_n$. And, multiplying by some arbitrary nonzero scalar yields $0 = \alpha_1 v_1 + \alpha_2 v_2 + \cdots + \alpha_n v_n$, where α_1 is guaranteed to be nonzero. Notice that testing each vector in W leads to this same equation. So, one test for linear dependence is the existence of some nonzero scalars which make the equation true.

Theorem 3.3.8

Let $W = \{v_1, v_2, \cdots, v_n\}$ be a subset of vector space $(V, +, \cdot)$. W is linearly dependent if and only if there exist scalars $\alpha_1, \alpha_2, \cdots, \alpha_n$, not all zero, such that

$$\alpha_1 v_1 + \alpha_2 v_2 + \cdots + \alpha_n v_n = 0. \tag{3.9}$$

Proof. ($\Rightarrow$) Suppose W is linearly dependent. Then some vector in W can be written as a linear combination of the other vectors in W. Without loss of generality[5], suppose $v_1 = a_2 v_2 + a_3 v_3 + \cdots + a_n v_n$. We have $0 = (-1)v_1 + a_2 v_2 + a_3 v_3 + \cdots + a_n v_n$ so that the linear dependence relation holds for scalars not all zero.

($\Leftarrow$) Suppose $\alpha_1 v_1 + \alpha_2 v_2 + \cdots + \alpha_n v_n = 0$ for scalars not all zero. Without loss of generality, suppose $\alpha_1 \neq 0$. Then, $v_1 = (-\alpha_2/\alpha_1)v_2 + \cdots + (-\alpha_n/\alpha_1)v_n$. Since v_1 is a linear combination of the vectors $v_2, v_3, \cdots, v_n$, W is linearly dependent. $\square$

Definition 3.3.9

Equation 3.9 is called the **linear dependence relation**. $\square$

Corollary 3.3.10

Let $X = \{v_1, v_2, \cdots, v_n\}$ be a subset of the vector space $(V, +, \cdot)$. X is linearly independent if and only if the linear dependence relation $\alpha_1 v_1 + \alpha_2 v_2 + \cdots + \alpha_n v_n = 0$ has only the trivial solution $\alpha_1 = \alpha_2 = \cdots = \alpha_n = 0$.

---

[5] We say "without loss of generality" here because although it seems like we singled out the specific vector v_1. If in fact it was a different vector that could be written as a linear combination of the others we could change the labeling so that the new v_1 is the vector that is written as a linear combination. This technique is useful because it allows us to simplify notation. We'll use it frequently throughout the text, so even if it seems a little mysterious at first, you'll see plenty of examples in the coming sections.

The linear dependence relation is *always* true if $\alpha_1 = \alpha_2 = \ldots = \alpha_n = 0$, but this tells us nothing about the linear dependence of the set. To determine linear independence, one must determine whether or not the linear dependence relation is true *only* when all the scalars are zero.

> **Lemma 3.3.11**
> Let $(V, +, \cdot)$ be a vector space. The set $\{v_1, v_2\} \subset V$ is linearly independent if and only if v_1 and v_2 are nonzero and v_1 is not a scalar multiple of v_2.

Proof. Let V be a vector space.
($\Rightarrow$) First, we will show that if if $\{v_1, v_2\}$ is linearly independent then $v_1 \neq k v_2$ for any scalar k. Suppose, to the contrary, that $\{v_1, v_2\}$ is linearly independent and that $v_1 = k v_2$ for some scalar k. Then $v_1 - k v_2 = 0$. So, by the definition of linear dependence, since the coefficient of v_1 is nonzero, $\{v_1, v_2\}$ is linearly dependent, contradicting our original assumption about linear dependence. $\rightarrow \leftarrow$ Therefore, if $\{v_1, v_2\}$ is linearly independent, v_1 is not a scalar multiple of v_2.
($\Leftarrow$) Now, we show that if v_1 is not a scalar multiple of v_2 then $\{v_1, v_2\}$ is linearly independent. We show this by proving the contrapositive instead. Assume that $\{v_1, v_2\}$ is linearly dependent. Then there are scalars α_1, α_2 not both zero so that $\alpha_1 v_1 + \alpha_2 v_2 = 0$. Suppose, without loss of generality, that $\alpha_1 \neq 0$. Then $v_1 = \frac{\alpha_2}{\alpha_1} v_2$. Thus, v_1 is a scalar multiple of v_2. Therefore, if v_1 is not a scalar multiple of v_2 then $\{v_1, v_2\}$ is linearly independent. $\qquad\square$

Example 3.3.12 Let us determine whether $\{x + 1, x^2 + 1, x^2 + x + 1\} \subseteq \mathcal{P}_2(\mathbb{R})$ is linearly dependent or independent. We start by setting up the linear dependence relation. We let

$$\alpha(x + 1) + \beta(x^2 + 1) + \gamma(x^2 + x + 1) = 0.$$

Now, we want to decide whether or not α, β, and γ must all be zero. Matching up like terms in the linear dependence relation leads to the system of equations

$$
\begin{aligned}
(x^2 \text{ term:}) \quad & 0 = \quad\ \beta + \gamma \\
(x \text{ term:}) \quad & 0 = \alpha \quad\ + \gamma. \\
(\text{constant term:}) \quad & 0 = \alpha + \beta + \gamma
\end{aligned}
$$

Using elimination, we get

$$
\begin{array}{ll}
0 = \quad\ \beta + \gamma & \quad 0 = \quad\ \beta + \gamma \\
0 = \alpha \quad\ + \gamma & \quad 0 = \alpha \quad\ + \gamma. \\
0 = \alpha + \beta + \gamma & \quad 0 = \alpha
\end{array}
\qquad \xrightarrow{E3 = -e_1 + e_3}
$$

The only solution to this system of equations is $\alpha = 0$, $\beta = 0$, $\gamma = 0$. This means that $\{x + 1, x^2 + 1, x^2 + x + 1\}$ is linearly independent. $\qquad\square$

Example 3.3.13 Now, let us determine the linear dependence of the set

$$\left\{ \begin{pmatrix} 1 & 3 \\ 1 & 1 \end{pmatrix}, \begin{pmatrix} 1 & 1 \\ 1 & -1 \end{pmatrix}, \begin{pmatrix} 1 & 2 \\ 1 & 0 \end{pmatrix} \right\}.$$

Again, we begin by setting up the linear dependence relation. Let

$$\begin{pmatrix} 0 & 0 \\ 0 & 0 \end{pmatrix} = \alpha \begin{pmatrix} 1 & 3 \\ 1 & 1 \end{pmatrix} + \beta \begin{pmatrix} 1 & 1 \\ 1 & -1 \end{pmatrix} + \gamma \begin{pmatrix} 1 & 2 \\ 1 & 0 \end{pmatrix}.$$

We want to find α, β, and γ so that this is true. Matching up entries, we get the following system of equations.

$$\begin{aligned} ((1,1) \text{ entry:}) \ 0 &= \ \alpha + \beta + \ \gamma \\ ((1,2) \text{ entry:}) \ 0 &= 3\alpha + \beta + 2\gamma \\ ((2,1) \text{ entry:}) \ 0 &= \ \alpha + \beta + \ \gamma \\ ((2,2) \text{ entry:}) \ 0 &= \ \alpha - \beta \end{aligned}$$

We again, use elimination, but this time, let us use a coefficient matrix and reduce it.

$$\begin{pmatrix} 1 & 1 & 1 & | & 0 \\ 3 & 1 & 2 & | & 0 \\ 1 & 1 & 1 & | & 0 \\ 1 & -1 & 0 & | & 0 \end{pmatrix} \xrightarrow[R_3=-r_1+r_3, R_4=-r_1+r_4]{R_2=-3r_1+r_2} \begin{pmatrix} 1 & 1 & 1 & | & 0 \\ 0 & -2 & -1 & | & 0 \\ 0 & 0 & 0 & | & 0 \\ 0 & -2 & -1 & | & 0 \end{pmatrix}$$

$$\xrightarrow[R_4=-r_2+r_4]{R_2=-\frac{1}{2}r_2} \begin{pmatrix} 1 & 1 & 1 & | & 0 \\ 0 & 1 & \frac{1}{2} & | & 0 \\ 0 & 0 & 0 & | & 0 \\ 0 & 0 & 0 & | & 0 \end{pmatrix} \xrightarrow{R_1=-r_2+r_1} \begin{pmatrix} 1 & 0 & \frac{1}{2} & | & 0 \\ 0 & 1 & \frac{1}{2} & | & 0 \\ 0 & 0 & 0 & | & 0 \\ 0 & 0 & 0 & | & 0 \end{pmatrix}$$

Thus, γ can be any real number and $\alpha = -\frac{1}{2}\gamma$ and $\beta = -\frac{1}{2}\gamma$. Thus there are infinitely many possible choices for α, β, and γ. Thus,

$$\left\{ \begin{pmatrix} 1 & 3 \\ 1 & 1 \end{pmatrix}, \begin{pmatrix} 1 & 1 \\ 1 & -1 \end{pmatrix}, \begin{pmatrix} 1 & 2 \\ 1 & 0 \end{pmatrix} \right\}$$

is a linearly dependent set. Indeed, we found that (choosing $\gamma = 2$)

$$\begin{pmatrix} 0 & 0 \\ 0 & 0 \end{pmatrix} = - \begin{pmatrix} 1 & 3 \\ 1 & 1 \end{pmatrix} - \begin{pmatrix} 1 & 1 \\ 1 & -1 \end{pmatrix} + 2 \begin{pmatrix} 1 & 2 \\ 1 & 0 \end{pmatrix}.$$

$\square$

The linear (in)dependence of the set of vectors given by the columns of an $n \times n$ matrix M helps to determine the nature of the solutions to the matrix equation $Mx = b$.

Theorem 3.3.14

Let M be an $n \times n$ matrix with columns $v_1, v_2, \ldots, v_n \in \mathbb{R}^n$. The matrix equation $Mx = b$ has a unique solution $x \in \mathbb{R}^n$ for every $b \in \mathbb{R}^n$ if, and only if, $\{v_1, v_2, \ldots, v_n\}$ is a linearly independent set.

Proof. ($\Rightarrow$) Suppose M is an $n \times n$ matrix with columns $v_1, v_2, \ldots, v_n$, and the matrix equation $Mx = b$ has a unique solution $x \in \mathbb{R}^n$ for every $b \in \mathbb{R}^n$. That is, b is uniquely written as

$b = Mw = w_1 v_1 + w_2 v_2 + \ldots + w_n v_n$. Now, suppose, by way of contradiction, that $\{v_1, v_2, \ldots, v_n\}$ is linearly dependent. Then there exist scalars $y_1, y_2, \ldots, y_n$, not all zero, such that $y_1 v_1 + y_2 v_2 + \ldots + y_n v_n = 0$. Then $w + y \neq w$ also solves the matrix equation $Mx = b$, a contradiction. Thus, $\{v_1, v_2, \ldots, v_n\}$ is linearly independent.

($\Leftarrow$) Suppose M is an $n \times n$ matrix with columns $v_1, v_2, \ldots, v_n$ that form a linearly independent set. Suppose, by way of contradiction, that there exists $b \in \mathbb{R}^n$ such that $Mx = b$ and $My = b$ for some $x \neq y$. Then $M(x - y) = 0$, or equivalently, $(x_1 - y_1)v_1 + \ldots + (x_n - y_n)v_n = 0$. Since $x \neq y$, $x_k - y_k \neq 0$ for some k, then $v_1, v_2, \ldots, v_n$ is linearly dependent, a contradiction. Thus, $Mx = b$ has a unique solution $x \in \mathbb{R}^n$ for every $b \in \mathbb{R}^n$. $\qquad\square$

Linear dependence of a set can also be determined by considering spans of proper subsets of the set.

> **Theorem 3.3.15**
> Let X be a subset of the vector space $(V, +, \cdot)$. X is linearly dependent if and only if there exists a proper subset U of X such that Span $U =$ Span X.

The proof is the subject of Exercise 15.

Example 3.3.16 Consider $X = \left\{ \begin{pmatrix} 1 \\ 0 \end{pmatrix}, \begin{pmatrix} 1 \\ 3 \end{pmatrix}, \begin{pmatrix} 0 \\ 1 \end{pmatrix}, \begin{pmatrix} 2 \\ 3 \end{pmatrix} \right\} \subseteq \mathbb{R}^2$. We notice that Span $X = \mathbb{R}^2$ and $U = \left\{ \begin{pmatrix} 1 \\ 0 \end{pmatrix}, \begin{pmatrix} 0 \\ 1 \end{pmatrix} \right\} \subset X$ with Span $U = \mathbb{R}^2$. Thus, X is linearly dependent. $\qquad\square$

Example 3.3.17 Consider

We have seen that

is such that Span $X =$ Span Y. Thus X is linearly dependent. $\qquad\square$

Building off of Theorem 3.3.15, it seems that if X is a linearly independent set and $X \subset Y$, but Span $Y \neq$ Span X, then there must be a vector v in Y so that $X \cup \{v\}$ is also linearly independent. Indeed, the next theorem gives a more precise result.

Theorem 3.3.18

Let X be a linearly independent subset of the vector space $(V, +, \cdot)$ and let $v \in V$ be a vector so that $v \notin$ Span X. Then $X \cup \{v\}$ is linearly independent.

Proof. Let $(V, +, \cdot)$ be a vector space, $X = \{x_1, x_2, \ldots, x_n\} \subset V$ be a linearly independent set of vectors, and let $v \in V$ be a vector not in the span of X. Now, suppose there are scalars $\alpha_1, \alpha_2, \ldots, \alpha_{n+1}$ so that

$$\alpha_1 x_1 + \alpha_2 x_2 + \ldots + \alpha_n x_n + \alpha_{n+1} v = 0.$$

Notice that $\alpha_{n+1} = 0$ for otherwise

$$v = -\frac{\alpha_1}{\alpha_{n+1}} x_1 - \frac{\alpha_2}{\alpha_{n+1}} x_2 - \ldots - \frac{\alpha_n}{\alpha_{n+1}} x_n \in \text{Span } X.$$

Therefore,

$$\alpha_1 x_1 + \alpha_2 x_2 + \ldots + \alpha_n x_n = 0.$$

Since X is linearly independent, $\alpha_1 = \alpha_2 = \ldots = \alpha_n = 0$. Therefore, $\{x_1, x_2, \ldots, x_n, v\}$ is a linearly independent set. $\square$

Example 3.3.19 Notice that $X = \{x^2 + 2, 2\} \subset \mathcal{P}_2$ is a linearly independent set because the vectors are not multiples of one another. Also, we can easily see that $x^2 + 3x - 2 \notin$ Span X. So, we have found an even larger linearly independent set, $X \cup \{x^2 + 3x - 2\} = \{x^2 + 2, 2, x^2 + 3x - 2\} \subset \mathcal{P}_2$. $\square$

3.3.3 Summary of Linear Dependence

As we have seen with all other concepts in this text, there is appropriate and inappropriate language surrounding the idea of linear (in)dependence. Below, we share the appropriate way to discuss the linear dependence of vectors and sets of vectors.

⋆⋆ **Watch Your Language!** Linear dependence is a property of a set of vectors, not a property of a vector. For a linearly dependent set $W = \{v_1, v_2, \cdots, v_n\}$, we have the following grammatically and mathematically correct statements:

✓ W is linearly (in)dependent.
✓ W is a linearly (in)dependent set.
✓ $\{v_1, v_2, \cdots, v_n\}$ is linearly (in)dependent.
✓ The vectors $v_1, v_2, \cdots, v_n$ are linearly (in)dependent.
✓ The vectors $v_1, v_2, \cdots, v_n$ form a linearly (in)dependent set.
✓ The columns (or rows) of a matrix, M, form a linearly (in)dependent set.

But it would be incorrect to say

✗ W has linearly (in)dependent vectors.
✗ $\{v_1, v_2, \cdots, v_n\}$ are linearly (in)dependent .
✗ The matrix, M, is linearly (in)dependent.

In general, to show that a set is linearly dependent, you need to exhibit only one linear dependency (i.e., you need only write one vector as a linear combination of the others, or equivalently, find one set of nonzero constants for which the linear dependence relation holds).

To show a set is linearly independent, on the other hand, requires that you show that no nontrivial solutions to the linear dependence relation exist; that is, you must show that it is not possible to write any of the vectors in the set as a linear combination of the others.

We summarize a few of the observations from the examples in this section that can help with the process of categorizing whether a set in an arbitrary vector space is linearly dependent or independent.

▶ **Determining if a set W is linearly dependent or linearly independent**

1. If W contains the zero vector, then W is linearly dependent.
2. If W contains a single vector and it is not zero, then W is linearly independent.
3. The empty set is linearly independent. (See Exercise 16.)
4. If W contains two vectors, and neither is zero, then W is linearly dependent if and only if one vector is a scalar multiple of the other.
5. If any element of W is a scalar multiple of another element of W, then W is linearly dependent.
6. If any element of W is a linear combination of other elements of W, then W is linearly dependent.
7. If $W = \{w_1, \ldots w_k\} \subset \mathbb{R}^n$, then we can row reduce the augmented matrix $[w_1 \ldots w_k \,|\, 0]$. If the corresponding system has only the trivial solution, then W is linearly independent. On the other hand, if the system has infinitely many solutions then W is linearly dependent. Moreover, the process of row reducing a matrix does not change linear dependencies between columns, so one can often easily identify linear dependencies from the row reduced matrix.
8. If $W = \{v_1, v_2, \cdots, v_n\}$ then W is linearly independent if

$$\alpha_1 v_1 + \alpha_2 v_2 + \cdots + \alpha_n v_n = 0$$

has *only* the trivial solution

$$\alpha_1 = \alpha_2 = \cdots = \alpha_n = 0.$$

3.3.4 Exercises

1. Use the linear dependence relation to determine whether the given set is linearly independent or linearly dependent.

(a) $\left\{ \begin{pmatrix} 1 \\ 1 \\ -1 \end{pmatrix}, \begin{pmatrix} 0 \\ -1 \\ -1 \end{pmatrix}, \begin{pmatrix} -1 \\ 2 \\ 1 \end{pmatrix}, \begin{pmatrix} 0 \\ 0 \\ 1 \end{pmatrix} \right\} \subseteq \mathbb{R}^3$

(b) $\{1, x, x^2\} \subseteq \mathcal{P}_2(\mathbb{R})$.
(c) $\{1, x + x^2, x^2\} \subseteq \mathcal{P}_2(\mathbb{R})$
(d) $\{1, 1 - x, 1 + x, 1 + x^2\} \subseteq \mathcal{P}_2(\mathbb{R})$.
(e) $\{1 + x, 1 - x, x\} \subseteq \mathcal{P}_2(\mathbb{R})$.

(f) $\left\{ \begin{pmatrix} 1 & 1 \\ 1 & 0 \end{pmatrix}, \begin{pmatrix} 0 & 1 \\ -1 & -1 \end{pmatrix}, \begin{pmatrix} -1 & 2 \\ 1 & 0 \end{pmatrix} \right\} \subseteq \mathcal{M}_{2\times2}(\mathbb{R})$

(g) $\{\sin x, \sin 2x, \sin 3x\} \subseteq \mathcal{F}(\mathbb{R})$, where $\mathcal{F}(\mathbb{R})$ is the vector space of functions on $[0, \pi]$.

(h)

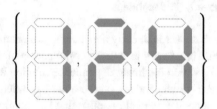

$\subseteq \mathcal{D}(Z_2)$ (the vector space of 7-bar LCD images).

2. Given the vector space W given by

$$W = \text{Span}$$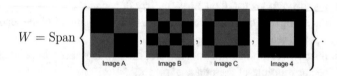

Can you write W as a span of fewer vectors? Justify.

3. Let $(V, +, \cdot)$ be a vector space with scalar field $\mathbb{R}$. Suppose the set of vectors $\{v_1, v_2, v_3, v_4\}$ is linearly independent. Determine if the following sets are linearly independent. Justify your answer. If not, remove as few vectors as you can so the resulting set is linearly independent.

   (a) $\{v_1, v_2\}$.
   (b) $\{v_1, v_2, v_3, v_4, v_1 - 2v_3\}$.
   (c) $\{v_1 + v_3, v_2 + v_4, v_3, v_4\}$.
   (d) $\{v_1 - 2v_2, v_2, v_3 - v_4 - v_2, v_4\}$.

4. True or False: Suppose $\{v_1, \ldots, v_k\}$ is a linearly dependent set of vectors in a vector space V, so that $v_1 \in \text{Span}\{v_2, \ldots, v_k\}$. Then there must exist $i \in \{2, \ldots, k\}$ so that $v_i \in \text{Span}\{v_1, \ldots, v_{i-1}, v_{i+1}, v_k\}$.

5. Given the linearly independent set of vectors $S = \{u, v, w\}$, show that the set of vectors $T = \{u + 2v, u - w, v + w\}$ is linearly independent and that $\text{Span}\, S = \text{Span}\, T$.

6. Given a linearly independent set S, use Exercise 5 to make a general statement about how to obtain a different linearly independent set of vectors T with $\text{Span}\, S = \text{Span}\, T$. Be careful to use accurate linear algebra language.

7. Show that $X = \{x^2 + 2, x - 2, 3x^2, 2x - 1, 1\}$ is a linearly dependent set. Now, find two different linearly independent subsets of X containing only three vectors.

8. Does the vector space $\{0\}$ have a linearly independent subset?

9. Find a set of four distinct vectors in $\mathbb{R}^3$ that is linearly dependent and for which no subset of three vectors is linearly independent.

10. Given a homogeneous system of three linear equations with three variables, show that the system has only the trivial solution whenever the corresponding coefficient matrix has linearly independent rows.

11. Consider the given heat states in $H_4(\mathbb{R})$ (Figure 3.7 in Section 3.4). Find a linearly independent set of four heat states.

12. Let $\mathcal{I}_4(\mathbb{R})$ be the vector space of 4×4 grayscale images. Show that Span $Y \neq \mathcal{I}_4(\mathbb{R})$ in Example 3.3.17.

13. Find four elements of D which form a linearly independent set in $\mathcal{D}(Z_2)$. See Example 2.4.17.

14. Determine whether $S = \{I_1, I_2, I_3\}$, where the I_n are given below, is a linearly independent set in the vector space of images with the given geometry.

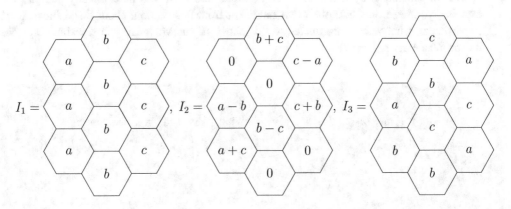

15. Prove Theorem 3.3.15.

16. Use Theorem 3.3.15 to determine if the given set is linearly dependent or linearly independent.

   (a) $W = \emptyset$.
   (b) $W = \{0\}$.
   (c) $W = \{1 + x, 1 - x, x\} \subseteq \mathcal{P}_2(\mathbb{R})$.
   (d)

$$W = \left\{ \text{[image]}, \text{[image]}, \text{[image]} \right\}.$$

17. Consider the 4×4 images of Example 3.3.17. Determine a linearly dependent set distinct from Y.

18. Consider the finite vector space, given in Section 2.4 Exercise 21, by

$$V = \left\{ a = \text{[image]}, \ b = \text{[image]}, \ c = \text{[image]}, \ d = \text{[image]} \right\}.$$

Define $\oplus$, vector addition, and $\odot$, scalar multiplication, according to the following tables:

$\oplus$	a	b	c	d
a	a	b	c	d
b	b	a	d	c
c	c	d	a	b
d	d	c	b	a

$\odot$	a	b	c	d
0	a	a	a	a
1	a	b	c	d

(a) List, if possible, one subset of V with two or more elements that is linearly independent. Justify.
(b) List, if possible, a set with three or fewer elements that is linearly dependent. Justify.

19. In Section 3.1, Exercise 28, we began combining equivalent statements. Continue combining equivalent statements by completing the theorem below. (If you have already completed that exercise, you need only add one more statement from this section, about linear independence. In particular, what vectors are related to the matrix A, and what can you say about their linear independence/dependence?)

Theorem 3.3.20
Let A be an $n \times n$ matrix and $b \in \mathbb{R}^n$. The following statements are equivalent.

- $Ax = 0$ has a unique solution.
-
-
-
-
-

Prove Theorem 3.3.20.

3.4   Basis and Dimension

We have now seen many examples of vector spaces and subspaces that can be described as the span of some smaller set of vectors. We have also seen that there is a lot of freedom in the choice of a spanning set. Moreover, two sets of very different sizes can span the same vector space. This suggests that larger sets contain redundant information in the form of "extra" vectors.

If a spanning set does contain "extra" vectors, the set is linearly dependent. We seek an efficient description of a vector space in terms of some spanning set. We see that a most efficient set must have two properties. First, it should be able to generate the vector space of interest (be a spanning set). Second, it should contain no unnecessary vectors (be linearly independent). Any set with these two properties is a maximally efficient spanning set for the vector space.

Consider, again, the problem in which an object is described by radiographic intensity values on a 4×4 grid. In Section 3.3, we found two spanning sets for the set of all objects that are linear combinations of the seven images from page 11. One was considered a "better" spanning set because it did not hold redundant information. In this section, we will explore sets that hold all the information we need to recreate a vector space and not more information than we need. We will also use these sets to define the *dimension* of a vector space.

3.4.1   Efficient Heat State Descriptions

Consider the set of six heat states $X = \{h_1, h_2, h_3, h_4, h_5, h_6\}$ shown in Figure 3.7 for $m = 4$. We are interested in two questions. First, can this set of heat states be used to describe (via linear combinations) all vectors in the vector space $(H_4(\mathbb{R}), +, \cdot)$ of heat states sampled in 6 places along a rod? Second,

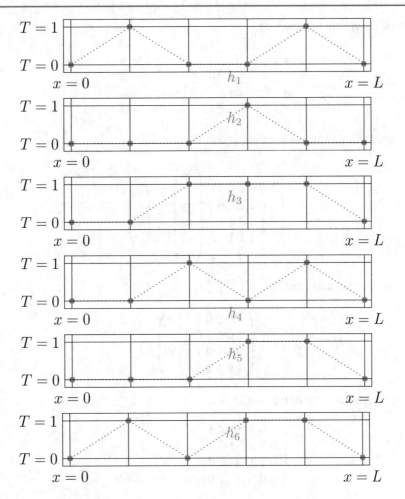

Fig. 3.7 The set $X = \{h_1, h_2, h_3, h_4, h_5, h_6\}$ of six heat states in $H_4(\mathbb{R})$.

if so, is this the most efficient spanning set? In answering these two questions we will use techniques from Sections 3.2 and 3.3.

Can this set of heat states be used to describe all vectors in the vector space $H_4(\mathbb{R})$? In other words, is Span $X = H_4(\mathbb{R})$? First note that X is a subset of $H_4(\mathbb{R})$. Thus, Span $X \subseteq H_4(\mathbb{R})$. We need only show that $H_4(\mathbb{R}) \subseteq$ Span X by showing that an arbitrary vector v in $H_4(\mathbb{R})$ is in Span X, i.e., it can be written as a linear combination of vectors in X.

For ease, we write heat states as $(m + 2)$-tuples of scalars (with first and last entries 0). For example, in our set X, $h_1 = (0, 1, 0, 0, 1, 0)$ and $h_2 = (0, 0, 0, 1, 0, 0)$. Consider arbitrary heat state $v = (0, a, b, c, d, 0) \in H_4(\mathbb{R})$, where $a, b, c, d \in \mathbb{R}$. We need to show that there exist scalars $\alpha_1, \cdots, \alpha_6$ such that

$$v = \alpha_1 h_1 + \alpha_2 h_2 + \cdots + \alpha_6 h_6.$$

Substituting for $v, h_1, \ldots, h_6$ above yields

$$(0, a, b, c, d, 0) = (0, \alpha_1 + \alpha_6, \alpha_3 + \alpha_4, \alpha_2 + \alpha_3 + \alpha_5 + \alpha_6, \alpha_1 + \alpha_3 + \alpha_4 + \alpha_5 + \alpha_6, 0)$$

with the equivalent system of equations

$$
\begin{array}{rcl}
\alpha_1 & +\alpha_6 &= a \\
\alpha_3 + \alpha_4 & &= b \\
\alpha_2 + \alpha_3 & +\alpha_5 + \alpha_6 &= c \\
\alpha_1 & +\alpha_3 + \alpha_4 + \alpha_5 + \alpha_6 &= d
\end{array}
$$

We can also write this system as the matrix equation

$$
\begin{pmatrix}
1 & 0 & 0 & 0 & 0 & 1 \\
0 & 0 & 1 & 1 & 0 & 0 \\
0 & 1 & 1 & 0 & 1 & 1 \\
1 & 0 & 1 & 1 & 1 & 1
\end{pmatrix}
\begin{pmatrix}
\alpha_1 \\
\alpha_2 \\
\alpha_3 \\
\alpha_4 \\
\alpha_5 \\
\alpha_6
\end{pmatrix}
=
\begin{pmatrix}
a \\
b \\
c \\
d
\end{pmatrix},
$$

with the equivalent augmented matrix

$$
\left(
\begin{array}{cccccc|c}
1 & 0 & 0 & 0 & 0 & 1 & a \\
0 & 0 & 1 & 1 & 0 & 0 & b \\
0 & 1 & 1 & 0 & 1 & 1 & c \\
1 & 0 & 1 & 1 & 1 & 1 & d
\end{array}
\right).
$$

The reduced row echelon form of this matrix is

$$
\left(
\begin{array}{cccccc|c}
1 & 0 & 0 & 0 & 0 & 1 & a \\
0 & 1 & 0 & -1 & 0 & 1 & a + c - d \\
0 & 0 & 1 & 1 & 0 & 0 & b \\
0 & 0 & 0 & 0 & 1 & 0 & -a - b + d
\end{array}
\right).
$$

Thus, we see that the system is consistent and has solutions for all values of a, b, c, and d. We have shown that $H_4(\mathbb{R}) \subseteq \text{Span } X$. Together with our previous findings, we conclude that X is indeed a spanning set for $H_4(\mathbb{R})$.

Is X the most efficient spanning set? In other words, do we *need* all six heat states in order to form a spanning set? Could we find a smaller subset of X that still forms a spanning set? If we examine the reduced row echelon form of the augmented matrix, we can write the solution set (in terms of free variables α_4 and α_6):

$$
\begin{aligned}
\alpha_1 &= a - \alpha_6 \\
\alpha_2 &= a + c - d + \alpha_4 - \alpha_6 \\
\alpha_3 &= b - \alpha_4 \\
\alpha_5 &= -a - b + d
\end{aligned}
$$

The solution set in parametric form is

$$
\left\{ \begin{pmatrix} \alpha_1 \\ \alpha_2 \\ \alpha_3 \\ \alpha_4 \\ \alpha_5 \\ \alpha_6 \end{pmatrix} = \begin{pmatrix} a \\ a+c-d \\ b \\ 0 \\ -a-b+d \\ 0 \end{pmatrix} + \alpha_4 \begin{pmatrix} 0 \\ 1 \\ -1 \\ 1 \\ 0 \\ 0 \end{pmatrix} + \alpha_6 \begin{pmatrix} -1 \\ -1 \\ 0 \\ 0 \\ 0 \\ 1 \end{pmatrix} \middle| \; \alpha_4, \alpha_6 \in \mathbb{R} \right\}.
$$

We see that we can write vector $v = (0, a, b, c, d, 0)$ by choosing *any* $\alpha_4, \alpha_6 \in \mathbb{R}$. In particular, we can choose $\alpha_4 = \alpha_6 = 0$ so that (arbitrary) v can be written in terms of h_1, h_2, h_3, and h_5, with coefficients $\alpha_1, \alpha_2, \alpha_3$, and α_5. That is $Y = \{h_1, h_2, h_3, h_5\}$ is a more efficient spanning set for $H_4(\mathbb{R})$ than X.

We also see that no proper subset of Y can span $H_4(\mathbb{R})$ because for arbitrary constants a, b, c, and d, we need all four coefficients $\alpha_1, \alpha_2, \alpha_3$, and α_5. This means that Y is, in some sense, a *smallest* spanning set. Another way to think about this is to recall that X is larger because it must contain redundant information. And Y, because it cannot be smaller, must not contain redundant information. Thus, we suspect that X is linearly dependent and Y is linearly independent. We next check these assertions.

Example 3.4.1 Show that X is linearly dependent. Notice that $h_1 + h_2 - h_6 = 0$. Thus, by Theorem 3.3.8, X is linearly dependent.   □

Example 3.4.2 Show that Y is linearly independent. From the solution set above, we see that the zero heat state can only be written as the trivial linear combination of the heat states in Y. That is, for $a = b = c = d = 0$ we have the unique solution $\alpha_1 = \alpha_2 = \alpha_3 = \alpha_5 = 0$. Thus, by Corollary 3.3.10, Y is linearly independent.   □

The results of the examples above indicate that there is no subset *of* X consisting of fewer than 4 heat states in $H_4(\mathbb{R})$ that spans $H_4(\mathbb{R})$. In fact there is no spanning set of $H_4(\mathbb{R})$ containing fewer than 4 heat states (see Exercise 15).

The relationship between minimal spanning sets and linear independence is explored in the next sections.

3.4.2 Basis

Linearly independent spanning sets for vector spaces, like the set of heat states $Y = \{h_1, h_2, h_3, h_5\} \subset H_4(\mathbb{R})$ that we found in the previous section, play an important role in linear algebra. We call such a set a basis.

Definition 3.4.3

A subset $\mathcal{B}$ of a vector space $(V, +, \cdot)$ is called a **basis** of V if Span $\mathcal{B} = V$ and $\mathcal{B}$ is linearly independent.   □

The first condition in the definition of a basis gives us that $\mathcal{B}$ is big enough to describe all of V ($\mathcal{B}$ is a spanning set for V) and the second condition says that $\mathcal{B}$ is not so big that it contains redundant information.

Example 3.4.4 The set $Y = \{h_1, h_2, h_3, h_5\}$ is a basis for $H_4(\mathbb{R})$.   □

It turns out that bases are not unique. Can you think of a different basis for $H_4(\mathbb{R})$ other than Y? We now consider bases for Euclidean spaces.

Example 3.4.5 A basis for $\mathbb{R}^3$ is

$$S = \left\{ \begin{pmatrix} 1 \\ 0 \\ 0 \end{pmatrix}, \begin{pmatrix} 0 \\ 1 \\ 0 \end{pmatrix}, \begin{pmatrix} 0 \\ 0 \\ 1 \end{pmatrix} \right\}.$$

First, we show that Span $S = \mathbb{R}^3$. Notice that Span $S \subseteq \mathbb{R}^3$ because arbitrary linear combinations of vectors in S are also vectors in $\mathbb{R}^3$. Next consider an arbitrary vector in $\mathbb{R}^3$,

$$v = \begin{pmatrix} a \\ b \\ c \end{pmatrix} \in \mathbb{R}^3.$$

We see that

$$v = a \begin{pmatrix} 1 \\ 0 \\ 0 \end{pmatrix} + b \begin{pmatrix} 0 \\ 1 \\ 0 \end{pmatrix} + c \begin{pmatrix} 0 \\ 0 \\ 1 \end{pmatrix} \tag{3.10}$$

which shows that v can be written as a linear combination of vectors in S. So, $v \in$ Span S and $\mathbb{R}^3 \subseteq$ Span S. Together we have Span $S = \mathbb{R}^3$.

Now, we show that S is linearly independent. The equation

$$\alpha \begin{pmatrix} 1 \\ 0 \\ 0 \end{pmatrix} + \beta \begin{pmatrix} 0 \\ 1 \\ 0 \end{pmatrix} + \gamma \begin{pmatrix} 0 \\ 0 \\ 1 \end{pmatrix} = \begin{pmatrix} 0 \\ 0 \\ 0 \end{pmatrix}$$

has unique solution

$$\begin{pmatrix} \alpha \\ \beta \\ \gamma \end{pmatrix} = \begin{pmatrix} 0 \\ 0 \\ 0 \end{pmatrix},$$

or, $\alpha = \beta = \gamma = 0$. Hence by the linear dependence relation test (Theorem 3.3.8), S is linearly independent. We showed both conditions for a basis hold for S, thus this set is a basis for $\mathbb{R}^3$.   □

The basis of the previous example is very natural and used quite often because, given a vector, we can easily see the linear combination of basis vectors that produces the vector (as in equation 3.10). This special basis is called the *standard basis* for $\mathbb{R}^3$. We also introduce notation for each of its vectors. We let e_1, e_2, and e_3 denote the three vectors in S, where

$$e_1 = \begin{pmatrix} 1 \\ 0 \\ 0 \end{pmatrix}, \quad e_2 = \begin{pmatrix} 0 \\ 1 \\ 0 \end{pmatrix}, \quad \text{and} \quad e_3 = \begin{pmatrix} 0 \\ 0 \\ 1 \end{pmatrix}.$$

In general, the standard basis for $\mathbb{R}^n$ is $\{e_1, e_2, \ldots, e_n\}$, where e_i is the vector with zeros in every position except the ith entry, which contains a one.

Example 3.4.6 Another basis for $\mathbb{R}^3$ is

$$\mathcal{B} = \left\{ \begin{pmatrix} 2 \\ 0 \\ 1 \end{pmatrix}, \begin{pmatrix} 1 \\ 1 \\ 0 \end{pmatrix}, \begin{pmatrix} 0 \\ 0 \\ 1 \end{pmatrix} \right\}.$$

By Exercise 7b in Section 3.2, we have that Span $\mathcal{B} = \mathbb{R}^3$. So, we need only show that $\mathcal{B}$ is linearly independent. Let

$$\alpha \begin{pmatrix} 2 \\ 0 \\ 1 \end{pmatrix} + \beta \begin{pmatrix} 1 \\ 1 \\ 0 \end{pmatrix} + \gamma \begin{pmatrix} 0 \\ 0 \\ 1 \end{pmatrix} = \begin{pmatrix} 0 \\ 0 \\ 0 \end{pmatrix}.$$

Then

$$\begin{pmatrix} 2\alpha + \beta \\ \beta \\ \alpha + \gamma \end{pmatrix} = \begin{pmatrix} 0 \\ 0 \\ 0 \end{pmatrix}.$$

Thus, by matching components, we see that $\alpha = \beta = \gamma = 0$. So, $\mathcal{B}$ is linearly independent and is therefore a basis for $\mathbb{R}^3$. □

The previous two examples make it clear that a basis for a vector space is not unique. Suppose we have a basis $\mathcal{B} = \{v_1, v_2, \cdots, v_n\}$ for some vector space V over $\mathbb{R}$. Then $\mathcal{B}' = \{\alpha v_1, v_2, \cdots, v_n\}$ is also a basis for V for *any* nonzero scalar $\alpha \in \mathbb{R}$.

Example 3.4.7 Which of the following sets of vectors in Figure 3.8 are bases for $\mathbb{R}^2$? Decide which you think are bases, and why, then check your answers below[6]. □

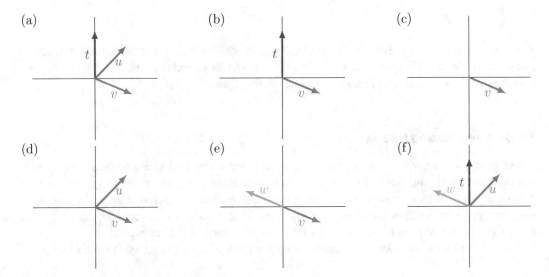

Fig. 3.8 Six sets of candidate basis vectors for $\mathbb{R}^2$. (See Example 3.4.7.)

---

[6] (b) and (d) are bases; (a), (e), and (f) are not linearly independent; and (c) and (e) do not span $\mathbb{R}^2$.

Example 3.4.8 A basis for $\mathcal{P}_2(\mathbb{R})$ is $\mathcal{S} = \{1, x, x^2\}$. In Exercise 1b in Section 3.3, we showed that $\mathcal{S}$ is linearly independent. And, Span $\mathcal{S} \subseteq \mathcal{P}_2(\mathbb{R})$. So we need to show that $\mathcal{P}_2(\mathbb{R}) \subseteq$ Span $\mathcal{S}$. Let $v = ax^2 + bx + c \in \mathcal{P}_2(\mathbb{R})$. Then v is a linear combination of $1, x$, and x^2. So $v \in$ Span $\mathcal{S}$. Thus Span $\mathcal{S} = \mathcal{P}_2(\mathbb{R})$. Therefore, $\mathcal{S}$ is a basis of $\mathcal{P}_2(\mathbb{R})$.   □

The basis $\mathcal{S} = \{1, x, x^2\}$ is called the *standard basis* for $\mathcal{P}_2(\mathbb{R})$, because, as with the standard basis for $\mathbb{R}^3$, one can trivially recover the linear combination of basis vectors that produces any polynomial in $\mathcal{P}_2(\mathbb{R})$.

Example 3.4.9 Let $\mathcal{B} = \{1, x + x^2, x^2\}$. In this example, we will show that $\mathcal{B}$ is also a basis for $\mathcal{P}_2$. In Exercise 1c in Section 3.3 we showed that $\mathcal{B}$ is linearly independent. So, we need only show that Span $\mathcal{B} = \mathcal{P}_2(\mathbb{R})$. This task reduces to showing that $\mathcal{P}_2(\mathbb{R}) \subseteq$ Span $\mathcal{B}$ since it is clear that Span $\mathcal{B} \subseteq \mathcal{P}_2(\mathbb{R})$. Let $v = ax^2 + bx + c \in \mathcal{P}_2(\mathbb{R})$. We want to find α, β, and γ so that

$$\alpha(1) + \beta(x + x^2) + \gamma(x^2) = ax^2 + bx + c.$$

Matching up like terms, we see that $\alpha = c$, $\beta = b$ and $\beta + \gamma = a$ or $\gamma = a - b$. That is, $v = c(1) + b(x) + (a - b)x^2 \in$ Span $\mathcal{B}$. Thus, $\mathcal{B}$ is a basis for $\mathcal{P}_2$.   □

Example 3.4.10 Consider $\mathcal{I}_4(\mathbb{R})$, the vector space of 4×4 grayscale images, and the set

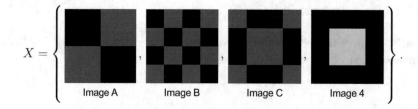

X is not a basis for $\mathcal{I}_4(\mathbb{R})$. The pixel-wise definitions of addition and scalar multiplication lead to a system of 16 equations in four unknowns when using the linear dependence relation. The system has only the trivial solution so X is linearly independent. But, Span $X \neq \mathcal{I}_4(\mathbb{R})$.

3.4.3 Constructing a Basis

While we can now appreciate that a basis provides an efficient way to describe all possible vectors in a vector space, it may not be clear just how to find a basis. For some common vector spaces, there are "standard" bases that are frequently convenient to use for applications. Some of these standard bases are listed in the table below. You should be able to verify that each of these does indeed represent a basis. Can you see why each is in fact called the *standard basis* of the corresponding vector space? Standard bases for image spaces, heat states, and other applications are explored in the Exercises.

Vector Space	Standard Basis
$(\mathbb{R}^n, +, \cdot)$	$\{e_1, e_2, \cdots, e_n\}$, where $e_i = \begin{pmatrix} \vdots \\ 0 \\ 1 \\ 0 \\ \vdots \end{pmatrix}$ with 1 in position i.
$(\mathcal{P}_n(\mathbb{R}), +, \cdot)$	$\{1, x, x^2, \cdots, x^n\}$
$(\mathcal{M}_{m \times n}(\mathbb{R}), +, \cdot)$	$\{M_{11}, M_{21}, \cdots, M_{mn}\}$, where M_{jk} is the matrix of all zeros except for a 1 in the jth row and k^{th} column.

If we do not have a standard basis or we wish to use a different basis, then we must construct one. There are two main methods for constructing a basis. We can either start with a spanning set, and selectively remove vectors until the set is linearly independent (but still spans the space), or we can start with a linearly independent subset of the space and selectively add vectors until the set spans the space (and is still linearly independent). Both methods make use of the definition of a basis and both will seem familiar.

Method #1: Spanning Set Reduction. Suppose we have a spanning set X for a vector space V. We might hope to find a basis $\mathcal{B}$ for V as a subset of X. If so, we must have Span $\mathcal{B} =$ Span $X = V$ *and* $\mathcal{B}$ is linearly independent. In other words, is there a linearly independent subset of X that has the same span as X? The following theorem answers this question.

Theorem 3.4.11
Let $(V, +, \cdot)$ be a vector space and X a finite subset of V such that Span $X = V$. Then, there exists a basis $\mathcal{B}$ for V such that $\mathcal{B} \subseteq X$.

Proof. Let $X = \{u_1, u_2, \cdots, u_k\}$ and Span $X = V$. We show that there must exist a linearly independent subset $\mathcal{B}$ of X such that Span $\mathcal{B} = V$.

If X is linearly independent then $\mathcal{B} = X$ is a basis for V. (This is true even if $X = \emptyset$. If this is the case, since Span $X = V$, it must be that $V = \{0\}$, so X is a basis of V.)

If X is not linearly independent, there exists a vector in X which can be written as a linear combination of the other vectors in X. Without loss of generality, suppose $u_k = \alpha_1 u_1 + \alpha_2 u_2 + \cdots + \alpha_{k-1} u_{k-1}$, and consider an arbitrary vector v in V. Observe:

$$v = a_1 u_1 + a_2 u_2 + \cdots + a_{k-1} u_{k-1} + a_k u_k$$
$$= a_1 u_1 + a_2 u_2 + \cdots + a_{k-1} u_{k-1} + a_k (\alpha_1 u_1 + \alpha_2 u_2 + \cdots + \alpha_{k-1} u_{k-1})$$
$$= (a_1 - a_k \alpha_1) u_1 + (a_2 - a_k \alpha_2) u_2 + \cdots + (a_{k-1} - a_k \alpha_{k-1}) u_{k-1}.$$

So, $v \in \operatorname{Span}\{u_1, u_2, \cdots, u_{k-1}\}$. In particular, $\operatorname{Span}\{u_1, u_2, \cdots, u_{k-1}\} = V$. Now, if this set is linearly independent then it is also a basis for V. If not, repeat the reduction step by eliminating any vector from the set which can be written as a linear combination of the remaining vectors. The process must terminate at some point because we started with a finite number of vectors, and the empty set of vectors is defined to be linearly independent. At the end of this process, we have a basis for V. $\square$

We now give several examples of this process at work.

Example 3.4.12 Find a basis for $\mathcal{P}_2(\mathbb{R})$ from among the set $X = \{2x^2, x, x^2 - 1, 3\}$.

We leave it to the reader to first verify that $\operatorname{Span} X = \mathcal{P}_2(\mathbb{R})$. Next, we find, if possible, any element of X which can be written as a linear combination of the remaining vectors. We see that $(3) = -3(x^2 - 1) + \frac{3}{2}(2x^2)$. So, $\operatorname{Span} X = \operatorname{Span}\{2x^2, x, x^2 - 1\}$. This remaining set is linearly independent (we leave it up to the reader to show this). Thus, $\mathcal{B} = \{2x^2, x, x^2 - 1\}$ is a basis for $\mathcal{P}_2(\mathbb{R})$.
$\square$

In the preceding example, we used the equation $(3) = -3(x^2 - 1) + \frac{3}{2}(2x^2)$ to show that the set is linearly dependent. This is equivalent to $3(x^2 - 1) = -3 + \frac{3}{2}(2x^2)$ and also to $(x^2 - 1) = -\frac{1}{3}3 + \frac{1}{2}(2x^2)$. So, we could, instead, have eliminated the vector $x^2 - 1$ to get a basis for $\mathcal{P}_2(\mathbb{R})$ consisting of $\{2x^2, x, 3\}$.

Example 3.4.13 Given the set $X = \{(4, 0), (1, 2), (3, 1), (1, 0)\} \subseteq \mathbb{R}^2$, find a basis for $\mathbb{R}^2$ that is a subset of X.

We leave it to the reader to first verify that $\operatorname{Span} X = \mathbb{R}^2$. Next, we eliminate any subset of vectors which can be written as linear combinations of remaining vectors. Notice that

$$(1, 2) = -\frac{5}{4}(4, 0) + 2(3, 1)$$

$$(1, 0) = \frac{1}{4}(4, 0) + 0(3, 1).$$

Thus, $\operatorname{Span} X = \operatorname{Span}\{(4, 0), (3, 1)\}$. Furthermore, since $\{(4, 0), (3, 1)\}$ is linearly independent, $\mathcal{B} = \{(4, 0), (3, 1)\}$ is a basis for $\mathbb{R}^2$. $\square$

In Example 3.4.13 above, there are many possible other choices of vectors to eliminate. What other combinations of vectors from X would also be bases?

In Examples 3.4.12 and 3.4.13, we recognized one vector that was a linear combination of the others. Let us look at how we can use the tool of matrix reduction to determine which vectors form a linearly independent set.

Example 3.4.14 Let us revisit Example 3.4.13. We can set up the linear dependence relation for the vectors in X as follows. Let α, β, γ, and δ be scalars so that

$$\alpha(4, 0) + \beta(1, 2) + \gamma(3, 1) + \delta(1, 0) = (0, 0).$$

We can then write the system of equations where each equation corresponds to a vector entry. That is, we have the following system of equations.

$$4\alpha + \beta + 3\gamma + \delta = 0$$
$$2\beta + \gamma = 0. \tag{3.11}$$

We can solve system (3.11) using matrix reduction as follows.

$$\begin{pmatrix} 4 & 1 & 3 & 1 & | & 0 \\ 0 & 2 & 1 & 0 & | & 0 \end{pmatrix} \sim \begin{pmatrix} 4 & -5 & 0 & 1 & | & 0 \\ 0 & 2 & 1 & 0 & | & 0 \end{pmatrix}. \tag{3.12}$$

It was easier to use columns 3 and 4 (corresponding to the variables γ and δ) to get "leading" 1's. The corresponding system of equations is

$$4\alpha - 5\beta + \delta = 0$$
$$2\beta + \gamma = 0.$$

So, α and β are free variables and $\delta = 5\beta - 4\alpha$ and $\gamma = -2\beta$. Putting this into the linear dependence relation tells us that, for any scalars α and β,

$$\alpha(4, 0) + \beta(1, 2) + (-2\beta)(3, 1) + (5\beta - 4\alpha)(1, 0) = (0, 0).$$

We can choose $\alpha = 1$ and $\beta = 0$ and we find that

$$(4, 0) - 4(1, 0) = (0, 0).$$

That is, $(4, 0)$ is a linear combination of $(1, 0)$. Now, if we choose $\alpha = 0$ and $\beta = 1$ we see that $(1, 2)$ is a linear combination of $(1, 0)$ and $(3, 1)$ as follows

$$(1, 2) - 2(3, 1) + 5(1, 0) = (0, 0).$$

Thus, Span $X = \text{Span}\{(1, 0), (3, 1)\}$. This is also indicated by the "leading" 1's in the reduced echelon form of the matrix given in (3.12). $\square$

Example 3.4.15 Consider the heat state example of Section 3.4.1. We constructed subset $Y = \{h_1, h_2, h_3, h_5\}$ from X by eliminating heat states that could be written as linear combinations of vectors in Y. Then, we showed in Example 3.4.2 that Y is linearly independent. Thus, Y is a basis for $H_4(\mathbb{R})$. $\square$

Method #2: Linearly Independent Set Augmentation. A second strategy for constructing a basis is to begin with a linearly independent subset and augment it, retaining linear independence, until it becomes a spanning set of the vector space of interest. This strategy is nearly identical to the strategy for building a spanning set discussed in Example 3.2.19. We simply modify the procedure to verify linear independence of the set as an additional test before accepting each new vector to the set.

To verify that this method does indeed work to find a basis, we refer the reader to the theorems in Section 3.4.4. We delay the verification because we need to first discuss the topic of dimension of a vector space. We can, however, use Theorem 3.3.18 to verify that if we have a linearly independent set that is not a spanning set, we can add a vector to obtain a new "larger" linearly independent set.

Example 3.4.17 Find a basis $\mathcal{B}$ for $\mathbb{R}^3$.

We begin with $\mathcal{B}_0 = \emptyset$. $\mathcal{B}_0$ is linearly independent but does not span $\mathbb{R}^3$.

We choose any vector from $\mathbb{R}^3$ to add to the current set $\mathcal{B}_0$ which is not in the span of the existing vectors. Since Span $\mathcal{B}_0 = \{0\}$, we can choose any nonzero vector. We will choose vector $b_1 = (0, 1, 2)^\mathsf{T}$. Let $\mathcal{B}_1 = \{(0, 1, 2)^\mathsf{T}\}$. $\mathcal{B}_1$ is linearly independent but does not span $\mathbb{R}^3$.

We continue by adding another vector from $\mathbb{R}^3$ that is not in the span of $\mathcal{B}_1$. In this case, we can add any arbitrary vector that is not a scalar multiple of $b_1 = (0, 1, 2)^\mathsf{T}$, say $b_2 = (1, 1, 1)^\mathsf{T}$. By Theorem 3.3.18, we know that $\mathcal{B}_2 = \{(0, 1, 2)^\mathsf{T}, (1, 1, 1)^\mathsf{T}\}$ is linearly independent. We next verify that $\mathcal{B}_2$ does not span all of $\mathbb{R}^3$. Recall that the vector $v = (x, y, z)^\mathsf{T}$ is in the span of b_1 and b_2 if and only if $v = \alpha b_1 + \beta b_2$ for some α and β in $\mathbb{R}$. Hence we row reduce the augmented matrix

$$\left(b_1 \; b_2 \middle| v\right) = \begin{pmatrix} 0 & 1 & \bigm| & x \\ 1 & 1 & \bigm| & y \\ 2 & 1 & \bigm| & z \end{pmatrix} \sim \begin{pmatrix} 1 & 0 & \bigm| & y - x \\ 0 & 1 & \bigm| & x \\ 0 & 0 & \bigm| & z - 2y + x \end{pmatrix}.$$

Hence the system is consistent if and only if $z - 2y + x = 0$; this equation describes a plane in $\mathbb{R}^3$, not all of $\mathbb{R}^3$. Hence we do not yet have a basis.

We continue by adding another vector from $\mathbb{R}^3$ which is not in the span of $\mathcal{B}_2$. We must find a vector $(a, b, c)^\mathsf{T}$ such that $(a, b, c)^\mathsf{T} \neq \alpha_1 (0, 1, 2)^\mathsf{T} + \alpha_2 (1, 1, 1)^\mathsf{T}$ for any scalars α_1 and α_2. One such vector is $(1, 0, 0)^\mathsf{T}$ because this choice leads to an inconsistent system of equations in coefficients α_1 and α_2. We have $\mathcal{B}_3 = \{(0, 1, 2)^\mathsf{T}, (1, 1, 1)^\mathsf{T}, (1, 0, 0)^\mathsf{T}\}$ which is linearly independent. One can also verify that Span $\mathcal{B}_3 = \mathbb{R}^3$.

Thus, $\mathcal{B}_3$ is a linearly independent spanning set for $\mathbb{R}^3$. The set $\mathcal{B} = \mathcal{B}_3$ is a basis for $\mathbb{R}^3$.   □

Example 3.4.18 Find a basis for the vector space of 7-bar LCD characters, $\mathcal{D}(Z_2)$.

Using the method of Linearly Independent Set Augmentation, we can see that a basis is constructed by the following set:

$$\mathcal{B} = \left\{ \text{⊟}, \text{⊟}, \text{⊟}, \text{⊟}, \text{⊟}, \text{⊟}, \text{⊟} \right\}$$

Each vector cannot be written as a linear combination of those that precede it. Once these seven characters are included in the set, it becomes a spanning set for $\mathcal{D}(Z_2)$. (You are asked to verify these claims in Exercise 36.)   □

3.4.4 Dimension

Looking back at the examples of previous sections, we see that different bases for the same vector space seem to have the same number of vectors. We saw two different bases for $\mathbb{R}^2$, each with two vectors, and two different bases for $\mathbb{R}^3$, each with three vectors. We saw several different bases for $\mathcal{P}_2(\mathbb{R})$, each with three vectors.

Since a basis is a minimal spanning set (see Exercise 31), it should not be surprising that any two bases for a vector space should consist of the same number of vectors. Is this always true? If so, then the number of basis vectors is an important property of the vector space itself. Since each basis vector is linearly independent with respect to all others, larger bases span richer vector spaces.

In the following theorem, we prove that, for a *finite-dimensional* vector space, all bases have the same number of elements.

Definition 3.4.19

A **finite-dimensional** vector space is one that can be spanned by a finite set of vectors.   □

Theorem 3.4.20
Let $(V, +, \cdot)$ be a finite-dimensional vector space with bases $\mathcal{B}_1 = \{v_1, v_2, \ldots, v_n\}$ and $\mathcal{B}_2 = \{u_1, u_2, \ldots, u_m\}$. Then $n = m$.

Proof. Suppose both $\mathcal{B}_1$ and $\mathcal{B}_2$ are bases for V consisting of n and m elements, respectively. We show that $m = n$ "by contradiction." That is, we will assume that $m \neq n$ and show that this leads to a logical inconsistency. Without loss of generality, suppose $m > n$. We will produce our contradiction by showing that $\mathcal{B}_2$ cannot be linearly independent (and therefore could not be a basis). We write each of the elements of $\mathcal{B}_2$ as a linear combination of elements of $\mathcal{B}_1$. Specifically, since $\mathcal{B}_2$ is a subset of V, we know that there exist $\alpha_{i,j}$ for $1 \leq i \leq m$ and $1 \leq j \leq n$ so that

$$
\begin{aligned}
u_1 &= \alpha_{1,1}v_1 + \alpha_{1,2}v_2 + \ldots + \alpha_{1,n}v_n \\
u_2 &= \alpha_{2,1}v_1 + \alpha_{2,2}v_2 + \ldots + \alpha_{2,n}v_n \\
&\;\;\vdots \\
u_m &= \alpha_{m,1}v_1 + \alpha_{m,2}v_2 + \ldots + \alpha_{m,n}v_n.
\end{aligned}
$$

To determine whether $\mathcal{B}_2$ is linearly independent, suppose there is a linear dependence relation

$$
\beta_1 u_1 + \beta_2 u_2 + \ldots + \beta_m u_m = 0.
$$

We solve for all possible $\beta_1, \beta_2, \ldots, \beta_m$ that satisfy this equation. If we replace $u_1, u_2, \ldots, u_n$ with the linear combinations above and rearrange terms, we get

$$
\begin{aligned}
&(\beta_1\alpha_{1,1} + \beta_2\alpha_{2,1} + \ldots + \beta_m\alpha_{m,1})v_1 \\
+&(\beta_1\alpha_{1,2} + \beta_2\alpha_{2,2} + \ldots + \beta_m\alpha_{m,2})v_2 \\
&\qquad\qquad\vdots \\
+&(\beta_1\alpha_{1,n} + \beta_2\alpha_{2,n} + \ldots + \beta_m\alpha_{m,n})v_n = 0.
\end{aligned}
$$

Since $\mathcal{B}_1$ is a basis, we get that the coefficients of $v_1, v_2, \ldots, v_n$ are all zero. That is

$$\beta_1\alpha_{1,1} + \beta_2\alpha_{2,1} + \ldots + \beta_m\alpha_{m,1} = 0 \tag{3.13}$$

$$\beta_1\alpha_{1,2} + \beta_2\alpha_{2,2} + \ldots + \beta_m\alpha_{m,2} = 0 \tag{3.14}$$

$$\vdots \tag{3.15}$$

$$\beta_1\alpha_{1,n} + \beta_2\alpha_{2,n} + \ldots + \beta_m\alpha_{m,n} = 0. \tag{3.16}$$

This is a system of equations in the variables $\beta_1, \beta_2, \ldots, \beta_m$ (recall that we want to know what possible values $\beta_1, \beta_2, \ldots, \beta_m$ can take on). We know that this system has a solution because it is homogeneous. But, because there are m variables and only n equations, this system must have infinitely many solutions. This means that $\mathcal{B}_2$ cannot be linearly independent and so it cannot be a basis.

Thus, if both $\mathcal{B}_1$ and $\mathcal{B}_2$ are bases of the same vector space, then $n = m$. $\qquad\square$

Because the number of elements in a basis is unique to the vector space, we give it a name.

Definition 3.4.21

Given a vector space $(V, +, \cdot)$ and a basis $\mathcal{B}$ for V with n elements, we say that the **dimension** of V is n and write $\dim V = n$. If V has no finite basis, we say that V is **infinite-dimensional**. $\quad\square$

When we proved Theorem 3.4.20, we actually showed that an n-dimensional vector space V has no linearly independent set with more than n elements. That is, if we have a set $\{u_1, u_2, \ldots, u_k\} \subseteq V$ and $k > n$, then we automatically know that the set is linearly dependent. This gives us another tool to make a quick check of linear dependence and may save us time.

Corollary 3.4.22

Let S be a k-element subset of n-dimensional vector space $(V, +, \cdot)$. If $k > n$, then S is linearly dependent.

Corollary 3.4.22 also justifies our earlier assertion that Method #2 for constructing bases (Linearly Independent Set Augmentation) works, at least in the setting of finite-dimensional vector spaces. Indeed, one iteratively adds vectors to a linearly independent set, each time maintaining linear independence. We continue this until we arrive at a set containing n linearly independent vectors. Corollary 3.4.22 guarantees that adding another vector beyond n will create a linearly dependent set. Hence, every vector in the space must lie in the span of these n vectors, so those n vectors form a basis.

The contrapositive of Corollary 3.4.22 is also true: If S is linearly independent, then $k \leq n$. So, then, a basis is the largest linearly independent set.

Lemma 3.4.23

Let S be a k-element subset of an n-dimensional vector space $(V, +, \cdot)$. If $k < n$ then S is not a spanning set.

Proof. Suppose $\mathcal{B} = \{v_1, v_2, \ldots, v_n\}$ is a basis for V and $S = \{s_1, s_2, \ldots, s_k\}$. We want to show that there is an element $v \in V$ with $v \notin \text{Span } S$. We will assume also that S spans V and look for a contradiction. We break this proof into two cases: Case 1: S is linearly independent and Case 2: S is linearly dependent.

Case 1: Suppose S is linearly independent. If S spans V then S is a basis for V, but by Theorem 3.4.20, $k = n$. Since $k < n$, we have found a contradiction.

Case 2: Suppose S is linearly dependent. Then some of the vectors in S can be written as linear combinations of the others. This means there is a subset $S' \subset S$ that is linearly independent and Span $S' =$ Span $S = V$. But then S', by definition is a basis. Again, this contradicts Theorem 3.4.20.

Thus, S cannot be a spanning set. $\qquad\square$

Perhaps even more surprisingly, in an n-dimensional vector space, *any* set of n linearly independent vectors is a spanning set, and hence a basis.

Theorem 3.4.24

Let $(V, +, \cdot)$ be an n-dimensional vector space and $S \subseteq V$. S is a basis for V if and only if S is linearly independent and contains exactly n elements.

Proof. ($\Rightarrow$) Suppose S is a basis for the n-dimensional vector space V. Then S is linearly independent and Span$(S) = V$. By Lemma 3.4.23, S must have at least n elements. Also, by Corollary 3.4.22, S cannot have more than n elements. Thus, S contains exactly n elements.

($\Leftarrow$) Now suppose S is a linearly independent set containing exactly n vectors from n-dimensional vector space V. If S is not a basis for V then S is not a spanning set of V and a basis for V would contain more than n vectors, a contradiction. Thus, S is a spanning set (and basis) for V. $\qquad\square$

In the next few examples, we illustrate how we find the dimension of various vector spaces.

Example 3.4.25 Because Y is a basis for $H_4(\mathbb{R})$ with 4 elements, $H_4(\mathbb{R})$ is 4-dimensional. $\qquad\square$

Example 3.4.26 Let

$$V = \left\{ \begin{pmatrix} x \\ y \\ z \end{pmatrix} \,\middle|\, x + y + z = 0, 2x + y - 4z = 0, 3x + 2y - 3z = 0 \right\}.$$

Notice that V is the solution set of a homogeneous system of equations. So, we first, show that V is a subspace of $\mathbb{R}^3$ (and therefore a vector space). We show next that V has dimension 1.

First, we will represent V as the span of a set of vectors, and check the linear dependence of this set. (If the spanning set is not linearly independent we can use Spanning Set Reduction to pare it down into a basis.) We first reduce the matrix corresponding to the system of equations

$$\begin{aligned} x + y + z &= 0 \\ 2x + y - 4z &= 0 \\ 3x + 2y - 3z &= 0. \end{aligned}$$

The reduction results in the reduced echelon form

$$\begin{pmatrix} 1 & 1 & 1 & 0 \\ 2 & 1 & -4 & 0 \\ 3 & 2 & -3 & 0 \end{pmatrix} \sim \begin{pmatrix} 1 & 0 & -5 & 0 \\ 0 & 1 & 6 & 0 \\ 0 & 0 & 0 & 0 \end{pmatrix}.$$

Thus, the solution is of the form

$$\begin{pmatrix} x \\ y \\ z \end{pmatrix} = z \begin{pmatrix} 5 \\ -6 \\ 1 \end{pmatrix},$$

where z can be any real number. This means we can rewrite V as below

$$V = \left\{ z \begin{pmatrix} 5 \\ -6 \\ 1 \end{pmatrix} \middle| z \in \mathbb{R} \right\} = \text{Span} \left\{ \begin{pmatrix} 5 \\ -6 \\ 1 \end{pmatrix} \right\}.$$

The set

$$\mathcal{B} = \left\{ \begin{pmatrix} 5 \\ -6 \\ 1 \end{pmatrix} \right\}$$

is linearly independent and spans V. Thus it is a basis for V. Since $\mathcal{B}$ has one element, V has dimension 1.   □

Example 3.4.27 Let $V = \left\{ ax^2 + bx + c \mid a + b - 2c = 0 \right\}$. We show that V is a subspace of $\mathcal{P}_2$. Indeed, $0x^2 + 0x + 0 \in V$ and if $v_1 = a_1 x^2 + b_1 x + c_1$ and $v_2 = v_1 = a_2 x^2 + b_2 x + c_2$ are vectors in V and α, β are scalars, then

$$a_1 + b_1 - 2c_1 = 0 \quad \text{and} \quad a_2 + b_2 - 2c_2 = 0.$$

Now,

$$\alpha v_1 + \beta v_2 = (\alpha a_1 + \beta a_2)x^2 + (\alpha b_1 + \beta b_2)x + \alpha c_1 + \beta c_2.$$

Also, we see that

$$\alpha a_1 + \beta a_2 + \alpha b_1 + \beta b_2 + 2\alpha c_1 + 2\beta c_2 = \alpha(a_1 + b_1 - 2c_1) + \beta(a_2 + b_2 - 2c_2)$$
$$= 0 + 0$$
$$= 0.$$

Thus, $\alpha v_1 + \beta v_2 \in V$. Therefore, V is a subspace of $\mathcal{P}_2$. Now, below, we show that V is 2-dimensional. Indeed, we can rewrite V:

$$V = \left\{ (2c - b)x^2 + bx + c \mid b, c \in \mathbb{R} \right\}$$
$$= \left\{ (-x^2 + x)b + (2x^2 + 1)c \mid b, c \in \mathbb{R} \right\}$$
$$= \text{Span}\{-x^2 + x, 2x^2 + 1\}.$$

Now, we can see that the elements of the set $\mathcal{B} = \{-x^2 + x, 2x^2 + 1\}$ are not scalar multiples of one another so $\mathcal{B}$ is linearly independent. Thus, $\mathcal{B}$ is a basis for V. Since $\mathcal{B}$ has two elements, V is 2-dimensional.   □

What we see is that, in order to find the dimension of a vector space, we need to find a basis and count the elements in the basis.

Example 3.4.28 Consider the vector space $\mathcal{I}_{512 \times 512 \times 512}$ of grayscale brain images represented as voxels (cubic pixels) in a $512 \times 512 \times 512$ array. Let $b_{i,j,k}$ be the brain image of all zero values except a 1 at array location (i, j, k). The set of images

$$\mathcal{B} = \left\{ b_{i,j,k} \mid 1 \leq i, j, k \leq 512 \right\}$$

is a basis for V with $512 \times 512 \times 512 = 136{,}839{,}168$ elements. Thus, V is a vector space of dimension $136{,}839{,}168$. □

Example 3.4.29 Consider the vector space V of color images from a 12-megapixel phone camera (see Section 2.4, Exercise 6). Images are represented on a rectangular grid with 3024 rows and 4032 columns, such as Figure 3.9. The color in each pixel is determined by the relative brightness of red, green, and blue light specified by three scalars.

Let $p_{i,j,k}$ be the image of all zero values except a 1 in the ith row, j^{th} column and k^{th} color (red,green,blue). The set of images

$$\mathcal{B} = \left\{ p_{i,j,k} \mid 1 \leq i \leq 3024, \ 1 \leq j \leq 4032, \ k = 1, 2, 3 \right\}$$

is a basis for V with $3024 \times 4032 \times 3 = 36{,}578{,}304$ elements. Thus, V is a vector space of dimension $36{,}578{,}304$. □

Fig. 3.9 An example of a 12-megapixel phone camera image with 3024-pixel rows, 4032-pixel columns, and three color channels (red, green, blue).

3.4.5   Properties of Bases

We began this section by considering $X = \{h_1, h_2, h_3, h_4, h_5, h_6\}$, a spanning set for $H_4(\mathbb{R})$, the vector space of heat states of $m = 4$. This means that any heat state in $H_4(\mathbb{R})$ can be written as a linear combination of these six representative heat states. This, in itself, provides a compact way to catalogue and represent heat states. But we went further by showing that $Y = \{h_1, h_2, h_3, h_5\}$ is also a spanning set and has fewer representative states. So, for efficient descriptions, Y makes a better catalogue. Then we showed that no catalogue can be smaller than Y. Any smallest such catalogue is called a basis, and the number of elements in a basis is called the dimension of the vector space that the basis spans.

However, we also saw that a basis for a vector space need not be unique. Much of our work with applications and their descriptions will center around good choices of bases. This choice can dramatically affect our ability to (a) interpret our results and descriptions and (b) efficiently perform computations.

There are some important questions we have yet to answer. Because we have seen that a vector space is easily described by very few vectors when it has a basis, we want to know that the basis exists. The good news is that every vector space has a basis. In the next section we address this question with a proof of the existence of a basis. We also, address the question of uniqueness of representing a vector as a linear combination of basis elements. Can we find more than one representation given a particular basis? This question is important because alternate representations might make it difficult to create algorithms that answer questions about the vector space.

Existence of a Basis

As just mentioned, it is important to know that every vector space has a basis. The next theorem and its proof give us this result. The idea is that, because of the way we defined a finite-dimensional vector spaces, we know that there is a finite spanning set on which we may apply the spanning set reduction algorithm to get a basis. We state the result here as a corollary to Theorem 3.4.11.

Corollary 3.4.30
Every finite-dimensional vector space has a basis.

Proof. Suppose V is a finite-dimensional vector space. Then V is spanned by some finite set X. By Theorem 3.4.11, there exists a basis for V that is a subset of X; hence V has a basis.   □

Most examples in this section have involved finite-dimensional vector spaces. We have considered vector spaces $\mathbb{R}^n$, $\mathcal{P}_n(\mathbb{R})$, $\mathcal{D}(Z_2)$, $\mathcal{M}_{m \times n}(\mathbb{R})$ as well as some image spaces, all of which are finite-dimensional.

Although we will mostly refrain from discussing infinite-dimensional vector spaces in the remainder of this book, we make a few observations here. One can show[7] that every infinite-dimensional vector

---

[7] The proof that every vector space (including infinite-dimensional vector spaces) has a basis requires a tool called Zorn's Lemma, which is beyond the scope of this course. The technique of Linearly Independent Set Augmentation will allow one to create larger and larger linearly independent sets, but it is not clear how to proceed with this process ad infinitum. Zorn's Lemma allows one to circumvent this issue (and also allows for uncountable bases, which a sequential application of Linearly Independent Set Augmentation would not accommodate).

space has a basis (that is, a maximal linearly independent subset) as well, though it may not be clear how to construct such a basis. Sometimes, however, a basis for an infinite-dimensional vector space can be readily constructed using the method of Linearly Independent Set Augmentation—extended to account for non-finite spanning sets. The following examples are two infinite-dimensional vector spaces for which we can easily construct infinite bases.

Example 3.4.31 (Infinite-Dimensional Vector Space) Consider the set of all polynomials $\mathcal{P}(\mathbb{R})$. Every polynomial has a finite number of terms and so, we begin listing simple 1-term polynomial vectors to create the standard basis $\{1, x, x^2, x^3, \cdots\}$. Although we do not formally prove it here, we note that for $k \in \mathbb{N}$, x^k cannot be represented as a finite sum of other 1-term polynomials; hence the set is linearly independent.   □

Example 3.4.32 (Infinite-Dimensional Vector Space). Consider $\mathcal{S}(\mathbb{R})$, the set of sequences of real numbers with a finite number of nonzero entries. In a very similar way, we can begin listing basis elements $\{s_1, s_2, s_3, \cdots\}$ as follows

$$s_1 = (1, 0, 0, 0, \ldots)$$
$$s_2 = (0, 1, 0, 0, \ldots)$$
$$s_3 = (0, 0, 1, 0, \ldots)$$
$$\vdots$$

where the kth vector in the basis is the sequence whose kth term is 1 and all other terms are 0. Again, though we will not prove it formally, s_k cannot be written as a linear combination of finitely many other sequences, so $\{s_1, s_2, s_3, \ldots\}$ is linearly independent. In addition, if a sequence is in $\mathcal{S}(\mathbb{R})$, it has finitely many nonzero entries, so it can be written as a linear combination of the vectors corresponding to its nonzero entries.   □

On the other hand, a space such as the vector space of all sequences of real numbers does not have such a simple basis. To see this, observe that the sequence vector $(1, 1, 1, \ldots)$ cannot be written as a linear combination of the vectors $s_1, s_2, s_3, \ldots$ (recall Definition 3.1.1). If we were to then add this vector, it still would not be a basis[8].

Uniqueness of Linear Combinations

The concept of basis evolved from finding efficient spanning sets used as representatives of all possible vectors in a vector space. However, we could think of a basis from a different viewpoint. We could instead look for a set $\mathcal{B}$ so that every vector in V has a *unique* representation as a linear combination of vectors in $\mathcal{B}$. Does such a set always exist? The answer, amazingly enough, is "yes."

Consider the vector space $\mathcal{I}_{108 \times 108}(\mathbb{R})$, the set of 108×108 grayscale images on a regular grid. Suppose we are interested in brain scan images in this space such as those shown in Figure 3.10. We could hypothetically take a brain scan of every living human and have a complete catalogue of images. This database would contain about 8 billion images. At first glance, it might seem reasonable to conclude that such a database would be sufficient for describing all possible brain images in $\mathcal{I}_{108 \times 108}(\mathbb{R})$.

However, there are three main problems with this conclusion.

---

[8] A typical proof to show that no such basis is found uses a diagonalization argument. We encourage the interested reader to look at a similar argument made by Cantor.

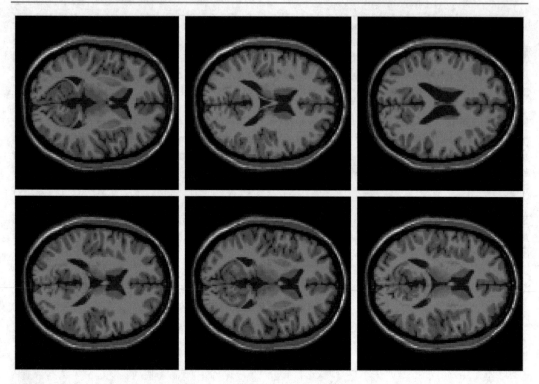

Fig. 3.10 Six example brain scan images in the vector space $\mathcal{I}_{108 \times 108}(\mathbb{R})$.

- It may still be possible for the brain image of a new person not to be in the span of the 8 billion existing images.
- Even if a new brain image is in this span, it may not be simply described in this catalogue.
- Even if a new brain image is in this span, it may not be *uniquely* described with this catalogue.

In fact, as few as 11,664 images are needed to generate (with linear combinations) the entire set of possible brain images. Moreover, every brain image is *uniquely* described using this catalogue. It turns out that any catalogue that is comprehensive (generates the entire space via linear combinations) and gives a unique representation of the brain images in the space $\mathcal{I}_{108 \times 108}(\mathbb{R})$ is in fact a basis for $\mathcal{I}_{108 \times 108}(\mathbb{R})$. This useful characterization is true more generally for vector spaces.

Theorem 3.4.33
Let $X = \{u_1, u_2, \cdots, u_n\}$ be a subset of vector space $(V, +, \cdot)$. Then X is a basis for V if and only if each vector in V is uniquely expressed as a linear combination of vectors in X.

Proof. Let $X = \{u_1, u_2, \cdots, u_n\}$ be a subset of vector space $(V, +, \cdot)$.

($\Rightarrow$) Suppose X is a basis for V and v is an arbitrary vector in V. Suppose, by way of contradiction, that v can be expressed as a linear combination of vectors in X in more than one way. In particular, suppose

$$v = \alpha_1 u_1 + \alpha_2 u_2 + \cdots + \alpha_n u_n,$$

and
$$v = \beta_1 u_1 + \beta_2 u_2 + \cdots + \beta_n u_n,$$

where $\beta_k \neq \alpha_k$ for some $1 \leq k \leq n$. Then we have

$$0 = v - v = (\alpha_1 - \beta_1)u_1 + (\alpha_2 - \beta_2)u_2 + \cdots + (\alpha_n - \beta_n)u_n.$$

At least one coefficient $(\alpha_k - \beta_k) \neq 0$ which implies that X is linearly dependent, a contradiction. So $\alpha_k = \beta_k$ for all k, and the expression of v as a linear combination of vectors in X is unique.

($\Leftarrow$) We will prove the contrapositive; that is, we will show that if X is not linearly independent, then there is not a unique representation for every vector in terms of elements of X. In other words, we will prove that if X is linearly dependent, then there exists a vector $v \in V$ that can be represented in two ways with as linear combinations of vectors from X. Toward this end, suppose that X is linearly dependent. Then there are scalars $\alpha_1, \alpha_2, ..., \alpha_n$, not all zero, so that

$$0 = \alpha_1 u_1 + \alpha_2 u_2 + \cdots + \alpha_n u_n.$$

Without loss of generality, suppose $\alpha_1 \neq 0$. Then

$$u_1 = -(\alpha_2/\alpha_1)u_2 - (\alpha_3/\alpha_1)u_3 - \cdots - (\alpha_n/\alpha_1)u_n$$
$$\text{and}$$
$$u_1 = 1 \cdot u_1.$$

In other words, the vector u_1 can be expressed as two distinct linear combinations of vectors in X. Since $X \subset V$, we have a vector, u_1 in V that can be expressed as two distinct linear combinations of vectors in X.

Therefore, if each vector in V is uniquely expressed as a linear combination of vectors in X, then X must be linearly independent. Also, since each vector can be expressed as a linear combination of vectors in X, X is a spanning set for V; hence X is a basis. $\square$

The following two examples show how this theorem can be useful in the vector spaces $\mathcal{P}_2(\mathbb{R})$ and $\mathcal{D}(Z_2)$.

Example 3.4.34 The vector space $\mathcal{P}_2(\mathbb{R})$ has dimension 3, so any basis has three elements. However, the three-element set $X = \{x^2 + x, x + 1, x^2 - 1\}$ is not a basis. We could show that X is linearly dependent. Or, we could notice that vector $x^2 + 2x + 1$ is not uniquely expressed in terms of the elements of X:

$$x^2 + 2x + 1 = 1(x^2 + x) + 1(x + 1) + 0(x^2 - 1),$$
$$\text{and}$$
$$x^2 + 2x + 1 = 2(x^2 + x) + 0(x + 1) - 1(x^2 - 1).$$ $\square$

Example 3.4.35 The set $\mathcal{B}$ in Example 3.4.18 is a basis for the vector space $\mathcal{D}(Z_2)$ because any character is uniquely expressed in terms of the basis vectors. To show this, let $\mathcal{B} = \{b_1, b_2, b_3, b_4, b_5, b_6, b_7\}$ in the order shown. Consider the representation of an arbitrary character, say $v = \alpha_1 b_1 + \cdots + \alpha_7 b_7$. We will show that there is a unique choice of scalar coefficients.

Notice that α_7 is entirely dependent on whether the lowest horizontal bar of v is green ($\alpha_7 = 1$) or not green ($\alpha_7 = 0$). Remember that the scalar field is $Z_2 = \{0, 1\}$ so we have uniquely determined α_7.

Next, notice that α_6 is determined by whether the lower-right vertical bar of v is green or not green *and* by the value of α_7. In particular, if this bar is green in v then we must have $1 = \alpha_6 + \alpha_7$. And if this bar is not green in v then we must have $0 = \alpha_6 + \alpha_7$. At each step we must be careful to perform all arithmetic in Z_2. So, we have uniquely determined α_6.

Next, notice that α_5 is determined by whether the lower-left vertical bar of v is green or not green *and* by the values of α_6 and α_7. In particular, if this bar is green in v then we must have $1 = \alpha_5 + \alpha_6 + \alpha_7$. And if this bar is not green in v then we must have $0 = \alpha_5 + \alpha_6 + \alpha_7$. So, we have uniquely determined α_5.

Continuing this process uniquely determines all coefficients $\alpha_1, \cdots, \alpha_7$. Thus, by Theorem 3.4.33, $\mathcal{B}$ is a basis for $\mathcal{D}(Z_2)$. $\square$

Theorem 3.3.14 allows us to derive the following key relationships between a basis for $\mathbb{R}^n$, linear combinations of vectors expressed in terms of this basis, and the matrix equation corresponding to such linear combinations.

Corollary 3.4.36

Let M be an $n \times n$ matrix with columns $v_1, v_2, \ldots, v_n$. Then, $B = \{v_1, v_2, \ldots, v_n\}$ is a basis for $\mathbb{R}^n$ if and only if $Mx = b$ has a unique solution $x \in \mathbb{R}^n$ for every $b \in \mathbb{R}^n$.

Proof. ($\Rightarrow$) Suppose M is an $n \times n$ matrix with columns $v_1, v_2, \ldots, v_n$, and $B = \{v_1, v_2, \ldots, v_n\}$ is a basis for $\mathbb{R}^n$. Then, B is linearly independent. Thus, by Theorem 3.3.14, $Mx = b$ has a unique solution $x \in \mathbb{R}^n$ for every $b \in \mathbb{R}^n$.

($\Leftarrow$) Suppose M is an $n \times n$ matrix with columns $v_1, v_2, \ldots, v_n$, and let $B = \{v_1, v_2, \ldots, v_n\}$. Further, suppose $Mx = b$ has a unique solution $x \in \mathbb{R}^n$ for every $b \in \mathbb{R}^n$. By Theorem 3.3.14, B is linearly independent. Thus, by Theorem 3.4.24, B is a basis for $\mathbb{R}^n$. $\square$

By Corollary 3.1.32, we can equivalently restate the previous result in terms of the matrix equation $Mx = 0$.

Corollary 3.4.37

Let M be an $n \times n$ matrix with columns $v_1, v_2, \ldots, v_n$. Then, $B = \{v_1, v_2, \ldots, v_n\}$ is a basis for $\mathbb{R}^n$ if and only if $Mx = 0$ has only the trivial solution $x = 0$.

We conclude with two examples using these corollaries for the case $n = 3$.

Example 3.4.38 Consider the matrix equation $Mx = b$ where

$$M = \begin{pmatrix} 1 & 1 & -2 \\ -2 & 1 & 1 \\ -1 & 2 & -1 \end{pmatrix} \quad \text{and} \quad b = \begin{pmatrix} -5 \\ 7 \\ 2 \end{pmatrix}.$$

Notice that

$$\begin{pmatrix} x \\ y \\ z \end{pmatrix} = \begin{pmatrix} -4 \\ -1 \\ 0 \end{pmatrix} \quad \text{and} \quad \begin{pmatrix} x \\ y \\ z \end{pmatrix} = \begin{pmatrix} 0 \\ 3 \\ 4 \end{pmatrix}$$

are both solutions to $Mx = b$. Thus, by Corollary 3.4.36 the set of column vectors of M is not a basis for $\mathbb{R}^3$. We can also say that, because there are three column vectors, we know that they must form a linearly dependent set. (Recall, Theorem 3.4.24.) □

Example 3.4.39 Consider the set of vectors in $\mathbb{R}^3$

$$B = \left\{ \begin{pmatrix} 1 \\ 0 \\ 1 \end{pmatrix}, \begin{pmatrix} 1 \\ 2 \\ 1 \end{pmatrix}, \begin{pmatrix} -1 \\ 1 \\ 1 \end{pmatrix} \right\}.$$

Consider also the matrix M whose columns are the vectors in B. We can solve the matrix equation $Mx = 0$ to determine whether or not B is a basis for $\mathbb{R}^3$. Using matrix reduction, we obtain the reduced echelon form of $[M|0]$ as follows

$$\begin{pmatrix} 1 & 1 & -1 & | & 0 \\ 0 & 2 & 1 & | & 0 \\ 1 & 1 & 1 & | & 0 \end{pmatrix} \sim \begin{pmatrix} 1 & 0 & 0 & | & 0 \\ 0 & 1 & 0 & | & 0 \\ 0 & 0 & 1 & | & 0 \end{pmatrix}.$$

Therefore $Mx = 0$ has only the trivial solution. Therefore, by Corollary 3.4.37, B is a basis for $\mathbb{R}^3$. □

3.4.6 Exercises

For Exercises 1 to 3, determine whether the set is a basis for $\mathcal{P}_2(\mathbb{R})$.

1. $\{1, x + x^2, x^2\}$
2. $\{x^2 - 1, 1 + x, 1 - x\}$
3. $\{1, 1 - x, 1 + x, 1 + x^2\}$

For Exercises 4 to 10, suppose $\{v_1, v_2, v_3, v_4\}$ is linearly independent. Determine whether the set is a basis for $\text{Span}\{v_1, v_2, v_3, v_4\}$. Justify your answer.

4. $\{v_1, v_2\}$
5. $\{v_1, v_2, v_3, v_4, v_1 - 2v_3\}$
6. $\{v_1 + v_2, v_3, v_4\}$
7. $\{v_1 + v_3, v_2 + v_4, v_3, v_4\}$
8. $\{v_1 + v_2 + v_3 + v_4, v_1 - v_2 + v_3 - v_4, v_1 - v_2 - v_3 + v_4, v_1 - v_2 - v_3 - v_4\}$
9. $\{v_1 - 2v_2, v_2, v_3 - v_4 - v_2, v_4\}$
10. $\{v_1 - v_2, v_1 + v_2, 2v_1 + v_2 - v_3, v_1 - v_2 - v_3 - 2v_4, v_3 - v_4\}$

For Exercises 11 to 14, decide whether or not $\mathcal{B}$ is a basis for the vector space V.

11. $\mathcal{B} = \left\{ \begin{pmatrix} 1 \\ 1 \\ 1 \end{pmatrix}, \begin{pmatrix} 1 \\ 2 \\ 3 \end{pmatrix}, \begin{pmatrix} 3 \\ 2 \\ 1 \end{pmatrix}, \begin{pmatrix} 0 \\ 0 \\ 1 \end{pmatrix} \right\}, V = \mathbb{R}^3$

12. $\mathcal{B} = \left\{ \begin{pmatrix} 1 \\ 1 \end{pmatrix}, \begin{pmatrix} 1 \\ 2 \end{pmatrix} \right\}, V = \mathbb{R}^2$

13. $\mathcal{B} = \left\{ \begin{pmatrix} 1 & 0 \\ 1 & 2 \end{pmatrix}, \begin{pmatrix} 1 & 2 \\ 3 & -1 \end{pmatrix}, \begin{pmatrix} 3 & 0 \\ 0 & 1 \end{pmatrix}, \begin{pmatrix} 1 & 0 \\ 0 & 0 \end{pmatrix} \right\}, V = \mathcal{M}_{2\times2}(\mathbb{R})$

14. $\mathcal{B} = \{x^2, x^2 + x, x^2 + x + 2\}, V = \mathcal{P}_2(\mathbb{R})$

15. Prove that there is no basis of $H_4(\mathbb{R})$ consisting of three or fewer heat states.

For Exercises 16 to 20 find a basis $\mathcal{B}$ (that is not the standard basis) for the given vector space. State the dimension of the vector space.

16. $\left\{ \begin{pmatrix} a & c \\ 3d & b \end{pmatrix} \middle| a + b + c - 2d = 0, a + 3b - 4c + d = 0, a - d + b = c \right\}$

17. $\{cx^2 + 3bx - 4a \mid a - b - 2c = 0\}$

18. $\mathcal{M}_{3\times2}(\mathbb{R})$

19. $\mathcal{P}_3(\mathbb{R})$

20. span $\left\{ \begin{pmatrix} 1 \\ 1 \\ 1 \end{pmatrix}, \begin{pmatrix} 1 \\ 2 \\ 3 \end{pmatrix}, \begin{pmatrix} 3 \\ 2 \\ 1 \end{pmatrix}, \begin{pmatrix} 0 \\ 0 \\ 1 \end{pmatrix} \right\}$

21. Given the set $\mathcal{B} = \{u, v, w\}$. Show that if $\mathcal{B}$ is a basis, then so is

$$\mathcal{B}' = \{u + 2v, u - w, v + w\}.$$

22. Using Exercise 21, make a general statement about how to get a basis from another basis. Be careful to use accurate linear algebra language.

For Exercises 23 to 26 provide specific examples (with justification) of the given scenario for vector space $V = \mathbb{R}^2$.

23. $\mathcal{B}_1$ is a nonstandard basis.

24. $\mathcal{B}_2$ is a nonstandard basis with $\mathcal{B}_2 \cap \mathcal{B}_1 = \emptyset$.

25. $W \subseteq V$ where W has three elements, Span $W = V$, and W is not a basis for V.

26. $W \subseteq V$ where W has three elements, Span $W \neq V$, and W is not a basis for V.

27. Does the vector space $\{0\}$ have a basis? If so, what is it? If not, show that it cannot have a basis.

28. Find a basis for $\mathcal{D}(Z_2)$ which has no elements in common with the basis of Example 3.4.18.

29. Determine whether $\mathcal{B} = \{I_1, I_2, I_3\}$, where the I_n are given below, is a basis for the vector space of images with the same geometric orientation as each of the I_n below.

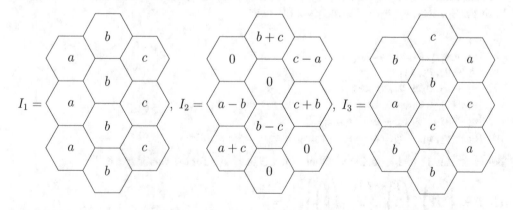

30. What is the dimension of the vector space of heat signatures given in Section 3.4.1?

31. Prove that a basis is a minimal spanning set. Specifically, suppose that B is a basis for vector space V over $\mathbb{R}$. Using just the definition of a basis, prove that if any vector is removed from B, the resulting set no longer spans V, and hence is not a basis for V.

32. Show that if $B = \{v_1, v_2, \cdots, v_n\}$ is a basis for vector space V over $\mathbb{R}$, then $B' = \{\alpha v_1, v_2, \cdots, v_n\}$ is also a basis for V for any nonzero scalar α.

33. Given a basis of some vector space, V given by $B = \{v_1, v_2, \cdots, v_n\}$. Determine whether $B = \{v_1 - v_n, v_2, \cdots, v_n\}$ is also a basis. Prove your result.

34. Prove that a subset X of a vector space V over $\mathbb{R}$ is a basis for V if and only if X is a maximal linear independent subset of V. In other words, prove that there is no superset of X (i.e., there is no strictly larger set containing X) that is linearly independent.

35. In Section 3.1, Exercise 28 and Section 3.3, Exercise 19, we began combining equivalent statements. Continue combining equivalent statements by completing the theorem below. (If you have already completed these exercises, you need only add one more statement from this section connecting these ideas to bases of $\mathbb{R}^n$. Otherwise, you can look at the exercise statements for hints about the earlier statements.)

Theorem 3.4.40

Let A be an $n \times n$ matrix and $b \in \mathbb{R}^n$. The following statements are equivalent.

- $Ax = 0$ has a unique solution.
-
-
-
-
-

Prove Theorem 3.4.40.

36. Verify that the given set B in Example 3.4.18 is a basis for the vector space of 7-bar LCD characters, $\mathcal{D}(Z_2)$. Specifically, prove the following.

    (a) Explain why each vector in B cannot be written as a linear combination of those that precede it.

    (b) Explain why B is a spanning set for $\mathcal{D}(Z_2)$.

37. Verify that the given set B in Example 3.4.28 is a basis for the vector space of $512 \times 512 \times 512$ arrays of real numbers.

For Exercises 38 to 40, determine whether or not W is a vector space. If not, justify. If so, prove it, find a basis for W, and determine the dimension of W.

38. $W = \left\{ y \mid y = \begin{pmatrix} 1 & 1 \\ 0 & 2 \end{pmatrix} x \text{ for } x \in \mathbb{R}^2 \right\}$.

39. $W = \left\{ \begin{pmatrix} 1 & 1 \\ 0 & 2 \end{pmatrix} + x \mid x \in \mathcal{M}_{2 \times 2} \right\}$.

40. $W = \left\{ x \mid y = \begin{pmatrix} 1 & 1 \\ 0 & 2 \end{pmatrix} x \text{ for } y \in \mathbb{R}^2 \right\}$.

41. Show that the space of continuous functions on $[0, 1]$ is not finite-dimensional.

3.5 Coordinate Spaces

We have now seen that a vector space or subspace can be most efficiently described in terms of a set of basis vectors and we understand that any vector in the space can be written as a unique linear combination of the basis vectors. That is, each vector in the space can be written as a linear combination of the basis vectors in exactly one way. This uniqueness suggests a very simple method for creating a cataloging system for vectors in the vector space. This system will help us represent vectors in a more standard way. Using the fact that the number of basis vectors is the dimension of the space, we are able to further describe the space in which these standardized vectors are located.

When considering the vector spaces of images, it becomes tedious to always need to write

Instead, because we have already found that Image 2 can be written uniquely as

we prefer a more compact way of expressing Image 2 as a particular linear combination of Images A, B, C and Image 4. In this section, we will describe a means to write image vectors (and any other vector) in a more compact and standardized way.

Suppose we have a basis $\mathcal{B} = \{u_1, u_2, \cdots, u_k\}$ for a k-dimensional vector space V. Consider an arbitrary vector v in V. Since $\mathcal{B}$ spans V, there exist coefficients $\alpha_1, \alpha_2, \cdots, \alpha_k$ such that $v = \alpha_1 u_1 + \alpha_2 u_2, \cdots, \alpha_k u_k$. And since $\mathcal{B}$ is linearly independent, the coefficients are unique – there is no other way to express v as a linear combination of $u_1, u_2, \cdots, u_k$. Thus, the coefficients $\alpha_1, \alpha_2, \cdots, \alpha_k$ are uniquely associated with the vector v. We will use this understanding to create the cataloging system.

3.5.1 Cataloging Heat States

Let's see how these ideas work in the vector space $H_4(\mathbb{R})$. We found in Section 3.4.1 that $Y = \{h_1, h_2, h_3, h_5\}$ formed a basis for the space of heat states $H_4(\mathbb{R})$ (See also Example 3.4.4). This basis is shown in Figure 3.11. Now, Y can be used to form a cataloging system for $H_4(\mathbb{R})$. To see how this works, consider the heat state v shown in Figure 3.12, chosen somewhat arbitrarily. We seek coefficients α_1, α_2, α_3, and α_4 so that

$$v = \alpha_1 h_1 + \alpha_2 h_2 + \alpha_3 h_3 + \alpha_4 h_5. \tag{3.17}$$

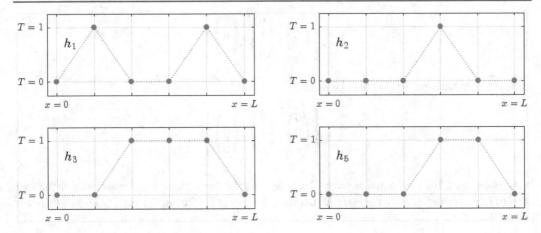

Fig. 3.11   Four heat state vectors which form a basis $Y = \{h_1, h_2, h_3, h_5\}$ for $H_4(\mathbb{R})$.

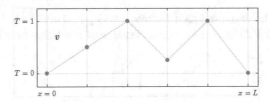

Fig. 3.12   Arbitrarily chosen heat state v in $H_4(\mathbb{R})$.

We rewrite Equation 3.17 as the system of equations

$$
\begin{aligned}
\alpha_1 & & & & &= 1/2 \\
& & \alpha_3 & & &= 1 \\
& \alpha_2 &+\alpha_3 &+\alpha_4 &= 1/4 \\
\alpha_1 & &+\alpha_3 &+\alpha_4 &= 1 &.
\end{aligned}
$$

The unique solution to the system of equations gives us that the scalars in Equation 3.17 are

$$
\alpha_1 = \frac{1}{2}, \alpha_2 = -\frac{1}{4}, \alpha_3 = 1, \text{ and } \alpha_4 = -\frac{1}{2}.
$$

These coefficients, along with the given basis Y, uniquely determine the heat state v. Any heat state can be completely and uniquely described by coefficients relative to a basis. We now lay out the path to express these vectors in terms of coordinates.

Definition 3.5.1

Given a basis $\mathcal{B}$ for the finite-dimensional vector space $(V, +, \cdot)$, we say that $\mathcal{B}$ is an **ordered basis** for V if a particular order is given to the elements of $\mathcal{B}$.   □

Let us consider an example to make sense out of this definition.

Example 3.5.2 We know that

$$S = \left\{ \begin{pmatrix} 1 \\ 0 \\ 0 \end{pmatrix}, \begin{pmatrix} 0 \\ 1 \\ 0 \end{pmatrix}, \begin{pmatrix} 0 \\ 0 \\ 1 \end{pmatrix} \right\}$$

is the standard basis for $\mathbb{R}^3$. We also know that

$$\hat{S} = \left\{ \begin{pmatrix} 0 \\ 0 \\ 1 \end{pmatrix}, \begin{pmatrix} 0 \\ 1 \\ 0 \end{pmatrix}, \begin{pmatrix} 1 \\ 0 \\ 0 \end{pmatrix} \right\}$$

is the same basis. But, we would not consider S and $\hat{S}$ to be the same *ordered* basis because the first and last basis vectors in S are the last and first basis vectors in $\hat{S}$. That is, the vectors are in a different *order*.   □

Definition 3.5.3

Let v be a vector in the finite-dimensional vector space $(V, +, \cdot)$ over $\mathbb{F}$ and let $\mathcal{B} = \{u_1, u_2, \cdots, u_k\}$ be an ordered basis for V. Then the **coordinate vector of v relative to $\mathcal{B}$** is

$$[v]_{\mathcal{B}} = \begin{pmatrix} \alpha_1 \\ \alpha_2 \\ \vdots \\ \alpha_k \end{pmatrix} \in \mathbb{F}^k,$$

where $\alpha_1, \alpha_2, \cdots, \alpha_k$ are the unique coefficients such that

$$v = \alpha_1 u_1 + \alpha_2 u_2 + \cdots + \alpha_k u_k.$$ □

Now, as long as we have an ordered basis for a vector space, *any* vector in the space is uniquely determined by its coordinate vector.

Example 3.5.4 Looking back at the example above with heat states, suppose we have the coordinate vector

$$[w]_{\mathcal{B}} = \begin{pmatrix} 1/2 \\ 1/2 \\ 1 \\ -1 \end{pmatrix},$$

with $\mathcal{B}$ defined above and $w \in H_4(\mathbb{R})$. This coordinate vector tells us that

$$w = (1/2)h_1 + (1/2)h_2 + (1)h_3 + (-1)h_5.$$

The reader can verify that the heat state, w is

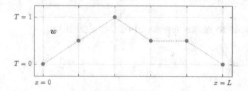

□

★★ **Watch Your Language!** Coordinate vectors $[v]_B$ are vectors in $\mathbb{F}^k$ for a given basis B of a k-dimensional vector space $(V, +, \cdot)$. Vectors v in V can be very abstract objects such as a grayscale image, a polynomial, a differentiable function, a 7-bar LCD image, a heat state, etc. However abstract they may be, they can be catalogued as vectors in $\mathbb{F}^k$ given an ordered basis.

✓ $[v]_B$ is a coordinate vector for the vector v.
✓ $[v]_B \in \mathbb{F}^k$.
✓ V is a k dimensional vector space.
✓ $v \in V$ is represented by $[v]_B \in \mathbb{F}^k$.

But it would be incorrect to say

✗ $[v]_B \in V$.
✗ $v \in \mathbb{R}^k$.
✗ $V = \mathbb{F}^k$.
✗ $v = [v]_B$.

3.5.2 Coordinates in $\mathbb{R}^n$

We are already familiar with coordinate representations of points in the xy-plane and xyz-space (which we call "3D space"). You might wonder if these coordinates are related to the coordinates that we are exploring now. Now, we connect these two notions of coordinates. In our typical 3D space, we talk about vectors that look like this: $v = \begin{pmatrix} x \\ y \\ z \end{pmatrix}$. When we say that the coordinates are x, y, and z, respectively, we mean that the vector points from the origin to a point that is x units horizontally, y units vertically, and z units up from the origin (as in Figure 3.13). If x, y, and z are coordinates relative to some ordered basis, then it is natural to ask what the chosen ordered basis is.

We can write

$$v = \begin{pmatrix} x \\ y \\ z \end{pmatrix} = x \begin{pmatrix} 1 \\ 0 \\ 0 \end{pmatrix} + y \begin{pmatrix} 0 \\ 1 \\ 0 \end{pmatrix} + z \begin{pmatrix} 0 \\ 0 \\ 1 \end{pmatrix}.$$

From this, we see that x, y, and z are the scalar weights associated with writing the vector as a linear combination of vectors in the standard basis for $\mathbb{R}^3$

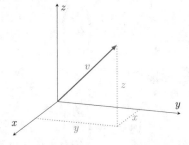

Fig. 3.13 The coordinate representation of a point v in 3D space with coordinates x, y, and z.

$$S = \left\{ \begin{pmatrix} 1 \\ 0 \\ 0 \end{pmatrix}, \begin{pmatrix} 0 \\ 1 \\ 0 \end{pmatrix}, \begin{pmatrix} 0 \\ 0 \\ 1 \end{pmatrix} \right\}.$$

That is to say, our usual interpretation of coordinates in 3D assumes that we are working in the standard (ordered) basis for $\mathbb{R}^3$.

Example 3.5.5 Now, let us consider a different ordered basis for $\mathbb{R}^3$:

$$\mathcal{B} = \left\{ v_1 = \begin{pmatrix} 1 \\ 1 \\ 1 \end{pmatrix}, v_2 = \begin{pmatrix} 1 \\ 0 \\ 1 \end{pmatrix}, v_3 = \begin{pmatrix} 0 \\ 0 \\ 1 \end{pmatrix} \right\}.$$

The coordinates of the vector $w = \begin{pmatrix} 1 \\ 2 \\ 3 \end{pmatrix}$ in the standard basis are 1, 2, and 3, respectively, but in this alternate basis, they are found by finding the scalars α_1, α_2, and α_3 so that

$$w = \begin{pmatrix} 1 \\ 2 \\ 3 \end{pmatrix} = \alpha_1 \begin{pmatrix} 1 \\ 1 \\ 1 \end{pmatrix} + \alpha_2 \begin{pmatrix} 1 \\ 0 \\ 1 \end{pmatrix} + \alpha_3 \begin{pmatrix} 0 \\ 0 \\ 1 \end{pmatrix}.$$

We find that $\alpha_1 = 2$, $\alpha_2 = -1$, and $\alpha_3 = 2$. (Be sure to check this.) So, we can represent the vector v in coordinates according to the basis $\mathcal{B}$ as

$$[w]_\mathcal{B} = \begin{pmatrix} 2 \\ -1 \\ 2 \end{pmatrix}.$$

□

The vector w in $\mathbb{R}^3$ is the vector that points from the origin to the regular cartesian grid location one unit along the x-axis, then two units along the y-axis direction, and then three units along the z-axis direction. $[w]_\mathcal{B}$ is the *representation* of v relative to basis $\mathcal{B}$. The vector w itself remains unchanged. See Figure 3.14.

If we are given bases for $\mathbb{R}^3$, $\mathcal{B}_1 = \{v_1, v_2, v_3\}$ and $\mathcal{B}_2 = \{u_1, u_2, u_3\}$, then if $w = \alpha_1 v_1 + \alpha_2 v_2 + \alpha_3 v_3$, we have

$$[w]_{\mathcal{B}_1} = \begin{pmatrix} \alpha_1 \\ \alpha_1 \\ \alpha_3 \end{pmatrix}.$$

And, if $w = \beta_1 u_1 + \beta_2 u_2 + \beta_3 u_3$, we have

$$[w]_{\mathcal{B}_2} = \begin{pmatrix} \beta_1 \\ \beta_1 \\ \beta_3 \end{pmatrix}.$$

w, $[w]_{\mathcal{B}_1}$, and $[w]_{\mathcal{B}_2}$ are different representations of the *same* vector.

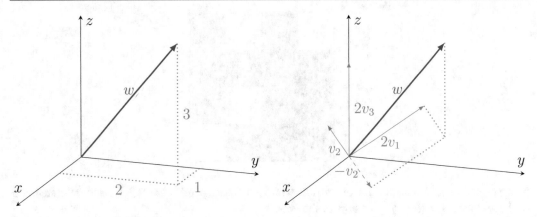

Fig. 3.14 The coordinate vectors $[w]_S$ (left) and $[w]_B$ (right) from Example 3.5.5.

Example 3.5.6 Let $w = \begin{pmatrix} 1 \\ -1 \\ 1 \end{pmatrix}$. We leave as an exercise to find $[w]_{B_1}$ and $[w]_{B_2}$, given

$$\mathcal{B}_1 = \left\{ \begin{pmatrix} 1 \\ 1 \\ 1 \end{pmatrix}, \begin{pmatrix} 1 \\ 0 \\ 1 \end{pmatrix}, \begin{pmatrix} 0 \\ 0 \\ 1 \end{pmatrix} \right\}$$

and

$$\mathcal{B}_2 = \left\{ \begin{pmatrix} 1 \\ 1 \\ 1 \end{pmatrix}, \begin{pmatrix} 0 \\ 1 \\ 1 \end{pmatrix}, \begin{pmatrix} 0 \\ 0 \\ 1 \end{pmatrix} \right\}.$$

See Exercise 3. □

3.5.3 Example Coordinates of Abstract Vectors

The following examples illustrate that abstract objects in a k-dimensional vector space V can be represented as coordinate vectors in $\mathbb{F}^k$, given an ordered basis for V. Coordinates present a useful and powerful tool for cataloging vectors and performing computations because their representations are familiar vectors in $\mathbb{F}^k$.

Example 3.5.7 Consider the 16-dimensional vector space of 4×4 grayscale images. Recall, this vector space is defined to have scalars in $\mathbb{R}$. Let us, now, see how we can catalog 4×4 grayscale images using vectors in the more familiar vector space, $\mathbb{R}^{16}$. Let v be the image

$$v = \quad$$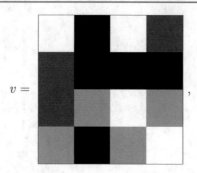

where black indicates a pixel value of zero and white indicates a pixel value of 3. Consider the standard (ordered) basis $\mathcal{B} = \{b_1, b_2, \cdots, b_{16}\}$ as shown:

$$\mathcal{B} = $$

Each element of $\mathcal{B}$ consists of all black pixels except for a single pixel with a value of 1. We have

$$v = 3b_1 + 1b_2 + 1b_3 + 2b_4 + 0b_5 + 0b_6 + 2b_7 + 0b_8$$
$$+ 3b_9 + 0b_{10} + 3b_{11} + 2b_{12} + 1b_{13} + 0b_{14} + 2b_{15} + 3b_{16},$$

so that

$$[v]_{\mathcal{B}} = (3\,1\,1\,2\,0\,0\,2\,0\,3\,0\,3\,2\,1\,0\,2\,3)^{\mathsf{T}}.$$

The coordinate vector $[v]_{\mathcal{B}}$ is in $\mathbb{R}^{16}$ and represents the 4×4 grayscale image v. In this standard ordered basis the coordinate values in $[v]_{\mathcal{B}}$ are the same as the pixel values in v (in a particular order). □

In the next example, we consider coordinate vectors in the less familiar vector space Z_2^k. The process for finding such coordinate vectors is the same.

Example 3.5.8 Let us consider the 7-dimensional vector space of 7-bar LCD images whose scalar set is Z_2. Consider the two different bases

$$\mathcal{B}_1 = $$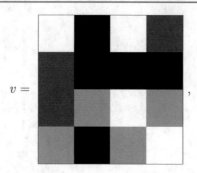

and

$$\mathcal{B}_2 = \left\{ \text{⊓}, \text{⊟}, \text{⊡}, \text{⊔}, \text{⊓}, \text{⊞}, \text{⊟} \right\}.$$

Let v be the "4" character. To find the coordinate vector $[v]_{\mathcal{B}_1}$, we write v as a linear combination of the basis elements in $\mathcal{B}_1$. Actually, v can be written as a linear combination of only the second, third, fourth, and sixth basis elements. Indeed,

$$v = \text{⊔} = \text{⊡} + \text{⊟} + \text{⊏} + \text{⊡}.$$

Therefore, $[v]_{\mathcal{B}_1} = \begin{pmatrix} 0 \\ 1 \\ 1 \\ 1 \\ 0 \\ 1 \\ 0 \end{pmatrix} \in Z_2^7$. Now, if we write v as a linear combination of the basis elements in

$\mathcal{B}_2$. We find

$$v = \text{⊔} = \text{⊡} + \text{⊓} + \text{⊟} + \text{⊞}.$$

Notice that $[v]_{\mathcal{B}_2} = \begin{pmatrix} 1 \\ 0 \\ 0 \\ 1 \\ 1 \\ 1 \\ 0 \end{pmatrix}$ is different than $[v]_{\mathcal{B}_1}$ even though they represent the same 7-bar LCD

character.

Remember, v is the LCD image of the "4" character, not a vector in Z_2^7. And, $[v]_{\mathcal{B}_1}$ and $[v]_{\mathcal{B}_2}$ are two different coordinate vector representations in Z_2^7, with respect to two different (ordered) bases, $\mathcal{B}_1$ and $\mathcal{B}_2$. $\qquad\square$

We provide the next example not only to practice finding coordinate vectors, but to also use a coordinate vector and an ordered basis to find the corresponding vector.

Example 3.5.9 Consider the subspace $V = \left\{ ax^2 + bx + c \mid a + b - 2c = 0 \right\} \subseteq \mathcal{P}_2(\mathbb{R})$. We saw in Example 3.4.27 that a basis for V is $\mathcal{B} = \{ -x^2 + x, \ 2x^2 + 1 \}$. We also know that $v = 3x^2 - 3x \in V$. This means that we can write v as a coordinate vector in $\mathbb{R}^2$. We do this by finding $\alpha_1, \alpha_2 \in \mathbb{R}$ so that

$$v = \alpha_1(-x^2 + x) + \alpha_2(2x^2 + 1).$$

Notice that

$$v = -3(-x^2 + x) + 0(2x^2 + 1).$$

Therefore,

$$[v]_{\mathcal{B}} = \begin{pmatrix} -3 \\ 0 \end{pmatrix} \in \mathbb{R}^2.$$

Working within the same subspace, we consider another coordinate vector,

$$[w]_{\mathcal{B}} = \begin{pmatrix} 2 \\ -1 \end{pmatrix} \in \mathbb{R}^2.$$

Using our understanding of the coordinate vector notation and which ordered basis is being used to create the coordinate vector, we write w as

$$w = (2)(-x^2 + x) + (-1)(2x^2 + 1) = -4x^2 + 2x - 1 \in V.$$

$\square$

In some instances, it is much easier to work in one coordinate space than in another. The following example shows how we can change a coordinate vector, according to one ordered basis, into another coordinate vector, according to a different ordered basis.

Example 3.5.10 Consider the subspace

$$V = \left\{ \begin{pmatrix} a & a+b \\ a-b & b \end{pmatrix} \middle| a, b \in \mathbb{R} \right\} \subseteq \mathcal{M}_{2\times2}(\mathbb{R}).$$

In order to catalog vectors in V, using coordinate vectors, we must first find a basis for V. Notice that we can write V as the span of the linearly independent vectors in

$$\mathcal{B}_1 = \left\{ \begin{pmatrix} 1 & 1 \\ 1 & 0 \end{pmatrix}, \begin{pmatrix} 0 & 1 \\ -1 & 1 \end{pmatrix} \right\}.$$

Indeed,

$$V = \left\{ a \begin{pmatrix} 1 & 1 \\ 1 & 0 \end{pmatrix} + b \begin{pmatrix} 0 & 1 \\ -1 & 1 \end{pmatrix} \middle| a, b \in \mathbb{R} \right\}$$

$$= \text{Span} \left\{ \begin{pmatrix} 1 & 1 \\ 1 & 0 \end{pmatrix}, \begin{pmatrix} 0 & 1 \\ -1 & 1 \end{pmatrix} \right\}.$$

Consider the vector w so that

$$[w]_{\mathcal{B}_1} = \begin{pmatrix} 1 \\ 2 \end{pmatrix}.$$

Then, we know that

$$w = \begin{pmatrix} 1 & 1 \\ 1 & 0 \end{pmatrix} + 2 \begin{pmatrix} 0 & 1 \\ -1 & 1 \end{pmatrix} = \begin{pmatrix} 1 & 3 \\ -1 & 2 \end{pmatrix} \in V.$$

Suppose we would like to change to a different coordinate system, relative to the ordered basis

$$\mathcal{B}_2 = \left\{ \begin{pmatrix} 1 & 2 \\ 0 & 1 \end{pmatrix}, \begin{pmatrix} 1 & 0 \\ 2 & -1 \end{pmatrix} \right\}.$$

(Check to see that you actually believe this is a basis of V.) That is, suppose we want to write $[w]_{\mathcal{B}_2}$, the coordinate vector according to the basis $\mathcal{B}_2$. Then, we need to find scalars α and β so that

$$w = \begin{pmatrix} 1 & 3 \\ -1 & 2 \end{pmatrix} = \alpha \begin{pmatrix} 1 & 2 \\ 0 & 1 \end{pmatrix} + \beta \begin{pmatrix} 1 & 0 \\ 2 & -1 \end{pmatrix}.$$

By inspection of the matrix entries, or using elimination on a system of equations, we see that $\alpha = \frac{3}{2}$ and $\beta = -\frac{1}{2}$. Therefore,

$$[w]_{\mathcal{B}_2} = \begin{pmatrix} 3/2 \\ -1/2 \end{pmatrix} \in \mathbb{R}^2.$$

$\square$

The next two examples provide more practice and clarity for the reader.

Example 3.5.11 Consider the subspace $V = \{ax^2 + (b-a)x + (a+b) \mid a, b \in \mathbb{R}\} \subseteq \mathcal{P}_2(\mathbb{R})$. Since

$$V = \{a(x^2 - x + 1) + b(x+1) \mid a, b \in \mathbb{R}\} = \mathrm{Span}\{x^2 - x + 1, x + 1\},$$

a basis for V is

$$\mathcal{B} = \{v_1 = x^2 - x + 1, v_2 = x + 1\}.$$

Thus, dim $V = 2$ and vectors in V can be represented by coordinate vectors in $\mathbb{R}^2$. Let $v = 3x^2 + x + 7$. We can write v in terms of $\mathcal{B}$ as $v = 3v_1 + 4v_2$ (check this). Therefore, $v \in V$ and the coordinate vector for v is $[v]_{\mathcal{B}} = \begin{pmatrix} 3 \\ 4 \end{pmatrix} \in \mathbb{R}^2$.

$\square$

Example 3.5.12 Now, consider the subspace $W = \left\{ \begin{pmatrix} \alpha & \beta \\ \gamma & \alpha+\beta+\gamma \end{pmatrix} \middle| \alpha, \beta, \gamma \in \mathbb{R} \right\} \subseteq \mathcal{M}_{2\times2}(\mathbb{R})$.
A basis for W is

$$\mathcal{B} = \left\{ \begin{pmatrix} 1 & 0 \\ 0 & 1 \end{pmatrix}, \begin{pmatrix} 0 & 1 \\ 0 & 1 \end{pmatrix}, \begin{pmatrix} 0 & 0 \\ 1 & 1 \end{pmatrix} \right\}.$$

Now consider the vector $w = \begin{pmatrix} 3 & 4 \\ -1 & 6 \end{pmatrix} \in \mathcal{M}_{2\times2}$. If $w \in W$, we can find the coordinate vector, $[w]_{\mathcal{B}} \in \mathbb{R}^3$. Let $v_1, v_2,$ and v_3 be the above basis vectors, respectively. Then we determine whether there are scalars $\alpha_1, \alpha_2, \alpha_3$ so that $w = \alpha_1 v_1 + \alpha_2 v_2 + \alpha_3 v_3$. That is, we determine whether there are scalars so that

$$\begin{pmatrix} 3 & 4 \\ -1 & 6 \end{pmatrix} = \alpha_1 \begin{pmatrix} 1 & 0 \\ 0 & 1 \end{pmatrix} + \alpha_2 \begin{pmatrix} 0 & 1 \\ 0 & 1 \end{pmatrix} + \alpha_3 \begin{pmatrix} 0 & 0 \\ 1 & 1 \end{pmatrix}.$$

Equating corresponding entries in the matrices on the left and right leads to the system of equations

$$3 = \alpha_1$$
$$4 = \alpha_2$$
$$-1 = \alpha_3$$
$$6 = \alpha_1 + \alpha_2 + \alpha_3.$$

Thus, we find that $\alpha_1 = 3$, $\alpha_2 = 4$, $\alpha_3 = -1$ telling us that $w \in W$ and has coordinate vector $[w]_{\mathcal{B}} = \begin{pmatrix} 3 \\ 4 \\ -1 \end{pmatrix}$.

Now, consider the vector $\tilde{w} = \begin{pmatrix} 3 & 3 \\ 4 & 4 \end{pmatrix} \in \mathcal{M}_{2 \times 2}(\mathbb{R})$. Following the same procedure as above, we can determine whether or not $[\tilde{w}]_{\mathcal{B}}$ exists. That is, we determine whether there are scalars so that

$$\begin{pmatrix} 3 & 3 \\ 4 & 4 \end{pmatrix} = \alpha_1 \begin{pmatrix} 1 & 0 \\ 0 & 1 \end{pmatrix} + \alpha_2 \begin{pmatrix} 0 & 1 \\ 0 & 1 \end{pmatrix} + \alpha_3 \begin{pmatrix} 0 & 0 \\ 1 & 1 \end{pmatrix}.$$

The corresponding system of equations is

$$3 = \alpha_1 \qquad\qquad\qquad \text{(Upper left entry)}$$
$$3 = \alpha_2 \qquad\qquad\qquad \text{(Upper right entry)}$$
$$4 = \alpha_3 \qquad\qquad\qquad \text{(Lower left entry)}$$
$$4 = \alpha_1 + \alpha_2 + \alpha_3 \qquad \text{(Lower right entry).}$$

Clearly, the system above has no solution. Therefore, $\tilde{w} \notin W$ and $[\tilde{w}]_{\mathcal{B}}$ does not exist. □

In the previous example, we were able to answer two questions with one set of steps. That is, we can determine whether a vector space contains a particular vector while computing the coordinate vector at the same time.

3.5.4 Brain Scan Images and Coordinates

One of our larger goals is to understand and determine how density distributions of the human head (brain scan images) produce a given set of radiographs. Thus far, we have focused on understanding both brain images and radiographic images as mathematical objects. We have arrived at an important place in which these objects can not only be arithmetically manipulated, but can be categorized and cataloged, as well. Consider the following stepping stones.

1. Vector Space Arithmetic. We can add (and subtract) images and multiply (scale) images by scalars.
2. Vector Space Closure. All possible images are contained in a vector space. Arithmetic operations on images lead only to images of the same class.
3. Subspace. Some subclasses of images retain all the properties of the larger vector space.
4. Linear Combination. Images can be expressed as simple weighted sums of other images. There are relationships among elements of sets of images.
5. Spanning Set. A small (usually finite) set of images can be used to characterize larger (possibly infinite) subspaces of images through linear combinations.
6. Linear Independence. Minimal spanning sets have the property of linear independence in which no image can be written as a linear combination of the others.

7. Basis. Minimal spanning sets of images comprise a most efficient catalog set for a vector space. Each image in the space is represented uniquely as a linear combination of basis images.

8. Coordinates. Using a given set of basis images, every image in a space is uniquely paired with a coordinate vector in $\mathbb{R}^n$.

This last item, coordinates, is a major breakthrough. Now, and only now, can we represent arbitrary abstract vectors (images, matrices, functions, polynomials, heat states, etc.) from finite-dimensional vector spaces as vectors in $\mathbb{R}^k$. Given an ordered basis for a (possibly very abstract) vector space, we can perform all mathematical operations in the equivalent (and much more familiar) coordinate space (see Exercise 8).

☞ **Path to New Applications**

When looking for a function that predicts results from collected data, researchers will use regression analysis. This method requires you to find a coordinate vector, with respect to chosen basis functions, that minimizes an error of fit. See Section 8.3.2 for more information about connections between linear algebra tools and regression analysis.

3.5.5 Exercises

1. Using the vector space V and basis $\mathcal{B}$ in Example 3.5.11, determine if the given vector is in V, and if so, find the coordinate vector.

    (a) $\left[5x^2 - 7x + 3\right]_{\mathcal{B}}$.
    (b) $\left[x^2 + 2\right]_{\mathcal{B}}$.
    (c) $\left[x^2 + 4x + 2\right]_{\mathcal{B}}$.

2. Let $X = \{(x, y, x - y, x + y) | x, y \in \mathbb{R}\}$. Find a basis, $\mathcal{B}$ for X. Determine if $v = (2, 3, 1, 5) \in X$. If so, find $[v]_{\mathcal{B}}$.

3. Complete Example 3.5.6.

4. Given vector space V with basis $\mathcal{B} = \{2x + 1, x^2, 2x^3\}$, find w when

    (a) $[w]_{\mathcal{B}} = \begin{pmatrix} 2 \\ 1 \\ 3 \end{pmatrix}$.

    (b) $[w]_{\mathcal{B}} = \begin{pmatrix} -1 \\ 0 \\ 3 \end{pmatrix}$.

5. Given $\mathcal{B} = \left\{ \begin{pmatrix} 1 & 2 \\ 1 & 1 \end{pmatrix}, \begin{pmatrix} 0 & 1 \\ 1 & 1 \end{pmatrix}, \begin{pmatrix} 0 & 0 \\ 1 & 1 \end{pmatrix} \right\}$, and $V = \text{Span}\,\mathcal{B}$,

    (a) Verify that $\mathcal{B}$ is indeed a basis for V.

    (b) Find the element $v \in V$ so that $[v]_{\mathcal{B}} = \begin{pmatrix} -1 \\ 0 \\ 2 \end{pmatrix}$.

(c) Find the element $v \in V$ so that $[v]_B = \begin{pmatrix} 0 \\ 0 \\ 1 \end{pmatrix}$.

6. Given $V = \text{Span}\{v_1, v_2, v_3, v_4\}$. Suppose also that $B_1 = \{v_1, v_2, v_3, v_4\}$ is linearly independent. Let $u = v_1 + 2v_2 - 3v_3 - v_4$.

   (a) Find $[u]_{B_1}$.
   (b) Show that $B_2 = \{v_1 - v_2, v_1 + v_2 + 3v_3 - v_4, v_2, v_3\}$, is a basis for V.
   (c) Find $[u]_{B_2}$.

7. Use the basis B_2 from Example 3.5.8 to find the following coordinate vectors.

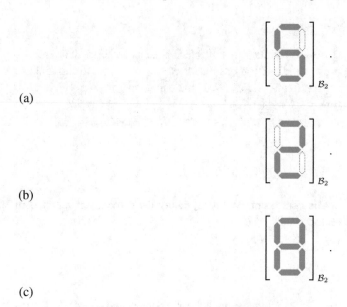

   (a)

   (b)

   (c)

8. Let v and w be vectors in finite-dimensional vector space V, α be a scalar, and B a basis for V. Prove the following.

   (a) $[\alpha v]_B = \alpha[v]_B$.
   (b) $[v + w]_B = [v]_B + [w]_B$.
   (c) The additive inverse of the coordinate vector for v is the coordinate vector of the additive inverse of v.
   (d) The coordinate vector for the additive identity vector is independent of choice of basis.

9. The standard ordered basis for the set of all 108×108 grayscale images is defined analogously to the standard ordered basis of Example 3.5.7. This basis is simple, easy to enumerate, and lends itself to straightforward coordinate vector decomposition. Suggest a different, potentially more useful, basis in the context of medical diagnosis of brain images.

10. Dava is training for a sprint triathlon consisting of a 3 mile run, followed by a 20 mile bike ride, and ending with a half mile swim. Recently, she has completed three training sessions with total times shown in the table below.

run (mi)	bike (mi)	swim (mi)	time (min)
2	2	½	35
0	10	1	60
5	0	1	65

Calculate Dava's expected triathlon time using two different methods.

(a) Compute the rates at which Dava can expect to perform in each event (minutes per mile) assuming that all times scale linearly with distance. Then determine her total expected time.

(b) Consider Dava's training session distances as vectors in $\mathbb{R}^3$. Show that this set of vectors is linearly independent. Using these vectors as a basis, find the linear combination of training runs that results in a vector of sprint triathlon distances. Then compute her expected time.

11. (**Data Classification.**) Suppose that $B = \{b_1, \ldots, b_n\}$ is a basis for $\mathbb{R}^n$. One can show, with appropriate definitions, that the $(n-1)$-dimensional subspace $S = \text{Span}\{b_1, \ldots, b_{n-1}\}$ (a hyperplane) separates $\mathbb{R}^n$ into two regions. Informally, when we say that a subspace separates $\mathbb{R}^n$ into two regions, we mean that any point not in this subspace lies on one side of the subspace or the other. Given two vectors v and w in $\mathbb{R}^n \setminus S$, find a method, using coordinates, to determine if v and w lie on the same side of the hyperplane. Explain.

☞ **Path to New Applications**

These ideas are useful in data classification. If we want to group data (vectors) into categories, we can look for subspaces that separate the data points that have different characteristics. (See Exercise 30 of Section 7.1.) For more information about data classification using linear algebra tools, see Section 8.3.2

Linear Transformations

4

In Chapter 1, we introduced several applications. In Chapter 2, we considered vector spaces of objects to be radiographed and their respective radiographs, vector spaces of images of a particular size and vector spaces of heat states. In these particular applications, we see that vectors from one vector space are transformed into vectors in the same or another vector spaces. For example, the heat along a rod is measured as a heat state and the process of the rod cooling takes one heat state to another. For the brain scan radiography application, object vectors are transformed into radiograph vectors as in Figure 4.1.

In this chapter, we will discuss the functions that transform vector spaces to vector spaces. We call such functions, transformations. We then consider the key property of transformations that is necessary so that we can apply linear algebra techniques. We will explore, in more detail, how heat states transform through heat diffusion. Using the result for heat state transformations and our knowledge of coordinate spaces we find a way to represent these transformations using matrices. We categorize linear transformations and their matrix representations by a number called the determinant. Finally, we explore ideal properties for the reconstruction of brain images from radiographs. We then explore these properties to learn more information about the vector spaces being transformed.

4.1 Explorations: Computing Radiographs and the Radiographic Transformation

In preceding chapters, we have seen how to model images as vectors (and the space of images as a vector space). We now focus on modeling the radiographic *process* that starts with a brain and produces an image of the brain.

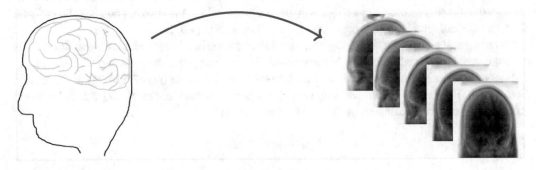

Fig. 4.1 In brain scan radiography, objects are transformed into radiographs.

© Springer Nature Switzerland AG 2022, corrected publication 2023
H. A. Moon et al., *Application-Inspired Linear Algebra*, Springer Undergraduate Texts in Mathematics and Technology, https://doi.org/10.1007/978-3-030-86155-1_4

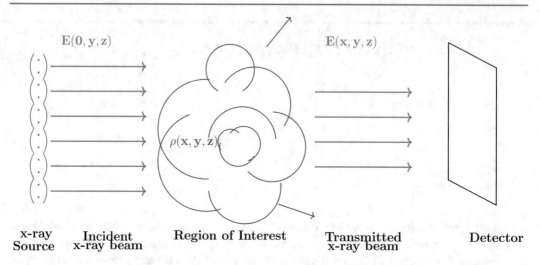

x-ray   Incident   Region of Interest   Transmitted   Detector
Source   x-ray beam   x-ray beam

Fig. 4.2 A schematic of the radiographic process. The function E measures the intensity of the X-ray beam. Here, the X-rays are traveling in the x direction.

We will need to make some further simplifying assumptions to model this process with linear algebra. We begin by describing some basic principles of the process and how they lead to a linear algebraic model of the radiographic process. Then, we explore how the radiographs on small objects are computed. We discuss the notation and setup needed to do these computations as well.

The material in this section is a no-frills sketch of the basic mathematical framework for radiography, and only lightly touches on physics concepts, since a deep understanding of physics principles is not necessary for understanding radiography. We refer the reader to Appendix A for more details on the process and the physics behind it.

4.1.1   Radiography on Slices

In transmission radiography such as that used in CAT scans, changes in an X-ray or other high-energy particle beam are measured and recorded after passing through an object of interest, such as a brain (Figure 4.2).

To further simplify the setup, we consider the problem one layer at a time. That is, we fix a height, and consider the slice of the object and the slice of detector at that height. In what follows we model the radiographic process restricted to a single height. At a given height, the slice of brain is 2-dimensional (2D), and the radiographic image of the slice is a 1-dimensional (1D) image.

To get the full picture, we "paste together" the lower dimensional objects and images; this models the radiographic process that takes a 3-dimensional (3D) brain and transforms it into a 2D image.

For our basic setup, the detector is divided into some fixed number (m) of "bins" (numbered, say, from 1 to m). For the kth bin, we denote the initial number of photons sent by p_k^0 and the total detected (after passing through the object) by p_k. Then it turns out that

$$p_k = p_k^0 e^{-s_k/\alpha},$$

where s_k is the total mass in the path of the kth bin portion of the beam, and α is a constant proportional to the bin area.

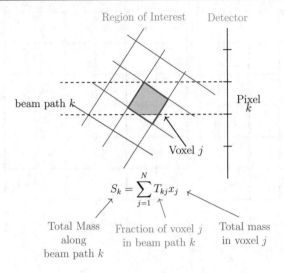

Fig. 4.3 Object space and radiograph space discretization.

We consider (a slice of) the region of interest to be subdivided into N cubic voxels (3D pixels). Let x_j be the mass in object voxel j and T_{kj} the fraction of voxel j in beam path k (see Figure 4.3). (Note that x_j is *not* related to the direction that the X-ray beams are traveling.) Then the mass along beam path k is

$$s_k = \sum_{j=1}^{N} T_{kj} x_j,$$

and the expected photon count p_k at radiograph pixel k is given by

$$p_k = p_k^0 e^{-\frac{1}{\alpha} \sum_{j=1}^{N} T_{kj} x_j},$$

or equivalently, we define b_k as

$$b_k \equiv \left(-\alpha \ln \frac{p_k}{p_k^0} \right) = \sum_{j=1}^{N} T_{kj} x_j.$$

What we have done here is replace photon counts p_k with the quantities b_k. This amounts to a variable change that allows us to formulate a matrix expression for the radiographic transformation

$$b = Tx.$$

So what have we accomplished? We are modeling slices of objects x as vectors in a vector space, in much the same way that we modeled images before. As before, we model slices of radiographic images b as vectors in another vector space. And, the calculations above produce a mathematical model of the radiographic transformation *that is given by matrix multiplication*. Moreover, the matrix does not depend on the specific vectors x or b, it *only* depends on the way the object and detectors are arranged. This means that we can determine this matrix before we ever produce a radiograph.

In the next section, we delve into the process of producing this matrix in a little more detail.

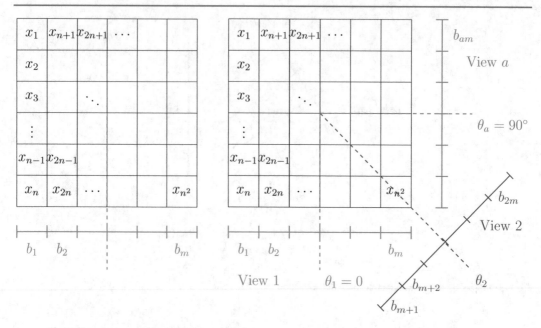

Fig. 4.4 Left: The geometry of a single-view radiographic transformation. Right: The geometry of a multiple-view radiographic transformation showing view 1, view 2, and view a (for some integer a).

4.1.2   Radiographic Scenarios and Notation

Keeping in mind that we are working with 2D slices of the object/region of interest. For example, a single-view radiographic setup consists of a 2D area of interest where the object will be placed, and a 1D screen onto which the radiograph will be recorded. A multiple-view radiographic setup consists of a single 2D area of interest experimentally designed so that radiographs of this area can be recorded for different locations about the object.

The geometry of a radiographic scenario is illustrated in Figure 4.4. Again, this is just for a single slice. The notation we use is as follows.

▶ Radiographic Scenario Notation

- Slice for region of interest: n by n array of voxels, where n is an even integer.
- Total number of voxels in each image slice is $N = n^2$.
- Each voxel has a width and height of 1 unit.
- For each radiographic view, we record m pixels of data, where m is an even integer. The center of each radiographic view "lines up with" the center of the object.
- The width of each pixel is $ScaleFac$. If $ScaleFac = 1$, then pixel width is the same as voxel width.
- Number of radiographic angles (views): a.
- Total number of pixels in the radiograph image: $M = am$
- Angle of the ith view (the angle of the line, connecting centers of the object and the radiograph, measured in degrees east of south): θ_i.
- Object mass at voxel j is x_j.
- Recorded radiograph value at pixel k is b_k.

In this exploration, we will be constructing matrix representations of the radiographic process for several different scenarios.

We consider object coordinate vectors $[x]_{\mathcal{B}_O}$ and radiographic data coordinate vectors $[b]_{\mathcal{B}_R}$, both using standard bases. Thus, objects are represented as vectors in $\mathbb{R}^N$ and radiographs as vectors in $\mathbb{R}^M$. What will be the size of the corresponding matrix that transforms an object vector $[x]_{\mathcal{B}_O}$ to its corresponding radiograph vector $[b]_{\mathcal{B}_R}$ through multiplication by the matrix? We call this matrix a matrix operator because the operation on objects is to multiply by the matrix. Read again the definition of the matrix operator from Section 4.1.1 (or Appendix A). Recalling the key point that particular values for $[x]$ and $[b]$ are *not necessary* for computing the matrix of the radiographic operator.

4.1.3   A First Example

Consider the setup pictured below. For this scenario, we have

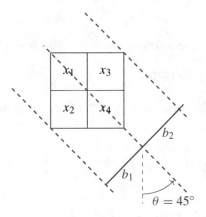

- Total number of voxels: $N = 4$ ($n = 2$).
- Total number of pixels: $M = m = 2$.
- $ScaleFac = \sqrt{2}$.
- Number of views: $a = 1$.
- Angle of the single view: $\theta_1 = 45°$.

Recalling that T_{kj} is the fraction of voxel j which projects perpendicularly onto pixel k, the matrix associated with this radiographic setup is

$$T = \begin{pmatrix} ^1/_2 & 1 & 0 & ^1/_2 \\ ^1/_2 & 0 & 1 & ^1/_2 \end{pmatrix}.$$

Be sure and check this to see if you agree. Hence, for *any* input vector $[x]$, the radiographic output is $[b] = T[x]$. Find the output when the object is the vector

$$[x] = \begin{pmatrix} 10 \\ 0 \\ 5 \\ 10 \end{pmatrix}.$$

For this simple example, it was easy to produce the matrix T "by hand." But in general, we will be radiographing much larger objects. Code that automates this process is in the MATLAB/OCTAVE file tomomap.m.

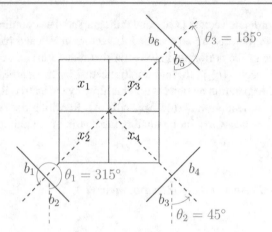

Fig. 4.5 Sketch of a radiographic scenario with three views at $\theta_1 = 315°$, $\theta_2 = 45°$, and $\theta_3 = 135°$.

4.1.4   Radiographic Setup Example

To illustrate how we identify the angles in a radiographic setup, consider the radiographic scenario below.

- Total number of voxels: $N = 4$ ($n = 2$).
- Number of pixels per radiographic view: $m = 2$.
- Number of views: $a = 3$.
- Total number of pixels: $M = am = 6$.
- Stretch factor for radiographic pixels: $ScaleFac = 1$.
- Angle of the views: $\theta_1 = 315°$, $\theta_2 = 45°$, and $\theta_3 = 135°$.

A sketch for this setup is found in Figure 4.5.

Using tomomap.m, we type

```
T = full(tomomap(2,2,[315,45,135],1))
```

With OCTAVE/MATLAB output:

```
T =

0.82843    0.50000    0.50000    0.00000
0.00000    0.50000    0.50000    0.82843
0.50000    0.82843    0.00000    0.50000
0.50000    0.00000    0.82843    0.50000
0.00000    0.50000    0.50000    0.82843
0.82843    0.50000    0.50000    0.00000
```

The reader should verify that this matrix represents the expected transformation.

4.1.5   Exercises

Some of the following exercises will ask you to use MATLAB or OCTAVE to compute radiographic transformations. You will need to first download the function tomomap.m from the IMAGEMath.org website. If you do not yet have access to personal or department computing resources, you can complete this assignment using OCTAVE-Online in a web browser.

Standard Bases for Radiographic Data

1. Consider the radiographic matrix operator T described in Appendix A and the example of Section 4.1.4. Suppose $\mathcal{B}_O$ is the standard ordered basis for the object space and $\mathcal{B}_R$ is the standard ordered basis for the radiograph space. The order for these bases is consistent with the description in this section.

    (a) Sketch and describe basis elements in $\mathcal{B}_O$.
    (b) Sketch and describe basis elements in $\mathcal{B}_R$.

With an ordered basis we are now able to represent both objects and radiographs as coordinate vectors in $\mathbb{R}^N$ and $\mathbb{R}^M$, respectively. This will make notation much simpler in future exercises.

Constructing Radiographic Transformation Matrices

2. Suppose you have the setup where

    • Height and width of the image in voxels: $n = 2$ (Total voxels $N = 4$)
    • Pixels per view in the radiograph: $m = 2$
    • $ScaleFac = 1$
    • Number of views: $a = 2$
    • Angle of the views: $\theta_1 = 0°, \theta_2 = 90°$

    (a) Sketch this setup.
    (b) Calculate the matrix associated with the setup.
    (c) Find the radiographs of the two objects below.

4	3		0	1
6	1		1	0

    (d) Represent the radiographs in the coordinate space you described in Exercise 1 above.

3. Suppose you have the setup where

    • Height and width of the image in voxels: $n = 2$ (Total voxels $N = 4$)
    • Pixels per view in the radiograph: $m = 2$
    • $ScaleFac = 1$
    • Number of views: $a = 2$
    • Angle of the views: $\theta_1 = 0°, \theta_2 = 45°$

    (a) Sketch this setup.
    (b) Calculate the matrix associated with the setup.
    (c) Repeat step (b) using the code tomomap.

4. Suppose you have the setup where

    • Height and width of the image in voxels: $n = 2$ (Total voxels $N = 4$)
    • Pixels per view in the radiograph: $m = 2$
    • $ScaleFac = \sqrt{2}$
    • Number of views: $a = 2$
    • Angle of the views: $\theta_1 = 45°, \theta_2 = 135°$

(a) Sketch this setup.
(b) Calculate the matrix associated with the setup.
(c) Repeat step (b) using the code **tomomap**.

5. Suppose you have the setup where

  - Height and width of the image in voxels: $n = 2$ (Total voxels $N = 4$)
  - Pixels per view in the radiograph: $m = 4$
  - $ScaleFac = \sqrt{2}/2$
  - Number of views: $a = 1$
  - Angle of the views: $\theta_1 = 45°$

(a) Sketch this setup.
(b) Calculate the matrix associated with the setup.
(c) Repeat step (b) using the code **tomomap**.

6. Suppose you have the setup where

  - Height and width of the image in voxels: $n = 4$ (Total voxels $N = 16$)
  - Pixels per view in the radiograph: $m = 2$
  - $ScaleFac = 1$
  - Number of views: $a = 2$
  - Angle of the views: $\theta_1 = 0°$, $\theta_2 = 45°$

(a) Sketch this setup.
(b) Calculate the matrix associated with the setup.
(c) Find the radiographs of images A, B, and C from Section 2.1 under this transformation.
(d) Repeat steps (b) and (c) using the code **tomomap**.

7. Suppose you have the setup where

  - Height and width of the image in voxels: $n = 4$ (Total voxels $N = 16$)
  - Pixels per view in the radiograph: $m = 4$
  - $ScaleFac = 1$
  - Number of views: $a = 3$
  - Angle of the views: $\theta_1 = 0°$, $\theta_2 = 25.5°$, and $\theta_3 = 90°$

(a) Sketch this setup.
(b) Calculate the matrix associated with the setup using **tomomap**.
(c) Find the radiographs of images A, B, and C from section 2.1 under this transformation.

8. A *block matrix* is a matrix of matrices. In other words, it is a large matrix that has been partitioned into sub-matrices. We usually represent the block matrix by drawing vertical and horizontal lines between the blocks. For example, the matrix

$$A = \begin{pmatrix} 1 & 0 & 1 & 0 \\ 0 & 1 & 0 & 1 \end{pmatrix}$$

can be considered as the block matrix

$$\left(\begin{array}{cc|cc} 1 & 0 & 1 & 0 \\ 0 & 1 & 0 & 1 \end{array} \right) = (I_2 | I_2).$$

(a) Choose one of the two-view radiographic setups from Exercises 2, 3, 5, or 6. Find the two matrices associated with each of the component single-view radiographic transformations. Compare these two matrices with the overall radiographic transformation matrix. What do you notice?

(b) Repeat this for the radiographic setup from Exercise 7.

(c) In general, the transformation associated with a multiple-view radiograph can be represented as a block matrix in a natural way, where the blocks represent the transformations associated with the component single-view radiographs. Suppose that you know that for a particular radiographic setup with k views, the individual views are represented by the matrices $T_1, \ldots, T_k$. What is the block matrix that represents the overall transformation T?

Radiographs of Linear Combinations of Objects.

Take the two objects in Exercise 2 to be x (left object) and y (right object). For each of the transformations in Exercises 2, 4, and 5, answer the following questions.

9. Determine the radiographs of the following objects.

   (a) $3x$

   (b) $0.5y$

   (c) $3x + 0.5y$

10. Generalize these observations to arbitrary linear combinations of object vectors. Write your conjecture(s) in careful mathematical notation.

4.2   Transformations

In Section 4.1, we became more familiar with the radiographic process. We identified two important features of this process. First, the process takes vectors in an object space to vectors in a radiograph space. Second, if we compute the radiograph of a linear combination of objects, this radiograph is a linear combination of the radiographs corresponding to those objects. That is, if x_1, x_2 are objects corresponding to radiographs b_1 and b_2, respectively, then the radiograph corresponding to $\alpha x_1 + x_2$, for some scalar α, is $\alpha b_1 + b_2$. At first, this property may not seem very useful or informative. However, suppose we have detailed knowledge about part of the object that is being radiographed. Suppose $x_{\exp}$ is the expected part of the object and we wish to discover how different the actual object, x, is from what is *expected*. One possible scenario is illustrated in Figure 4.6 where a known (or expected) object (at left) is similar to the actual (unknown) object (at right). The circle represents an unknown or unexpected part of the true object. We find the unexpected object as part of the difference object $x_{\text{diff}} = x - x_{\exp}$. Because we know the radiographic transformation, T, we can produce an expected radiograph $b_{\exp} = T x_{\exp}$. We then attempt to recover the radiograph that corresponds to the difference object. That is:

$$b = Tx = T(x_{\exp} + x_{\text{diff}}) = b_{\exp} + b_{\text{diff}},$$

so that the difference object is one that produces the difference radiograph $T x_{\text{diff}} = b_{\text{diff}}$. While this problem is mathematically equivalent to solving $Tx = b$, the significant knowledge of $x_{\exp}$ can help make the problem more tractable. For example, the difference image may be in a subspace of relatively small dimension, or the process of finding x_{diff} may be less prone to uncertainties.

A similar scenario arises when comparing radiographs of the "same" object taken at different times. Suppose we radiograph the object, $x_{\exp}$ as in Figure 4.6 (on the left) and find that the radiograph is $b_{\exp}$. But weeks later, we radiograph the same object (or so we think) and we get a radiograph that is

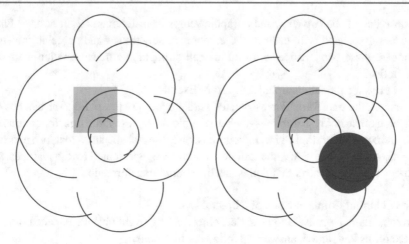

Fig. 4.6 Two possible objects. Left: An expected scenario x_{exp}. Right: The actual scenario x.

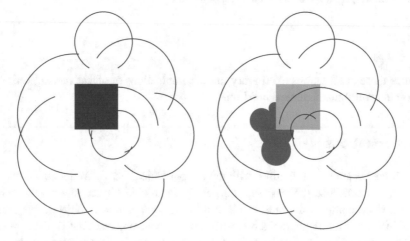

Fig. 4.7 Two objects with more mass than might be expected, indicating more density within the object being radiographed.

1.3 times the radiograph b_{exp}. This could mean that the object now looks more like one of the objects we see in Figure 4.7. We notice that the object density is proportionally larger, possibly meaning the object became more dense (as indicated by a darker color on the left in Figure 4.7) or another object is present along with the object we expected (represented by the figure on the right). Again, the change in the object is given by x_{diff}.

In this section, we will explore the properties of transformations such as radiographic transformations that possess the property that linear combinations of inputs result in the same linear combination of outputs. We call such transformations *linear*.

In order to avoid cumbersome notation, we will write vector spaces $(V, +, \cdot)$ as V, leaving to the reader the responsibility of understanding that each vector space is defined with specific operations of vector addition and scalar multiplication. Each vector space is also defined over a field, of which the reader must also infer from the context. For example, the declarative

"Let $(V, +, \cdot)$ and $(W, \oplus, \odot)$ be vector spaces over the field $\mathbb{F}$ and let $a, b \in \mathbb{F}$."

will be shortened to

"Let V and W be vector spaces and let a and b be scalars."

without loss of meaning.

4.2.1 Transformations are Functions

Simply stated, a transformation is a function whose domain and codomain are vector spaces. The point is that the word transformation is not referring to a new concept, rather a very familiar concept with the added context of vector spaces.

Definition 4.2.1

A function $T : V \to W$ is called a **transformation** if the **domain**, V, and the **codomain**, W, are vector spaces. □

Not all functions are transformations. Consider the function whose input is the date and time and whose output is the shirt color of the person closest to you. This function is only defined for dates and times for which you were living—this domain is not a vector space. The function outputs could be considered RGB values in $\mathbb{R}^3$ (a vector space). However, the closest person may not be wearing a shirt, in which case an acceptable function output is "no color" which is not an element of $\mathbb{R}^3$.

Example 4.2.2 Let A be the set of all possible angles measured counterclockwise from a particular ray in $\mathbb{R}^2$. The set A is a vector space. Let $T : A \to \mathbb{R}$ be the transformation defined by $T(a) = \cos(a)$. Notice that $T(a)$ is defined for all a in the domain A. We also know that $-1 \leq \cos(a) \leq 1$ for all $a \in A$. But, we still write that $T : A \to \mathbb{R}$ because $[-1, 1] \subseteq \mathbb{R}$. In this case, T transforms vectors from the space of angles into vectors in the space of real numbers. □

Definition 4.2.3

The **range** of a function (and therefore any transformation) $T : V \to W$ is the set of all codomain vectors y for which there exists a domain vector x such that $T(x) = y$.

□

Definition 4.2.3 is equivalently written:

Let $T : V \to W$. The range of T is the set $\{y \in W \mid T(x) = y \text{ for some } x \in V\}$.

It is not necessary for a transformation to have the potential to output all possible vectors in the codomain. The transformation outputs need only be vectors in the codomain. In Example 4.2.2, the transformation has codomain $\mathbb{R}$ and range $[-1, 1]$. This particular example tells us that the range of a transformation need not be a vector space.

4.2.2 Linear Transformations

As we discussed in the introduction to this section, some transformations have the interesting and useful property of preserving linear combinations. That is, transforming a linear combination of domain vectors is equivalent to first transforming the individual domain vectors and then computing the appropriate linear combination in the codomain.

> **Definition 4.2.4**

Let V and W be vector spaces. We say that $T : V \to W$ is a **linear transformation** if for every pair of vectors $v_1, v_2 \in V$ and every scalar α,

$$T(\alpha \cdot v_1 + v_2) = \alpha \cdot T(v_1) + T(v_2). \tag{4.1}$$

Equation 4.1 is called the *linearity condition* for transformations. Not all transformations are linear, though many familiar examples are. In fact, the example from Section 4.1 led us to the discussion of linear transformations and is, therefore, our first example.

Example 4.2.5 Let V_O be the vector space of objects and V_R be the space of radiographs. In Section 4.1, we found that $T : V_O \to V_R$ is a transformation so that if we have two objects $x_1, x_2 \in V_O$ whose corresponding radiographs are $T(x_1) = b_1$ and $T(x_2) = b_2$, then for scalar α, $T(\alpha x_1 + x_2) = \alpha b_1 + b_2$. Thus, radiographic transformations are examples of linear transformations. $\qquad\square$

We now consider a few examples to help understand how we determine whether or not a transformation is linear.

Example 4.2.6 Consider the vector space $\mathcal{F} = \{f : \mathbb{R} \to \mathbb{R} \mid f \text{ is continuous}\}$ defined with the standard operations of addition and scalar multiplication on functions (see Section 2.3). Consider fixed scalar a and define $T_a : \mathcal{F} \to \mathbb{R}$ to be the transformation defined by $T_a(f) = f(a)$. T_a is a linear transformation.

Proof. Let $a \in \mathbb{R}$ and define T_a as above. We will show that T_a satisfies the linearity condition by considering arbitrary functions $f, g \in \mathcal{F}$ and an arbitrary scalar $\alpha \in \mathbb{R}$. Then

$$\begin{aligned} T_a(\alpha f + g) &= (\alpha f + g)(a) \\ &= \alpha f(a) + g(a) \\ &= \alpha T_a(f) + T_a(g). \end{aligned}$$

Therefore, by Definition 4.2.4, T_a is linear. $\qquad\square$

Example 4.2.7 Consider the transformation $T : \mathcal{M}_{2\times3}(\mathbb{R}) \to \mathbb{R}^2$ defined as follows:

$$T \begin{pmatrix} a & b & c \\ d & f & g \end{pmatrix} = \begin{pmatrix} a+b+c \\ d+f+g \end{pmatrix}.$$

We can show that T is a linear transformation, again by testing the linearity condition. Consider arbitrary vectors $v_1, v_2 \in \mathcal{M}_{2\times3}(\mathbb{R})$ and an arbitrary scalar α. We will show that

$$T(\alpha v_1 + v_2) = \alpha T(v_1) + T(v_2).$$

Let

$$v_1 = \begin{pmatrix} a_1 & b_1 & c_1 \\ d_1 & f_1 & g_1 \end{pmatrix} \quad \text{and} \quad v_2 = \begin{pmatrix} a_2 & b_2 & c_2 \\ d_2 & f_2 & g_2 \end{pmatrix}$$

for some scalars $a_1, b_1, c_1, d_1, f_1, g_1, a_2, b_2, c_2, d_2, f_2, g_2 \in \mathbb{R}$. Then we know that

$$T(\alpha v_1 + v_2) = T\begin{pmatrix} \alpha a_1 + a_2 & \alpha b_1 + b_2 & \alpha c_1 + c_2 \\ \alpha d_1 + d_2 & \alpha f_1 + f_2 & \alpha g_1 + g_2 \end{pmatrix}$$

$$= \begin{pmatrix} \alpha a_1 + a_2 + \alpha b_1 + b_2 + \alpha c_1 + c_2 \\ \alpha d_1 + d_2 + \alpha f_1 + f_2 + \alpha g_1 + g_2 \end{pmatrix}$$

$$= \begin{pmatrix} \alpha(a_1 + b_1 + c_1) + (a_2 + b_2 + c_2) \\ \alpha(d_1 + f_1 + g_1) + (d_2 + f_2 + g_2) \end{pmatrix}$$

$$= \begin{pmatrix} \alpha(a_1 + b_1 + c_1) \\ \alpha(d_1 + f_1 + g_1) \end{pmatrix} + \begin{pmatrix} (a_2 + b_2 + c_2) \\ (d_2 + f_2 + g_2) \end{pmatrix}$$

$$= \alpha \begin{pmatrix} a_1 + b_1 + c_1 \\ d_1 + f_1 + g_1 \end{pmatrix} + \begin{pmatrix} a_2 + b_2 + c_2 \\ d_2 + f_2 + g_2 \end{pmatrix}$$

$$= \alpha T(v_1) + T(v_2).$$

Therefore, T maps linear combinations to linear combinations with the same scalars. So, T is a linear transformation. □

Because linear transformations preserve linear combinations, we can relate this idea back to matrix multiplication examples from Section 3.1.

Example 4.2.8 Let M be the matrix defined by

$$M = \begin{pmatrix} 2 & 3 & 1 \\ 1 & 2 & 1 \end{pmatrix}.$$

Define the transformation $T : \mathbb{R}^3 \to \mathbb{R}^2$ by

$$T\begin{pmatrix} a \\ b \\ c \end{pmatrix} = M\begin{pmatrix} a \\ b \\ c \end{pmatrix}.$$

Recall that multiplying by M will result in a linear combination of the columns of M. Therefore, we can rewrite T as

$$T\begin{pmatrix} a \\ b \\ c \end{pmatrix} = a\begin{pmatrix} 2 \\ 1 \end{pmatrix} + b\begin{pmatrix} 3 \\ 2 \end{pmatrix} + c\begin{pmatrix} 1 \\ 1 \end{pmatrix}.$$

This form of T inspires us to believe that it is a linear transformation. Indeed, let $x, y \in \mathbb{R}^2$. Then,

$$x = \begin{pmatrix} a \\ b \\ c \end{pmatrix} \text{ and } y = \begin{pmatrix} d \\ e \\ f \end{pmatrix},$$

for some scalars $a, b, c, d, e, f \in \mathbb{R}$. Therefore, if α is also a scalar, then

$$T(\alpha x + y) = M(\alpha x + y)$$

$$= M \begin{pmatrix} \alpha a + d \\ \alpha b + e \\ \alpha c + f \end{pmatrix}$$

$$= (\alpha a + d) \begin{pmatrix} 2 \\ 1 \end{pmatrix} + (\alpha b + e) \begin{pmatrix} 3 \\ 2 \end{pmatrix} + (\alpha c + f) \begin{pmatrix} 1 \\ 1 \end{pmatrix}$$

$$= \begin{pmatrix} 2(\alpha a + d) \\ 1(\alpha a + d) \end{pmatrix} + \begin{pmatrix} 3(\alpha b + e) \\ 2(\alpha b + e) \end{pmatrix} + \begin{pmatrix} 1(\alpha c + f) \\ 1(\alpha c + f) \end{pmatrix}$$

$$= \begin{pmatrix} 2\alpha a \\ \alpha a \end{pmatrix} + \begin{pmatrix} 3\alpha b \\ 2\alpha b \end{pmatrix} + \begin{pmatrix} \alpha c \\ \alpha c \end{pmatrix} + \begin{pmatrix} 2d \\ d \end{pmatrix} + \begin{pmatrix} 3e \\ 2e \end{pmatrix} + \begin{pmatrix} f \\ f \end{pmatrix}$$

$$= \alpha \left(a \begin{pmatrix} 2 \\ 1 \end{pmatrix} + b \begin{pmatrix} 3 \\ 2 \end{pmatrix} + c \begin{pmatrix} 1 \\ 1 \end{pmatrix} \right) + \left(d \begin{pmatrix} 2 \\ 1 \end{pmatrix} + e \begin{pmatrix} 3 \\ 2 \end{pmatrix} + f \begin{pmatrix} 1 \\ 1 \end{pmatrix} \right)$$

$$= \alpha M \begin{pmatrix} a \\ b \\ c \end{pmatrix} + M \begin{pmatrix} d \\ e \\ f \end{pmatrix}$$

$$= \alpha T(x) + T(y).$$

Thus, T satisfies the linearity condition and is, therefore, a linear transformation. $\square$

Example 4.2.9 In a first class in algebra, we learn that $f : \mathbb{R} \to \mathbb{R}$, defined by $f(x) = mx + b$ for $m, b \in \mathbb{R}$, is called a "linear function." If f is a linear transformation, then it should satisfy the linearity condition.

Let $x, y \in \mathbb{R}$ and let α be a scalar. Then

$$\alpha f(x) + f(y) = \alpha(mx + b) + my + b.$$
$$= m(\alpha x + y) + b + \alpha b$$
$$= f(\alpha x + y) + \alpha b.$$

Notice that $\alpha f(x) + f(y) = f(\alpha x + y)$ only when $b = 0$. Thus, in general, f is *not* a linear transformation. The reason that f is called a linear function is because the graph of f is a line. However, f does not satisfy the definition of linear, and we do not consider this function to be linear. Mathematicians call functions of the form $f(x) = mx + b$ affine functions. $\square$

Definition 4.2.10

Let V and W be vector spaces. Transformation $T : V \to W$ is an **affine transformation** if there exists constant vector $b \in W$ such that $\hat{T}(x) = T(x) - b$ is a linear transformation. $\square$

Example 4.2.11 Consider the vector space V (from Example 3.4.29) of color images that can be created by a 12 megapixel phone camera. Suppose you have the image on the left in Figure 4.8 and you want to lighten it to show more details of the cat you photographed (such as the image on the right.) You can apply a transformation $T : V \to V$ to the left image, I_{dark}. When adjusting the brightness of such an image, we add more white to the whole image (so as not to change contrast). That is, we add a flat image (all pixel intensities are equal). The transformation $T : V \to V$ used to brighten an image $I \in V$ is given by $T(I) = I + \alpha \mathbf{1}$, where $\mathbf{1}$ is the image of constant value 1 and α is a scalar brightening factor.

Fig. 4.8 Left: A 2448 × 3264 (approximately 8 megapixel) color image taken by a 12 megapixel phone camera. Right: Same image lightened by adding a flat image.

Fig. 4.9 A better lightening of the left image in Figure 4.8.

T is not a linear transformation. Indeed, if I_1 and I_2 are images in V, then $T(I_1 + I_2) = I_1 + I_2 + \alpha\mathbf{1}$, but $T(I_1) + T(I_2) = I_1 + \alpha\mathbf{1} + I_2 + \alpha\mathbf{1} = I_1 + I_2 + 2\alpha\mathbf{1}$.

Notice, on the right in Figure 4.8, that such a transformation leaves us with a poorly lightened image. Another transformation that can be performed on I_{dark}, to allow for better contrast, is to apply a nonlinear transformation such as $T_2 : V \to V$ defined pixelwise by $T_2((p_{i,j,k})) = (255 * \sqrt{p_{i,j,k}/255})$, where $I_{\text{dark}} = (p_{i,j,k})$ for $1 \leq i \leq 2448$, $1 \leq j \leq 3264$, and $1 \leq k \leq 3$). The resulting image is found in Figure 4.9. T_2 is not a linear transformation and this is easily seen by the square root, since $\sqrt{a + b} \neq \sqrt{a} + \sqrt{b}$. $\qquad\square$

The domain and codomain of linear (and nonlinear) transformations are often the same vector space. We see this in the image brightening example. Consider two more examples.

Example 4.2.12 Let us consider the vector space $\mathcal{D}(Z_2)$ of 7-bar LCD images given in Example 2.4.17. Consider, also, the transformation $T : \mathcal{D}(Z_2) \to \mathcal{D}(Z_2)$ defined by adding a vector to itself. That is, for $x \in \mathcal{D}(Z_2)$, $T(x) = x + x$. Notice that T maps every vector to 0, the zero vector. We can see this by considering the vector space properties of $\mathcal{D}(Z_2)$ and the scalar field properties of Z_2:

$$T(x) = x + x = 1 \cdot x + 1 \cdot x = (1 + 1) \cdot x = 0 \cdot x = 0.$$

Some specific examples are:

$$T\begin{pmatrix}\boxed{0}\end{pmatrix} = \boxed{0} + \boxed{0} = \boxed{8},$$

$$T\begin{pmatrix}\boxed{5}\end{pmatrix} = \boxed{5} + \boxed{5} = \boxed{8},$$

$$T\begin{pmatrix}\boxed{H}\end{pmatrix} = \boxed{H} + \boxed{H} = \boxed{8}.$$

Let $x, y \in \mathcal{D}(Z_2)$ and let $\alpha \in Z_2$, then

$$\begin{aligned}
T(\alpha \cdot x + y) &= (\alpha \cdot x + y) + (\alpha \cdot x + y) \\
&= \alpha \cdot x + \alpha \cdot x + y + y \\
&= \alpha \cdot (x + x) + (y + y) \\
&= \alpha T(x) + T(y).
\end{aligned}$$

Because the linearity condition holds for arbitrary domain vectors and arbitrary scalars, T is a linear transformation.   □

In Example 4.2.12, we introduced the zero transformation, the transformation $T : V \to W$ that maps all vectors in V to $0 \in W$ (the additive identity vector in W).

Definition 4.2.13

Let V and W be vector spaces and $T : V \to W$. We say that T is the **zero transformation** if $T(v) = 0 \in W$ for all $v \in V$.   □

Example 4.2.14 Consider the vector space $\mathbb{R}^2$ viewed as a 2D coordinate grid with x (horizontal) and y (vertical) axes. All vectors in $\mathbb{R}^2$ can then be viewed using arrows that originate at the origin and terminate at a point indicated by the coordinates of the vector. Using this picture of $\mathbb{R}^2$, we consider the transformation $T : \mathbb{R}^2 \to \mathbb{R}^2$ that is a combination of three other transformations: $T_1 : \mathbb{R}^2 \to \mathbb{R}^2$ that first rotates the vector 90° counterclockwise about the origin, next $T_2 : \mathbb{R}^2 \to \mathbb{R}^2$ reflects it across the line $y = x$, and finally, $T_3 : \mathbb{R}^2 \to \mathbb{R}^2$ reflects it over the x-axis. That is $T(x) = T_3(T_2(T_1(x)))$ for all $x \in \mathbb{R}^2$. For example,

$$\begin{pmatrix}1\\2\end{pmatrix} \overset{T_1}{\mapsto} \begin{pmatrix}-2\\1\end{pmatrix} \overset{T_2}{\mapsto} \begin{pmatrix}1\\-2\end{pmatrix} \overset{T_3}{\mapsto} \begin{pmatrix}1\\2\end{pmatrix}.$$

(Here, we use the notation "$a \overset{f}{\mapsto} b$" to indicate that $f(a) = b$. We read it as "a maps to b.") This example suggests that this transformation maps a vector onto itself. We can show that this is, indeed, true. Let $x = \begin{pmatrix}x_1\\x_2\end{pmatrix}$, then we have

$$T(x) = T \begin{pmatrix} x_1 \\ x_2 \end{pmatrix}$$

$$= T_3 \left(T_2 \left(T_1 \begin{pmatrix} x_1 \\ x_2 \end{pmatrix} \right) \right)$$

$$= T_3 \left(T_2 \begin{pmatrix} -x_2 \\ x_1 \end{pmatrix} \right)$$

$$= T_3 \begin{pmatrix} x_1 \\ -x_2 \end{pmatrix}$$

$$= \begin{pmatrix} x_1 \\ x_2 \end{pmatrix}$$

$$= x.$$

This transformation is also linear. Let $x, y \in \mathbb{R}^2$ and let $\alpha \in \mathbb{R}$. Then $T(\alpha x + y) = \alpha x + y = \alpha T(x) + T(y)$ and the linearity condition holds. $\quad\square$

Definition 4.2.15

Let V be a vector space and $T : V \to V$. We say that T is the **identity transformation** if $T(v) = v$ for all $v \in V$. $\quad\square$

Another very useful transformation is one that transforms a vector space into the corresponding coordinate space. We know that, given a basis for n-dimensional vector space V, we are able to represent any vector $v \in V$ as a coordinate vector in the vector space $\mathbb{R}^n$. Suppose $\mathcal{B} = \{v_1, v_2, \ldots, v_n\}$ is a basis for V. Recall that we find the coordinate vector $[v]_\mathcal{B}$ by finding the scalars, $\alpha_1, \alpha_2, \ldots, \alpha_n$, that make the linear combination $v = \alpha_1 v_1 + \alpha_2 v_2 + \ldots + \alpha_n v_n$, giving us

$$[v]_\mathcal{B} = \begin{pmatrix} \alpha_1 \\ \alpha_2 \\ \vdots \\ \alpha_n \end{pmatrix} \in \mathbb{R}^n.$$

The transformation that maps a vector to its coordinate vector is called the *coordinate transformation*. Our next theorem shows that the coordinate transformation is linear.

Theorem 4.2.16

Let V be an n-dimensional vector space and $\mathcal{B}$ an ordered basis for V. Then $T : V \to \mathbb{R}^n$ defined by $T(v) = [v]_\mathcal{B}$ is a linear transformation.

Proof. Suppose V is a vector space and $\mathcal{B}$ an ordered basis for V. Let $u, v \in V$ and α be a scalar. Then there are scalars $a_1, a_2, \ldots, a_n, b_1, b_2, \ldots, b_n$ so that

$$[u]_{\mathcal{B}} = \begin{pmatrix} a_1 \\ a_2 \\ \vdots \\ a_n \end{pmatrix} \text{ and } [v]_{\mathcal{B}} = \begin{pmatrix} b_1 \\ b_2 \\ \vdots \\ b_n \end{pmatrix}.$$

Notice that

$$T(\alpha u + v) = [\alpha u + v]_{\mathcal{B}}$$

$$= \begin{pmatrix} \alpha a_1 + b_1 \\ \alpha a_2 + b_2 \\ \vdots \\ \alpha a_n + b_n \end{pmatrix}$$

$$= \alpha \begin{pmatrix} a_1 \\ a_2 \\ \vdots \\ a_n \end{pmatrix} + \begin{pmatrix} b_1 \\ b_2 \\ \vdots \\ b_n \end{pmatrix}$$

$$= \alpha [u]_{\mathcal{B}} + [v]_{\mathcal{B}}$$

$$= \alpha T(u) + T(v).$$

Thus, the linearity condition holds and T is a linear transformation. □

⋆⋆ Watch Your Language!

It is important to notice the subtle difference between an everyday function and a transformation. A function need not have vector space domain and codomain, whereas a transformation requires both to be vector spaces. Linear transformations are particular transformations that preserve linear combinations. It is correct to say

✓ The range of a transformation is the set of all outputs.
✓ Linear transformations satisfy the linearity property.
✓ Linear transformations preserve linear combinations.

But it would be incorrect to say

✗ The codomain of a transformation is the set of all outputs.

4.2.3   Properties of Linear Transformations

The transformations T_1, T_2, and T_3 from Example 4.2.14 are all linear (the proofs are left to the reader). This leads us to question whether it is always true if we compose linear transformations, that we get a linear transformation. What about other typical function operations on linear transformations? Do they also lead to linear transformations? These questions are answered in this section. First, let us define some of the common operations on transformations.

Definition 4.2.17

Let U and V be vector spaces and let $T_1 : V \to U$ and $T_2 : V \to U$ be transformations. We define the **transformation sum**, $T_1 + T_2 : V \to U$ by

$$(T_1 + T_2)(v) = T_1(v) + T_2(v)$$

for every $v \in V$. We define the **transformation difference**, $T_1 - T_2 : V \to U$ by

$$(T_1 - T_2)(v) = T_1(v) + (-1) \cdot T_2(v)$$

for every $v \in V$.

□

Example 4.2.18 Consider the transformations $T_1 : \mathbb{R}^2 \to \mathbb{R}^2$ and $T_2 : \mathbb{R}^2 \to \mathbb{R}^2$ given by

$$T_1 \begin{pmatrix} x \\ y \end{pmatrix} = \begin{pmatrix} x - y \\ x + y \end{pmatrix}, \text{ and } T_2 \begin{pmatrix} x \\ y \end{pmatrix} = \begin{pmatrix} x \\ 2y + x \end{pmatrix}.$$

We have

$$(T_1 + T_2) \begin{pmatrix} x \\ y \end{pmatrix} = \begin{pmatrix} x - y \\ x + y \end{pmatrix} + \begin{pmatrix} x \\ 2y + x \end{pmatrix} = \begin{pmatrix} x - y + x \\ x + y + 2y + x \end{pmatrix} = \begin{pmatrix} 2x - y \\ 2x + 3y \end{pmatrix}.$$

As a particular realization, notice that

$$T_1 \begin{pmatrix} 1 \\ 2 \end{pmatrix} = \begin{pmatrix} -1 \\ 3 \end{pmatrix}, \quad T_2 \begin{pmatrix} 1 \\ 2 \end{pmatrix} = \begin{pmatrix} 1 \\ 5 \end{pmatrix}, \quad T_1 \begin{pmatrix} 1 \\ 2 \end{pmatrix} + T_2 \begin{pmatrix} 1 \\ 2 \end{pmatrix} = \begin{pmatrix} 0 \\ 8 \end{pmatrix},$$

and, therefore,

$$(T_1 + T_2) \begin{pmatrix} 1 \\ 2 \end{pmatrix} = \begin{pmatrix} 0 \\ 8 \end{pmatrix}.$$

□

Definition 4.2.19

Let U, V, and W be vector spaces and let $T_1 : V \to W$ and $T_2 : W \to U$ be transformations. We define the **composition transformation**

$$T_2 \circ T_1 : V \to U$$

by $(T_2 \circ T_1)(v) = T_2(T_1(v))$ for every $v \in V$.

□

Multiple sequential compositions of transformations, such as we saw in Example 4.2.14, are commonly written without extraneous parenthetical notation. In the example, we have $T = T_3 \circ (T_2 \circ T_1)$, but we write simply $T = T_3 \circ T_2 \circ T_1$.

Example 4.2.20 Consider the transformations $T_1 : \mathbb{R}^2 \to \mathbb{R}^2$ given by

$$T_1 \begin{pmatrix} x \\ y \end{pmatrix} = \begin{pmatrix} 3x \\ y - x \end{pmatrix}$$

and $T_2 : \mathbb{R}^2 \to \mathbb{R}$ given by

$$T_2 \begin{pmatrix} a \\ b \end{pmatrix} = a + b.$$

Then

$$(T_2 \circ T_1)\begin{pmatrix} x \\ y \end{pmatrix} = T_2\left(T_1\begin{pmatrix} x \\ y \end{pmatrix}\right) = T_2\begin{pmatrix} 3x \\ y - x \end{pmatrix} = 3x + y - x = 2x + y.$$

$\square$

Considering the examples of combinations of transformations that we have encountered, it is useful to know whether these new transformations preserve linearity. Is the sum of linear transformations also linear? Is the composition of linear transformations also linear? The answer is yes and is presented in the following theorem.

Theorem 4.2.21

Let U, V, and W be vector spaces and let $T_1 : V \to W$, $T_2 : V \to W$, and $T_3 : W \to U$ be linear transformations. Then the following transformations are linear:

1. $T_1 + T_2 : V \to W$,
2. $T_1 - T_2 : V \to W$,
3. $\alpha T_1 : V \to W$, for any scalar α,
4. $T_3 \circ T_1 : V \to U$.

The reader is asked to provide a proof in Exercise 39.

In examples like Example 4.2.9, we might wonder why something whose graph is a line, is not called linear. Here, we explore what went wrong with affine functions. When looking at our work, we see that, in order for f to be linear, b and αb need to be equal for all scalars α. This only occurs when $b = 0$, that is when $f(x) = mx$. In that case,

$$\begin{aligned} f(\alpha x + y) &= m(\alpha x + y) \\ &= \alpha(mx) + my \\ &= \alpha f(x) + f(y). \end{aligned}$$

That means, $f : \mathbb{R} \to \mathbb{R}$ defined by $f(x) = mx$ is linear. This brings to light a very important property of linear transformations.

Theorem 4.2.22

Let V and W be vector spaces. If $T : V \to W$ is a linear transformation, then $T(0_V) = 0_W$, where 0_V and 0_W are the zero vectors of V and W, respectively.

Proof. Let V and W be vector spaces and let $T : V \to W$ be a linear transformation. We know that, for any scalar α, $\alpha 0_V = 0_V$. So,

$$T(0_V) = T(\alpha 0_V) = \alpha T(0_V). \tag{4.2}$$

In particular, when $\alpha = 0$ we see that $T(0_V) = 0_W$. $\square$

Theorem 4.2.22 gives us a quick check to see whether a transformation is *not* linear. If $T(0_V) \neq 0_W$ then T is not linear. The converse is not necessarily true. The fact that $T(0_V) = 0_W$ is not a sufficient test for linearity. Consider the following examples.

Example 4.2.23 Consider again the vector space $\mathcal{D}(Z_2)$. Define $T : \mathcal{D}(Z_2) \to \mathbb{R}$ to be the transformation that counts the number of "lit" bars. For example,

$$T \begin{pmatrix} \end{pmatrix} = 5 \text{ and } T \begin{pmatrix} \end{pmatrix} = 7.$$

Notice that $0 \in \mathcal{D}(Z_2)$ is the LCD vector with no bars "lit." And so,

$$T(0) = T \begin{pmatrix} \end{pmatrix} = 0.$$

Therefore, $0 \in \mathcal{D}(Z_2)$ maps to $0 \in \mathbb{R}$. But, T is not a linear transformation. We can see this by considering the following example. Notice that,

$$T \begin{pmatrix} + \end{pmatrix} = T \begin{pmatrix} \end{pmatrix} = 5,$$

but

$$T \begin{pmatrix} \end{pmatrix} + T \begin{pmatrix} \end{pmatrix} = 5 + 2 \neq 5.$$

Thus, T is not linear. $\square$

Example 4.2.24 Consider again the image brightening Example 4.2.11. The transformation is $T(I) = I + \alpha \mathbf{1}$. The zero image, represented as all black pixels, is transformed into a nonzero image of constant intensity α. Thus, for any $\alpha \neq 0$, T is not linear. $\square$

Linear transformations have the special property that they can be defined in terms of their action on a basis for the domain. First, we define what it means for two transformations to be equal, then show two important results.

Definition 4.2.25

Let V and W be vector spaces. We say that two transformations $T_1 : V \to W$ and $T_2 : V \to W$ are **equal**, and write $\mathbf{T_1} = \mathbf{T_2}$, if $T_1(x) = T_2(x)$ for all $x \in V$.   □

We have seen an example of two equal transformations. Recall that the transformation in Example 4.2.14 was written as the composition of three transformations and as the identity transformation. The next theorem begins the discussion about how linear transformations act on basis elements of the domain space.

Theorem 4.2.26

Let V and W be vector spaces and suppose $\{v_1, v_2, \cdots, v_n\}$ is a basis for V. Then for $\{w_1, w_2, \cdots, w_n\} \subseteq W$, there exists a unique linear transformation $T : V \to W$ such that $T(v_k) = w_k$ for $k = 1, 2, \cdots, n$.

▶ **Note about the proof:** When proving the existence of something in mathematics, we need to be able to find and name the object which we are proving exists. Watch as we do that very thing in the following proof.

Proof. Suppose V and W are vector spaces and $\mathcal{B} = \{v_1, v_2, \cdots, v_n\}$ a basis for V. Let $x \in V$. Then there exist unique scalars $a_1, a_2, \dots, a_n$ such that $x = a_1 v_1 + a_2 v_2 + \cdots + a_n v_n$. Also, suppose $w_1, w_2, \cdots, w_n$ are arbitrary vectors in W. Define $T : V \to W$ by

$$T(x) = a_1 w_1 + a_2 w_2 + \cdots + a_n w_n.$$

Now, since v_k is uniquely written as a linear combination of $v_1, v_2, \dots, v_n$ using scalars

$$a_i = \begin{cases} 0 & \text{if } i \neq k \\ 1 & \text{if } i = k, \end{cases}$$

we know that $T(v_k) = w_k$ for $k = 1, 2, \dots, n$. We next show that T is linear and that T is unique.

Linearity

To show that T is linear, we consider $y, z \in V$ and scalar α, and show that

$$T(\alpha y + z) = \alpha T(y) + T(z).$$

Because $\mathcal{B}$ is a basis of V, there exist scalars b_k and c_k, for $k = 1, 2, \dots, n$ so that $y = b_1 v_1 + b_2 v_2 + \cdots + b_n v_n$ and $z = c_1 v_1 + c_2 v_2 + \dots + c_n v_n$. We have

$$T(\alpha y + z) = T\left(\alpha \sum_{k=1}^{n} b_k v_k + \sum_{k=1}^{n} c_k v_k \right)$$

$$= T\left(\sum_{k=1}^{n} (\alpha b_k v_k + c_k v_k) \right)$$

$$= T \left(\sum_{k=1}^{n} (\alpha b_k + c_k) v_k \right)$$

$$= \sum_{k=1}^{n} (\alpha b_k + c_k) w_k$$

$$= \alpha \sum_{k=1}^{n} b_k w_k + \sum_{k=1}^{n} c_k w_k$$

$$= \alpha T(y) + T(z).$$

Therefore, by Definition 4.2.4, T is a linear transformation.

Uniqueness

To show that T is unique we will show that, for any other linear transformation $U : V \to W$ with the property that $U(v_k) = w_k$ for $k = 1, 2, \ldots, n$, it must be true that $U = T$. Let $x \in V$, then there are unique scalars $a_1, a_2, \ldots, a_n$ so that

$$x = a_1 v_1 + a_2 v_2 + \ldots + a_n v_n.$$

Therefore, by linearity of U,

$$U(x) = U \left(\sum_{k=1}^{n} a_k v_k \right) = \sum_{k=1}^{n} a_k U(v_k) = \sum_{k=1}^{n} a_k w_k = T(x).$$

Thus, by definition of equal transformations, $U = T$. $\square$

Theorem 4.2.26 tells us that, there is always one and only one linear transformation that maps basis elements, of an n-dimensional vector space, to n vectors of our choosing in another vector space. The following corollary makes clear an important result that follows directly.

Corollary 4.2.27
Let V and W be vector spaces and suppose $\{v_1, v_2, \cdots, v_n\}$ is a basis for V. If $T : V \to W$ and $U : V \to W$ are linear transformations such that $T(v_k) = U(v_k)$ for $k = 1, 2, \cdots, n$, then $T = U$.

Proof. Suppose V and W are vector spaces and $\{v_1, v_2, \cdots, v_n\}$ is a basis for V. Let T, U be linear transformations defined as above. Let $w_k = T(v_k)$ for $k = 1, 2, \cdots, n$ with $w_k \in W$. Then, by supposition, $U(v_k) = w_k$. So by Theorem 4.2.26, $T = U$. $\square$

Corollary 4.2.27 shows us that the action of a transformation on a basis for the domain fully determines the action the transformation will have on *any* domain vector. There cannot be two different linear transformations that have the same effect on a basis for the domain.

Consider the radiographic transformation of Section 4.1.4. The domain is the vector space of 2×2 voxel objects and the codomain is the vector space of 2 pixel radiographs. The transformation is fully defined by its action on a basis for the domain. Suppose we consider the standard basis

$$B = \left\{ \begin{array}{|c|c|} \hline 1 & 0 \\ \hline 0 & 0 \\ \hline \end{array}, \begin{array}{|c|c|} \hline 0 & 0 \\ \hline 1 & 0 \\ \hline \end{array}, \begin{array}{|c|c|} \hline 0 & 1 \\ \hline 0 & 0 \\ \hline \end{array}, \begin{array}{|c|c|} \hline 0 & 0 \\ \hline 0 & 1 \\ \hline \end{array} \right\}.$$

The action of the transformation on these basis vectors yields four radiographs:

$$T(B) = \left\{ \tfrac{1}{2}\diagup\diagup , \begin{array}{c}0\\1\end{array}\diagup\diagup , \begin{array}{c}1\\0\end{array}\diagup\diagup , \tfrac{1}{2}\diagup\diagup \right\}.$$

No other *linear* transformation transforms the vectors of B into the vectors in $T(B)$. No other information about T is needed in order to determine its action on arbitrary vectors. As a specific example, suppose

$$x = \begin{array}{|c|c|} \hline 2 & 3 \\ \hline 1 & 1 \\ \hline \end{array},$$

and we wish to find the radiograph b that results in transforming x under the transformation T. Using the fact that T is linear, we have

$$b = T(x)$$

$$= T\left(\begin{array}{|c|c|} \hline 2 & 3 \\ \hline 1 & 1 \\ \hline \end{array} \right)$$

$$= T\left(2 \cdot \begin{array}{|c|c|} \hline 1 & 0 \\ \hline 0 & 0 \\ \hline \end{array} + 1 \cdot \begin{array}{|c|c|} \hline 0 & 0 \\ \hline 1 & 0 \\ \hline \end{array} + 3 \cdot \begin{array}{|c|c|} \hline 0 & 1 \\ \hline 0 & 0 \\ \hline \end{array} + 1 \cdot \begin{array}{|c|c|} \hline 0 & 0 \\ \hline 0 & 1 \\ \hline \end{array} \right)$$

$$= 2 \cdot T\left(\begin{array}{|c|c|} \hline 1 & 0 \\ \hline 0 & 0 \\ \hline \end{array} \right) + 1 \cdot T\left(\begin{array}{|c|c|} \hline 0 & 0 \\ \hline 1 & 0 \\ \hline \end{array} \right) + 3 \cdot T\left(\begin{array}{|c|c|} \hline 0 & 1 \\ \hline 0 & 0 \\ \hline \end{array} \right) + 1 \cdot T\left(\begin{array}{|c|c|} \hline 0 & 0 \\ \hline 0 & 1 \\ \hline \end{array} \right)$$

$$= 2 \cdot \left(\tfrac{1}{2}\diagup \right) + 1 \cdot \left(\begin{array}{c}0\\1\end{array}\diagup \right) + 3 \cdot \left(\begin{array}{c}1\\0\end{array}\diagup \right) + 1 \cdot \left(\tfrac{1}{2}\diagup \right)$$

$$= \tfrac{5}{2}\, \tfrac{9}{2}\diagup .$$

All we needed to know is how T acts on a basis for the object space as well as the coordinates of the given object in the same basis. The reader should compare this result carefully with the discussion in Section 4.1.4. These ideas will be explored more completely in Section 4.4.

4.2.4 Exercises

For Exercises 1 through 13, determine which of the given transformations are linear. For each, provide a proof or counterexample as appropriate.

1. Define M to be the 3×2 matrix

$$M = \begin{pmatrix} 1 & 3 \\ 1 & 1 \\ 1 & 2 \end{pmatrix}$$

and define $f : \mathbb{R}^2 \to \mathbb{R}^3$ by $f(v) = Mv$

2. Define

$$M = \begin{pmatrix} 1 & 3 \\ 2 & -2 \end{pmatrix}$$

and $f : \mathbb{R}^2 \to \mathbb{R}^2$ by $f(v) = M \begin{pmatrix} 1 \\ 3 \end{pmatrix} + v$.

3. Define M as in Exercise 2 and define $T : \mathbb{R}^2 \to \mathbb{R}^2$ by $T(v) = Mv$.
4. Define $f : \mathbb{R}^3 \to \mathbb{R}^2$ by $f(v) = Mv + x$, where

$$M = \begin{pmatrix} 1 & 2 & 1 \\ 1 & 2 & 1 \end{pmatrix} \quad \text{and} \quad x = \begin{pmatrix} 1 \\ 0 \end{pmatrix}$$

5. Let $T : \mathcal{P}_2 \to \mathbb{R}$ be defined by $T(ax^2 + bx + c) = a + b + c$.
6. Define $\mathcal{F} : V \to \mathcal{P}_1$, where

$$V = \{ax^2 + (3a - 2b)x + b \mid a, b \in \mathbb{R}\} \subseteq \mathcal{P}_2.$$

by

$$\mathcal{F}(ax^2 + (3a - 2b)x + b) = 2ax + 3a - 2b.$$

7. Define $\mathcal{G} : \mathcal{P}_2 \to \mathcal{M}_{2 \times 2}$ by

$$\mathcal{G}(ax^2 + bx + c) = \begin{pmatrix} a & a - b \\ c - 2 & c + 3a \end{pmatrix}.$$

8. Define $h : V \to \mathcal{P}_1$, where

$$V = \left\{ \begin{pmatrix} a & b & c \\ 0 & b - c & 2a \end{pmatrix} \middle| \, a, b, c \in \mathbb{R} \right\} \subseteq \mathcal{M}_{2 \times 3}$$

by

$$h \begin{pmatrix} a & b & c \\ 0 & b - c & 2a \end{pmatrix} = ax + c.$$

9. Let

$$\mathcal{I} = \left\{ I = \begin{vmatrix} & 3a & \\ -b & 0 & 2a \\ c & 3c & b \end{vmatrix} \; \middle| \; a, b, c \in \mathbb{R} \right\}.$$

And define $f : \mathcal{I} \to \mathcal{P}_2$ by

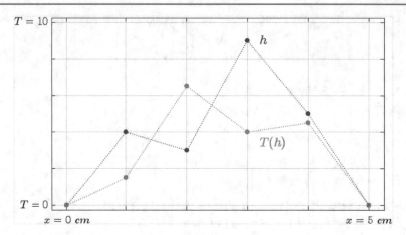

Fig. 4.10 Example of averaging heat state transformation.

$$f(I) = ax^2 + (b+c)x + (a+c).$$

10. Define $f : \mathcal{M}_{2\times 2} \to \mathbb{R}^4$ by

$$f \begin{pmatrix} a & b \\ c & d \end{pmatrix} = \begin{pmatrix} a \\ b \\ c \\ d \end{pmatrix}.$$

11. Define $f : \mathcal{P}_2 \to \mathbb{R}^2$ by

$$f(ax^2 + bx + c) = \begin{pmatrix} a+b \\ a-c \end{pmatrix}.$$

12. Let $\mathcal{H}_4$ be the set of all possible heat states sampled every 1 cm along a 5 cm long rod. Define a function $T : \mathcal{H}_4 \to \mathcal{H}_4$ by replacing each value (which does not correspond to an endpoint) with the average of its neighbors. The endpoint values are kept at 0. An example of T is shown in Figure 4.10.

13. Define $\mathcal{O}: \mathcal{D}(Z_2) \to Z_2$ by $\mathcal{O}(d) = 1$ if d is an LCD digit with at least one lit bar, and $\mathcal{O}(d) = 0$ otherwise.

For Exercises 14-27, you will explore the geometry of a linear transformation. Let $\square_1$ be a square that resides in the positive quadrant of $\mathbb{R}^2$, has sides aligned with the x- and y-axes, a vertex at the origin, and side length equal to 1. (It might be useful to know the result of Exercise 32: a linear transformation T maps line segments to line segments, so the image of a line segment whose endpoints are a and b is the line segment with endpoints $T(a)$ and $T(b)$.)

14. Draw $\square_1$ and write the vectors corresponding to each vertex.

15. Consider the function $S : \mathbb{R}^2 \to \mathbb{R}^2$, defined as

$$S \begin{pmatrix} x \\ y \end{pmatrix} = \begin{pmatrix} 2 & 0 \\ 0 & 1 \end{pmatrix} \begin{pmatrix} x \\ y \end{pmatrix}.$$

Verify that S is a linear transformation. Use S to transform $\square_1$. This means that you will find $S(x, y)$ for each vertex on $\square_1$. What did S do to $\square_1$?

16. Consider the function $R_1 : \mathbb{R}^2 \to \mathbb{R}^2$, defined as

$$R_1 \begin{pmatrix} x \\ y \end{pmatrix} = \begin{pmatrix} 1 & 0 \\ 0 & -1 \end{pmatrix} \begin{pmatrix} x \\ y \end{pmatrix}.$$

Verify that R_1 is a linear transformation. Use R_1 to transform $\square_1$. What does R_1 do to $\square_1$?

17. Consider the function $R_2 : \mathbb{R}^2 \to \mathbb{R}^2$, defined as

$$R_2 \begin{pmatrix} x \\ y \end{pmatrix} = \begin{pmatrix} 1 & 0 \\ 0 & -1 \end{pmatrix} \begin{pmatrix} x \\ y \end{pmatrix}.$$

Verify that R_2 is a linear transformation. Use R_2 to transform $\square_1$. What does R_2 do to $\square_1$?

18. Consider the function $T : \mathbb{R}^2 \to \mathbb{R}^2$, defined as

$$T \begin{pmatrix} x \\ y \end{pmatrix} = \begin{pmatrix} 1 & 0 \\ 1 & 0 \end{pmatrix} \begin{pmatrix} x \\ y \end{pmatrix}.$$

Verify that T is a linear transformation. Use T to transform $\square_1$. What does T do to $\square_1$?

19. Consider the function $\tilde{T} : \mathbb{R}^2 \to \mathbb{R}^2$, defined as

$$\tilde{T} \begin{pmatrix} x \\ y \end{pmatrix} = \begin{pmatrix} 2 & 0 \\ 2 & 0 \end{pmatrix} \begin{pmatrix} x \\ y \end{pmatrix}.$$

Verify that $\tilde{T}$ is a linear transformation. Use $\tilde{T}$ to transform $\square_1$. What does $\tilde{T}$ do to $\square_1$?

20. Consider the function $S_2 : \mathbb{R}^2 \to \mathbb{R}^2$, defined as

$$S_2 \begin{pmatrix} x \\ y \end{pmatrix} = \begin{pmatrix} 1 & 1 \\ 0 & 1 \end{pmatrix} \begin{pmatrix} x \\ y \end{pmatrix}.$$

Verify that S_2 is a linear transformation. Use S_2 to transform $\square_1$. What does S_2 do to $\square_1$?

21. Consider the function $S_3 : \mathbb{R}^2 \to \mathbb{R}^2$, defined as

$$S_3 \begin{pmatrix} x \\ y \end{pmatrix} = \begin{pmatrix} 2 & 0 \\ 2 & 2 \end{pmatrix} \begin{pmatrix} x \\ y \end{pmatrix}.$$

Verify that S_3 is a linear transformation. Use S_3 to transform $\square_1$. What does S_3 do to $\square_1$?

22. Consider the function $R_2 : \mathbb{R}^2 \to \mathbb{R}^2$, defined as

$$R_2 \begin{pmatrix} x \\ y \end{pmatrix} = \frac{1}{\sqrt{2}} \begin{pmatrix} 1 & -1 \\ 1 & 1 \end{pmatrix} \begin{pmatrix} x \\ y \end{pmatrix}.$$

Verify that R_2 is a linear transformation. Use R_2 to transform $\square_1$. What does R_2 do to $\square_1$?

23. Consider the function $R_3 : \mathbb{R}^2 \to \mathbb{R}^2$, defined as

$$R_3 \begin{pmatrix} x \\ y \end{pmatrix} = \frac{1}{\sqrt{2}} \begin{pmatrix} 1 & 1 \\ -1 & 1 \end{pmatrix} \begin{pmatrix} x \\ y \end{pmatrix}.$$

Verify that R_3 is a linear transformation. Use R_3 to transform $\square_1$. What does R_3 do to $\square_1$?

24. Consider the function $R_4 : \mathbb{R}^2 \to \mathbb{R}^2$, defined as

$$R_4 \begin{pmatrix} x \\ y \end{pmatrix} = \frac{\sqrt{3}}{2} \begin{pmatrix} -1 & 1 \\ -1 & -1 \end{pmatrix} \begin{pmatrix} x \\ y \end{pmatrix}.$$

Verify that R_4 is a linear transformation. Use R_4 to transform $\square_1$. What does R_4 do to $\square_1$?

25. Describe, geometrically, the transformations $S \circ T$ and $T \circ S$. What do they do to $\square_1$?
26. Describe, geometrically, the transformations $R_2 \circ S$ and $S \circ R_2$. What do they do to $\square_1$?
27. Create transformations that will transform $\square_1$ in the following ways:

   (a) Stretch $\square_1$ horizontally by a factor of 5 and vertically by a factor of 7.
   (b) Rotate $\square_1$ counterclockwise by an angle of 30°.
   (c) Transform $\square_1$ to a vertical line segment of length 2.
   (d) Transform $\square_1$ to a horizontal line segment of length 3.
   (e) Reflect $\square_1$ over the $x-$axis.

For Exercises 28-31, define $\square_{(x_0, y_0)}$ to be $\square_1$ (defined for Exercises 14-27) translated up a distance y_0 and right a distance x_0. (If y_0 is negative, $\square_1$ is actually translated down. Similarly, if x_0 is negative, $\square_1$ is translated left.)

28. Draw $\square_{(1,1)}$, $\square_{(-1,1)}$, and $\square_{(-2,-2)}$.
29. Describe what happens to each of $\square_{(1,1)}$, $\square_{(-1,1)}$, and $\square_{(-2,-2)}$ when you transform them by R_4 in Exercise 24.
30. Describe what happens to each of $\square_{(1,1)}$, $\square_{(-1,1)}$, and $\square_{(-2,-2)}$ when you transform them by T in Exercise 18.
31. Describe what happens to each of $\square_{(1,1)}$, $\square_{(-1,1)}$, and $\square_{(-2,-2)}$ when you transform them by S in Exercise 15.
32. We can parameterize a line segment connecting two points in $\mathbb{R}^2$ as follows

$$\ell(\lambda) = \lambda v + (1 - \lambda)u,$$

where u and v are the vectors corresponding to the two points and $0 \leq \lambda \leq 1$.

   (a) Find the vectors corresponding to $\ell(0)$ and to $\ell(1)$.
   (b) Let $T : \mathbb{R}^2 \to \mathbb{R}^2$ be a linear transformation. Show that T maps the points on the line segment to the points on another line segment in $\mathbb{R}^2$.

33. Consider the vector space of functions

$$\mathcal{D}^1(\mathbb{R}) = \{f : \mathbb{R} \to \mathbb{R} \mid f \text{ is continuous and } f' \text{ is continuous}\}.$$

Show that $T : \mathcal{D}^1(\mathbb{R}) \to \mathcal{F}$ defined by $T(f) = f'$ is linear. Here, $\mathcal{F}$ is the vector space of functions given in Example 2.4.1.

34. Consider the space of functions

$$\mathcal{R}(\mathbb{R}) = \{f : \mathbb{R} \to \mathbb{R} \mid f \text{ is integrable on } [a, b]\}.$$

Show that $T : \mathcal{R}(\mathbb{R}) \to \mathbb{R}$ defined by $T(f) = \int_a^b f(x) \, dx$ is linear.

35. Using Theorem 4.2.22, show that the function $f : \mathbb{R} \to \mathbb{R}$, defined by $f(x) = 3x - 7$, is not linear.
36. Consider, again, the vector space of 7-bar LCD images, $\mathcal{D}(Z_2)$ from Example 2.4.17.

   (a) Show that if we have a transformation $T : \mathcal{D}(Z_2) \to V$, where V is a vector space with the same scalar set Z_2, then T is linear if $T(x + y) = T(x) + T(y)$ for all $x, y \in \mathcal{D}(Z_2)$.

(b) Use Part (a) to show that if $T : \mathcal{D}(Z_2) \to \mathcal{D}(Z_2)$ is the transformation that flips the digits upside down, then T is linear. Some example transformations are as follows:

$$T\left(\text{⸤digit⸥}\right) = \text{⸤digit⸥} \quad \text{and} \quad T\left(\text{⸤digit⸥}\right) = \text{⸤digit⸥}$$

$$T\left(\text{⸤digit⸥} + \text{⸤digit⸥}\right) = T\left(\text{⸤digit⸥}\right) = \text{⸤digit⸥}.$$

37. Let $\mathcal{J}_n(\mathbb{R})$ be the vector space of histograms with n ordered bins with real values. Consider the "re-binning" transformation $T \colon \mathcal{J}_{12}(\mathbb{R}) \to \mathcal{J}_6(\mathbb{R})$ defined by $T(J) = K$, which re-bins data by adding the contents of bin pairs. That is, the value in the first bin of K is the sum of the values of the first two bins of J, the value of the second bin of K is the sum of the values of the third and fourth bins of J, etc. An example is shown in Figure 4.11. Show that T is a linear transformation.

38. Consider $\mathcal{I}_{512 \times 512}(\mathbb{R})$, the vector space of 512×512 grayscale radiographic images, and $\mathcal{J}_{256}(\mathbb{R})$, the vector space of 256-bin histograms. Suppose $T \colon \mathcal{I}_{512 \times 512}(\mathbb{R}) \to \mathcal{J}_{256}(\mathbb{R})$ is the transformation which creates a histogram of the intensity values in a radiograph. More precisely, suppose $h = T(b)$ for some radiograph b and let h_k indicate the value for the kth histogram bin. The action of T is defined as follows: h_1 is the number of pixels in b with value less than one; h_{256} is the number of pixels in b with value greater than or equal to 255; otherwise, h_k is the number of pixels in b with value greater than or equal to $k - 1$ and less than k. Determine if T is a linear transformation.

39. Let U, V, and W be vector spaces and let $T_1 : V \to W$, $T_2 : V \to W$, and $T_3 : W \to U$ be linear transformations. (See Theorem 4.2.21.) Prove that the following transformations are linear.

(a) $T_1 + T_2 : V \to W$,
(b) $T_1 - T_2 : V \to W$,
(c) $\alpha T_1 : V \to W$, for any scalar α,
(d) $T_3 \circ T_1 : V \to U$.

40. Prove that the zero transformation and the identity transformation are both linear.

41. Using the concepts from this section, prove that the zero transformation and the identity transformation are both unique.

42. Let V and W be vector spaces, $S \subseteq V$, and $T \colon V \to W$ a linear transformation. Define $T(S) = \{T(s) \mid s \in S\}$. Prove that S is a subspace of V if, and only if, $T(S)$ is a subspace of W.

43. Let V be an n-dimensional vector space with basis $\mathcal{B}$. Define $S \colon \mathbb{R}^n \to V$ to be the inverse coordinate transformation: for all $u \in \mathbb{R}^n$, $S(u) = v$ where $[v]_\mathcal{B} = u$. Prove that S is linear.

44. Describe a scenario in which the difference transformation $(T_1 - T_2)(x) = T_1(x) + (-1)T_2(x)$ could be of interest in a radiography setting.

45. Suggest another image brightening transformation (see Example 4.2.11) that can improve contrast in both dark and light areas of an image. Is your transformation linear?

46. Find a transformation $T \colon V \to W$ for which $T(x + y) \neq T(x) + T(y)$ for all $x, y \in V$, but $T(\alpha x) = \alpha T(x)$ for all $x \in V$ and all scalars α.

47. Consider the vector space of 7-bar LCD images, $\mathcal{D}(Z_2)$ and ordered basis

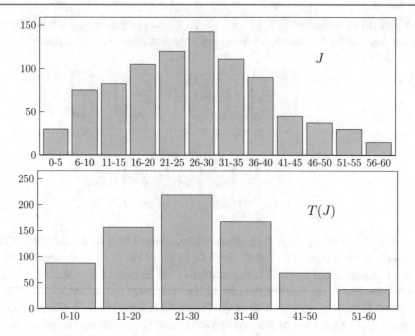

Fig. 4.11 Example histogram $J \in \mathcal{J}_{12}(\mathbb{R})$ and the result of the re-binning transformation $K = T(J) \in \mathcal{J}_6(\mathbb{R})$.

$$B = \left\{ \boxed{8}, \boxed{8}, \boxed{8}, \boxed{9}, \boxed{8}, \boxed{8}, \boxed{8} \right\}.$$

Define $T : \mathcal{D}(\mathbb{Z}_2) \to \mathcal{D}(\mathbb{Z}_2)$ by $T(b_k) = d_k$, where b_k is the kth element of B and d_k is the kth element of

$$D = \left\{ \boxed{8}, \boxed{8}, \boxed{8}, \boxed{8}, \boxed{8}, \boxed{8}, \boxed{8} \right\}.$$

Find $T\left(\boxed{8} \right)$. Is D a basis for $\mathcal{D}(\mathbb{Z}_2)$?

4.3   Explorations: Heat Diffusion

Recall in Chapter 1, we introduced the application of diffusion welding. A manufacturing company uses the process of diffusion welding to adjoin several smaller rods into a single longer rod. The diffusion welding process leaves the final rod heated to various temperatures along the rod with the ends of the rod held at a fixed relatively cool temperature T_0. At regular intervals along the rod, a machine records the temperature difference (from T_0) obtaining a set of values which we call a heat state. We assume that the rod is thoroughly insulated except at the ends so that the major mechanism for heat loss is diffusion through the ends of the rod.

We want to explore this application further. Suppose we have a rod of length L with ends at $x = a$ and $x = b$. Let $T(x, t)$ be the temperature of the rod at position $x \in [a, b]$ and time t. Since the ends of the rod are kept at a fixed temperature T_0, we have $T(a, t) = T(b, t) = T_0$. Define a function $f : [a, b] \times \mathbb{R} \to \mathbb{R}$ that measures the difference in temperature from the temperature at the ends of the rod at time t. That is, $f(x, t) = T(x, t) - T(a, t)$. Notice that $f(a, t) = f(b, t) = 0$. Even though f measures a temperature difference, we will often call the quantity $f(x, t)$ the temperature of the rod at position x and time t.

The quantity $f(x, t)$ varies with respect to position x and evolves in time. As time progresses, the heat will spread along the rod changing the temperature distribution. We can imagine that after a very long time the heat will diffuse along the rod until the temperature is uniform, $\lim_{t \to \infty} f(x, t) = 0$. We can even predict some details on *how* the heat will diffuse. Consider the illustration in Figure 4.12. The green curve shows a possible temperature profile at some time t. The magenta curve shows a temperature profile a short time Δt later. We notice that the diffusion will follow the following trends.

1. Rod locations where the temperature is higher than the surrounding local area will begin to cool. We reason that warm regions would not get warmer unless there is some heat being added to the rod at that point. The red (downward) arrow in Figure 4.12 indicates that the warm area begins to cool.

2. Rod locations where the temperature is lower than the surrounding local area will begin to warm. We reason that cool regions would not get colder unless there is some heat being removed from the rod at that point. The blue (upward) arrow in Figure 4.12 indicates that the cool area begins to warm.

3. Suppose we have two equally warm regions of the rod (e.g., locations x_1 and x_2 in Figure 4.12). Location x_1 has relatively cool areas very nearby, while location x_2 does not. The temperature at x_1 will cool faster than at x_2 because heat is more quickly transferred to the nearby cool regions. Geometrically, we observe that sharply varying temperature differences disappear more quickly than slowly varying temperature differences.

4. The long-term behavior (in time) is that temperatures smooth out and become the same. In this case, temperatures approach the function $f(x, t) = 0$.

4.3.1 Heat States as Vectors

It turns out that we can use linear algebra to describe this long-term behavior! In Section 2.4, we defined the finite-dimensional vector space of *heat states*. We will use our knowledge of linear algebra to compute the heat state, at any later time, as the heat is redistributed along the rod. In physics, the process is known as heat diffusion. The linear algebra formulation is known as heat state evolution. We will remind the reader about our initial discussion on heat states from Section 2.4.1 by repeating some of the same discussion points here.

In the introduction to this section, we modeled the temperature profile along a bar by a continuous function $f(x, t)$, which we call the heat signature of the bar. We discretize such a heat signature $f(x, t)$ (in position) by sampling the temperature at m locations along the bar. For any fixed time, these discretized heat signatures are called **heat states**. If we space the m sampling locations equally, then for $\Delta x = \frac{L}{m+1} = \frac{b-a}{m+1}$, we can choose the sampling locations to be $a + \Delta x, a + 2\Delta x, \dots, a + m\Delta x$. (We do not need to sample the temperature at the end points because the temperature at the ends is fixed at T_0.) Then, the heat state has the coordinate vector (according to a standard basis) given by the following vector u in $\mathbb{R}^{m+2}$.

$$u = [0, u_1, u_2, \dots, u_m, 0] = [f(a), f(a + \Delta x), f(a + 2\Delta x), \dots, f(a + m\Delta x), f(b)],$$

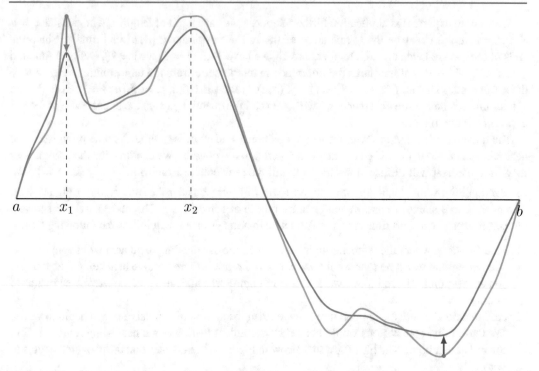

Fig. 4.12 Example 1D temperature profile with higher temperatures on the left end of the rod and lower temperatures on the right. Arrows show the temperature trend predictions at local extrema. The arrows point to a curve showing a temperature profile a short time later.

where we have temporarily suppressed the time dependence for notational clarity. Notice that $f(a) = f(b) = 0$, where $b = a + (m + 1)\Delta x$. These are the familiar heat states that we have seen before. Also, if $u_j = f(x)$ for some $x \in [a, b]$ then $u_{j+1} = f(x + \Delta x)$ and $u_{j-1} = f(x - \Delta x)$. The figure below shows, at some fixed time, a continuous heat signature on the left and the same heat signature with sampling points, to create the heat state, marked on the right.

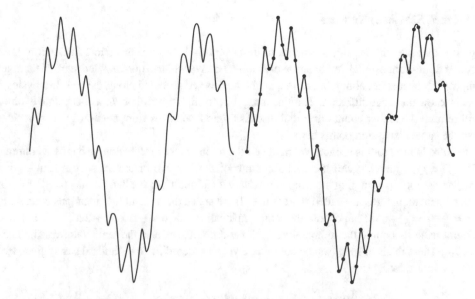

Recall the operations of addition and scalar multiplication in this vector space.

1. We defined scalar multiplication in the usual component-wise fashion. Scalar multiplication results in a change in amplitude only. In the illustration below, the blue heat state is 2 times the red heat state. Heat states appear below as continuous curves, but are actually made up of finitely many (thousands of) points.

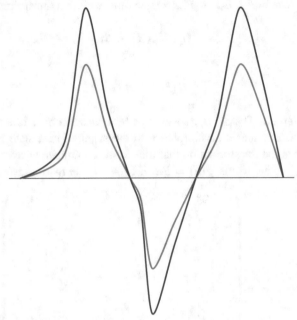

2. We defined vector addition in the usual component-wise fashion. Addition can result in changes in both amplitude and shape. In the illustration below, the red heat state is the sum of the blue and green heat states.

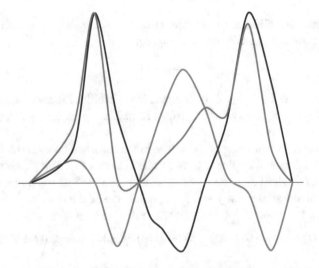

4.3.2   Heat Evolution Equation

The discrete heat evolution equation and time evolution transformation given in this section are derived in Appendix B.

Given a heat state at time t, $h(t) \in \mathcal{H}_m(\mathbb{R})$, the time evolution transformation

$$U : \mathcal{H}_m(\mathbb{R}) \to \mathcal{H}_m(\mathbb{R})$$

is defined by

$$U(h(t)) = h(t + \Delta t)$$

for discrete-time step Δt. That is, U transforms a heat state at time t to a heat state at time $t + \Delta t$. We can work in the coordinate space relative to the standard heat state basis $\mathcal{B}$, where $u = [h]_\mathcal{B}$ and $g : \mathbb{R}^m \to \mathbb{R}^m$ is the transformation that transforms coordinate vectors to coordinate vectors by $g([h]_\mathcal{B}) = E[h]_\mathcal{B}$; in other words, $g(u) = Eu$. E is the $m \times m$ matrix given by

$$E = \begin{pmatrix} 1 - 2\delta & \delta & 0 & \cdots & & & 0 \\ \delta & 1 - 2\delta & \delta & & & & \\ 0 & \delta & \ddots & \ddots & & & \vdots \\ & & & & & \delta & 0 \\ \vdots & & \ddots & \ddots & \delta & 0 \\ & & & & \delta & 1 - 2\delta & \delta \\ 0 & & \cdots & & 0 & \delta & 1 - 2\delta \end{pmatrix}, \tag{4.3}$$

where $\delta \equiv \frac{\Delta t}{(\Delta x)^2}$. Notice that E has nonzero entries on the main diagonal and on both adjacent diagonals. All other entries in E are zero. In this coordinate space,

$$u(t + \Delta t) = Eu(t).$$

Here, we note that we need $0 < \delta \leq \frac{1}{4}$ for computational stability. Since Δx is fixed, we need to take small enough time steps Δt to satisfy this inequality. Hence, as we let $\Delta x \to 0$, we are also implicitly forcing $\Delta t \to 0$.

It is useful to consider the meaning of the values in the rows and columns of E. For example, we might wonder how to interpret the values in a particular column of E. The jth column shows how the heat at time t distributes to heat at time $t + \Delta t$ at location j in the heat state. Fraction $(1 - 2\delta)$ of the heat at location j remains at location j and fraction δ of the heat moves to each of the two nearest neighboring locations $j + 1$ and $j - 1$. So, away from the end points:

$$u_j(t) = \delta u_{j-1}(t + \Delta t) + (1 - 2\delta)u_j(t + \Delta t) + \delta u_{j+1}(t + \Delta t).$$

We can similarly interpret the values in a particular row of E. The jth row shows from where the heat at time $t + \Delta t$ came. We have

$$u_j(t + \Delta t) = \delta u_{j-1}(t) + (1 - 2\delta)u_j(t) + \delta u_{j+1}(t).$$

In particular, fraction $1 - 2\delta$ of the heat at location j was already at location j and fraction δ came from each of the two nearest neighbors for this location.

We also notice that all but the first and last columns (and rows) of E sum to 1. How can we interpret this observation? Based on the above discussion, we see that this guarantees that no heat is lost from the system *except at the end points*. Heat is only redistributed (diffused) at points that are away from the end points and is diffused with some loss at the end points. At each iteration, fraction δ of the heat at $u_1(t)$ and at $u_m(t)$ is lost out the ends of the rod.

4.3.3 Exercises

For Exercises 1 through 8, consider the heat state diffusion transformation E given in Equation 4.3. Suppose we know the heat state $u(0)$ at time $t = 0$. If we want to find the heat state k time steps in the future, $u(k\Delta t)$, we compute $u(k\Delta t) = E^k u(0)$.

1. What is the equation for the heat state 2 steps in the future? 1000 time steps in the future?
2. Find an explicit expression for E^2.
3. Does it look like computing E^k is an easy way to find the heat state at some time far in the future (for example, $k = 1000$ time steps away)?
4. Pick your favorite nontrivial vector $u(0) \in \mathbb{R}^4$ and compute $u(1) = Eu(0)$, $u(2) = Eu(1)$ and $u(3) = Eu(2)$.
5. Does it look like computing $u(k)$ (see Exercise 4) is an easy way to find the heat state at some time, $k \gg 1$, far in the future?
6. Suppose the heat diffusion matrix had the form

$$G = \begin{pmatrix} a_1 & b_1 & c_1 & d_1 \\ 0 & b_2 & c_2 & d_2 \\ 0 & 0 & c_3 & d_3 \\ 0 & 0 & 0 & d_4 \end{pmatrix}.$$

    G is an upper triangular. How would the computations for the iterative process $u(k) = G^k u(0)$ compare to the heat diffusion process governed by E? Would they be simpler or not?
7. Clearly, the computations would be much easier if E was the identity matrix or if E was a matrix of all zeros. Why would we not care to discuss an iterative process defined with these matrices?
8. If you had to perform repeated matrix multiplication operations to compute $u(k\Delta t)$ for large k, what matrix characteristics would you prefer a diffusion operator to have?
9. Verify, for $m = 6$, that multiplication by D_2 results in the same heat state as using the formula given in Equation B.1 from Appendix B.
10. Following similar reasoning as in Equation (B.1) shows that the discretization of the time derivative can be approximated at the jth sampling point on the rod by

$$\frac{\partial}{\partial t} u_j(t) \approx \frac{u_j(t + \Delta t) - u_j(t)}{\Delta t}.$$

4.3.4 Extending the Exploration: Application to Image Warping

Recall, in Chapter 1, we briefly discussed the application of Image Warping. The idea is that we want to create a video that transitions smoothly from one image to the next. We see such a transition in Figure 4.13.

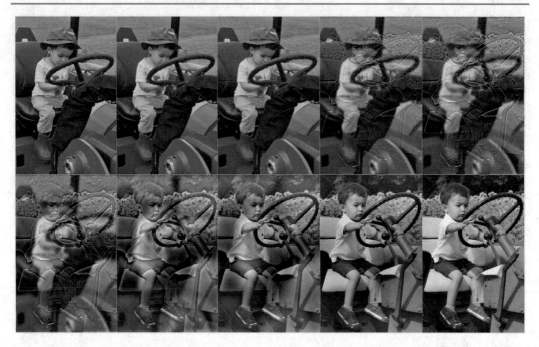

Fig. 4.13 Ten images in the warping sequence that begins with the a young boy on a tractor and ends with the same boy a little older.

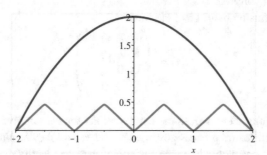

Fig. 4.14 Two heat signatures v (smooth curve) and w (jagged curve). We will evolve v into w.

In this part of the heat diffusion exploration, we will consider the process of image warping by first looking at the process on 1D heat signatures and then discussing how this extends to 2D images.

Morphing 1D heat signatures

Suppose that we start with two heat signatures, v and w, and we want to heat evolve one into the other. One way to do this is to heat evolve *the difference* between the two heat signatures $v - w$.

In the following, we will explore this process with the heat signatures v and w in Figure 4.14.

1. On a new set of axes, sketch the difference $v - w$.
2. Use your intuition about heat diffusion to sketch some (at least 3) future heat signatures that might arise from evolving $v - w$ for different lengths of time. (Add these to your sketch from the previous part). Define a family of heat signatures x_t, where $x_0 = v - w$ is the initial heat signature, and x_t is the heat signature after time t. The heat signatures in your drawings that occur earlier in the evolution are associated with lower values of t than the ones that occur later. Label your heat signatures with this notation, indicating the order based on t.

3. Add a sketch of the limiting heat signature, x_∞, that results from evolving $v - w$.

4. Now discuss how the sketches you made might help you to morph the function v into the function w. (Hint: Consider the related family of heat signatures, where for $t \geq 0$, $z_t := w + x_t$. On a new set of axes, sketch the graphs of z_t that correspond to each of the x_t you sketched. Also include z_0 and z_∞ corresponding to x_0 and x_∞. What is the result of morphing z_0 to z_∞?)

Observations

5. The family of functions z_t shows a heat signature morphing into another heat signature. What do you notice about this morphing? In particular, what types of features disappear and appear?

6. Another method for morphing heat signature v into heat signature w is to first diffuse v and then follow this with the inverse of the diffusion of w. How is this different from what we obtained above by diffusing the signature $v - w$? (You may produce sketches of the heat evolutions of v and w to support your discussion.)

Morphing Images

We now consider the task of morphing a 2D grayscale image into another image[1].

Such an image can be thought of as consisting of pixels in a rectangular array with intensity values assigned to each pixel: white pixels have high intensity (heat) and black pixels have low intensity (heat). As in the 1D case, we will assume the boundary intensities are zero. Boundary pixels are those on the border of the rectangle.

The principles behind heat diffusion in two dimensions are similar to those we have been discussing in one dimension. In particular,

- Places where the temperature is higher than the surrounding area will cool.
- Places where the temperature is cooler than the surrounding area will warm.
- As the heat diffuses, sharpest temperature differences disappear first.
- The long-term behavior is that temperatures smooth out approaching 0 heat everywhere.

Hence, if we take an image and heat diffuse it, the final heat state of the image will be the black (zero) image. We define a morphing between image V and image W as in the 1D case above: by diffusing the difference $V - W$ and considering the associated family X_t as before.

7. In Exercise 17 of Section 2.3, we determined that as long as we have images of the same size, we can perform pixelwise addition. The process of heat diffusion can be modeled for the difference of images, but in two dimensions.

8. Given an $m \times n$ image (at total of mn pixels), what additional properties should we require in order to consider the image as a heat state? In particular, what should be true at the boundary of heat state images?

9. It turns out that one can construct a matrix $\tilde{E}$ so that the heat evolution operator $T : \mathcal{I}_{m \times n} \to \mathcal{I}_{m \times n}$ on images can written as $T(V) = \tilde{E} V$.

10. Above, we observed that the diffusion matrix for 1D heat states sends heat from warmer regions to cooler regions, raising the temperature there. How can we think of the pixel intensities for a difference image as heat values? With this thought, what analogous ideas can we apply to a difference image so that warmer regions lose heat to cooler regions?

---

[1] Recall that most images we encounter are color images, but up to this point of the text, we have considered only grayscale images. Morphing a color image is done by considering the red, green, and blue components of the image separately and then combining them.

11. Consider the structure of the 1D heat diffusion matrix E. What similarities would E have with $\tilde{E}$? Describe the features of such a matrix $\tilde{E}$. (It may help to consider the interpretation of the rows and columns of E found on Page 208.)

12. (**Challenge Question**) Fix a basis (say the standard basis) for the space of 4×4 heat state images and determine the matrix $\tilde{E}$.

4.4 Matrix Representations of Linear Transformations

In our studies of the radiographic transformation and of the heat state evolution transformation, we have modeled the transformations with the matrix multiplication operation. In reality, brain images, heat states, and radiographs are somewhat abstract objects which do not lend themselves to matrix operations. However, we now have tools for numerical operations based on coordinate vectors in familiar vector spaces ($\mathbb{R}^n$). Between these coordinate spaces, matrix operators serve as proxies for the transformations of interest. In this section, we will solidify these general ideas. We need to know which types of transformations can be written in terms of a matrix multiplication proxy. We also want to keep sight of our radiography goal: does an inverse transformation exist, and if so, what is its relationship to the proxy matrix?

The main result of this section is the fact that every *linear* transformation between finite-dimensional vector spaces can be uniquely represented as a left matrix multiplication operation between coordinate spaces. Consider the illustrative graphic in Figure 4.15. We can think of a linear transformation $T : V \to W$ as a composition of three transformations: $T = T_3 \circ T_2 \circ T_1$. Consider two vectors spaces V and W with $\dim(V) = n$ and $\dim(W) = m$. The first transformation, T_1, maps $v \in V$ to its coordinate vector $[v]_{B_V} \in \mathbb{R}^n$. The next transformation, T_2, maps the coordinate vector $[v]_{B_V}$ to the coordinate vector $[w]_{B_W}$. And finally, T_3 maps the coordinate vector $[w]_{B_W}$ to the vector $w \in W$. Instead of directly computing $w = T(v)$, our explorations with radiography and heat state evolution have been with the coordinate space transformation $[w]_{B_W} = T_2([v]_{B_V}) = M[v]_{B_V}$, where $M \in \mathcal{M}_{m \times n}(\mathbb{R})$. The transformation T_2 is shown in red to indicate that we have yet to answer questions concerning its existence. The coordinate transformation T_1 exists and is linear. The inverse coordinate transformation T_3 exists and is linear (see Exercise 43 of Section 4.2). Table 4.1 shows some of the correspondences that we have already used. So far, we have been able to find a matrix M so that $T(v) = Mx$ for $x = [v]_{B_V}$ in the coordinate space of V relative to some basis B_V. We found that Mx is the coordinate vector of $T(v)$ relative to a basis of W, say B_W. But is it always possible to find such a matrix M? Figure 4.16 shows in red a summary of the transformations for which we do not yet know their existence. In Section 4.2, we showed that the coordinate transformations, S_1 and S_3 exist and are linear. We do not yet know under what conditions, if any, ensure the existence of the transformations S or S_2. The transformation S (if it exists) is an inverse transformation in that $S \circ T$ is the identity transformation on V and $T \circ S$ is the identity transformation on W. The question of the existence of S is equivalent to understanding the existence of S_2 because $S = S_1 \circ S_2 \circ S_3$. In this section, we will discuss the answer about the existence of T_2 in Figure 4.15 and build some tools to answer the remaining questions.

4.4.1 Matrix Transformations between Euclidean Spaces

In the situation where both the domain and the codomain are euclidean spaces, the situation of Figure 4.15 is simpler. We saw, in Section 4.2, that multiplication by any $m \times n$ matrix is a linear transformation from $\mathbb{R}^n$ to $\mathbb{R}^m$. It turns out that *every* linear transformation between $\mathbb{R}^n$ and $\mathbb{R}^m$ can be represented as multiplication by an $m \times n$ matrix.

Table 4.1   Three examples of vectors, corresponding coordinate vectors, and matrix representations of a transformation. The radiography example is that of Section 4.1.4. The heat diffusion example is for $m = 4$ with a diffusion parameter $\delta = \frac{1}{4}$. The polynomial differentiation example is for transformation $T : \mathcal{P}_2(\mathbb{R}) \to \mathcal{P}_2(\mathbb{R})$ defined by $T(p(x)) = p'(x)$. In all cases, we use the standard bases $\mathcal{B}_V$ and $\mathcal{B}_W$.

	Radiography	Heat Diffusion	Polynomial Differentiation
v			$3x^2 - 2x + 3$
$[v]_{\mathcal{B}_V}$	$\begin{pmatrix} 1 \\ 3 \\ 2 \\ 4 \end{pmatrix}$	$\begin{pmatrix} 4 \\ 3 \\ 9 \\ 5 \end{pmatrix}$	$\begin{pmatrix} 3 \\ -2 \\ 3 \end{pmatrix}$
M	$\begin{pmatrix} 1/2 & 1 & 0 & 1/2 \\ 1/2 & 0 & 1 & 1/2 \end{pmatrix}$	$\begin{pmatrix} 1/2 & 1/4 & 0 & 0 \\ 1/4 & 1/2 & 1/4 & 0 \\ 0 & 1/4 & 1/2 & 1/4 \\ 0 & 0 & 1/4 & 1/2 \end{pmatrix}$	$\begin{pmatrix} 0 & 0 & 0 \\ 2 & 0 & 0 \\ 0 & 1 & 0 \end{pmatrix}$
$[w]_{\mathcal{B}_W}$	$\begin{pmatrix} 11/2 \\ 9/2 \end{pmatrix}$	$\begin{pmatrix} 11/4 \\ 19/4 \\ 13/2 \\ 19/4 \end{pmatrix}$	$\begin{pmatrix} 0 \\ 6 \\ -2 \end{pmatrix}$
w			$6x - 2$

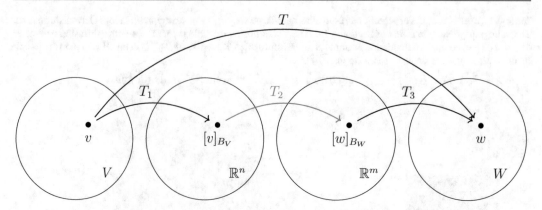

Fig. 4.15 Illustration of the equivalence of linear transformation T with the composition of two coordinate transformations T_1 and T_3 and one matrix product $T_2\left([v]_{B_V}\right) = M[v]_{B_V} = [w]_{B_W}$.

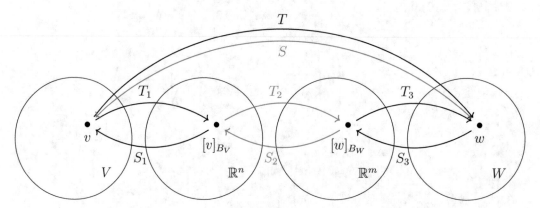

Fig. 4.16 Illustration of the equivalence of linear transformation T and a possible inverse transformation S. Compare to Figure 4.15.

Definition 4.4.1

Given the linear transformation $T : \mathbb{R}^n \to \mathbb{R}^m$, we say that $M \in \mathcal{M}_{m \times n}(\mathbb{R})$ is the **matrix representation** for the transformation T if $T(x) = Mx$ for every $x \in \mathbb{R}^n$. $\square$

Lemma 4.4.2

Let $T : \mathbb{R}^n \to \mathbb{R}^m$ be a linear transformation. Then there exists a unique $m \times n$ matrix A such that $T(x) = Ax$ for all $x \in \mathbb{R}^n$.

Proof. Let $T : \mathbb{R}^n \to \mathbb{R}^m$ be a linear transformation. By Theorem 4.2.26, we know that T is uniquely determined by how T transforms the basis elements. Consider the $m \times n$ matrix A

$$A = \begin{pmatrix} | & | & & | \\ T(e_1) & T(e_2) & \cdots & T(e_n) \\ | & | & & | \end{pmatrix}.$$

whose columns are the outputs of the n standard basis elements of $\mathbb{R}^n$, written with respect to the standard basis on $\mathbb{R}^m$. One can verify that $Ae_i = T(e_i)$ for $i = 1, \ldots, n$. (See Exercise 24.) Hence, by Corollary 4.2.27, $T(x) = Ax$ for all $x \in \mathbb{R}^n$, and so A is a matrix representation of T.

To see that A is the only matrix that represents T, we recognize that any matrix B that represents T must map (via multiplication) the standard basis of $\mathbb{R}^n$ to the same vectors that T maps them to. Hence, the ith column of B must be $T(e_i)$ (written with respect to the standard basis.)   □

Example 4.4.3 Consider the linear transformation $T : \mathbb{R}^3 \to \mathbb{R}^2$ given by

$$T \begin{pmatrix} x \\ y \\ z \end{pmatrix} = \begin{pmatrix} x + y \\ 2z + y - x \end{pmatrix}$$

The matrix representation of T has columns $T(e_1) = \begin{pmatrix} 1 \\ -1 \end{pmatrix}$, $T(e_2) = \begin{pmatrix} 1 \\ 1 \end{pmatrix}$, and $T(e_3) = \begin{pmatrix} 0 \\ 2 \end{pmatrix}$. We verify with a computation that

$$\begin{pmatrix} 1 & 1 & 0 \\ -1 & 1 & 2 \end{pmatrix} \begin{pmatrix} x \\ y \\ z \end{pmatrix} = \begin{pmatrix} x + y \\ 2z + y - x \end{pmatrix}$$

as desired.   □

4.4.2 Matrix Transformations

Suppose we have two vector spaces V and W. Let V be n-dimensional with basis $\mathcal{B}_V$ and W be m-dimensional with basis $\mathcal{B}_W$. Suppose we are given a linear transformation $T : V \to W$. We are interested in finding a way to transform vectors from V to W, possibly taking a new path using matrix multiplication. Recall that the transformation $T_1 : V \to \mathbb{R}^n$ defined by $T_1(v) = [v]_{\mathcal{B}_V}$ is linear (see Theorem 4.2.16). Let T_3 be the transformation that takes coordinate vectors in $\mathbb{R}^m$ back to their corresponding vectors in W. We know that T_3 is a linear transformation (see Exercise 43 of Section 4.2). We also know that we can multiply vectors in $\mathbb{R}^n$ by $m \times n$ matrices to get vectors in $\mathbb{R}^m$. We want to find $M \in \mathcal{M}_{m \times n}(\mathbb{R})$ so that

$$T_2 : \mathbb{R}^n \to \mathbb{R}^m \text{ by } T_2(x) = Mx \text{ for all } x \in \mathbb{R}^n$$

and so that

$$T_2([v]_{\mathcal{B}_V}) = M[v]_{\mathcal{B}_V} = [T(v)]_{\mathcal{B}_W} \text{ for all } v \in V.$$

That is, we want T_2 to transform $[v]_{\mathcal{B}_V}$ into $[w]_{\mathcal{B}_W}$ in the same way T transforms v into w. (See Figure 4.15).

First, we answer our central questions concerning the existence of S_2 and of S, as well as their relationship to T and the coordinate transformations. We will show that if T is linear, then the matrix representation $M = [T]_{\mathcal{B}_V}^{\mathcal{B}_W}$ not only exists but is uniquely determined by T, $\mathcal{B}_V$ and $\mathcal{B}_W$. As you read through the next two lemmas and their proofs, use Figure 4.16 to keep the transformations organized in your mind.

Lemma 4.4.4
Given vector spaces V and W and a linear transformation $T : V \to W$. If $\dim V = n$ and $\dim W = m$, then there exists a linear transformation $T_2 : \mathbb{R}^n \to \mathbb{R}^m$ so that for all $v \in V$

$$T_2([v]_{\mathcal{B}_V}) = [T(v)]_{\mathcal{B}_W}. \tag{4.4}$$

for any bases $\mathcal{B}_V$ and $\mathcal{B}_W$ of V and W, respectively.

Proof. Suppose $T : V \to W$ linear. Suppose also that $\mathcal{B}_V$ and $\mathcal{B}_W$ are bases for V and W. Let $v \in V$ and define $w = T(v) \in W$. We will show that there is a linear transformation $T_2 : \mathbb{R}^m \to \mathbb{R}^n$ satisfying (4.4) above. We know, by Exercise 43 of Section 4.2 that there exist a coordinate transformation

$$S_3 : W \to \mathbb{R}^m \text{ so that } S_3(w) = [w]_{\mathcal{B}_W}$$

and an inverse coordinate transformation

$$S_1 : \mathbb{R}^n \to V \text{ so that } S_1([v]_{\mathcal{B}_V}) = v.$$

By Theorem 4.2.21 and the linearity of T, we know that $S_3 \circ T \circ S_1$ is also linear. Now, we need only show that if we define $T_2 = S_3 \circ T \circ S_1$, then T_2 satisfies (4.4) above. Using Definition 4.2.25, we have

$$\begin{aligned}
T_2([v]_{\mathcal{B}_V}) &= (S_3 \circ T \circ S_1)([v]_{\mathcal{B}_V}) \\
&= S_3(T(S_1([v]_{\mathcal{B}_V}))) \\
&= S_3(T(v)) \\
&= S_3(w) \\
&= [w]_{\mathcal{B}_W}.
\end{aligned}$$

Therefore, T_2 is a linear transformation that satisfies (4.4). $\square$

We continue establishing the existence of each of the transformations in Figure 4.16.

Lemma 4.4.5
Let V and W be vector spaces with $\dim V = n$ and $\dim W = m$. Suppose there is a linear transformation $T : V \to W$. Then, there exists a linear transformation

$$S : W \to V \text{ with } S(T(v)) = v, \text{ for all } v \in V$$

if and only if there exists a linear transformation

$$S_2 : \mathbb{R}^m \to \mathbb{R}^n \text{ with } S_2([T(v)]_{\mathcal{B}_W}) = [v]_{\mathcal{B}_V}, \text{ for all } v \in V.$$

Proof. Suppose V and W are vector spaces with dim $V = n$ and dim $W = m$. Suppose $T : V \to W$ is linear. Suppose, also, that $v \in V$. Define $w = T(v) \in W$. We've shown that there exist linear coordinate transformations $T_1 : V \to \mathbb{R}^n$ and $S_3 : \mathbb{R}^m \to W$. We also know that there exist linear inverse coordinate transformations $S_1 : \mathbb{R}^n \to V$ and $T_3 : \mathbb{R}^m \to W$.

($\Rightarrow$) First, suppose S, as described above, exists and define $S_2 = T_1 \circ S \circ T_3$. By Theorem 4.2.21, we know that S_2 is linear. We need only show that $S_2([w]_{\mathcal{B}_W}) = [v]_{\mathcal{B}_V}$. We know that

$$S_2([w]_{\mathcal{B}_W}) = (T_1 \circ S \circ T_3)([w]_{\mathcal{B}_W})$$
$$= T_1(S(T_3([w]_{\mathcal{B}_W}))).$$

Thus, $S_2 = T_1 \circ S \circ T_3$ is a linear transformation with $S_2([w]_{\mathcal{B}_W}) = [v]_{\mathcal{B}_V}$.

($\Leftarrow$) Now, suppose S_2, as described above, exists and define $S = S_1 \circ S_2 \circ S_3$. Again, we know that, by Theorem 4.2.21, S is linear. Now, we know that

$$S(w) = (S_1 \circ S_2 \circ S_3)(w)$$
$$= S_1(S_2(S_3(w)))$$
$$= S_1(S_2([w]_{\mathcal{B}_W}))$$
$$= S_1([v]_{\mathcal{B}_V})$$
$$= v.$$

Therefore, $S = S_1 \circ S_2 \circ S_3$ is a linear transformation so that $S(w) = v$. $\square$

Theorem 4.4.6

Let V and W be finite-dimensional vector spaces with bases $\mathcal{B}_V$ and $\mathcal{B}_W$, respectively. Let $T : V \to W$. If T is linear then there exists unique matrix $M = [T]_{\mathcal{B}_V}^{\mathcal{B}_W}$ so that $M[v]_{\mathcal{B}_V} = [T(v)]_{\mathcal{B}_W}$ for all $v \in V$.

Proof. Suppose V and W are finite-dimensional vector spaces with bases $\mathcal{B}_V = \{y_1, y_2, \cdots, y_n\}$ and $\mathcal{B}_W = \{z_1, z_2, \cdots, z_m\}$, respectively. Let $T : V \to W$ be linear. Let $v \in V$ and define $w = T(v) \in W$. We will show that there exists a unique matrix M, such that $M[v]_{\mathcal{B}_V} = [w]_{\mathcal{B}_W}$. Suppose $v \in V$ and $w \in W$. Then, we can express v and w, uniquely, as linear combinations of basis vectors, that is, there are scalars a_k and c_j so that

$$v = \sum_{k=1}^{n} a_k y_k \quad \text{and} \quad w = \sum_{j=1}^{m} c_j z_j.$$

Now, since T is linear, we have

$$\sum_{j=1}^{m} c_j z_j = w$$

$$= T(v)$$

$$= T\left(\sum_{k=1}^{m} a_k y_k\right)$$

$$= \sum_{k=1}^{n} a_k T(y_k).$$

Since the codomain of T is W, we can express each vector $T(y_k)$ as a linear combination of basis vectors in B_W. That is, there are scalars b_{jk} so that

$$T(y_k) = \sum_{j=1}^{m} b_{jk} z_j.$$

Therefore, we have

$$\sum_{j=1}^{m} c_j z_j = \sum_{j=1}^{m} a_j \left(\sum_{k=1}^{n} b_{jk}\right) z_j = \sum_{j=1}^{m} \left(\sum_{k=1}^{n} a_k b_{jk}\right) z_j.$$

So $c_j = \sum_{k=1}^{n} b_{jk} a_k$, for all $j = 1, 2, \ldots, m$. We can then, combine these unique coefficient relationships as the matrix equation

$$\begin{pmatrix} c_1 \\ c_2 \\ \vdots \\ c_m \end{pmatrix} = \begin{pmatrix} b_{11} & b_{12} & \cdots & b_{1n} \\ b_{21} & b_{22} & \cdots & b_{2n} \\ \vdots & \vdots & \ddots & \vdots \\ b_{m1} & b_{m2} & \cdots & b_{mn} \end{pmatrix} \begin{pmatrix} a_1 \\ a_2 \\ \vdots \\ a_n \end{pmatrix}.$$

Thus, we have

$$[w]_{\mathcal{B}_W} = \begin{pmatrix} c_1 \\ c_2 \\ \vdots \\ c_m \end{pmatrix} = M \begin{pmatrix} a_1 \\ a_2 \\ \vdots \\ a_n \end{pmatrix} = M[v]_{\mathcal{B}_V},$$

where

$$M = \begin{pmatrix} b_{11} & b_{12} & \cdots & b_{1n} \\ b_{21} & b_{22} & \cdots & b_{2n} \\ \vdots & \vdots & \ddots & \vdots \\ b_{m1} & b_{m2} & \cdots & b_{mn} \end{pmatrix}.$$

Notice that the entries in M are determined by the basis vectors $\mathcal{B}_V$ and $\mathcal{B}_W$, not the specific vectors v or w. Thus, M is unique for the given bases. $\square$

The unique matrix given in Theorem 4.4.6 is the matrix representation of T.

Definition 4.4.7

Let V and W be vector spaces with dim $V = n$ and dim $W = m$. Given the linear transformation $T : V \to W$, we say that the $m \times n$ matrix M is the **matrix representation** of T with respect to bases $\mathcal{B}_V$ and $\mathcal{B}_W$ if $[T(x)]_{\mathcal{B}_W} = M[x]_{\mathcal{B}_V}$ for every $x \in V$.

$\square$

It is common to indicate the matrix representation M of a linear transformation $T : V \to W$ by $M = [T]_{\mathcal{B}_V}^{\mathcal{B}_W}$, where $\mathcal{B}_V$ and $\mathcal{B}_W$ are bases for V and W, respectively. If V and W are the same vector spaces, and we choose to use the same basis $\mathcal{B}$, then we typically write $M = [T]_{\mathcal{B}}$ to indicate $M = [T]_{\mathcal{B}}^{\mathcal{B}}$.

The proof of Theorem 4.4.6 suggests a method for constructing a matrix representation for a linear transformation between finite-dimensional vector spaces, given a basis for each. The kth column of M is the coordinate vector, relative to $\mathcal{B}_W$, of the transformed k^{th} basis vector of $\mathcal{B}_V$. That is, $M_k = [T(y_k)]_{\mathcal{B}_W}$.

Corollary 4.4.8

Let V and W be vector spaces with ordered bases $\mathcal{B}_V = \{y_1, y_2, \ldots, y_n\}$ and $\mathcal{B}_W$, respectively. Also, let $T : V \to W$ be linear. Then the matrix representation $M = [T]_{\mathcal{B}_V}^{\mathcal{B}_W}$ is given by

$$
M = \left(\begin{array}{cccc}
| & | & & | \\
[T(y_1)]_{\mathcal{B}_W} & [T(y_2)]_{\mathcal{B}_W} & \cdots & [T(y_n)]_{\mathcal{B}_W} \\
| & | & & |
\end{array} \right), \tag{4.5}
$$

where $[T(y_k)]_{\mathcal{B}_W}$ is the kth column of M.

Proof. The proof follows directly from the proof of Theorem 4.4.6. $\square$

Example 4.4.9 Let $V = \{ax^2 + bx + (a + b) \mid a, b \in \mathbb{R}\}$ and let $W = \mathcal{M}_{2 \times 2}(\mathbb{R})$. Consider the transformation $T : V \to W$ defined by

$$
T(ax^2 + bx + (a + b)) = \begin{pmatrix} a & b - a \\ a + b & a + 2b \end{pmatrix}.
$$

The reader can verify that T is linear. We seek a matrix representation, M, of T. Recall, that the columns of the matrix representation M of T are coordinate vectors in W according to some basis $\mathcal{B}_W$. Thus, we must choose a basis for $W = \mathcal{M}_{2 \times 2}(\mathbb{R})$. We must also apply the transformation T to basis elements for a basis of V. Therefore, we must also choose a basis for V. Now, $T : V \to \mathcal{M}_{2 \times 2}(\mathbb{R})$, so we will choose $\mathcal{B}_W$ to be the standard basis for $\mathcal{M}_{2 \times 2}(\mathbb{R})$. To find the basis $\mathcal{B}_V$, we write V as the span of linearly independent vectors. In fact, we can see that

$$
V = \{ax^2 + bx + (a + b) \mid a, b \in \mathbb{R}\} = \text{Span}\left\{x^2 + 1, x + 1\right\}.
$$

So we define the basis $\mathcal{B}_V = \{x^2 + 1, x + 1\}$. Since V is a 2D space, the corresponding coordinate space is $\mathbb{R}^2$. The coordinate space for the 4-dimensional space, W is $\mathbb{R}^4$. Thus, the matrix representation M is a 4×2 matrix. We also want M to act like T. That is, we want $[T(v)]_{\mathcal{B}_W} = M[v]_{\mathcal{B}_V}$. We need to determine to where the basis elements of V get mapped. By the definition of T, we have

$$T(x^2 + 1) = \begin{pmatrix} 1 & -1 \\ 1 & 1 \end{pmatrix}$$

and

$$T(x + 1) = \begin{pmatrix} 0 & 1 \\ 1 & 2 \end{pmatrix}.$$

And, the corresponding coordinate vectors in $\mathbb{R}^4$ are

$$[T(x^2 + 1)]_{\mathcal{B}_W} = \left[\begin{pmatrix} 1 & -1 \\ 1 & 1 \end{pmatrix} \right]_{\mathcal{B}_W} = \begin{pmatrix} 1 \\ 1 \\ -1 \\ 1 \end{pmatrix}$$

and

$$[T(x + 1)]_{\mathcal{B}_W} = \left[\begin{pmatrix} 0 & 1 \\ 1 & 2 \end{pmatrix} \right]_{\mathcal{B}_W} = \begin{pmatrix} 0 \\ 1 \\ 1 \\ 2 \end{pmatrix}.$$

According to Corollary 4.4.8,

$$M = \begin{pmatrix} 1 & 0 \\ 1 & 1 \\ -1 & 1 \\ 1 & 2 \end{pmatrix}.$$

We can (and should) check that the transformation $\tilde{T} : \mathbb{R}^2 \to \mathbb{R}^4$ defined by $\tilde{T}(x) = Mx$ transforms the coordinate vectors in the same way T transforms vectors. Let us test our matrix representation on $v = 2x^2 + 4x + 6$. We know that $v \in V$ because it corresponds to the choices $a = 2$ and $b = 4$. According to the definition for T, we get

$$T(v) = T(2x^2 + 4x + (2 + 4)) = \begin{pmatrix} 2 & 4 - 2 \\ 2 + 4 & 2 + 2(4) \end{pmatrix} = \begin{pmatrix} 2 & 2 \\ 6 & 10 \end{pmatrix}.$$

Notice that $v = 2(x^2 + 1) + 4(x + 1)$. So

$$[v]_{\mathcal{B}_V} = \begin{pmatrix} 2 \\ 4 \end{pmatrix}.$$

Finally, we check that multiplying $[v]_{\mathcal{B}_V}$ by M gives $[T(v)]_{\mathcal{B}_W}$. Indeed

$$M[v]_{\mathcal{B}_V} = \begin{pmatrix} 1 & 0 \\ 1 & 1 \\ -1 & 1 \\ 1 & 2 \end{pmatrix} \begin{pmatrix} 2 \\ 4 \end{pmatrix}$$

$$= \begin{pmatrix} 2 \\ 6 \\ 2 \\ 10 \end{pmatrix}$$

$$= \left[\begin{pmatrix} 2 & 2 \\ 6 & 10 \end{pmatrix} \right]_{\mathcal{B}_W}$$

$$= [T(v)]_{\mathcal{B}_W}.$$

We can check this for a general case by using an arbitrary vector in $v \in V$. Let, $v = ax^2 + bx + a + b$ for some $a, b \in \mathbb{R}$. Then

$$T(v) = \begin{pmatrix} a & b - a \\ a + b & a + 2b \end{pmatrix}.$$

We see that

$$M[v]_{\mathcal{B}_V} = \begin{pmatrix} 1 & 0 \\ 1 & 1 \\ -1 & 1 \\ 1 & 2 \end{pmatrix} \begin{pmatrix} a \\ b \end{pmatrix}$$

$$= \begin{pmatrix} a \\ a + b \\ b - a \\ a + 2b \end{pmatrix}$$

$$= \left[\begin{pmatrix} a & b - a \\ a + b & a + 2b \end{pmatrix} \right]_{\mathcal{B}_W}$$

$$= [T(v)]_{\mathcal{B}_W}.$$

Thus, as expected $[T(v)]_{\mathcal{B}_W} = M[v]_{\mathcal{B}_V}$. $\qquad \square$

Next, we apply this procedure to a small radiographic transformation. In Section 4.1, we constructed some matrix representations for radiographic transformations with a few object voxels and radiograph pixels. In the next example, we consider a similar scenario, constructing the transformation T using Corollary 4.4.8.

Example 4.4.10 Consider the following radiographic scenario.

- Height and width of the image in voxels: $n = 2$ (Total voxels $N = 4$)
- Pixels per view in the radiograph: $m = 2$
- *ScaleFac* $= \sqrt{2}$
- Number of views: $a = 2$
- Angle of the views: $\theta_1 = 0°, \theta_2 = 135°$

There are two radiographic views of two pixels each and the object space has four voxels as shown here:

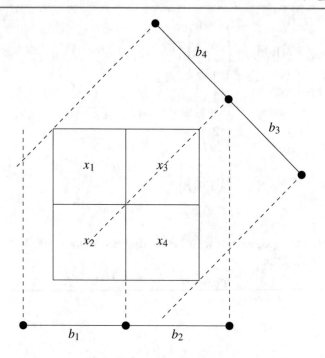

We will use the standard bases for both the object and radiograph spaces. That is,

$$\mathcal{B}_O = \left\{ \begin{array}{|c|c|} \hline 1 & 0 \\ \hline 0 & 0 \\ \hline \end{array}, \begin{array}{|c|c|} \hline 0 & 0 \\ \hline 1 & 0 \\ \hline \end{array}, \begin{array}{|c|c|} \hline 0 & 1 \\ \hline 0 & 0 \\ \hline \end{array}, \begin{array}{|c|c|} \hline 0 & 0 \\ \hline 0 & 1 \\ \hline \end{array} \right\},$$

$$\mathcal{B}_R = \left\{ \quad , \quad , \quad , \quad \right\}.$$

The columns of the matrix representation of the transformation are found using Corollary 4.4.8 as follows.

$$\left[T\left(\begin{array}{|c|c|} \hline 1 & 0 \\ \hline 0 & 0 \\ \hline \end{array} \right) \right]_{\mathcal{B}_R} = \left[\quad \right]_{\mathcal{B}_R} = \begin{pmatrix} 1 \\ 0 \\ 0 \\ 1 \end{pmatrix},$$

$$\left[T\left(\begin{array}{|c|c|} \hline 0 & 0 \\ \hline 1 & 0 \\ \hline \end{array} \right) \right]_{\mathcal{B}_R} = \left[\quad \right]_{\mathcal{B}_R} = \begin{pmatrix} 1 \\ 0 \\ 1/2 \\ 1/2 \end{pmatrix},$$

$$\left[T\left(\begin{array}{|c|c|} \hline 0 & 1 \\ \hline 0 & 0 \\ \hline \end{array}\right)\right]_{\mathcal{B}_R} = \left[\begin{array}{c} {}^{1/2} \\ {}^{1/2} \\ \hline 0 \quad 1 \end{array}\right]_{\mathcal{B}_R} = \begin{pmatrix} 0 \\ 1 \\ {}^{1/2} \\ {}^{1/2} \end{pmatrix},$$

$$\left[T\left(\begin{array}{|c|c|} \hline 0 & 0 \\ \hline 0 & 1 \\ \hline \end{array}\right)\right]_{\mathcal{B}_R} = \left[\begin{array}{c} 0 \\ 1 \\ \hline 0 \quad 1 \end{array}\right]_{\mathcal{B}_R} = \begin{pmatrix} 0 \\ 1 \\ 1 \\ 0 \end{pmatrix}.$$

Thus, the matrix representation of the radiographic transformation T is

$$[T]_{\mathcal{B}_O}^{\mathcal{B}_R} = \begin{pmatrix} 1 & 1 & 0 & 0 \\ 0 & 0 & 1 & 1 \\ 0 & {}^{1/2} & {}^{1/2} & 1 \\ 1 & {}^{1/2} & {}^{1/2} & 0 \end{pmatrix}.$$

□

★★ **Watch Your Language!** The matrix representation is not unique, but rather depends upon chosen bases. Let V and W be vector spaces with bases $\mathcal{B}_V$ and $\mathcal{B}_W$, respectively. Suppose $T : V \to W$ is a linear transformation with matrix representation M. We list key relationships and the corresponding linear algebra language used to discuss such relationships. (Pay attention to which statements are correct to say and which are incorrect.)

$$\mathbf{T}(\mathbf{v}) = \mathbf{w}.$$

✓ T transforms v to w.
✓ w is the transformation of v under T.
✓ T of v is w.
✗ T times v is w.

$$\mathbf{M}\,[\mathbf{v}]_{\mathcal{B}_V} = [\mathbf{w}]_{\mathcal{B}_W}.$$

✓ The coordinate vector of v is transformed to the coordinate vector of w through multiplication by the matrix M.
✓ The coordinate vector of w is the transformation of the coordinate vector of v using multiplication by the matrix M.
✓ M times the coordinate vector of v is the coordinate vector of w.
✗ M transforms the coordinate vector of v to the coordinate vector of w.
✗ M transforms v to w.

$$\mathbf{M} = [\mathbf{T}]_{\mathcal{B}_V}^{\mathcal{B}_W}.$$

✓ M is a matrix representation of the linear transformation T.

✓ M is the matrix representation of the linear transformation T corresponding to the bases $\mathcal{B}_V$ and $\mathcal{B}_W$.
(In appropriate context, we often say, "M is the matrix representation of T.")

✗ M is equal to T.

✗ M is the linear transformation that transforms vectors in V to vectors in W in the same way T does.

4.4.3   Change of Basis Matrix

Consider brain images represented in coordinate space $\mathbb{R}^N$ relative to a basis $\mathcal{B}_0$. Perhaps this basis is the standard basis for brain images. Now suppose that we have another basis $\mathcal{B}_1$ for the space of brain images for which v_{431} is a brain image strongly correlated with disease X. If a brain image x is represented as a coordinate vector $[x]_{\mathcal{B}_0}$, it may be relatively simple to perform necessary calculations, but it may be more involved to diagnose if disease X is present. However, the 431st coordinate of $[x]_{\mathcal{B}_1}$ tells us directly the relative contribution of v_{431} to the brain image. Ideas such as this inspire the benefits of being able to quickly change our coordinate system.

Let $T : \mathbb{R}^n \to \mathbb{R}^n$ be the change of coordinates transformation from ordered basis $\mathcal{B} = \{b_1, b_2, \ldots, b_n\}$ to ordered basis $\tilde{\mathcal{B}} = \{\tilde{b}_1, \tilde{b}_2, \ldots, \tilde{b}_n\}$. We represent the transformation as a matrix $M = [T]_{\mathcal{B}}^{\tilde{\mathcal{B}}}$. The key idea is that a change of coordinates *does not change the vectors themselves, only their coordinate representation*. Thus, T must be the identity transformation. Using Corollary 4.4.8, we have

$$[T]_{\mathcal{B}}^{\tilde{\mathcal{B}}} = [I]_{\mathcal{B}}^{\tilde{\mathcal{B}}}$$

$$= \begin{pmatrix} | & | & & | \\ [I(b_1)]_{\tilde{\mathcal{B}}} & [I(b_2)]_{\tilde{\mathcal{B}}} & \cdots & [I(b_n)]_{\tilde{\mathcal{B}}} \\ | & | & & | \end{pmatrix}$$

$$= \begin{pmatrix} | & | & & | \\ [b_1]_{\tilde{\mathcal{B}}} & [b_2]_{\tilde{\mathcal{B}}} & \cdots & [b_n]_{\tilde{\mathcal{B}}} \\ | & | & & | \end{pmatrix}.$$

Definition 4.4.11

Let $\mathcal{B}$ and $\tilde{\mathcal{B}}$ be two ordered bases for a vector space V. The matrix representation $[I]_{\mathcal{B}}^{\tilde{\mathcal{B}}}$ for the transformation changing coordinate spaces is called a **change of basis matrix**.

□

The kth column of the change of basis matrix is the coordinate representation of the kth basis vector of $\mathcal{B}$ relative to the basis $\tilde{\mathcal{B}}$.

Example 4.4.12 Consider an ordered basis $\mathcal{B}$ for $\mathbb{R}^3$ given by

$$\mathcal{B} = \left\{ v_1 = \begin{pmatrix} 1 \\ 1 \\ 0 \end{pmatrix}, v_2 = \begin{pmatrix} 1 \\ 0 \\ 1 \end{pmatrix}, v_3 = \begin{pmatrix} 1 \\ 1 \\ 1 \end{pmatrix} \right\}.$$

Find the change of basis matrix M from the standard basis $\mathcal{B}_0$ for $\mathbb{R}^3$ to $\mathcal{B}$. We have

$$M = \begin{pmatrix} | & | & | \\ [e_1]_\mathcal{B} & [e_2]_\mathcal{B} & [e_2]_\mathcal{B} \\ | & | & | \end{pmatrix}.$$

We can find $[e_1]_\mathcal{B}$ by finding scalars a, b, c so that

$$e_1 = av_1 + bv_2 + cv_3.$$

Solving the corresponding system of equations, we get $a = 1, b = 1, c = -1$. So,

$$[e_1]_\mathcal{B} = \begin{pmatrix} 1 \\ 1 \\ -1 \end{pmatrix}.$$

Similarly, we find that

$$[e_2]_\mathcal{B} = \begin{pmatrix} 0 \\ -1 \\ 1 \end{pmatrix} \text{ and } [e_3]_\mathcal{B} = \begin{pmatrix} -1 \\ 0 \\ 1 \end{pmatrix}.$$

Thus,

$$M = \begin{pmatrix} 1 & 0 & -1 \\ 1 & -1 & 0 \\ -1 & 1 & 1 \end{pmatrix}.$$

Now, we can write any given coordinate vector (with respect to the standard basis) $v = \begin{pmatrix} x \\ y \\ z \end{pmatrix}$, as a coordinate vector in terms of $\mathcal{B}$ as

$$[v]_\mathcal{B} = M[v]_{\mathcal{B}_0}$$

$$= \begin{pmatrix} 1 & 0 & -1 \\ 1 & -1 & 0 \\ -1 & 1 & 1 \end{pmatrix} \begin{pmatrix} x \\ y \\ z \end{pmatrix}$$

$$= x \begin{pmatrix} 1 \\ 1 \\ -1 \end{pmatrix} + y \begin{pmatrix} 0 \\ -1 \\ 1 \end{pmatrix} + z \begin{pmatrix} -1 \\ 0 \\ 1 \end{pmatrix}$$

$$= \begin{pmatrix} x - z \\ x - y \\ -x + y + z \end{pmatrix}.$$

□

Often, when we are working with several transformations and several bases, it is helpful to be able to combine the matrix representations of the transformations with respect to the bases. Consider the following theorems which combine matrix representations of multiple linear transformations. The proof of each follow from the properties of matrix multiplication and the definition of the matrix representation. The first theorem shows that matrix representations of linear transformations satisfy linearity properties themselves.

Theorem 4.4.13
Let $T, U : V \to W$ be linear, α a scalar, and V and W be finite-dimensional vector spaces with ordered bases $\mathcal{B}$ and $\tilde{\mathcal{B}}$, respectively. Then

(a) $[T + U]_{\mathcal{B}}^{\tilde{\mathcal{B}}} = [T]_{\mathcal{B}}^{\tilde{\mathcal{B}}} + [U]_{\mathcal{B}}^{\tilde{\mathcal{B}}}$

(b) $[\alpha T]_{\mathcal{B}}^{\tilde{\mathcal{B}}} = \alpha [T]_{\mathcal{B}}^{\tilde{\mathcal{B}}}.$

Proof. Suppose V and W are finite-dimensional vector spaces with bases $\mathcal{B}$ and $\tilde{\mathcal{B}}$, respectively. Suppose $T, U : V \to W$ are linear.

(a) We show that $[T + U]_{\mathcal{B}}^{\tilde{\mathcal{B}}} [x]_{\mathcal{B}} = \left([T]_{\mathcal{B}}^{\tilde{\mathcal{B}}} + [U]_{\mathcal{B}}^{\tilde{\mathcal{B}}} \right) [x]_{\mathcal{B}}$ for arbitrary $x \in V$ so that by Definition 4.2.25, $[T + U]_{\mathcal{B}}^{\tilde{\mathcal{B}}} = [T]_{\mathcal{B}}^{\tilde{\mathcal{B}}} + [U]_{\mathcal{B}}^{\tilde{\mathcal{B}}}$. Let $y = (T + U)(x)$, $y_T = T(x)$ and $y_U = U(x)$. We have $y = (T + U)(x) = T(x) + U(x) = y_T + y_U$. Because coordinate transformations are linear, $[y]_{\tilde{\mathcal{B}}} = [y_T + y_U]_{\tilde{\mathcal{B}}} = [y_T]_{\tilde{\mathcal{B}}} + [y_U]_{\tilde{\mathcal{B}}}$. And finally,

$$[T + U]_{\mathcal{B}}^{\tilde{\mathcal{B}}} [x]_{\mathcal{B}} = [y]_{\tilde{\mathcal{B}}} = [y_T]_{\tilde{\mathcal{B}}} + [y_U]_{\tilde{\mathcal{B}}} = [T]_{\mathcal{B}}^{\tilde{\mathcal{B}}} [x]_{\mathcal{B}} + [U]_{\mathcal{B}}^{\tilde{\mathcal{B}}} [x]_{\mathcal{B}}.$$

(b) Similar to part (a), let α be any scalar and $y = (\alpha T)(x)$ and $y_T = T(x)$. We have

$$[\alpha T]_{\mathcal{B}}^{\tilde{\mathcal{B}}} [x]_{\mathcal{B}} = [y]_{\tilde{\mathcal{B}}} = \alpha [y_T]_{\tilde{\mathcal{B}}} = \alpha [T]_{\mathcal{B}}^{\tilde{\mathcal{B}}} [x]_{\mathcal{B}},$$

so that $[\alpha T]_{\mathcal{B}}^{\tilde{\mathcal{B}}} = a [T]_{\mathcal{B}}^{\tilde{\mathcal{B}}}.$

□

Theorem 4.4.13 suggests that, given finite-dimensional vector spaces V and W, the set of all transformations with domain V and codomain W might also be a vector space over $\mathbb{R}$. We leave the exploration of this suggestion to Exercise 23.

The next theorem shows that matrix representations of compositions of linear transformations behave as matrix multiplication operations (in appropriate bases representations).

> **Theorem 4.4.14**
> Let $T : V \to W$ and $U : W \to X$ be linear, $u \in V$. Let V, W, and X be finite-dimensional vector spaces with ordered bases $\mathcal{B}$, $\mathcal{B}'$, and $\tilde{\mathcal{B}}$, respectively. Then
>
> $$[U \circ T]_{\mathcal{B}}^{\tilde{\mathcal{B}}} = [U]_{\mathcal{B}'}^{\tilde{\mathcal{B}}} \, [T]_{\mathcal{B}}^{\mathcal{B}'} \, .$$

The proof of this theorem is similar to the proof of Theorem 4.4.13 and is the subject of Exercise 19.

4.4.4 Exercises

For Exercises 1 through 12, find the matrix representation, $M = [T]_{\mathcal{B}_V}^{\mathcal{B}_W}$, of given linear transformation $T : V \to W$, using the given bases $\mathcal{B}_V$ and $\mathcal{B}_W$. Be sure you can verify the given bases are indeed bases.

1. Let V be the space of objects with 4 voxels and W the space of radiographs with 4 pixels. Define $T : V \to W$ as

$$T\left(\begin{array}{|c|c|} \hline x_1 & x_3 \\ \hline x_2 & x_4 \\ \hline \end{array} \right) = \begin{array}{|c|} \hline x_1 + x_2 \\ \hline x_3 + x_4 \\ \hline \frac{2}{3}x_1 + x_2 + \frac{2}{3}x_4 \\ \hline \frac{1}{3}x_1 + x_3 + \frac{1}{3}x_4 \\ \hline \end{array} \, .$$

Let $\mathcal{B}_V$ and $\mathcal{B}_W$ be the standard bases for the object and radiograph spaces, respectively.

2. Consider the same transformation as in Exercise 1. Use the standard basis for the object space. Use the following basis for the radiograph space.

$$\mathcal{B}_V = \left\{ \begin{array}{|c|} \hline 1 \\ \hline 0 \\ \hline 0 \\ \hline 0 \\ \hline \end{array} \, , \begin{array}{|c|} \hline 1 \\ \hline 1 \\ \hline 0 \\ \hline 0 \\ \hline \end{array} \, , \begin{array}{|c|} \hline 1 \\ \hline 1 \\ \hline 1 \\ \hline 0 \\ \hline \end{array} \, , \begin{array}{|c|} \hline 1 \\ \hline 1 \\ \hline 1 \\ \hline 1 \\ \hline \end{array} \right\} \, .$$

3. Consider the same transformation as in Exercise 1. Use the standard basis for the radiograph space. Use the following basis for the object space.

$$\mathcal{B}_V = \left\{ \begin{array}{|c|c|} \hline 1 & 1 \\ \hline 1 & 0 \\ \hline \end{array} \, , \begin{array}{|c|c|} \hline 1 & 0 \\ \hline 1 & 1 \\ \hline \end{array} \, , \begin{array}{|c|c|} \hline 1 & 1 \\ \hline 0 & 1 \\ \hline \end{array} \, , \begin{array}{|c|c|} \hline 0 & 1 \\ \hline 1 & 1 \\ \hline \end{array} \right\}$$

4. Let $V = \mathcal{M}_{2 \times 2}(\mathbb{R})$ and $W = \mathbb{R}^4$. Define T as

$$T\begin{pmatrix} a & b \\ c & d \end{pmatrix} = \begin{pmatrix} a \\ b \\ a+b \\ c-d \end{pmatrix}.$$

Let $B_V = \left\{ \begin{pmatrix} 1 & 1 \\ 1 & 1 \end{pmatrix}, \begin{pmatrix} 1 & 1 \\ 1 & 0 \end{pmatrix}, \begin{pmatrix} 1 & 1 \\ 0 & 0 \end{pmatrix}, \begin{pmatrix} 1 & 0 \\ 0 & 0 \end{pmatrix} \right\}$ and let B_W be the standard basis for $\mathbb{R}^4$.

5. Let $V = W = \mathcal{P}_2(\mathbb{R})$ and T defined by $T(ax^2 + bx + c) = cx^2 + ax + b$, and let $B_V = B_W = \{x^2 + 1, x - 1, 1\}$.

6. Let $V = W = \mathcal{P}_2(\mathbb{R})$ and T defined by $T(ax^2 + bx + c) = (a+b)x^2 - b + c$ and let B_V and B_W be the standard basis for $\mathcal{P}_2(\mathbb{R})$.

7. Let $V = H_4(\mathbb{R})$ and $W = \mathbb{R}^4$ with T defined by $T(v) = [v]_Y$, where Y is the basis given in Example 3.4.15 and let B_W be the standard basis for $\mathbb{R}^4$.

8. Let $V = W = \mathcal{D}(Z_2)$ and T defined by $T(x) = x + x$, where $B_V = B_W$ is the basis given in Example 3.4.18.

9. Consider the transformation of Exercise 12 in Section 4.2. Let $B_V = B_W$ be the basis given in Example 3.4.15.

10. Let $V = \mathcal{P}_3(\mathbb{R})$ with the basis $\mathcal{B} = \{x^3, x^2 + 1, x + 1, 1\}$ and $W = \mathcal{P}_2$ with the standard basis. Define T as $T(ax^3 + bx^2 + cx + d) = 3ax^2 + 2bx + c$.

11. Let $V = \mathbb{R}^3$ with the standard basis and $W = \mathbb{R}^3$ with basis

$$\mathcal{B} = \left\{ \begin{pmatrix} 1 \\ 0 \\ 0 \end{pmatrix}, \begin{pmatrix} 1 \\ 1 \\ 1 \end{pmatrix}, \begin{pmatrix} 0 \\ 0 \\ 1 \end{pmatrix} \right\}$$

and define T as

$$T\begin{pmatrix} x \\ y \\ z \end{pmatrix} = \begin{pmatrix} x+y \\ y-z \\ 0 \end{pmatrix}.$$

12. Let $V = \mathcal{M}_{3 \times 2}(\mathbb{R})$ with the standard basis and $W = \mathcal{M}_{2 \times 2}(\mathbb{R})$ with the basis

$$\mathcal{B} = \left\{ \begin{pmatrix} 1 & 1 \\ 0 & 0 \end{pmatrix}, \begin{pmatrix} 0 & 0 \\ 1 & 1 \end{pmatrix}, \begin{pmatrix} 1 & 0 \\ 0 & 1 \end{pmatrix}, \begin{pmatrix} 0 & 1 \\ 0 & 0 \end{pmatrix} \right\}$$

and define T as

$$T\begin{pmatrix} a_{11} & a_{12} \\ a_{21} & a_{22} \\ a_{31} & a_{32} \end{pmatrix} = \begin{pmatrix} a_{11} & a_{12} + a_{22} \\ a_{21} + a_{31} & a_{32} \end{pmatrix}$$

For Exercises 13-17, choose common vector spaces V and W and a linear transformation $T : V \to W$ for which M is the matrix representation of T when using the standard bases for V and W. Check your answers with at least two examples.

13. $M = \begin{pmatrix} 3 & 2 \\ 2 & 1 \end{pmatrix}$

14. $M = \begin{pmatrix} 1 & 1 & 1 \\ 2 & 2 & 1 \end{pmatrix}$

15. $M = \begin{pmatrix} -1 & 0 \\ 2 & -1 \\ 3 & 0 \end{pmatrix}$

16. $M = \begin{pmatrix} 1 & 0 & 0 & 1 \\ 0 & 2 & 1 & 1 \end{pmatrix}$

17. $M = \begin{pmatrix} -1 & 1 & -1 & 1 \\ 1 & 2 & 1 & 2 \\ 2 & 2 & 0 & 0 \end{pmatrix}$

Additional Exercises.

18. Consider the space of 7-bar LCD images. Find the change of basis matrix $[I]_{B_1}^{B_2} \in \mathcal{M}_{7\times 7}(\mathbb{Z}_2)$ where

$$B_1 = \left\{ \begin{array}{cccccc} \end{array} \right\},$$

and

$$B_2 = \left\{ \begin{array}{cccccc} \end{array} \right\}.$$

Show that your matrix satisfies the identity relation on the "five" digit:

$$[I]_{B_1}^{B_2} \left[\begin{array}{c} \end{array} \right]_{B_1} = \left[\begin{array}{c} \end{array} \right]_{B_2}.$$

19. Prove Theorem 4.4.14.

20. Can the matrix $M = \begin{pmatrix} 2 & 1 & 0 \\ 0 & 1 & 0 \\ 1 & 0 & 1 \end{pmatrix}$ be the matrix representation of the identity transformation on a vector space of your choosing? Can M be the matrix representation of the zero transformation on a vector space of your choosing? Explain.

21. A regional economic policy group is trying to understand relationships between party affiliation of elected officials and economic health. The table shows a summary of recent election data and economic health indicators. Let $v_k \in \mathbb{R}^3$ represents the percentage of voters aligning with three political parties in election year k. Let $w_k \in \mathbb{R}^2$ be the corresponding economic indicator values two years after the election in year k. Can economic health prediction be modeled as a linear transformation $T : \mathbb{R}^3 \to \mathbb{R}^2$?

year	v_k	w_k
$k = 5$	(40, 45, 10)	(30, 25)
$k = 10$	(45, 35, 25)	(40, 35)
$k = 15$	(35, 40, 20)	(35, 30)

22. Find a basis for $H_2(\mathbb{R})$ for which the matrix representation of the heat diffusion transformation is a diagonal matrix. Repeat for $H_3(\mathbb{R})$. Would these bases be useful when computing the time evolution of a given heat state?

23. Given V and W with dim $V = n$ and dim $W = m$. Show that

$$\mathcal{T} = \{T : V \to W\}$$

is a vector space. What is the dimension of $\mathcal{T}$?

24. Show that if $T : \mathbb{R}^n \to \mathbb{R}^m$ is a linear transformation with $T(x) = Ax$ for some $m \times n$ matrix. Then the ith column of A is Ae_i.

4.5 The Determinants of a Matrix

In this section, we investigate the determinant, a number that describes the geometry of a linear transformation from $\mathbb{R}^n \to \mathbb{R}^n$.

We begin by considering transformations from $\mathbb{R}^2 \to \mathbb{R}^2$, since these are the most simple to visualize. Recall, in Section 4.2 Exercises 14-27, that we showed that any linear transformation T from $\mathbb{R}^2$ to $\mathbb{R}^2$ maps the unit square onto a (possibly degenerate) parallelogram that is determined by the two vectors $T(e_1)$ and $T(e_2)$. The vertices of the parallelogram are 0, $T(e_1)$, $T(e_2)$, and $T(e_1) + T(e_2)$. We now develop this idea further.

Example 4.5.1 Consider the linear transformation $T : \mathbb{R}^2 \to \mathbb{R}^2$ be given by $T(x) = Ax$, where $A = \begin{pmatrix} 1 & 0 \\ 0 & 2 \end{pmatrix}$. Notice, in Figure 4.17, that $T(e_1) = e_1$ and $T(e_2) = 2e_2$, so the unit square is mapped to a rectangle that is one unit wide and two units tall, twice the area of the original square.

Next, what happens if we choose some small number δ and consider the smaller square determined by the vectors δe_1 and δe_2? This smaller square has area δ^2 and its image is the rectangle determined by the vectors $T(\delta e_1) = \delta T(e_1) = \delta e_1$ and $T(\delta e_2) = \delta T(e_2) = 2\delta e_2$. So the image of the smaller square has area $2\delta^2$; the transformation has again doubled the area.

Moreover, even if we translate the square, the area of the image doubles: consider the smaller square in the previous paragraph translated by $\begin{pmatrix} a \\ b \end{pmatrix}$ (so the vertices are (a, b), $(a + \delta, b)$, $(a + \delta, b + \delta)$, and $(a, b + \delta)$). The image of the square under the map T has vertices

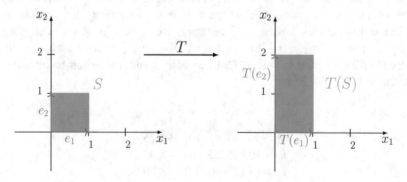

Fig. 4.17 The transformation T stretches the unit square vertically by a factor of 2.

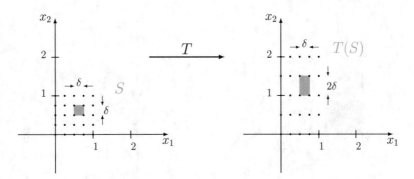

Fig. 4.18 The transformation T stretches the δ square vertically by a factor of 2.

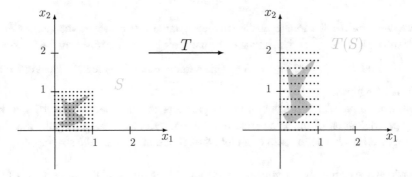

Fig. 4.19 The transformation T stretches the "blob" vertically by a factor of 2.

$$T\begin{pmatrix} a \\ b \end{pmatrix} = \begin{pmatrix} a \\ 2b \end{pmatrix}, \quad T\begin{pmatrix} a + \delta \\ b \end{pmatrix} = \begin{pmatrix} a + \delta \\ 2b \end{pmatrix},$$

$$T\begin{pmatrix} a + \delta \\ b + \delta \end{pmatrix} = \begin{pmatrix} a + \delta \\ 2b + 2\delta \end{pmatrix}, \text{ and } T\begin{pmatrix} a \\ b + \delta \end{pmatrix} = \begin{pmatrix} a \\ 2b + 2\delta \end{pmatrix}.$$

These are the vertices of a rectangle of width δ and height 2δ and lower left corner at the point (a, b). In this case as well, the square of area δ^2 has been transformed into a rectangle of area $2\delta^2$ (as can be seen in Figure 4.18).

So apparently, T doubles the area of squares (whose sides are parallel to the coordinate axes). What about other shapes? Suppose we start with the irregular set (a "blob") in $\mathbb{R}^2$ and we want to calculate the image of the blob under T. We can imagine overlaying a fine square grid over the blob; the image of any small grid square doubles under the transformation T, so the area of the blob must also double; see Figure 4.19.

This is consistent with our understanding that the linear transformation T preserves distances in the e_1-direction but stretches distances in the e_2-direction by 2. $\qquad\square$

Example 4.5.2 Consider the linear transformation $T : \mathbb{R}^2 \to \mathbb{R}^2$ be given by $T(x) = Ax$, where $A = \begin{pmatrix} 1 & 0 \\ 1 & 1 \end{pmatrix}$. Notice that $T(e_1) = \begin{pmatrix} 1 \\ 1 \end{pmatrix}$ and $T(e_2) = e_2$, so the unit square is mapped to the parallelogram shown in Figure 4.20. This parallelogram also has unit area.

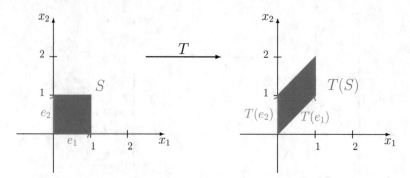

Fig. 4.20   The transformation T transforms the unit square to a parallelogram of the same area.

As before, we consider the image of the smaller square determined by the vectors δe_1 and δe_2. This smaller square has area δ^2 and its image is the parallelogram determined by the vectors $T(\delta e_1) = \delta T(e_1) = \delta \begin{pmatrix} 1 \\ 1 \end{pmatrix}$ and $T(\delta e_2) = \delta T(e_2) = \delta e_2$. This parallelogram has area δ^2, equal to the area of the square with side lengths δ.

Also, following a parallel argument as in the previous example, any square of side length δ is mapped to a parallelogram of the same area. In addition, by the same grid argument as before, this transformation will not change the area of a blob, though it does "shear" the shape.   □

In the previous two examples, we considered the area of the image of the unit square $T(S)$, and this number, in each case, was the amount that each transformation increased area. It turns out that any linear transformation from $\mathbb{R}^2$ to $\mathbb{R}^2$ will change areas of sets in $\mathbb{R}^2$ by a fixed multiplicative amount, independent of the set! And this factor can be determined just by looking at where the transformation sends the unit square!

Example 4.5.3   For each of the following matrices A, consider the linear transformation $T : \mathbb{R}^2 \to \mathbb{R}^2$ defined by $T(x) = Ax$. Find the factor by which each transformation increases area. That is, find the area of the image of the unit square under each of these transformations. (Hint: sketch the image of the unit square in each case.)

(a) $\begin{pmatrix} 1 & -1/2 \\ 0 & 1/2 \end{pmatrix}$

(b) $\begin{pmatrix} -1 & 1 \\ 1 & 1 \end{pmatrix}$

(c) $\begin{pmatrix} 1/2 & 1/2 \\ -1/2 & 1/2 \end{pmatrix}$

(d) $\begin{pmatrix} -1/2 & 1 \\ 1/2 & 0 \end{pmatrix}$

(e) $\begin{pmatrix} 2 & 1 \\ 4 & 2 \end{pmatrix}$

□

It turns out that there is a formula that will compute the area distortion factor of a linear transformation from $\mathbb{R}^2$ to $\mathbb{R}^2$, given the matrix representation $A = \begin{pmatrix} a & b \\ c & d \end{pmatrix}$ of the transformation. Specifically, the

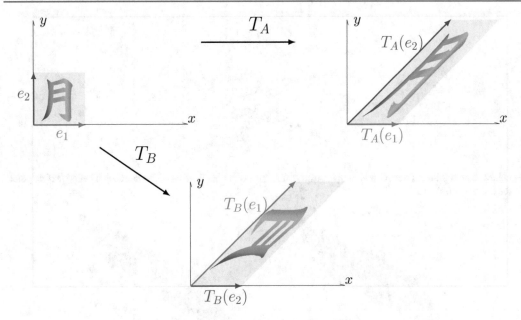

Fig. 4.21 A change in sign in the determinant shows as a "flip" of the image.

quantity $a \cdot d - b \cdot c$ is called the *determinant* of A, and its absolute value $|a \cdot d - b \cdot c|$ is the factor by which area is distorted.

Let's check this for the linear transformation in Example 4.5.1. We find that the determinant of $A = \begin{pmatrix} 1 & 0 \\ 0 & 2 \end{pmatrix}$ is $1 \cdot 2 - 0 \cdot 0 = 2$. This corresponds to our discovery that this linear transformation doubles area.

Similarly, for the linear transformation in Example 4.5.2, the determinant of $A = \begin{pmatrix} 1 & 0 \\ 1 & 1 \end{pmatrix}$ is $1 \cdot 1 - 0 \cdot 1 = 1$, which reflects the fact that this transformation did not change area.

Now, for each matrix $\begin{pmatrix} a & b \\ c & d \end{pmatrix}$ in Example 4.5.3 above, calculate the quantity $a \cdot d - b \cdot c$. You should see that indeed the absolute value of the determinant $|a \cdot d - b \cdot c|$ corresponds to the area distortion factor in each case.

So we have a good geometric interpretation for the absolute value of the determinant of a 2×2 matrix, but what about the sign of this quantity? Both of the matrices $A = \begin{pmatrix} 1 & 2 \\ 0 & 2 \end{pmatrix}$ and $B = \begin{pmatrix} 2 & 1 \\ 2 & 0 \end{pmatrix}$ (from parts (a) and (d) of Example 4.5.3) distort area by a factor of 2, but their determinants have the opposite sign; the determinant of A is 2 while the determinant of B is -2. In the following example we explore the geometry behind the difference in sign of the determinant for these two maps.

Example 4.5.4 Consider, again, the transformations T_A and T_B from above. We will apply these transformations to an image of the kanji for "moon" as seen in Figure 4.21. The image of the unit square is the same parallelogram in both cases but the sides come from different sides of the unit square. In particular, notice that in the linear transformation associated with B, the image of the unit square (and the kanji) has been "flipped over," but that in the linear transformation associated with A, the image of the unit square (and the kanji) has not. We say that T_A *preserves orientation* and T_B *reverses orientation*. □

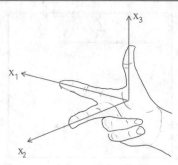

Fig. 4.22 The right-hand rule: Orientation is positive if x_3 points in the direction of the thumb and negative if x_3 points in the opposite direction.

Fig. 4.23 Left: Positive orientation.   Right: Negative orientation.

For a linear transformation from $\mathbb{R}^n$ to $\mathbb{R}^n$, we can analogously consider the volume expansion factor and whether the transformation is orientation preserving or reversing. We now describe the generalizations of these two geometric ideas for $n = 3$.[2]

A linear transformation $T : \mathbb{R}^3 \to \mathbb{R}^3$ maps a 3D cube to a (possibly degenerate) 3D parallelepiped whose corners are the images of the original cube's corners. We easily see that the factor by which T changes volume is just the volume of the image parallelepiped (because the volume of the unit cube is 1) and whose sign measures whether the orientation of the parallelepiped is different from that of the cube. Here, the idea of orientation is a little more complicated, but can still be visualized in the following way.

We will define whether T preserves or reverses orientation by first defining the orientation of a triple, or ordered set of (linearly independent) vectors. If three vectors x_1, x_2, and x_3 in $\mathbb{R}^3$ are linearly independent, then x_1 and x_2 span a plane, which separates $\mathbb{R}^3$ into two pieces, and x_3 must lie on one side or the other of this plane. Which side x_3 lies on determines whether or not the triple (x_1, x_2, x_3) has positive orientation. By convention we use the "right-hand rule:" if you point the index finger of your right hand in the direction of x_1 and the second finger toward x_2, then your thumb points to the positive orientation side of the plane (see Figure 4.22). If x_3 lies on this side of the plane then the triple (x_1, x_2, x_3) has positive orientation; if x_3 lies on the other side then the set has negative orientation. (See Figure 4.23.)

The 3 edges of the cube emanating from the coordinate origin correspond to the unit vectors e_1, e_2, and e_3, and the triple (e_1, e_2, e_3) has positive orientation. Hence, we can check the corresponding edges of the image of the cube: $T(e_1)$, $T(e_2)$, and $T(e_3)$. If the triple $(T(e_1), T(e_2), T(e_3))$ has positive orientation then we say that the transformation *preserves orientation*, and if it has negative orientation

---

[2] Although both volume expansion and orientation can be extended to dimensions higher than 3, we do not include them here.

(in other words, the orientation of the parallelepiped has "flipped" from the original orientation of the cube) then we say that the transformation *reverses orientation*.

The *determinant* of a matrix, which we will define algebraically in the next section, is geometrically interpreted as the product of the volume expansion factor and ± 1 (depending on whether the map preserves or reverses orientation). This geometric intuition can yield additional insights, so we encourage readers to keep both perspectives in mind.

4.5.1 Determinant Calculations and Algebraic Properties

Now that we have a geometric interpretation for the determinant of a square matrix, we will focus on an algebraic definition of the determinant. Indeed our geometric intuition suggests some of the algebraic properties that the determinant should possess. We give two motivating examples before proceeding with the definition.

As a map from $\mathbb{R}^n$ to $\mathbb{R}^n$, a diagonal matrix maps the unit n-cube to an n-dimensional rectangular parallelepiped whose side lengths are equal to the (absolute values of) the diagonal entries of the matrix. Hence, we would want our definition to assign the determinant of a diagonal matrix to be the product of the diagonal entries.

Also, suppose that two $n \times n$ matrices A and B are identical in every position except along the kth row, where the entries of B are α times the entries of A for some $\alpha > 0$. We can compare the parallelepipeds that are the images of the unit n-cube under T_A and T_B. The parallelepiped from T_B is the same as the parallelepiped from T_A except that under T_B all edges stretched by α times more in the direction e_k. This results in the parallelepiped from T_B having α times the volume of the one from T_A.

We now define the determinant as a function that assigns a real number to every matrix and satisfies the following properties. As you read the properties, we hope the reader will check that they are consistent with our geometric understanding of the determinant.

Definition 4.5.5

Let $\mathcal{M}_{n \times n}$ be the set of all $n \times n$ matrices. We define the determinant as the function det : $\mathcal{M}_{n \times n} \to \mathbb{R}$ with the following properties. For a number α (real or complex) and $n \times n$ matrices, $A, B \in \mathcal{M}_{n \times n}$,

- If A is in echelon form, then $\det(A)$ is the product of the diagonal elements.
- If B is obtained by performing the row operation, $R_k = \alpha r_j + r_k$ on A (replacing a row by the sum of the row and a multiple of another row), then $\det(B) = \det(A)$.
- If B is obtained by performing the row operation, $R_k = \alpha r_k$ on A (replacing a row by a multiple of the row), then $\det(B) = \alpha \det(A)$.
- If B is obtained by performing the row operation, $R_k = r_j$ and $R_j = r_k$ on A (interchanging two rows), then $\det(B) = -1 \cdot \det(A)$.

$\square$

Notation. It is common, when context permits, to write $|M|$ and mean $\det(M)$.

There are two important points to consider about this definition. First, you may notice that this definition only explicitly states the determinant of a matrix if it is in echelon form. However, we can row reduce any matrix to echelon form through elementary row operations, and the remaining properties all show how the determinant changes as these row operations are applied. Hence, we can use Definition 4.5.5 to find the determinant of a matrix by keeping track of how the determinant changes at each step of a matrix reduction and working backwards from the determinant of the row-reduced determinant.

Second, since there are many ways to row reduce a matrix. This means that we must show that *any* sequence of row operations that reduces the matrix to echelon form will change the determinant by the same amount, and that hence there is no ambiguity[3] in our definition. We will show this in Theorem 4.5.10 below, but before doing this we will give three examples illustrating how to use the definition to compute determinants.

Example 4.5.6 Find the determinant of

$$A = \begin{pmatrix} 1 & 1 & 1 \\ 2 & -1 & 3 \\ -1 & 1 & -2 \end{pmatrix}.$$

Our goal is to reduce A to echelon form all the while keeping track of how the determinant changes according the properties listed above. We can build a table to keep track of our operations, matrices and determinants.

Row Operations	Matrix	Determinant
	$\begin{pmatrix} 1 & 1 & 1 \\ 2 & -1 & 3 \\ -1 & 1 & -2 \end{pmatrix}$	$\det(A)$
$\xrightarrow[R_3=r_1+r_3]{R_2=-2r_1+r_2}$	$\begin{pmatrix} 1 & 1 & 1 \\ 0 & -3 & 1 \\ 0 & 2 & -1 \end{pmatrix}$	$\det(A)$
$\xrightarrow{R_2=r_2+r_3}$	$\begin{pmatrix} 1 & 1 & 1 \\ 0 & -1 & 0 \\ 0 & 2 & -1 \end{pmatrix}$	$\det(A)$
$\xrightarrow[R_3=2r_2+r_3]{R_2=-r_2}$	$\begin{pmatrix} 1 & 1 & 1 \\ 0 & 1 & 0 \\ 0 & 0 & -1 \end{pmatrix}$	$-\det(A)$
$\xrightarrow{R_3=-r_3}$	$\begin{pmatrix} 1 & 1 & 1 \\ 0 & 1 & 0 \\ 0 & 0 & 1 \end{pmatrix}$	$\det(A)$

Now, using the first property of Definition 4.5.5, we find that

$$\det \begin{pmatrix} 1 & 1 & 1 \\ 0 & 1 & 0 \\ 0 & 0 & 1 \end{pmatrix} = 1.$$

---

[3] In mathematical terminology, we say that the determinant is *well-defined*.

Matching the determinant of the reduced matrix with the far right column, we find that $\det(A) = 1$. Recall that geometrically, this tells us that the linear transformation T_A preserves volume (and orientation). $\qquad\square$

Example 4.5.7   Find the determinant of

$$A = \begin{pmatrix} 2 & 2 & 2 \\ 1 & 0 & 1 \\ -2 & 2 & -4 \end{pmatrix}.$$

Again, we reduce A to echelon form all the while keeping track of how the determinant changes.

Row Operations	Matrix	Determinant
	$\begin{pmatrix} 2 & 2 & 2 \\ 1 & 0 & 1 \\ -2 & 2 & -4 \end{pmatrix}$	$\det(A)$
$\xrightarrow{R_1=\frac{1}{2}r_1}$	$\begin{pmatrix} 1 & 1 & 1 \\ 1 & 0 & 1 \\ -2 & 2 & -4 \end{pmatrix}$	$\frac{1}{2}\det(A)$
$\xrightarrow[R_3=2r_1+r_3]{R_2=-r_1+r_2}$	$\begin{pmatrix} 1 & 1 & 1 \\ 0 & -1 & 0 \\ 0 & 4 & -2 \end{pmatrix}$	$\frac{1}{2}\det(A)$
$\xrightarrow[R_3=4r_2+r_3]{R_2=-r_2}$	$\begin{pmatrix} 1 & 1 & 1 \\ 0 & 1 & 0 \\ 0 & 0 & -2 \end{pmatrix}$	$-\frac{1}{2}\det(A)$
$\xrightarrow{R_3=-\frac{1}{2}r_3}$	$\begin{pmatrix} 1 & 1 & 1 \\ 0 & 1 & 0 \\ 0 & 0 & 1 \end{pmatrix}$	$\frac{1}{4}\det(A)$

Now, using the first property in Definition 4.5.5, we see that

$$\det \begin{pmatrix} 1 & 1 & 1 \\ 0 & 1 & 0 \\ 0 & 0 & 1 \end{pmatrix} = 1.$$

Again, matching the determinant of the reduced matrix with the far right column gives us that $\frac{1}{4}\det(A) = 1$ and so $\det(A) = 4$.

We conclude that the transformation T_A expands volume by a factor of 4, while preserving orientation. $\qquad\square$

Example 4.5.8 Find the determinant of

$$A = \begin{pmatrix} 1 & 1 & -1 \\ 1 & 2 & 1 \\ 2 & 3 & 0 \end{pmatrix}.$$

Again, we reduce A to echelon form all the while keeping track of how the determinant changes.

Row Operations	Matrix	Determinant
	$\begin{pmatrix} 1 & 1 & -1 \\ 1 & 2 & 1 \\ 2 & 3 & 0 \end{pmatrix}$	$\det(A)$
$\xrightarrow{R_2 = -r_1 + r_2}$	$\begin{pmatrix} 1 & 1 & -1 \\ 0 & 1 & 2 \\ 2 & 3 & 0 \end{pmatrix}$	$\det(A)$
$\xrightarrow{R_3 = -2r_1 + r_3}$	$\begin{pmatrix} 1 & 1 & -1 \\ 0 & 1 & 2 \\ 0 & 1 & 2 \end{pmatrix}$	$\det(A)$
$\xrightarrow{R_3 = -r_2 + r_3}$	$\begin{pmatrix} 1 & 1 & -1 \\ 0 & 1 & 2 \\ 0 & 0 & 0 \end{pmatrix}$	$\det(A)$

Now, using the first property in Definition 4.5.5, we see that

$$\det \begin{pmatrix} 1 & 1 & -1 \\ 0 & 1 & 2 \\ 0 & 0 & 0 \end{pmatrix} = 0.$$

Finally, matching the determinant of the reduced matrix with the far right column gives us that $\det(A) = 0$. Geometrically, we conclude that the map T_A maps the unit cube to a degenerate parallelepiped; one that has zero volume.   □

In the preceding example, the reduced echelon form of A has a zero along the diagonal and therefore the determinant is zero. Here we present this result more generally.

Theorem 4.5.9
Let A be an $n \times n$ matrix. If A is row equivalent to a matrix with a row of zeros, then $\det(A) = 0$. Conversely, if $\det(A) = 0$, then A is row equivalent to a matrix with a row of zeros.

The proof is Exercise 32. We will use this fact in several of the proofs that follow.

We now return to the issue of showing that the determinant of a matrix does not depend on which sequence of row operations are used to reduce the matrix. In order to do this, it will be helpful to use the language of matrix reduction using elementary matrices from Section 2.2.3. Recall that a matrix reduction can be performed through multiplication by elementary matrices. Recall also that the elementary matrix for a row operation is obtained by performing that row operation to the identity matrix.

We will consider determinants of the three types of elementary matrices: (1) multiplication of a row by a nonzero number, (2) addition of a multiple of a row to another, and (3) interchanging rows.

By Definition 4.5.5,

- the determinant of an elementary matrix E that multiplies a row by a nonzero number a is $\det(E) = a$;
- the determinant of an elementary matrix E that adds a multiple of a row to another row is $\det(E) = 1$; and
- the determinant of an elementary matrix E that switches two rows is $\det(E) = -1$.

With this notational tool, we are ready to show that our definition for the determinant of a matrix is well defined; that is, the calculation of the determinant of a matrix using Definition 4.5.5 does not depend on the sequence of row operations used to reduce the matrix.

Theorem 4.5.10
Let A be an $n \times n$ matrix. Then $\det A$ is uniquely determined by Definition 4.5.5.

Proof. Let us consider the matrix A whose reduced echelon form is A'. We have two cases to consider: (1) $A' \neq I$ and (2) $A = I$. In the first case, we know that A' has at least one row of all zeros and therefore, by Definition 4.5.5, $\det(A) = 0$. Since the reduced echelon form of a matrix is unique, the determinant for case (1) is always 0.

Next, let us consider the case where A reduces to I. This means that we have k elementary matrices $E_1, E_2, \ldots E_k$ so that $I = E_k \ldots E_2 E_1 A$. Suppose also that the row operation corresponding to elementary matrix E_ℓ applies a multiple α_ℓ to the determinant calculation so that $1 = \alpha_1 \alpha_2 \ldots \alpha_k \det(A)$. Now, suppose we find another sequence of row operations to reduce A to I corresponding to m elementary matrices $\tilde{E}_1, \tilde{E}_2, \ldots, \tilde{E}_m$ so that $I = \tilde{E}_m \ldots \tilde{E}_2 \tilde{E}_1 A$ and with $\tilde{E}_p$ applying a multiple β_p to the determinant calculation so that so that $1 = \beta_1 \beta_2 \ldots \beta_m \det(A)$. Notice that, all row operations can be undone:

- Multiplying row i by a nonzero constant, α is undone by multiplying row i by $\frac{1}{\alpha}$.
- Adding α times row i to row j is undone by adding $-\alpha$ times row i to row j.
- Changing the order of two rows is undone by changing them back.

Therefore, there are elementary matrices $E_1', E_2', \ldots, E_k'$ that undo the row operations corresponding to elementary matrices $E_1, E_2, \ldots, E_k$. In this case, E_ℓ' applies a multiple of $\frac{1}{\alpha_\ell}$ to the determinant calculation so that

$$\det(A) = \frac{1}{\alpha_1 \alpha_2 \ldots \alpha_k} \cdot 1.$$

Now, we see that

$$E_1' E_2' \ldots E_k' I = A.$$

Therefore, we can perform row operations on A to get back to I as follows.

$$(\tilde{E}_m \ldots \tilde{E}_2 \tilde{E}_1)(E_1' E_2' \ldots E_k' I) = (\tilde{E}_m \ldots \tilde{E}_2 \tilde{E}_1)A = I.$$

Therefore, by Definition 4.5.5,

$$\det(I) = \frac{\beta_1 \beta_2 \ldots \beta_m}{\alpha_1 \alpha_2 \ldots \alpha_k}.$$

But, since $\det(I) = 1$, we have that $\beta_1 \beta_2 \ldots \beta_m = \alpha_1 \alpha_2 \ldots \alpha_k$ and $\det(A)$ is uniquely defined.   □

In addition to reassuring us that Definition 4.5.5 is unambiguous, the proof above also shows that if A is any $n \times n$ matrix and E is any $n \times n$ elementary matrix, then we have

$$\det(EA) = \det(E)\det(A).$$

In fact, it is always the case that $\det(AB) = \det(A)\det(B)$. We present this result as a theorem.

> **Theorem 4.5.11**
> Let A and B be $n \times n$ matrices. Then $\det(AB) = \det(A)\det(B)$.

Similar to the proof of Theorem 4.5.10, the proof uses elementary matrices corresponding to A and B and also considers cases depending on the reduced echelon form of A and B.

Proof. Let A' denote the reduced echelon form of A and let B' denote the reduced echelon form of B. Then we can perform row operations on A' through multiplication by n elementary matrices to get to A. That is, there are elementary matrices E_i, $i = 1, 2, \ldots, n$ so that

$$A = E_1 E_2 \ldots E_n A'$$

And, we can perform row operations on B' through multiplication by k elementary matrices to get to B. So, there are elementary matrices $\tilde{E}_j$, $j = 1, 2, \ldots, k$ so that

$$B = \tilde{E}_1 \tilde{E}_2 \ldots \tilde{E}_k B'.$$

By the definition of determinant, we see that $\det(A) = \det(E_1)\det(E_2)\ldots\det(E_n)\det(A')$ and $\det(B) = \det(\tilde{E}_1)\det(\tilde{E}_2)\ldots\det(\tilde{E}_n)\det(B')$. Now, we consider cases (1) A' and B' are both I and (2) at least one of A' or B' is not I.

(1) In the case where both $A' = I$ and $B' = I$, we have

$$AB = (E_1 E_2 \ldots E_n A')(\tilde{E}_1 \tilde{E}_2 \ldots \tilde{E}_k B') = E_1 E_2 \ldots E_n \tilde{E}_1 \tilde{E}_2 \ldots \tilde{E}_k.$$

This means that AB reduces to I. Again, by the definition of the determinant,

$$\det(AB) = (\det(E_1)\det(E_2)\ldots\det(E_n))(\det(\tilde{E}_1)\det(\tilde{E}_2)\ldots\det(\tilde{E}_n)\det(I)) = \det(A)\det(B).$$

(2) Now, we consider the case where at least one of A' or B' is not the identity matrix. Suppose, first that $B' \neq I$. Then we see that AB reduces to $A'B'$ through the matrix multiplication

$$(E_1 E_2 \ldots E_n A')(\tilde{E}_1 \tilde{E}_2 \ldots \tilde{E}_k B') = E_1 E_2 \ldots E_n A' \tilde{E}_1 \tilde{E}_2 \ldots \tilde{E}_k B'.$$

Therefore, by the definition of the determinant,

$$\det(AB) = (\det(E_1) \det(E_2) \ldots \det(E_n))(\det(\tilde{E}_1) \det(\tilde{E}_2) \ldots \det(\tilde{E}_n) \det(B')) = \det(A) \det(B).$$

In fact, $\det(B') = 0$ because it has a row of all zeros. Similar investigations lead to $\det(AB) = 0$ for the case when A' is not the identity. (See Exercise 23.)

$\square$

We now present another method for computing determinants. This method is called *cofactor expansion*.

Lemma 4.5.12
The determinant of a 1×1 matrix M is just the value of its entry. That is, if $M = [a]$, then $\det(M) = a$.

The proof immediately follows from the fact that M is already in echelon form.
We use the determinant of a 1×1 matrix to find $\det(M) = |M|$ for any $n \times n$ matrix M using the following iterative computation.

Theorem 4.5.13
Let $A = (a_{i,j})$ be an $n \times n$ matrix, $n \geq 2$.

- (Expansion along ith row) For any i so that $1 \leq i \leq n$,

$$|A| = \sum_{j=1}^{n} (-1)^{i+j} a_{i,j} |M_{i,j}|.$$

- (Expansion along jth column) For any j so that $1 \leq j \leq n$,

$$|A| = \sum_{i=1}^{n} (-1)^{i+j} a_{i,j} |M_{i,j}|.$$

Here, $M_{i,j}$ is the sub-matrix of A where the ith row and jth column have been removed.

We omit the proof, but encourage the curious reader to consider how you might prove this theorem.[4] Let us use the cofactor expansion along the first row to compute the determinant of a 2×2 matrix.

Example 4.5.14 Let $A = \begin{pmatrix} a_{1,1} & a_{1,2} \\ a_{2,1} & a_{2,2} \end{pmatrix}$, then

$$|A| = a_{1,1}|M_{1,1}| - a_{1,2}|M_{1,2}|.$$

But $M_{1,1} = [a_{2,2}]$ and $M_{1,2} = [a_{2,1}]$ so we have $|A| = a_{1,1}a_{2,2} - a_{1,2}a_{2,1}$. Happily, we have recovered the familiar formula for the determinant of a 2×2 matrix. $\square$

If n is larger than 2, this process is iterative – the determinant of each submatrix $M_{i,j}$ is determined by another cofactor expansion. This process is repeated until the sub-matrices are 2×2.

We begin with a computation for a general 3×3 and then a general 4×4 matrix. We will follow these computations with the an example computing the determinant of a specific matrix.

Example 4.5.15 First, we consider an example when $n = 3$. Let $A = \begin{pmatrix} a_{1,1} & a_{1,2} & a_{1,3} \\ a_{2,1} & a_{2,2} & a_{2,3} \\ a_{3,1} & a_{3,2} & a_{3,3} \end{pmatrix}$, then

$$|A| = a_{1,1} \begin{vmatrix} a_{2,2} & a_{2,3} \\ a_{3,2} & a_{3,3} \end{vmatrix} - a_{1,2} \begin{vmatrix} a_{2,1} & a_{2,3} \\ a_{3,1} & a_{3,3} \end{vmatrix} + a_{1,3} \begin{vmatrix} a_{2,1} & a_{2,2} \\ a_{3,1} & a_{3,2} \end{vmatrix}.$$

From here, we use the formula for the determinant of a 2×2 matrix. Thus

$$|A| = a_{1,1}(a_{2,2}a_{3,3} - a_{3,2}a_{2,3}) - a_{1,2}(a_{2,1}a_{3,3} - a_{2,3}a_{3,1}) + a_{1,3}(a_{2,1}a_{3,2} - a_{2,2}a_{3,1}).$$

$\square$

We continue by finding the determinant of a general 4×4 matrix.

Example 4.5.16 Let $A = \begin{pmatrix} a_{1,1} & a_{1,2} & a_{1,3} & a_{1,4} \\ a_{2,1} & a_{2,2} & a_{2,3} & a_{2,4} \\ a_{3,1} & a_{3,2} & a_{3,3} & a_{3,4} \\ a_{4,1} & a_{4,2} & a_{4,3} & a_{4,4} \end{pmatrix}$, then, expanding about the first row gives

$$|A| = a_{1,1} \begin{vmatrix} a_{2,2} & a_{2,3} & a_{2,4} \\ a_{3,2} & a_{3,3} & a_{3,4} \\ a_{4,2} & a_{4,3} & a_{4,4} \end{vmatrix} - a_{1,2} \begin{vmatrix} a_{2,1} & a_{2,3} & a_{2,4} \\ a_{3,1} & a_{3,3} & a_{3,4} \\ a_{4,1} & a_{4,3} & a_{4,4} \end{vmatrix}$$

$$+ a_{1,3} \begin{vmatrix} a_{2,1} & a_{2,2} & a_{2,4} \\ a_{3,1} & a_{3,2} & a_{3,4} \\ a_{4,1} & a_{4,2} & a_{4,4} \end{vmatrix} - a_{1,4} \begin{vmatrix} a_{2,1} & a_{2,2} & a_{2,3} \\ a_{3,1} & a_{3,2} & a_{3,3} \\ a_{4,1} & a_{4,2} & a_{4,3} \end{vmatrix}.$$

From here, we finish by employing a cofactor expansion formula for the determinant of a 3×3 matrix and then the formula for a 2×2 matrix (as in Example 4.5.15 and Example 4.5.14 above). $\square$

---

[4] Hint: Can you show that a row expansion satisfies the properties of Definition 4.5.5? What about a column expansion?

In the previous two examples, we expanded about the first row. If it better suits us, we can expand about any other row or any column, and we will arrive at the same number. We demonstrate with the following specific example.

Example 4.5.17 Compute the determinant of the given matrix.

$$A = \begin{pmatrix} 2 & 2 & 2 \\ 1 & 0 & 1 \\ -2 & 2 & -4 \end{pmatrix}.$$

First, as before, we can expand along the first row. Then

$$\begin{aligned}
|A| &= 2\begin{vmatrix} 0 & 1 \\ 2 & -4 \end{vmatrix} - 2\begin{vmatrix} 1 & 1 \\ -2 & -4 \end{vmatrix} + 2\begin{vmatrix} 1 & 0 \\ -2 & 2 \end{vmatrix} \\
&= 2(0(-4) - 1(2)) - 2(1(-4) - 1(-2)) + 2(1(2) - 0(-2)) \\
&= -4 + 4 + 4 \\
&= 4.
\end{aligned}$$

We will now perform a cofactor expansion along the second column.

$$\begin{aligned}
|A| &= -2\begin{vmatrix} 1 & 1 \\ -2 & -4 \end{vmatrix} + 0\begin{vmatrix} 2 & 2 \\ -2 & -4 \end{vmatrix} - 2\begin{vmatrix} 2 & 2 \\ 1 & 1 \end{vmatrix} \\
&= -2(1(-4) - 1(-2)) + 0 - 2(2(1) - 2(1)) \\
&= -2(-4 - (-2)) \\
&= 4.
\end{aligned}$$

Both calculations yield the same result (indeed, expanding around any row or column must give the same result). However, the second calculation is a little bit simpler because we only needed to compute 2 of the 3 subdeterminants. We often choose the row/column to perform the expansion to be the one containing the most zeros. □

The following example further illustrates the idea of choosing a convenient row or column about which to expand.

Example 4.5.18 Let A be the matrix

$$A = \begin{pmatrix} 1 & 5 & -3 & 9 \\ 1 & 0 & 1 & 7 \\ 0 & 1 & 0 & 0 \\ 3 & 2 & 5 & -2 \end{pmatrix}$$

Compute the $\det(A)$.

Here we notice all the zeros in the third row and choose to expand about row 3, giving us one term in the first step, requiring a determinant of one 3×3 matrix.

$$|A| = -1 \begin{vmatrix} 1 & -3 & 9 \\ 1 & 1 & 7 \\ 3 & 5 & -2 \end{vmatrix}$$

$$= -1 \left(1 \begin{vmatrix} 1 & 7 \\ 5 & -2 \end{vmatrix} - 1 \begin{vmatrix} -3 & 9 \\ 5 & -2 \end{vmatrix} + 3 \begin{vmatrix} -3 & 9 \\ 1 & 7 \end{vmatrix} \right)$$

$$= -1 \left((1 \cdot (-2) - 5 \cdot 7) - 1((-3)(-2) - 5 \cdot 9) + 3((-3) \cdot 7 - 9 \cdot 1) \right)$$

$$= -(-37 - (-39) + 3(-30)) = 88.$$

Such choices, as we made here, can simplify the calculations. There are many choices and some might seem easier. We encourage the reader to attempt other choices of columns or rows about which to expand in order to become comfortable choosing a less tedious path.   □

Cofactor expansion gives us an easy way to prove the following useful property of the determinant.

Theorem 4.5.19
Given an $n \times n$ matrix, A, $\det(A^T) = \det(A)$.

The proof is Exercise 29.

Our primary motivation in this section has been to explore determinants as a means of understanding the geometry of linear transformations. However, as some of the proofs here suggest, determinants are intimately connected with other linear algebraic concepts, such as span. We will develop these connections more in future chapters, but to whet your appetite, we close this section with a theorem relating the determinant to the set of solutions to the matrix equation $Ax = b$.

Theorem 4.5.20
Let A be an $n \times n$ matrix. The matrix equation $Ax = 0$ has one solution if and only if $\det(A) \neq 0$.

The proof is Exercise 35.

Exercises

For Exercises 1-4, draw the image of the unit square under the transformation $T(x) = Mx$ and calculate the area multiplier determined by the transformation T.

1. $M = \begin{pmatrix} 2 & 0 \\ 0 & 5 \end{pmatrix}$

2. $M = \begin{pmatrix} -1 & 0 \\ 0 & 1 \end{pmatrix}$

3. $M = \begin{pmatrix} 1 & 1 \\ 2 & 2 \end{pmatrix}$

4. $M = \begin{pmatrix} 3 & 2 \\ 1 & 1 \end{pmatrix}$

For Exercises 5-9, draw the image of the unit cube under the (3D) transformation defined by $T(x) = Mx$ and calculate the volume multiplier determined by the transformation T.

5. $M = \begin{pmatrix} 2 & 0 & 0 \\ 0 & 5 & 0 \\ 0 & 0 & 3 \end{pmatrix}$

6. $M = \begin{pmatrix} -1 & 0 & 0 \\ 0 & 1 & 0 \\ 0 & 0 & 1 \end{pmatrix}$

7. $M = \begin{pmatrix} 1 & 1 & 1 \\ 2 & 2 & 1 \\ 1 & 1 & 0 \end{pmatrix}$

8. $M = \begin{pmatrix} 1 & 1 & 1 \\ 2 & 2 & 2 \\ 3 & 3 & 3 \end{pmatrix}$

9. $M = \begin{pmatrix} 1 & 2 & 0 \\ 0 & 1 & 0 \\ 0 & 0 & 1 \end{pmatrix}$

Find the determinant of the matrices in Exercises 10-22.

10. $\begin{pmatrix} 2 & 2 \\ 1 & 3 \end{pmatrix}$

11. $\begin{pmatrix} 4 & 3 \\ 1 & -4 \end{pmatrix}$

12. $\begin{pmatrix} 5 & 10 \\ 1 & 6 \end{pmatrix}$

13. $\begin{pmatrix} 1 & 2 \\ 1/2 & 3 \end{pmatrix}$

14. $\begin{pmatrix} 1 & 1 \\ 1 & 2 \end{pmatrix}$

15. $\begin{pmatrix} 2 & 2 \\ 2 & 4 \end{pmatrix}$

16. $\begin{pmatrix} 1 & 6 \\ 3 & 9 \end{pmatrix}$

17. $\begin{pmatrix} 1 & 2 \\ -2 & -4 \end{pmatrix}$

18. $\begin{pmatrix} 5 & 1 \\ 10 & 6 \end{pmatrix}$

19. $\begin{pmatrix} 1 & 1 & 1 \\ 1 & -1 & 1 \\ -1 & -1 & 1 \end{pmatrix}$

20. $\begin{pmatrix} 5 & 5 & 5 \\ 5 & -5 & 5 \\ -5 & -5 & 5 \end{pmatrix}$

21. $\begin{pmatrix} 1 & 1 & -2 \\ 2 & -1 & 2 \\ 2 & 1 & -4 \end{pmatrix}$

22. $\begin{pmatrix} 1 & 2 & 2 \\ 1 & -1 & 1 \\ -2 & 2 & -4 \end{pmatrix}$

23. Complete the proof of Theorem 4.5.11 by showing that if the reduced echelon form of a matrix A has a row of all zeros, then the reduce echelon form of AB also has a row of all zeros for any matrix B. How does this complete the proof?

24. Using the method following Definition 4.5.5, show that if

$$A = \begin{pmatrix} a & b \\ c & d \end{pmatrix}$$

then $\det(\alpha A) = \alpha^2 \det(A)$.

25. Using the method following Definition 4.5.5, show that if

$$A = \begin{pmatrix} a_{11} & a_{12} & a_{13} \\ a_{21} & a_{22} & a_{23} \\ a_{31} & a_{32} & a_{33} \end{pmatrix}.$$

then $\det(\alpha A) = \alpha^3 \det(A)$.

26. Given an $n \times n$ matrix A, show that $\det(\alpha A) = \alpha^n \det A$.

27. Using the method following Definition 4.5.5, show that if

$$A = \begin{pmatrix} a & b \\ c & d \end{pmatrix}$$

then $\det(A) = \det(A^{\mathsf{T}})$.

28. Using the method following Definition 4.5.5, show that if

$$A = \begin{pmatrix} a_{11} & a_{12} & a_{13} \\ a_{21} & a_{22} & a_{23} \\ a_{31} & a_{32} & a_{33} \end{pmatrix}.$$

then $\det(A) = \det(A^{\mathsf{T}})$.

29. Prove Theorem 4.5.19: given an $n \times n$ matrix A, show that $\det(A) = \det A^{\mathsf{T}}$.

30. Let A be an $n \times n$ matrix, and let B be the matrix obtained by multiplying the kth column of A by a factor α. Give both a geometric explanation and an algebraic explanation for the equation $\det(B) = \alpha \det(A)$.

31. Using the method following Definition 4.5.5, show that

$$\det \begin{pmatrix} a & b \\ c & d \end{pmatrix} = ad - bc.$$

(We calculated this determinant using cofactor expansion in the text, so you are being asked to use the row reduction method for this calculation.)

32. Prove Theorem 4.5.9: the determinant of an $n \times n$ matrix A is zero if and only if the reduced echelon form of A has a row of all zeros.

33. Suppose an $n \times n$ matrix A has determinant zero. Geometrically, we conclude that the map T_A maps the unit cube to a degenerate parallelepiped; that is, a parallelepiped of zero volume.

(a) What can you conclude about the span of the set $\{Ae_1, Ae_2, \ldots, Ae_n\}$?
(b) What can you conclude about the linear independence or dependence of the set $\{Ae_1, Ae_2, \ldots, Ae_n\}$?

[Hint: it may be helpful to look at an example, such as Example 4.5.8.]

34. **True or False**: If A and B are two $n \times n$ matrices then $\det(A + B) = \det(A) + \det(B)$. If true, give a proof, and if false, give an example showing that it is false.

35. Prove Theorem 4.5.20: Given an $n \times n$ matrix A, the matrix equation $Ax = 0$ has one solution if and only if $\det(A) \neq 0$.

4.6 Explorations: Re-Evaluating Our Tomographic Goal

In Section 4.1 we learned how to construct radiographs from objects given a known radiographic scenario. This first example of a transformation is a function which takes object vectors (in the object vector space) and outputs radiograph vectors (in the radiograph vector space). We observed that these types of transformations have the important property that linear combinations are preserved: $T(\alpha x + \beta y) = \alpha T(x) + \beta T(y)$. In Section 4.2 we formalized this idea by defining linear transformation between vector spaces. This linearity property is an important tool in analyzing radiographs and understanding the objects of which they are images. In this section, we continue our examination of radiographic transformations.

4.6.1 Seeking Tomographic Transformations

It is quite natural, but unfortunately incorrect, to assume that once the action of a transformation is known that we can follow this action backwards to find the object that produced a given radiograph. Here we begin to explore why this backward action is not so straightforward.

For any radiographic scenario, we have a transformation $T : V \to W$, where V is the vector space of objects and W is the vector space of radiographs. Our ultimate goal is to find a transformation $S : W \to V$ which "undoes" the action of T, that is, we seek S such that $S \circ T$ is the identity transformation. This idea is illustrated in the following table. Radiography is the process of obtaining a radiograph b. Tomography is the process which seeks to reverse the process by finding an object x which provides radiograph b.

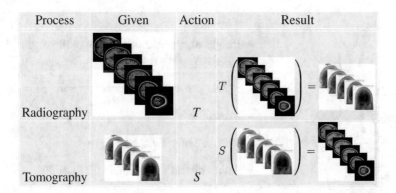

Notice that $(S \circ T)(x) = S(T(x)) = S(b) = x$; thus, $S \circ T$ is the identity transformation $I : V \to V$. We do not yet have any guarantee that such a transformation S exists. We can only explore the possibility of S by carefully understanding T. Two key questions rise to the top.

1. Could two different objects produce the same radiograph?
If so, then S may have no way to determine which object is the correct object.

2. Is it possible to have a radiograph which could not possibly come from any object?
If so, then S may be unable to point to any reasonable object *at all*.

In our explorations, we will make use of the following definitions.

Definition 4.6.1

Consider a vector space $(V, +, \cdot)$ of objects to be radiographed and a vector space $(W, +, \cdot)$ of radiographs. Suppose there exists a radiographic transformation $T : V \to W$. Then,

1. The **zero object** is the object $0 \in V$ whose voxel intensities are all zero.
2. The **zero radiograph** is the radiograph $0 \in W$ whose pixel intensities are all zero.
3. A **nonzero object** is an object $x \in V$ for which at least one voxel intensity is nonzero.
4. A **nonzero radiograph** is a radiograph $b \in W$ for which at least one pixel intensity is nonzero.
5. An **invisible object** is an object $x \in V$ which produces the zero radiograph.
6. A **possible radiograph** is a radiograph $b \in W$ for which there exists an object $x \in V$ such that $b = T(x)$.
7. We say that two vectors (objects or radiographs) are **identical** if all corresponding (voxel or pixel) intensities are equal.
8. We say that two vectors (objects or radiographs) are **distinct** if at least one pair of corresponding (voxel or pixel) intensities are not equal.

$\square$

4.6.2 Exercises

Consider the three radiographic scenarios shown in Figures 4.24-4.26 which were previously examined in Section 4.1. Exercises 1 through 10 can be applied to one or more of these scenarios.

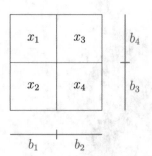

- Height and width of the image in voxels: $n = 2$ (Total voxels $N = 4$)

- Pixels per view in the radiograph: $m = 2$

- $ScaleFac = 1$

- Number of views: $a = 2$

- Angle of the views: $\theta_1 = 0°, \theta_2 = 90°$

Fig. 4.24 Tomographic Scenario A. Objects are in the vector space of 2×2 grayscale images. Radiographs are in the vector space of 2 views each with 2 pixels and the geometry as shown.

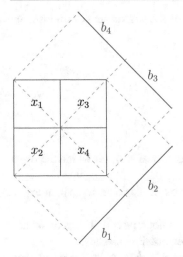

- Height and width of the image in voxels: $n = 2$ (Total voxels $N = 4$)

- Pixels per view in the radiograph: $m = 2$

- $ScaleFac = \sqrt{2}$

- Number of views: $a = 2$

- Angle of the views: $\theta_1 = 45°, \theta_2 = 135°$

Fig. 4.25   Tomographic Scenario B. Objects are in the vector space of 2×2 grayscale images. Radiographs are in the vector space of 2 views each with 2 pixels and the geometry as shown.

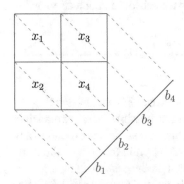

- Height and width of the image in voxels: $n = 2$ (Total voxels $N = 4$)

- Pixels per view in the radiograph: $m = 4$

- $ScaleFac = \sqrt{2}/2$

- Number of views: $a = 1$

- Angle of the views: $\theta_1 = 45°$

Fig. 4.26   Tomographic Scenario C. Objects are in the vector space of 2×2 grayscale images. Radiographs are in the vector space of 1 view with 4 pixels and the geometry as shown.

Carefully and completely, using linear algebra language, answer the questions. Be sure and provide examples and justifications to support your conclusions.

1. Is it possible for distinct objects to produce identical radiographs?
2. Are there nonzero invisible objects for this transformation?
3. Are there radiographs that cannot be the result of the transformation of *any* object? In other words, are there radiographs which are not possible?

The next three questions consider the deeper implications of your previous conclusions. Be creative and use accurate linear algebra language.

4. If possible, choose distinct objects that produce identical radiographs and subtract them. What is special about the resulting object?
5. Describe, using linear algebra concepts, the set of all invisible objects. Formulate a mathematical statement particular to the given transformation.
6. Similarly, describe the set of possible radiographs.

The next four questions ask you to dig deeper into the structure of the vector spaces themselves and how they relate to the transformation.

7. Show that the set of all invisible objects is a subspace of V.
8. Give a basis for the subspace of all invisible objects.
9. Show that the set of all possible radiographs is a subspace of W.
10. Give a basis for the set of all possible radiographs.

Additional questions.

11. Construct a radiographic scenario, using the same object space as Scenario A, for which there are no invisible objects.
12. Construct a radiographic scenario, using the same object space as Scenario A, for which every radiograph is a possible radiograph.
13. How might it be possible, in a brain scan application, to obtain a radiograph $b \in W$ that could not possibly be the radiograph of any brain $x \in V$? Give at least three possible reasons.
14. Make conjectures about and discuss the potential importance of a brain object $x \in V$ that contains negative intensities.
15. Discuss the potential importance of knowing which objects are in the subspace of objects invisible to brain scan radiography.

4.7   Properties of Linear Transformations

In Section 4.6, we saw that certain properties of linear transformations are crucial to understanding our ability to perform a tomographic process. In particular, we found that it is possible for (a) two distinct brain objects to produce identical radiographs and (b) real radiographic data not to correspond to any possible brain object. We want to know if it is possible for an abnormal brain to produce the same radiograph as a normal brain. Figure 4.27 shows two brain images which produce *identical* radiographs under the 30-view radiographic scenario described in Appendix A. We notice differences in the density variations (shown in darker gray in the left image) across the brain images which are invisible to the radiographic process. This means that if we found the difference in these brain images, the difference would be a nonzero invisible vector in the vector space of objects.

In addition, we want to understand the effects that noise or other measurement artifacts have on the ability to accurately determine a likely brain object. In these cases, we want to be able to recover the same brain object as if noise was never present. In later discussions, we will understand why noise can present a challenge (but not an insurmountable challenge) for recovering a meaningful brain image.

To understand a linear process between two vector spaces, we need to thoroughly understand these properties of the transformation.

4.7.1   One-To-One Transformations

Again, in Figure 4.27, we see that the radiographic transformation can map two distinct objects (domain vectors) to the same radiograph (codomain vector). It is important to recognize whether a transformation does this. If not and we have an output from the transformation, we can track the output vector back to its input vector. This would be ideal when trying to reconstruct brain images. Let us, now, talk about these ideal transformations.

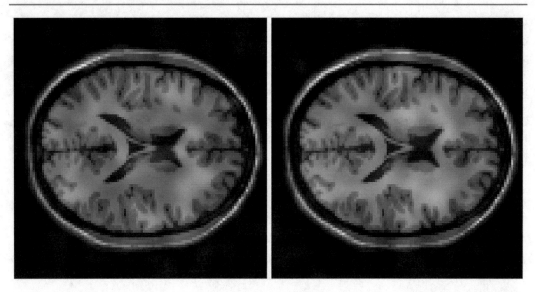

Fig. 4.27 Two different brain images that produce the same radiograph in a particular 30-view radiographic geometry.

Definition 4.7.1

Let V and W be vector spaces and $u, v \in V$. If, for a transformation $T : V \to W$,

$$T(u) = T(v) \text{ implies that } u = v,$$

then we say that T is **one-to-one** or that T is a **one-to-one transformation**.   □

In Definition 4.7.1, we do not require one-to-one transformations to be linear. A one-to-one transformation guarantees that distinct codomain vectors "came from" distinct domain vectors. For example, a one-to-one radiographic transformation guarantees that a given radiograph corresponds to one, and only one, brain object.

Consider the schematic pictures in Figure 4.28. The left schematic illustrates the action of a transformation, T, from a set with six elements, $V = \{v_1, v_2, v_3, v_4, v_5, v_6\}$, to a set with seven elements,

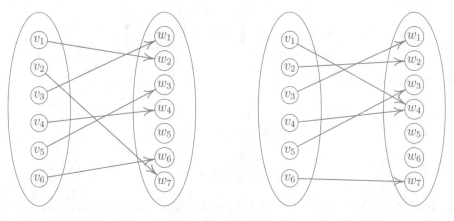

Fig. 4.28 Schematic illustrations of a one-to-one transformation (left) and a transformation that is not one-to-one (right).

$W = \{w_1, w_2, w_3, w_4, w_5, w_6, w_7\}$. For example, $T(v_1) = w_2$. This transformation is one-to-one because if $T(v_k) = w_i$ and $T(v_j) = w_i$, then $v_k = v_j$ for all w_i. That is, there is, at most, only one arrow pointing to each of the w_i. The right schematic illustrates the action of a different transformation $\hat{T}$ on the same sets. In this case, the transformation is not one-to-one because $\hat{T}(v_1) = \hat{T}(v_4)$ with $v_1 \neq v_4$. This is visually seen by two arrows pointing to w_4.

Example 4.7.2 Consider the radiographic transformation of Scenario A, call it T_A, described in Figure 4.24. The radiograph defined by $b_1 = b_2 = b_3 = b_4 = 1$ corresponds to at least two objects: for example,

$$T_A \left(\begin{array}{c|c} 1 & 0 \\ \hline 0 & 1 \end{array} \right) = \begin{array}{|c|} \hline 1 \\ \hline 1 \\ \hline 1 \\ \hline 1 \\ \hline \end{array}$$

$$T_A \left(\begin{array}{c|c} 1/2 & 1/2 \\ \hline 1/2 & 1/2 \end{array} \right) = \begin{array}{|c|} \hline 1 \\ \hline 1 \\ \hline 1 \\ \hline 1 \\ \hline \end{array}$$

Thus, this transformation is not one-to-one.   □

Example 4.7.3 Let $T : \mathbb{R} \to \mathbb{R}$ be defined by $T(x) = 5x$. Every vector, y, in the range of T, corresponds to a unique domain vector $x = y/5$. That is, if $T(x_1) = T(x_2)$ then $5x_1 = 5x_2$. So $x_1 = x_2$. Thus, T is one-to-one.   □

Example 4.7.4 Let $T : \mathbb{R}^2 \to \mathbb{R}^3$ be defined by $T(x) = Ax$, for the following matrix A.

$$A = \begin{pmatrix} 1 & 0 \\ 1 & 2 \\ 1 & 1 \end{pmatrix}.$$

This transformation is one-to-one. To see this, suppose that there are two vectors $u = \begin{pmatrix} u_1 \\ u_2 \end{pmatrix}$ and $v = \begin{pmatrix} v_1 \\ v_2 \end{pmatrix}$ such that $Au = Av$. Then we have

$$Au = \begin{pmatrix} u_1 \\ u_1 + 2u_2 \\ u_1 + u_2 \end{pmatrix} = Av = \begin{pmatrix} v_1 \\ v_1 + 2v_2 \\ v_1 + v_2 \end{pmatrix}.$$

But then we must have that

$$u_1 = v_1,$$
$$u_1 + 2u_2 = v_1 + 2v_2,$$
$$\text{and}$$
$$u_1 + u_2 = v_1 + v_2.$$

So we conclude that $u_1 = v_1$ and $u_2 = v_2$. This means that $u = v$ and so T is one-to-one.   □

Example 4.7.5 Let $T : \mathbb{R}^2 \to \mathbb{R}^3$ be defined by $T(x) = Ax$, for the following matrix A.

$$A = \begin{pmatrix} 1 & -2 \\ 0 & 0 \\ -1 & 2 \end{pmatrix}.$$

This transformation is not one-to-one. We might try the same method as in the previous example. Suppose that there are two vectors

$$u = \begin{pmatrix} u_1 \\ u_2 \end{pmatrix} \text{ and } v = \begin{pmatrix} v_1 \\ v_2 \end{pmatrix}$$

such that $Au = Av$. Then we have

$$\begin{pmatrix} u_1 - 2u_2 \\ 0 \\ -u_1 + 2u_2 \end{pmatrix} = \begin{pmatrix} v_1 - 2v_2 \\ 0 \\ -v_1 + 2v_2 \end{pmatrix}.$$

This means that u_1, u_2, v_1, and u_2 must satisfy

$$u_1 - 2u_2 = v_1 - 2v_2$$

and

$$-u_1 + 2u_2 = -v_1 + 2v_2.$$

But these two equations are multiples of each other, and both say that $u_1 - 2u_2 = v_1 - 2v_2$. Therefore, we cannot conclude $u = v$. Indeed, for $u = \begin{pmatrix} 0 \\ 0 \end{pmatrix}$ and $v = \begin{pmatrix} 2 \\ 1 \end{pmatrix}$, $Au = Av$ and we see that T is not one-to-one. Many other vectors u and v could act as examples to show that T is not one-to-one. $\square$

Example 4.7.6 Let $T : \mathcal{P}_2(\mathbb{R}) \to \mathcal{M}_{2\times3}(\mathbb{R})$ be defined by

$$T(ax^2 + bx + c) = \begin{pmatrix} a & b & c \\ 0 & 0 & 0 \end{pmatrix}.$$

Notice that T is a linear transformation. Indeed, let $v = a_2x^2 + a_1x + a_0$ and $u = b_2x^2 + b_1x + b_0$, for coefficients $a_k, b_k \in \mathbb{R}, k = 0, 1, 2$. Then for a scalar $\alpha \in \mathbb{R}$,

$$\begin{aligned}
T(\alpha v + u) &= T\left((\alpha a_2 + b_2)x^2 + (\alpha a_1 + b_1)x + (\alpha a_0 + b_0)\right) \\
&= \begin{pmatrix} \alpha a_2 + b_2 & \alpha a_1 + b_1 & \alpha a_0 + b_0 \\ 0 & 0 & 0 \end{pmatrix} \\
&= \begin{pmatrix} \alpha a_2 & \alpha a_1 & \alpha a_0 \\ 0 & 0 & 0 \end{pmatrix} + \begin{pmatrix} b_2 & b_1 & b_0 \\ 0 & 0 & 0 \end{pmatrix} \\
&= \alpha \begin{pmatrix} a_2 & a_1 & a_0 \\ 0 & 0 & 0 \end{pmatrix} + \begin{pmatrix} b_2 & b_1 & b_0 \\ 0 & 0 & 0 \end{pmatrix} \\
&= \alpha T(v) + T(u).
\end{aligned}$$

Therefore, T is a linear transformation. Next, we show that T is one-to-one. Suppose $T(v) = T(u)$. We will show that $u = v$. Notice that $T(v) = T(u)$ means

$$\begin{pmatrix} a_2 & a_1 & a_0 \\ 0 & 0 & 0 \end{pmatrix} = \begin{pmatrix} b_2 & b_1 & b_0 \\ 0 & 0 & 0 \end{pmatrix}.$$

Thus,

$$a_2 = b_2, a_1 = b_1, \text{ and } a_0 = b_0$$

Thus, $v = u$. □

Example 4.7.7 Let $T : \mathbb{R}^2 \to \mathcal{P}_1(\mathbb{R})$ be defined by

$$T \begin{pmatrix} a \\ b \end{pmatrix} = ax + b.$$

We want to determine whether or not T is one-to-one. Suppose

$$T \begin{pmatrix} a \\ b \end{pmatrix} = T \begin{pmatrix} c \\ d \end{pmatrix}.$$

Then

$$ax + b = cx + d.$$

Matching up like terms gives us that $a = c$ and $b = d$. That is

$$\begin{pmatrix} a \\ b \end{pmatrix} = \begin{pmatrix} c \\ d \end{pmatrix}.$$

So, T is one-to-one. □

Example 4.7.8 Let $T : \mathcal{J}_{12}(\mathbb{R}) \to \mathbb{R}$ (the transformation from the space of histograms with 12 bins to the set of real numbers) be defined as: $T(J)$ is the sum of the values assigned to all bins. T is not one-to-one. We can understand this idea because simply knowing the sum of the values does not allow us to uniquely describe the histogram that produced the sum. □

Example 4.7.9 The identity transformation, $I : V \to V$, is one-to-one. Indeed, for any $v \in V$, the only vector $u \in V$, for which $I(u) = v$, is $u = v$. □

Example 4.7.10 The zero transformation, $0 : V \to W$ may or may not be one-to-one. See Exercise 40 for more discussion. □

4.7.2   Properties of One-To-One Linear Transformations

In this section, we will explore connections between one-to-one linear transformations and linear independence. In fact, we show linear independence is preserved by linear transformations that are one-to-one. But first, we introduce some notation.

Definition 4.7.11

Let V and W be vector spaces, let $S \subseteq V$, and let $T : V \to W$ be a transformation. We define the set

$$T(S) = \{T(s) \mid s \in S\} \subseteq W.$$

That is, $T(S)$ is the set of all vectors being mapped to by elements of S. We call this set the **image of S** under T. □

Example 4.7.12 Consider the linear tranformation $T : \mathcal{P}_2(\mathbb{R}) \to \mathbb{R}^3$ defined by

$$T(ax^2 + bx + c) = (c, b, a + b).$$

Let $S = \{1, x, x^2\} \subseteq \mathcal{P}_2(\mathbb{R})$. We have

$$T(S) = \left\{T(1), T(x), T(x^2)\right\} = \{(1, 0, 0), (0, 1, 1), (0, 0, 1)\}.$$

□

Example 4.7.13 Suppose $S = \emptyset$ Then $T(S)$ is the set of all vectors in W mapped to by vectors in S. Since, there are no vectors in S, there can be no vectors in $T(S)$. That is, $T(S) = \emptyset$. □

Example 4.7.14 Consider the vector space V of 7-bar LCD characters. Let $T : V \to V$ be the one-to-one linear transformation which is described by a reflection about the horizontal axis of symmetry. Let

$$S = \left\{ \text{⬛} , \text{⬛} \right\}, \quad \text{then} \quad T(S) = \left\{ \text{⬛} , \text{⬛} \right\}.$$

□

The next Lemma is a statement about how nested sets behave under linear transformations. This will be useful as we work toward our goal of understanding how linearly independent sets behave under one-to-one transformations.

Lemma 4.7.15

Let V and W be vector spaces, and let $T : V \to W$ be a transformation. Suppose S_1 and S_2 are subsets of V such that $S_1 \subseteq S_2 \subseteq V$. Then $T(S_1) \subseteq T(S_2)$.

Proof. Suppose V and W are vector spaces and T is defined as above. Suppose also that $S_1 \subseteq S_2 \subseteq V$. We will show that if $w \in T(S_1)$ then $w \in T(S_2)$. Suppose $w \in T(S_1)$. Then there exists $v \in S_1$ such that $T(v) = w$. Since $S_1 \subseteq S_2$, we know that $v \in S_2$ and, therefore, $T(v) \in T(S_2)$. So $w \in T(S_2)$. Thus, $T(S_1) \subseteq T(S_2)$. □

Now, we have the tools to verify that linear independence is preserved under one-to-one transformations.

Lemma 4.7.16

Let V and W be vector spaces. Let $S = \{v_1, v_2, \ldots, v_m\}$ be a linearly independent set in V, possibly empty. If $T : V \to W$ is a one-to-one linear transformation, then $T(S) = \{T(v_1), T(v_2), \ldots, T(v_m)\}$ is linearly independent.

Proof. Let V and W be vector spaces. Suppose $T : V \to W$ is a one-to-one linear transformation. First, note that if $S = \emptyset$ then $T(S) = \emptyset$ which is linearly independent.

Now, assume that S is not empty and that S is linearly independent. We want to show that $\{T(v_1), T(v_2), \ldots, T(v_m)\}$ is linearly independent. Let $\alpha_1, \alpha_2, \ldots, \alpha_m$ be scalars and consider the linear dependence relation

$$\alpha_1 T(v_1) + \alpha_2 T(v_2) + \ldots + \alpha_m T(v_m) = 0. \tag{4.6}$$

We wish to show that $\alpha_1 = \alpha_2 = \ldots = \alpha_m = 0$. Because T is linear, we can rewrite the left side of (4.6) to get

$$T(\alpha_1 v_1 + \alpha_2 v_2 + \ldots + \alpha_n v_m) = 0.$$

But, since $T(0) = 0$, we can rewrite the right side of (4.6) to get

$$T(\alpha_1 v_1 + \alpha_2 v_2 + \ldots + \alpha_m v_m) = T(0).$$

Now, since T is one-to-one, we know that

$$\alpha_1 v_1 + \alpha_2 v_2 + \ldots + \alpha_m v_m = 0.$$

Finally, since S is linearly independent

$$\alpha_1 = \alpha_2 = \ldots = \alpha_m = 0.$$

Thus, $\{T(v_1), T(v_2), \ldots, T(v_m)\}$ is linearly independent. $\square$

Example 4.7.17 In Example 4.7.12 the transformation T is one-to-one because

$$T(ax^2 + bx + c) = T(dx^2 + ex + f) \text{ implies that } (c, b, a + b) = (f, e, d + e).$$

Therefore, it must be true that $c = f$, $b = e$, and, so, $a = d$. Lemma 4.7.16 guarantees that the linearly independent set $\{1, x, x^2\} \subseteq \mathcal{P}_2(\mathbb{R})$ is mapped to a linearly independent set, namely, $\{(1, 0, 0), (0, 1, 1), (0, 0, 1)\} \subseteq \mathbb{R}^3$. $\square$

Since linearly independent sets in V form bases for subspaces of V, Lemma 4.7.16 along with Exercise 42 from Section 4.2 lead us to question how this affects the dimensions of these subspaces. The following lemma answers this question.

> **Lemma 4.7.18**
>
> Let V and W be vector spaces. Let $U \subseteq V$ be a subspace of dimension m. If $T : V \to W$ is a one-to-one linear transformation, then $T(U) \subseteq W$ is also a subspace of dimension m.

Proof. Let V and W be vector spaces with $U \subseteq V$, an m-dimensional subspace of V. Let $S = \{v_1, v_2, \ldots, v_m\}$ be a basis for U and suppose $T : V \to W$ is a one-to-one linear transformation. We will show that $T(S)$ is a basis for the subspace $T(U)$. That is, we need $T(S)$ to be linearly independent and to span $T(U)$. By Lemma 4.7.16, we already know $T(S)$ is linearly independent. So, we need only show that $T(S)$ spans $T(U)$.

Let $w \in T(U)$. Then, by Definition 4.7.11, there is a $u \in U$ so that $T(u) = w$. Since $u \in U$, there are scalars $\alpha_1, \alpha_2, \ldots, \alpha_m$ so that

$$u = \alpha_1 v_1 + \alpha_2 v_2 + \ldots + \alpha_m v_m.$$

So, by linearity of T

$$\begin{aligned} w &= T(\alpha_1 v_1 + \alpha_2 v_2 + \ldots + \alpha_m v_m) \\ &= \alpha_1 T(v_1) + \alpha_2 T(v_2) + \ldots + \alpha_m T(v_m). \end{aligned}$$

Thus, $w \in \operatorname{Span} T(S)$. Therefore, $T(U) \subseteq \operatorname{Span}\{T(v_1), T(v_2), \ldots, T(v_m)\} = \operatorname{Span} T(S)$. Since $T(S) \subseteq T(U)$, $\operatorname{Span}(T(S)) = T(U)$. Thus, $T(S)$ is a basis for $T(U)$.

$\square$

The close connection between one-to-one transformations and linear independence leads to the following important theorem.

> **Theorem 4.7.19**
>
> Let V and W be finite-dimensional vector spaces and let $\mathcal{B} = \{v_1, v_2, \ldots, v_n\}$ be a basis for V. The linear transformation $T : V \to W$ is one-to-one if and only if $T(\mathcal{B}) = \{T(v_1), T(v_2), \ldots, T(v_n)\}$ is a linearly independent set in W.

Proof. Let V and W be finite-dimensional vector spaces and let $\mathcal{B}$ be a basis for V. ($\Rightarrow$) Linear independence of $T(\mathcal{B})$ follows directly from Lemma 4.7.16.

($\Leftarrow$) Now suppose that $\{T(v_1), T(v_2), \ldots, T(v_n)\} \subset W$ is linearly independent. We want to show that T is one-to-one. Let $u, v \in V$ so that $T(u) = T(v)$. Then, $T(u) - T(v) = 0$. Thus, $T(u - v) = 0$. Since $u, v \in V$, there are scalars $\alpha_1, \alpha_2, \ldots, \alpha_n$ and $\beta_1, \beta_2, \ldots, \beta_n$ so that

$$u = \alpha_1 v_1 + \alpha_2 v_2 + \ldots + \alpha_n v_n \quad \text{and} \quad v = \beta_1 v_1 + \beta_2 v_2 + \ldots + \beta_n v_n.$$

Thus,

$$T((\alpha_1 - \beta_1)v_1 + (\alpha_2 - \beta_2)v_2 + \ldots + (\alpha_n - \beta_n)v_n) = 0.$$

This leads us to the linear dependence relation

$$(\alpha_1 - \beta_1)T(v_1) + (\alpha_2 - \beta_2)T(v_2) + \ldots + (\alpha_n - \beta_n)T(v_n) = 0.$$

Since $\{T(v_1), T(v_2), \ldots, T(v_n)\}$ is linearly independent, we know that

$$\alpha_1 - \beta_1 = \alpha_2 - \beta_2 = \ldots = \alpha_n - \beta_n = 0.$$

That is, $u = v$. Therefore, T is one-to-one. $\square$

Example 4.7.20 Consider $\mathcal{B} = \left\{1, x, x^2, x^3\right\}$, the standard basis for $\mathcal{P}_3(\mathbb{R})$. Let $T : \mathcal{P}_3(\mathbb{R}) \to \mathcal{P}_3(\mathbb{R})$ be defined by $T(f(x)) = f'(x)$. We have $T(\mathcal{B}) = \left\{0, 1, 2x, 3x^2\right\}$ which is linearly dependent. Thus, T is not one-to-one. $\square$

Example 4.7.21 Consider Example 4.7.6. Using the basis

$$\mathcal{B} = \left\{1 + x + x^2, 1 + x, 1\right\}$$

for $\mathcal{P}_2(\mathbb{R})$ we have

$$T(\mathcal{B}) = \left\{\begin{pmatrix} 1 & 1 & 1 \\ 0 & 0 & 0 \end{pmatrix}, \begin{pmatrix} 0 & 1 & 1 \\ 0 & 0 & 0 \end{pmatrix}, \begin{pmatrix} 0 & 0 & 1 \\ 0 & 0 & 0 \end{pmatrix}\right\}.$$

Since $T(\mathcal{B})$ is linearly independent, T is one-to-one. $\square$

Now, for vector spaces V and W, let us discuss the connection between one-to-one transformations, $T : V \to W$ and the dimensions of V and W.

Theorem 4.7.22

Let V and W be finite-dimensional vector spaces. If $T : V \to W$ is a one-to-one linear transformation, then dim $V \le$ dim W.

Proof. Let V and W be vector spaces with bases $\mathcal{B}_V$ and $\mathcal{B}_W$, respectively. Suppose also that dim $V = n$ and dim $W = m$. Now, let $T : V \to W$ be a one-to-one linear transformation. By Theorem 4.7.19, we know that since $\mathcal{B}_V$ is a basis for V, $\mathcal{B}_V$ has n elements and $T(\mathcal{B}_V)$ is a linearly independent set in W with n elements. Using Corollary 3.4.22, we know that $\mathcal{B}_W$ has m elements and any linearly independent set in W must have at most m elements. Therefore, $n \le m$. So, dim $V \le$ dim W. $\square$

The reader should revisit each example of a one-to-one linear transformations in this section and verify the conclusion of Theorem 4.7.22. That is, you should check that the dimension of the codomain in each of these examples is at least as large as the dimension of the domain.

4.7.3 Onto Linear Transformations

Next, we consider whether a transformation has the ability to map to every codomain vector. In the radiograph sense, this would mean that every radiograph is possible (see Definition 4.6.1).

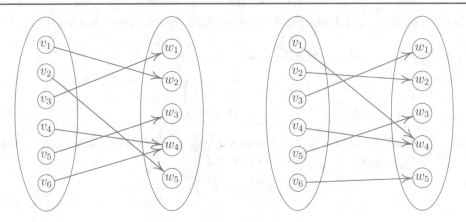

Fig. 4.29 Schematic illustrations of two different onto transformations.

Definition 4.7.23

Let V and W be vector spaces. We say that the transformation $T : V \to W$ **maps onto** W if for every vector $w \in W$ there exists $v \in V$ so that $T(v) = w$.

In the case that a transformation T maps onto its codomain, we say that T is an **onto** transformation.

$\square$

Definition 4.7.23 does not require an onto transformation to be linear.

Consider the schematic pictures in Figure 4.29. The left schematic illustrates the action of a transformation, T, from a vector space of six vectors, $V = \{v_1, \cdots, v_6\}$, to a vector space of five vectors, $W = \{w_1, \cdots, w_5\}$. For example, $T(v_1) = w_2$. This transformation is onto because every w_k can be written as $w_k = T(v_j)$ for some v_j. The right schematic illustrates the action of a different transformation on the same vector spaces. This transformation is also onto. But, neither transformation is one-to-one. If we consider the schematic pictures in Figure 4.28, we can see that neither transformation is onto because, in the left schematic, there are no vectors in the domain that map to w_5 in the codomain and, in the right schematic, we see that the equations

$$T(x) = w_5 \text{ and } T(y) = w_6$$

have no solutions.

Let us consider some examples in which we determine whether a transformation is onto.

Example 4.7.24 Consider the radiographic transformation of Scenario A described in Figure 4.24 and discussed in Example 4.7.2. There is no possible object which can produce the radiograph defined by $b_1 = b_2 = b_3 = 0$, $b_4 = 1$. Therefore, this transformation is not onto. $\square$

In the schematics above, we see pictures of possible scenarios where a transformation may be one-to-one and not onto or one that may be onto, but not one-to-one. Let us consider actual examples of these cases.

Example 4.7.25 Let $T : \mathcal{P}_2(\mathbb{R}) \to \mathbb{R}$ be defined by $T(ax^2 + bx + c) = a + b + c$. We know that T is linear (See Exercise 5). Let $u = x^2$ and $v = x$, then $u, v \in \mathcal{P}_2(\mathbb{R})$ and $u \neq v$, but $T(u) = T(v)$. Therefore, T is not one-to-one.

Now, for any $w \in \mathbb{R}$ we know that $wx^2 \in \mathcal{P}_2(\mathbb{R})$ and $T(wx^2) = w$. Therefore, T is onto. $\square$

Now, we know that there is a linear transformation that is onto, but not one-to-one.

Example 4.7.26 Let $T : \mathbb{R}^2 \to \mathcal{M}_{3\times2}(\mathbb{R})$ be defined by

$$T\begin{pmatrix} a \\ b \end{pmatrix} = \begin{pmatrix} a & -b \\ b & a+b \\ 0 & -a \end{pmatrix}.$$

We want to know if T is one-to-one and/or onto. First, we determine whether T is one-to-one. Let $u = (a, b)^T$ and $v = (c, d)^T$ and suppose that $T(u) = T(v)$:

$$T\begin{pmatrix} a \\ b \end{pmatrix} = T\begin{pmatrix} c \\ d \end{pmatrix},$$

then

$$\begin{pmatrix} a & -b \\ b & a+b \\ 0 & -a \end{pmatrix} = \begin{pmatrix} c & -d \\ d & c+d \\ 0 & -c \end{pmatrix}.$$

Matching up entries, gives us $a = c, -b = -d, b = d, a+b = c+d, 0 = 0$, and $-c = -d$, with unique solution $a = c$ and $b = d$. Thus, $u = v$ and T is one-to-one.

Next, we determine whether T maps onto $\mathcal{M}_{3\times2}$. Notice that,

$$w = \begin{pmatrix} 0 & 0 \\ 0 & 0 \\ 1 & 0 \end{pmatrix} \in \mathcal{M}_{3\times2}(\mathbb{R}).$$

But, there is no $v \in \mathbb{R}^2$ so that $T(v) = w$ because no vector in the range of T has a nonzero number as the lower left entry. Thus, T is not onto.   □

We now consider examples we saw in Section 4.7.1.

Example 4.7.27 As in Example 4.7.7, let $T : \mathbb{R}^2 \to \mathcal{P}_1(\mathbb{R})$ be defined by

$$T\begin{pmatrix} a \\ b \end{pmatrix} = ax + b.$$

If we pick $w \in \mathcal{P}_1$, then $w = ax + b$ for some $a, b \in \mathbb{R}$. And, if we let

$$v = \begin{pmatrix} a \\ b \end{pmatrix} \in \mathbb{R}^2$$

then $T(v) = w$. Thus, T is onto.   □

Example 4.7.28 The identity transformation $I : V \to V$ maps onto V because for every vector $y \in V$ there exists a vector $x \in V$ such that $I(x) = y$, namely, $x = y$.   □

Example 4.7.29 The zero transformation may or may not be onto. See Exercise 41.   □

4.7.4   Properties of Onto Linear Transformations

In this section, we explore linear transformations $T : V \to W$ that map onto W and what we know about the spaces V and W. We found that if T is a one-to-one linear transformation then the dimension of V is at most the dimension of W. We will consider a similar theorem about onto transformations here.

Lemma 4.7.30
Let V and W be finite-dimensional vector spaces, $T : V \to W$ be an onto linear transformation, and $S = \{w_1, w_2, \ldots, w_m\} \subseteq W$. If S is linearly independent, then there is a linearly independent set $U = \{v_1, v_2, \ldots, v_m\} \subseteq V$ such that $T(U) = S$.

Proof. Suppose $S \subset W$ is linearly independent and $T : V \to W$ is linear and onto. Since T maps onto W, there exists $U = \{v_1, v_2, \ldots, v_m\} \subset V$ so that $T(U) = S$. We will show that U is linearly independent. Let $\alpha_1, \alpha_2, \ldots, \alpha_m$ be scalars so that

$$\alpha_1 v_1 + \alpha_2 v_2 + \ldots + \alpha_m v_m = 0.$$

Then, by linearity,

$$T(\alpha_1 v_1 + \alpha_2 v_2 + \ldots \alpha_m v_m) = 0.$$

That is,

$$\alpha_1 T(v_1) + \alpha_2 T(v_2) + \ldots \alpha_m T(v_m) = 0.$$

But since $T(U) = S$, we have
$$\alpha_1 w_1 + \alpha_2 w_2 + \ldots \alpha_m w_m = 0.$$

Finally, since S is linearly independent, $\alpha_1 = \alpha_2 = \ldots = \alpha_m = 0$. Thus, U is linearly independent. $\square$

Example 4.7.31 Consider the linear transformation $T : \mathbb{R}^3 \to \mathbb{R}^2$ defined by $T(x, y, z) = (x, y)$. T is onto because any vector $(x, y) \in \mathbb{R}^2$ can be mapped to by some vector in $\mathbb{R}^3$. For example $(x, y, 4) \mapsto (x, y)$ for all $x, y \in \mathbb{R}$. Let $S = \{(1, 0), (1, 5)\}$. Since S is linearly independent, Lemma 4.7.30 tells us that there exists a linearly independent set $U \subseteq V$ such that $T(U) = S$. One possibility is $U = \{(1, 0, 17), (1, 5, 0)\}$. $\square$

Unlike Lemma 4.7.16, we do not claim that the converse is true (see Exercise 46). Lemma 4.7.30 tells us that a linearly independent set in V maps to a basis of W through an onto linear transformation. This is more formally written in the following corollary.

Corollary 4.7.32
Let V and W be finite-dimensional vector spaces and let $T : V \to W$ be a linear transformation that maps onto W. If $\mathcal{B}_W$ is a basis for W then there exists a linearly independent set $U \subset V$ so that $T(U) = \mathcal{B}_W$.

Proof. Let V and W be vector spaces with $\dim V = n$ and $\dim W = m$ and let $T : V \to W$ be an onto linear transformation. Suppose $\mathcal{B}_W$ is a basis for W. We will find a set of vectors in V that map to $\mathcal{B}_W$ and is linearly independent. Notice that if $m = 0$, then $\mathcal{B}_W = \emptyset$ and $U = \emptyset \subseteq V$ is a linearly independent set so that $T(U) = \mathcal{B}_W$. Now, suppose $m > 0$.

Assume that

$$\mathcal{B}_W = \{w_1, w_2, \ldots w_m\}$$

and consider the set

$$U = \{v \in V \mid T(v) \in \mathcal{B}_W\}.$$

We know that, because T maps onto w, there are vectors $v_1, v_2, \ldots, v_m$ so that $T(v_i) = w_i$, $i = 1, 2, \ldots, m$. That is,

$$U = \{v_1, v_2, \ldots, v_m\}.$$

Let $\alpha_1, \alpha_2, \ldots, \alpha_m$ be scalars and consider the linear dependence relation

$$\alpha_1 v_1 + \alpha_2 v_2 + \ldots + \alpha_m v_m = 0. \tag{4.7}$$

We will show that $\alpha_1 = \alpha_2 = \ldots = \alpha_m$. We know that because T is a function, (4.7) gives us that

$$T(\alpha_1 v_1 + \alpha_2 v_2 + \ldots + \alpha_m v_m) = T(0).$$

Therefore, because T is linear, we have

$$\alpha_1 T(v_1) + \alpha_2 T(v_2) + \ldots + \alpha_m T(v_m) = 0.$$

But, by the definition of U, we have

$$\alpha_1 w_1 + \alpha_2 w_2 + \ldots + \alpha_m w_m = 0.$$

But, since $\mathcal{B}_W$ is a basis, and therefore linearly independent, we know that

$$\alpha_1 = \alpha_2 = \ldots = \alpha_m = 0.$$

Therefore, U is linearly independent. $\square$

Using Corollary 4.7.32, we can discuss how the dimensions of V and W are related.

Theorem 4.7.33
Let V and W be finite-dimensional vector spaces. Let $T : V \to W$ be a linear transformation that maps onto W. Then $\dim V \geq \dim W$.

Proof. Let $T : V \to W$ be onto and linear. Also, let $\dim V = n$ and $\dim W = m$. We know that if $\mathcal{B}_W$ is a basis of W then there are m elements in $\mathcal{B}_W$. By Corollary 4.7.32, there is a linearly independent subset $S \subseteq V$ with m elements. By Corollary 3.4.22, we know that $m \leq n$. Thus, $\dim W \leq \dim V$. $\square$

4.7.5 Summary of Properties

At this point, we have collected many properties that link linear transformation that are one-to-one and onto to the domain and codomain spaces. We summarize them here.

▶ **Summary of Transformation Properties.** Suppose V and W are finite-dimensional vector spaces and that $T : V \to W$ is a linear transformation. We know that

1. If T preserves linear independence, then T is one-to-one.
2. If T maps some subset of V to a spanning set of W, then T is onto. (Exercise 39)

We know that if T is one-to-one then

3. T preserves linear independence.
4. T maps a basis of V to a basis of some subspace of W.
5. $\dim V \leq \dim W$.

We know that if T is onto then

6. A basis for W is mapped to by some linearly independent set in V.
7. $\dim W \leq \dim V$.

4.7.6 Bijections and Isomorphisms

In Sections 4.7.1 and 4.7.3, we have seen that some linear transformations are both one-to-one and onto. The actions of transformations with these properties are the simplest to understand. This statement requires evidence. So, let's, first, consider what this means in the radiographic sense.

A radiographic scenario, with an onto transformation, has the following properties.

- Every radiograph is a possible radiograph.
- For every radiograph there exists some brain image which can produce it through the transformation.
- If $T : V \to W$, then $T(V) = W$.
- If $T : V \to W$ and β a basis for V, then Span $T(\beta) = W$.

A radiographic scenario with a one-to-one transformation has the following properties.

- Distinct brain images produce distinct radiographs.
- Each possible radiograph is the transformation of exactly one brain image.
- If $T : V \to W$, then $\dim T(V) = \dim V$.

Taken together, a radiographic scenario with a transformation that is both one-to-one and onto has the key property that every possible radiographic image can be traced back to exactly one brain image. Radiographic transformations with this property guarantee, for every possible radiograph, the existence of a *unique* brain image which produces the given radiograph even though we do not yet know *how* to find it.

Definition 4.7.34

We say that a linear transformation, $T : V \to W$, is an **isomorphism** if T is both one-to-one and onto.

□

Whenever we have two vector spaces V and W with an isomorphism, T, between them, we know that each vector in V corresponds directly to a vector in W. By "correspond," we mean that the vector, v, in V really acts the same way in V as $T(v)$ acts in W. Essentially, these spaces are the same. We formalize this idea in the following definition.

Definition 4.7.35

Let V and W be vector spaces and let $T : V \to W$ be an isomorphism. Then we say that V is **isomorphic** to W and we write

$$V \cong W.$$

$\square$

Example 4.7.36 Combining Examples 4.7.7 and 4.7.27, we see that the transformation $T : \mathbb{R}^2 \to \mathcal{P}_1(\mathbb{R})$, defined by

$$T \begin{pmatrix} a \\ b \end{pmatrix} = ax + b$$

is an isomorphism. That means that $\mathbb{R}^2 \cong \mathcal{P}_1(\mathbb{R})$. According to our comments above, this means that elements of $\mathcal{P}_1(\mathbb{R})$ "act" like their corresponding elements in $\mathbb{R}^2$. $\square$

We already knew that $\mathbb{R}^2$ and $\mathcal{P}_1(\mathbb{R})$ act similarly because the coordinate space of $\mathcal{P}_1(\mathbb{R})$ is $\mathbb{R}^2$. We can actually show that the coordinate transformation is, indeed, an isomorphism. (See Theorem 4.7.43 below.)

4.7.7   Properties of Isomorphic Vector Spaces

Determining whether two finite-dimensional vector spaces are isomorphic seems to hinge on finding an isomorphism between them. However, reconsider the isomorphism of Example 4.7.36 giving us that $\mathcal{P}_1(\mathbb{R})$ is isomorphic to $\mathbb{R}^2$, $\mathcal{P}_1(\mathbb{R}) \cong \mathbb{R}^2$. And, $\dim \mathcal{P}_1(\mathbb{R}) = \dim \mathbb{R}^2$. This is not a coincidence. We will find that the only requirement for two finite-dimensional vector spaces to be isomorphic is that they have the same dimension.

Theorem 4.7.37

Let V and W be (finite-dimensional) vector spaces. Then, $V \cong W$ if and only if $\dim V = \dim W$.

Proof. ($\Rightarrow$) Suppose that $V \cong W$. Then there exists an isomorphism $T : V \to W$. By Theorem 4.7.22, we know that

$$\dim V \leq \dim W.$$

Now, by Theorem 4.7.33, we know that

$$\dim V \geq \dim W.$$

Therefore, we conclude that

$$\dim V = \dim W.$$

($\Leftarrow$) Now, suppose that $\dim V = \dim W = n$. Suppose also that a basis for V is $\mathcal{B}_V = \{v_1, v_2, \ldots, v_n\}$ and a basis for W is $\mathcal{B}_W = \{w_1, w_2, \ldots, w_n\}$. By Theorem 4.2.26, we can define $T : V \to W$ to be the linear transformation so that

$$T(v_1) = w_1, T(v_2) = w_2, \ldots, T(v_n) = w_n.$$

We will show that T is an isomorphism. We know that if $w \in W$, then

$$w = \alpha_1 w_1 + \alpha_2 w_2 + \ldots + \alpha_n w_n$$

for some scalars $\alpha_1, \alpha_2, \ldots, \alpha_n$. We also know that there is a vector $v \in V$, defined by

$$v = \alpha_1 v_1 + \alpha_2 v_2 + \ldots + \alpha_n v_n \in V,$$

since T is linear, we have

$$
\begin{aligned}
T(v) &= T(\alpha_1 v_1 + \alpha_2 v_2 + \ldots + \alpha_n v_n) \\
&= \alpha_1 T(v_1) + \alpha_2 T(v_2) + \ldots + \alpha_n T(v_n) \\
&= \alpha_1 w_1 + \alpha_2 w_2 + \ldots + \alpha_n w_n \\
&= w.
\end{aligned}
$$

Therefore, T maps onto W.

Now, suppose that

$$v = \alpha_1 v_1 + \alpha_2 v_2 + \ldots + \alpha_n v_n \text{ and } u = \beta_1 v_1 + \beta_2 v_2 + \ldots + \beta_n v_n$$

are vectors in V satisfying $T(v) = T(u)$. We will show that $v - u = 0$. We know that

$$
\begin{aligned}
0 &= T(v) - T(u) \\
&= T(v - u) \\
&= (\alpha_1 v_1 + \alpha_2 v_2 + \ldots + \alpha_n v_n - (\beta_1 v_1 + \beta_2 v_2 + \ldots + \beta_n v_n)) \\
&= T((\alpha_1 - \beta_1)v_1 + (\alpha_2 - \beta_2)v_2 + \ldots + (\alpha_n - \beta_n)v_n) \\
&= (\alpha_1 - \beta_1)T(v_1) + (\alpha_2 - \beta_2)T(v_2) + \ldots + (\alpha_n - \beta_n)T(v_n) \\
&= (\alpha_1 - \beta_1)w_1 + (\alpha_2 - \beta_2)w_2 + \ldots + (\alpha_n - \beta_n)w_n.
\end{aligned}
$$

Now, since $\mathcal{B}_W$ is a basis for W, we know that

$$\alpha_1 - \beta_1, \alpha_2 - \beta_2, \ldots, \alpha_n - \beta_n = 0.$$

That is to say $u - v = 0$. Therefore, $u = v$ and T is one-to-one. Since T is both one-to-one and onto, T is an isomorphism and, therefore, $V \cong W$. $\square$

Theorem 4.7.37 suggests that bases map to bases through an isomorphism as stated in the following theorem.

Theorem 4.7.38

Let V and W be finite-dimensional vector spaces. Suppose $\mathcal{B}_V$ is a basis for V and let $T : V \to W$ be a linear transformation. Then, T is an isomorphism if and only $T(\mathcal{B}_V)$ is also a basis of W.

Proof. Suppose V and W are finite-dimensional bases. Suppose also that $T : V \to W$ is an isomorphism. We know from Theorem 4.7.37, that dim $V =$ dim W, say that dim $V = n$. Now, define $\mathcal{B}_V = \{v_1, v_2, \ldots, v_n\}$ to be a basis for V. By Theorem 3.4.24, we need only show that $T(\mathcal{B}_V)$ is linearly independent. But, Lemma 4.7.16 tells us that since $\mathcal{B}_V$ is linearly independent, so is $T(\mathcal{B}_V)$. Therefore, $T(\mathcal{B}_V)$ is a basis of W. $\square$

Corollary 4.7.39

Let $(V, +, \cdot)$ be a vector space of dimension n with scalars from $\mathbb{F}$. Then V is isomorphic to $\mathbb{F}^n$.

This corollary is true for any scalar field $\mathbb{F}$, for example: $\mathbb{R}, \mathbb{C}, Z_2$, etc. This corollary suggests to us the idea that the most complicated abstract vector spaces (such as image spaces, heat state spaces, 7-bar LCD character spaces, polynomial spaces, etc.) are isomorphic to the simplest, most familiar spaces (such as $\mathbb{R}^n$). This is an indication of things to come; perhaps any vector space can be viewed as $\mathbb{R}^n$ through the right lens. After all, two vector spaces are isomorphic if their elements have a one-to-one relationship.

Let us take a moment to return to the discussion at the beginning of Section 4.5. In this discussion, we discussed matrix transformations $T(x) = Mx$, on a shape in $\mathbb{R}^2$. We found that $\det(M)$ is the factor by which the area of the shape changes after transforming it by T. But, in Exercise 18 of Section 4.2 we considered the transformation $T : \mathbb{R}^2 \to \mathbb{R}^2$, defined as

$$T \begin{pmatrix} x \\ y \end{pmatrix} = \begin{pmatrix} 1 & 0 \\ 1 & 0 \end{pmatrix} \begin{pmatrix} x \\ y \end{pmatrix}.$$

In Figure 4.30, we see that T transforms the unit square to a line segment connecting $(0, 0)$ and $(1, 0)$. In fact, no matter what shape we transform with T, we always get a line segment along the x-axis. Clearly, we have lost information and are not able to reconstruct the original shape. Following the same ideas in the previous discussion, we calculate

$$\det \begin{pmatrix} 1 & 0 \\ 1 & 0 \end{pmatrix} = 0.$$

This matches what we saw at the beginning of Section 4.5 because the area is reduced to 0. More important is the information we just gained here. A matrix whose determinant is zero results in a transformation that is not an isomorphism. We can see this because the transformation results in an image with smaller dimension. That is, $\dim(T(\mathbb{R}^n)) < n$. As we move forward, we will find this to be a very useful and quick check when determining whether or not we can get an exact radiographic reconstruction. Because this will prove to be very useful, we will put it in a box.

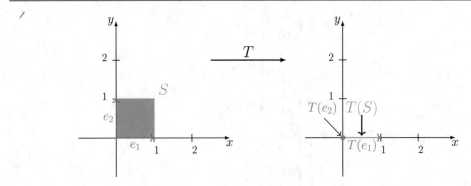

Fig. 4.30 The transformation T transforms S to a line segment.

Fact 4.7.40

Let M be an $n \times n$ matrix and define a transformation $T : \mathbb{R}^n \to \mathbb{R}^n$ by $T(x) = Mx$. Then $\det(M) = 0$ means that T is not an isomorphism. In fact, if $\det(M) \neq 0$, we will find that T is an isomorphism.

4.7.8 Building and Recognizing Isomorphisms

The proof of Theorem 4.7.37 suggests a tool for creating isomorphisms (if they exist). If we can define a linear operator $T : V \to W$ which maps a basis for V to a basis for W, then T is an isomorphism between V and W.

Example 4.7.41 Let $V = \mathcal{M}_{2 \times 3}$ and $W = \mathcal{P}_5$. We know that $V \cong W$ because both are 6-dimensional vector spaces. Indeed, a basis for V is

$$
\mathcal{B}_V = \left\{ \begin{pmatrix} 1\,0\,0 \\ 0\,0\,0 \end{pmatrix}, \begin{pmatrix} 0\,1\,0 \\ 0\,0\,0 \end{pmatrix}, \begin{pmatrix} 0\,0\,1 \\ 0\,0\,0 \end{pmatrix}, \right.
$$
$$
\left. \begin{pmatrix} 0\,0\,0 \\ 1\,0\,0 \end{pmatrix}, \begin{pmatrix} 0\,0\,0 \\ 0\,1\,0 \end{pmatrix}, \begin{pmatrix} 0\,0\,0 \\ 0\,0\,1 \end{pmatrix}, \right\}
$$

and a basis for W is

$$
\mathcal{B}_W = \{1, x, x^2, x^3, x^4, x^5\}.
$$

Thus, we can create an isomorphism T that maps V to W. Using Theorem 4.7.37 we define T as follows

$$T \begin{pmatrix} 1\,0\,0 \\ 0\,0\,0 \end{pmatrix} = 1,$$

$$T \begin{pmatrix} 0\,1\,0 \\ 0\,0\,0 \end{pmatrix} = x,$$

$$T \begin{pmatrix} 0\,0\,1 \\ 0\,0\,0 \end{pmatrix} = x^2,$$

$$T \begin{pmatrix} 0\,0\,0 \\ 1\,0\,0 \end{pmatrix} = x^3,$$

$$T \begin{pmatrix} 0\,0\,0 \\ 0\,1\,0 \end{pmatrix} = x^4,$$

$$T \begin{pmatrix} 0\,0\,0 \\ 0\,0\,1 \end{pmatrix} = x^5.$$

If we have any vector $v \in V$, we can find to where T maps it in W. Since $v \in V$, we know there are scalars, a, b, c, d, e, f so that

$$v = a \begin{pmatrix} 1\,0\,0 \\ 0\,0\,0 \end{pmatrix} + b \begin{pmatrix} 0\,1\,0 \\ 0\,0\,0 \end{pmatrix} + c \begin{pmatrix} 0\,0\,1 \\ 0\,0\,0 \end{pmatrix}$$
$$+ d \begin{pmatrix} 0\,0\,0 \\ 1\,0\,0 \end{pmatrix} + e \begin{pmatrix} 0\,0\,0 \\ 0\,1\,0 \end{pmatrix} + f \begin{pmatrix} 0\,0\,0 \\ 0\,0\,1 \end{pmatrix}).$$

That is,

$$v = \begin{pmatrix} a\,b\,c \\ d\,e\,f \end{pmatrix}.$$

Thus, since T is linear,

$$T(v) = a(1) + b(x) + c(x^2) + d(x^3) + e(x^4) + f(x^5) = a + bx + cx^2 + dx^3 + ex^4 + fx^5.$$

$\square$

Example 4.7.42 Consider the radiographic transformation of Scenario A described in Figure 4.24. We have already seen that the transformation is neither one-to-one nor onto. However, the dimensions of the object and radiograph spaces are equal. Theorem 4.7.37 tells us that these spaces are isomorphic. And by Definition 4.7.35 there exists an isomorphism between the two spaces. But,the important thing to note is that not every linear transformation between the two spaces is an isomorphism. $\square$

Consider $\mathcal{I}_{4\times4}(\mathbb{R})$, the space of 4×4 grayscale images. Since vectors in this vector space can be tedious to draw with any accuracy, let's consider their representation in $\mathbb{R}^{16}$. We choose to attempt this because these two vector spaces are isomorphic. *How do we know?*

Let's reconsider Example 3.5.7. We found that the image v, given by

$$v = $$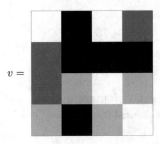

where black indicates a pixel value of zero and white indicates a pixel value of 3, could be represented as a coordinate vector $[v]_{\mathcal{B}_{\mathcal{I}}}$, where

$$\mathcal{B}_{\mathcal{I}} = \left\{ \text{} \right\}.$$

We found, $[v]_{\mathcal{B}_{\mathcal{I}}} = (3\ 1\ 1\ 2\ 0\ 0\ 2\ 0\ 3\ 0\ 3\ 2\ 1\ 0\ 2\ 3)^T \in \mathbb{R}^{16}$.

If we let $\mathcal{B}_{\mathcal{I}} = \{b_1, b_2, \cdots, b_{16}\}$ and suppose $\mathcal{E}_{16} = \{e_1, e_2, \cdots, e_{16}\}$ is the standard basis for $\mathbb{R}^{16}$, then we can define the linear transformation $T : \mathcal{I}_{4 \times 4}(\mathbb{R}) \to \mathbb{R}^{16}$ as $T(b_k) = e_k$, for $k = 1, 2, \ldots, 16$. That is, by Theorem 4.7.37, T is the isomorphism that maps a vector to its coordinate representation: $v \mapsto [v]_{\mathcal{B}}$.

Theorem 4.7.43

Let V be an n-dimensional vector space with ordered basis $\mathcal{B}$, and let $x \in V$. Then the coordinate transformation $x \mapsto [x]_{\mathcal{B}}$ is an isomorphism from V to $\mathbb{R}^n$.

Proof. From Theorem 4.2.26 we can define the transformation T by $T(b_k) = e_k$, for $k = 1, 2, \ldots, n$, where $\{e_1, e_2, \cdots, e_n\}$ is the standard basis for $\mathbb{R}^n$. Using the proof of Theorem 4.7.37, we know that T is an isomorphism. $\square$

4.7.9 Inverse Transformations

The exploration of transformations that are one-to-one and/or onto has led us to consider the possibility of recovering a domain vector from a codomain vector through understanding the properties of the given transformation. For example, we might wonder if we can recover a brain image from a radiograph by understanding the properties of the radiographic transformation. This section explores that possibility.

Definition 4.7.44

Let V and W be vector spaces and $T : V \to W$. We say that T is **invertible** if there exists a transformation $S : W \to V$, called the **inverse** of T, with the property

$$S \circ T = I_V \text{ and } T \circ S = I_W.$$

If such a transformation exists, we write

$$S = T^{-1}.$$

$\square$

Example 4.7.45 Consider the linear transformation $T : \mathbb{R} \to \mathbb{R}$ defined by $T(x) = 3x + 2$. The inverse transformation $S : \mathbb{R} \to \mathbb{R}$ is defined by $S(y) = \frac{1}{3}(y - 2)$, because

$$S(T(x)) = S(3x + 2) = \frac{1}{3}((3x + 2) - 2) = x,$$

and

$$T(S(y)) = T\left(\frac{1}{3}(y - 2)\right) = 3\left(\frac{1}{3}(y - 2)\right) + 2 = y.$$

$\square$

Example 4.7.46 Consider the polynomial differentiation transformation $T : \mathcal{P}_2(\mathbb{R}) \to \mathcal{P}_1(\mathbb{R})$ defined by $T(ax^2 + bx + c) = 2ax + b$. We can see that the polynomial integration operator $U : \mathcal{P}_1(\mathbb{R}) \to \mathcal{P}_2(\mathbb{R})$, defined by $U(dx + e) = \frac{1}{2}dx^2 + ex$, is not the inverse transformation of T. We see that $U(T(ax^2 + bx + c)) = ax^2 + bx \neq ax^2 + bx + c$ for all $c \neq 0$. But, $T(U(dx + e)) = dx + e$. The issue here is that there are infinitely many quadratic polynomials whose derivative are all the same. $\square$

Example 4.7.47 Consider, $\mathcal{D}(Z_2)$, the vector space of 7-bar LCD images, and the transformation $T : \mathcal{D}(Z_2) \to \mathcal{D}(Z_2)$ defined by

$$T(x) = x + \boxed{\text{[LCD image]}}.$$

In this case, $T^{-1} = T$ because

$$T(T(x)) = T\left(x + \boxed{\text{[LCD image]}}\right)$$

$$= \left(x + \boxed{\text{[LCD image]}}\right) + \boxed{\text{[LCD image]}}$$

$$= x + \left(\boxed{\text{[LCD image]}} + \boxed{\text{[LCD image]}}\right)$$

$$= x + \boxed{\text{[LCD image]}}$$

$$= x.$$

$\square$

Example 4.7.48 The identity transformation $I : V \to V$ is its own inverse, $I^{-1} = I$. □

It is not surprising that invertible transformations are isomorphisms.

Theorem 4.7.49
Let V and W be finite-dimensional vector spaces and let $T : V \to W$ be a linear transformation. Then T is invertible if and only if T is an isomorphism.

Proof. Let V and W be finite-dimensional vector spaces. Suppose $T : V \to W$ is linear. ($\Rightarrow$) First, we show that if T is invertible, then T is an isomorphism. Suppose T is invertible. Then there exists an inverse transformation S. We will show that T is both one-to-one and onto.

(Onto) Consider arbitrary $w \in W$. Define $v = S(w)$, then

$$T(v) = T(S(w)) = I(w) = w.$$

Thus, there is a $v \in V$ so that $T(v) = w$. Therefore, T maps onto W.

(One-to-one) Let $u, v \in V$ be vectors with $T(u) = T(v)$. We will show that this implies $u = v$. Because S is a function, we know that

$$u = I(u) = S(T(u)) = S(T(v)) = I(v) = v.$$

Thus, T is one-to-one.

($\Leftarrow$) Now we show that if T is an isomorphism, then T is invertible. Suppose $\dim V = n$. Since T is one-to-one, Theorems 4.7.19 and 4.7.38 give that if $\mathcal{B}_V = \{v_1, v_2, \ldots, v_n\}$ is a basis for V then $T(\mathcal{B}_V) = \mathcal{B}_W$ is a basis of W. Then

$$\mathcal{B}_W = \{w_1 = T(v_1), w_2 = T(v_2), \ldots, w_n = T(v_n)\}.$$

By Theorem 4.2.26, we can define a linear transformation $S : W \to V$ by

$$S(w_1) = v_1, S(w_2) = v_2, \ldots, S(w_n) = v_n.$$

Now, let $u \in V$ and $z \in W$ be vectors. We know that there are scalars, $\alpha_1, \alpha_2, \ldots, \alpha_n$ and $\beta_1, \beta_2, \ldots, \beta_n$ so that

$$u = \alpha_1 v_1 + \alpha_2 v_2 + \ldots + \alpha_n v_n \text{ and } z = \beta_1 w_1 + \beta_2 w_n + \ldots + \beta_n w_n.$$

We also know, by Theorem 4.2.21, that $T \circ S$ and $S \circ T$ are linear. We will show that $(T \circ S)(z) = z$ and $(S \circ T)(u) = u$. We have, using linearity and the definitions of $\mathcal{B}_W$ and S,

$$(T \circ S)(z) = T(S(z))$$
$$= T(S(\beta_1 w_1 + \beta_2 w_n + \ldots + \beta_n w_n))$$
$$= \beta_1 T(S(w_1)) + \beta_2 T(S(w_n)) + \ldots + \beta_n T(S(w_n))$$
$$= \beta_1 T(v_1) + \beta_2 T(v_2) + \ldots + \beta_n T(v_n)$$
$$= \beta_1 w_1 + \beta_2 w_2 + \ldots + \beta_n w_n$$
$$= z.$$

We also have

$$(S \circ T)(u) = S(T(u))$$
$$= S(T(\alpha_1 v_1 + \alpha_2 v_2 + \ldots + \alpha_n v_n))$$
$$= \alpha_1 S(T(v_1)) + \alpha_2 S(T(v_2)) + \ldots + \alpha_n S(T(v_n))$$
$$= \beta_1 S(w_1) + \beta_2 S(w_2) + \ldots + \beta_n S(w_n)$$
$$= \beta_1 v_1 + \beta_2 v_2 + \ldots + \beta_n v_n$$
$$= u.$$

Therefore, $S = T^{-1}$ and so T is invertible. $\qquad\qquad\square$

We have noticed that a linear transformation between isomorphic vector spaces may not be an isomorphism and therefore it will not be invertible. The next corollary helps complete this discussion by stating that invertible linear transformations only exist between vector spaces of the same dimension (between isomorphic vector spaces).

Corollary 4.7.50
Let V and W be finite-dimensional vector spaces and $T : V \to W$ linear. If T is invertible, then $\dim V = \dim W$.

Proof. The result follows directly from Theorem 4.7.49 and Theorem 4.7.37. $\qquad\square$

Example 4.7.51 The radiographic scenarios of Figures 4.24-4.26 feature transformations from the vector space of 2×2 images (dimension 4) to vector spaces of radiographic images (each of dimension 4). Yet, none of the transformations is invertible. Each transformation is neither one-to-one nor onto. $\qquad\square$

To complete the introduction to inverses, we present, in the next theorem, some useful properties.

Theorem 4.7.52
Let $T : V \to W$ be an invertible linear transformation. Then

(a) $(T^{-1})^{-1} = T$,
(b) T^{-1} is linear, and
(c) T^{-1} is an isomorphism.

The proof of this theorem is the subject of Exercises 53, 54, and 55.

4.7.10 Left Inverse Transformations

In general, radiographic transformations are not transformations between vector spaces of equal dimension, so we cannot expect invertibility. If such a transformation were invertible, and if we could determine the details of the inverse transformation, then for any brain image x with radiograph b we have $T(x) = b$ and more importantly $x = T^{-1}(b)$.

However, invertibility is actually *more* than we require. We really only hope to recover a brain image from a given radiograph, one that is the output of a radiographic transformation[5]. That is, suppose V is the space of brains and W is the space of radiographs and that we have a radiographic transformation $T : V \to W$ and another transformation $S : W \to V$ such that $S(T(x)) = x$ for all brains $x \in V$. Then we recover the brain x by $x = S(T(x)) = S(b)$ from the radiograph $b = T(x)$. In this section, we discuss this "almost inverse" transformation.

Definition 4.7.53

Let V and W be vector spaces and $T : V \to W$. Then $S : W \to V$, is called a **left inverse** of T if $S \circ T = I_V$. $\qquad\square$

If a transformation has a left inverse then a domain object can be uniquely recovered from the codomain object to which it maps.

Example 4.7.54 The integration transformation U from Example 4.7.46 has a left inverse transformation, the differentiation transformation T. That is, given a polynomial $p \in \mathcal{P}_1(\mathbb{R})$, we have, for $p = ax + b$

$$
\begin{aligned}
(T \circ U)(p) &= T(U(p)) \\
&= T(U(ax + b)) \\
&= T\left(\frac{a}{2}x^2 + bx\right) \\
&= ax + b \\
&= p.
\end{aligned}
$$

So, $T(U(p)) = p$ for all $p \in \mathcal{P}_1(\mathbb{R})$. $\qquad\square$

The following theorem tells us one way in which we can identify a linear transformation that has a left inverse.

Theorem 4.7.55

Let $T : V \to W$ be a linear transformation. Then T is one-to-one if and only if T has a left inverse.

---

[5] Later, we will discuss outputs that are the result of a corrupted radiographic process. That is, a radiograph that, due to noise, is not in the range of the radiographic transformation.

The proof can be extracted from relevant parts of the proof of Theorem 4.7.49. See also Exercise 56.

Example 4.7.56 Consider $\mathcal{H}_3(\mathbb{R})$, the vector space of histograms with 3 bins, and the transformation $T : \mathcal{H}_3(\mathbb{R}) \to \mathbb{R}^4$ defined by

$$T(H) = \begin{pmatrix} h_1 + h_2 + h_3 \\ h_1 - h_3 \\ h_2 - h_1 \\ h_3 - h_2 \end{pmatrix}$$

where H is a histogram (vector) and $h_1, h_2, h_3 \in \mathbb{R}$ are the three ordered values assigned to the histogram bins. We are interested in whether or not T has a left inverse. If so, then any vector in $\mathbb{R}^4$ (which is the result of the transformation of a histogram) uniquely defines that histogram.

We can determine if such a left inverse transformation exists by determining whether or not T is one-to-one. Consider two vectors $T(H)$ and $T(J)$ given by

$$T(H) = \begin{pmatrix} h_1 + h_2 + h_3 \\ h_1 - h_3 \\ h_2 - h_1 \\ h_3 - h_2 \end{pmatrix} \quad \text{and} \quad T(J) = \begin{pmatrix} j_1 + j_2 + j_3 \\ j_1 - j_3 \\ j_2 - j_1 \\ j_3 - j_2 \end{pmatrix}$$

If $T(H) = T(J)$, it is a matter of some algebra to show that indeed, $h_1 = j_1$, $h_2 = j_2$, and $h_3 = j_3$. That is, $H = J$. Thus, T is one-to-one and a left inverse transformation exists. $\qquad\square$

Corollary 4.7.57

Let $T : V \to W$ be a linear transformation. If T has a left inverse, then $\dim V \le \dim W$.

The proof follows directly from Theorems 4.7.22 and 4.7.55.
Now before we complete this section we discuss the language introduced.

$\star\star$ **Watch Your Language!** We have discussed several topics in this section. It is important, when discussing these topics, to recognize what the terminology is describing: vector spaces, transformations, or vectors. Be careful not to apply, for example, a term describing transformations to vector spaces. Here are appropriate uses of the terminology in this section.

✓ $T : V \to W$ maps V onto W.
✓ $T : V \to W$ is a one-to-one transformation.
✓ $T : V \to W$ is an isomorphism.
✓ V is isomorphic to W.
✓ V and W are isomorphic vector spaces.
✓ $T^{-1} : W \to V$ is the inverse of T.
✓ If T^{-1} is the inverse of T and $T(x) = y$, we can solve for x by applying the transformation, T^{-1} to both sides to get $T^{-1}(T(x)) = T^{-1}(y) = x$.

It is inappropriate to say the following.

✗ $T : V \to W$ maps $v \in V$ onto $w \in W$.

✗ V and W are one-to-one to each other.
✗ $T : V \rightarrow W$ is isomorphic.
✗ V and W are isomorphisms.
✗ V and W are equal/equivalent/the same.
✗ T^{-1} is the reciprocal of T.
✗ If T^{-1} is the inverse of T and $T(x) = y$, we can solve for x by dividing both sides by T to get $x = T^{-1}(y)$.

In a typical radiographic scenario, we wish to obtain high-resolution images in our brain image space, say V. We also want to achieve this goal without requiring excessive radiographic data, taking as few images as possible for our radiographic image space, say W. That is we hope that dim $W \ll$ dim V (that is, dim W is much less than dim V). Unfortunately, that also means that the transformation will not be one-to-one and we will not have a left inverse. This means that we require more linear algebra tools to accomplish our ultimate tomographic goals.

4.7.11   Exercises

1. Let $T : \mathbb{R}^2 \rightarrow \mathbb{R}^2$ be defined by $T(x) = Ax$, where A is the following matrix:

$$\begin{pmatrix} 1 & 0 \\ -1 & 1 \end{pmatrix}$$

   (a) Sketch the image of the unit square (vertices at $(0, 0)$, $(1, 0)$, $(1, 1)$, and $(0, 1)$) under the transformation T.
   (b) Based on your sketch, determine whether T is one to one, and whether T is onto. Your answer should include a brief discussion of your reasoning, based on the geometry of the transformation.
   (c) Prove your conjecture about whether T is one-to-one using the definition of a one-to-one map.
   (d) Prove your conjecture about whether T is onto using the definition of an onto map.

2. Let $T : \mathbb{R}^2 \rightarrow \mathbb{R}^2$ be defined by $T(x) = Ax$, where A is the following matrix:

$$\begin{pmatrix} 1/2 & -1/2 \\ -1 & 1 \end{pmatrix}$$

   (a) Sketch the image of the unit square (vertices at $(0, 0)$, $(1, 0)$, $(1, 1)$, and $(0, 1)$) under the transformation T.
   (b) Based on your sketch, determine whether T is one to one, and whether T is onto. Your answer should include a brief discussion of your reasoning, based on the geometry of the transformation.
   (c) Prove your conjecture about whether T is one-to-one using the definition of a one-to-one map.
   (d) Prove your conjecture about whether T is onto using the definition of an onto map.

For the Exercises 3-11 (Recall the linear transformations from Section 4.2 Exercises 1 to 12.), determine whether the transformation is one-to-one and/or onto. Prove your answer.

3. Let $T : \mathbb{R}^4 \rightarrow \mathbb{R}^3$ be defined by $T(x) = Ax$, where A is the following matrix:

$$\begin{pmatrix} 1 & -2 & 0 & 0 \\ -3 & 0 & -1 & 1 \\ 0 & -6 & -1 & 1 \end{pmatrix}$$

4. Define $f : \mathbb{R}^3 \to \mathbb{R}^2$ by $f(v) = Mv + x$, where

$$M = \begin{pmatrix} 1 & 2 & 1 \\ 1 & 2 & 1 \end{pmatrix} \quad \text{and} \quad x = \begin{pmatrix} 1 \\ 0 \end{pmatrix}.$$

5. Define $\mathcal{F} : V \to \mathcal{P}_1$, where

$$V = \{ax^2 + (3a - 2b)x + b \mid a, b \in \mathbb{R}\} \subseteq \mathcal{P}_2.$$

by

$$\mathcal{F}(ax^2 + (3a - 2b)x + b) = 2ax + 3a - 2b.$$

6. Define $\mathcal{G} : \mathcal{P}_2 \to \mathcal{M}_{2 \times 2}$ by

$$\mathcal{G}(ax^2 + bx + c) = \begin{pmatrix} a & a - b \\ c - 2 & c + 3a \end{pmatrix}.$$

7. Define $h : V \to \mathcal{P}_1$, where

$$V = \left\{ \begin{pmatrix} a & b & c \\ 0 & b - c & 2a \end{pmatrix} \middle| a, b, c \in \mathbb{R} \right\} \subseteq \mathcal{M}_{2 \times 3}$$

by

$$h \begin{pmatrix} a & b & c \\ 0 & b - c & 2a \end{pmatrix} = ax + c.$$

8. Let

$$\mathcal{I} = \left\{ \begin{array}{|c|c|} \hline & 3a & \\ \hline -b & & 2a \\ \hline & 0 & \\ \hline c & & b \\ \hline & 3c & \\ \hline \end{array} \middle| a, b, c \in \mathbb{R} \right\}.$$

And define $f : \mathcal{I} \to \mathcal{P}_2$ by

$$f(I) = ax^2 + (b + c)x + (a + c).$$

9. Define $f : \mathcal{M}_{2 \times 2} \to \mathbb{R}^4$ by

$$f \begin{pmatrix} a & b \\ c & d \end{pmatrix} = \begin{pmatrix} a \\ b \\ c \\ d \end{pmatrix}.$$

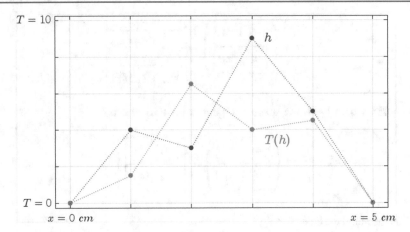

Fig. 4.31 Example of averaging heat state transformation.

10. Define $f : \mathcal{P}_2 \rightarrow \mathbb{R}^2$ by

$$f(ax^2 + bx + c) = \begin{pmatrix} a + b \\ a - c \end{pmatrix}.$$

11. Let $\mathcal{H}_4$ be the set of all possible heat states sampled every 1 cm along a 5 cm long rod. Define a function $T : \mathcal{H}_4 \rightarrow \mathcal{H}_4$ by replacing each value (which does not correspond to an endpoint) with the average of its neighbors. The endpoint values are kept at 0. An example of T is shown in Figure 4.31.

For Exercises 12-23, determine whether a transformation described can be created. If so, create it and prove that the transformation you created satisfies the description. If not, prove that no such transformation exists.

12. An onto transformation that maps from $\mathbb{R}^3$ to $\mathbb{R}^2$.
13. A one-to-one transformation that maps from $\mathbb{R}^2$ to $\mathbb{R}^3$.
14. An onto transformation that maps from $\mathbb{R}^3$ to $\mathcal{P}_1$.
15. A one-to-one transformation that maps from $\mathbb{R}^3$ to $\mathcal{P}_1$.
16. A one-to-one transformation that maps $\mathbb{R}^2$ to $\mathcal{M}_{2\times3}$.
17. An onto transformation that maps $\mathbb{R}^2$ to $\mathcal{M}_{2\times3}$.
18. A one-to-one transformation that maps $\mathcal{M}_{2\times3}$ to $\mathbb{R}^2$.
19. An onto transformation that maps $\mathcal{M}_{2\times3}$ to $\mathbb{R}^2$.
20. An onto transformation that maps $\mathcal{P}_1$ to $V = \{(x, y, z) | x + y + z = 0\}$.
21. A one-to-one transformation that maps $\mathcal{P}_1$ to $V = \{(x, y, z) | x + y + z = 0\}$.
22. A one-to-one transformation that maps $\mathcal{D}(Z_2)$ (the vector space of 7-bar LCD images) to H_7 (the vector space of heat states sampled 7 times along a rod).
23. An onto transformation that maps $\mathcal{D}(Z_2)$ (the vector space of 7-bar LCD images) to H_7 (the vector space of heat states sampled 7 times along a rod).

For Exercises 24-33 determine whether or not the transformation is an isomorphism, then determine whether it has an inverse. If not, determine whether it has a left inverse. Justify your conclusion.

24. $T : V_{\mathcal{O}} \rightarrow V_{\mathcal{R}}$, where $V_{\mathcal{O}}$ is the space of objects with 4 voxels and $V_{\mathcal{R}}$ is the space of radiographs with 4 pixels and

$$T\left(\begin{array}{|c|c|} \hline x_1 & x_3 \\ \hline x_2 & x_4 \\ \hline \end{array}\right) = \begin{array}{|c|} \hline x_1 + x_2 \\ \hline x_3 + x_4 \\ \hline \frac{2}{3}x_1 + x_2 + \frac{2}{3}x_4 \\ \hline \frac{1}{3}x_1 + x_3 + \frac{1}{3}x_4 \\ \hline \end{array}$$

25. $T : V_{\mathcal{O}} \to \mathbb{R}^4$, where $V_{\mathcal{O}}$ is the space of objects with 4 voxels and

$$T\left(\begin{array}{|c|c|} \hline x_1 & x_3 \\ \hline x_2 & x_4 \\ \hline \end{array}\right) = \left[\begin{array}{|c|c|} \hline x_1 & x_3 \\ \hline x_2 & x_4 \\ \hline \end{array}\right]_{\mathcal{B}},$$

   where $\mathcal{B}$ is the standard basis for $V_{\mathcal{O}}$.

26. $T : \mathcal{M}_{2\times 2} \to \mathbb{R}^4$ defined by $T\begin{pmatrix} a & b \\ c & d \end{pmatrix} = \begin{pmatrix} a \\ b \\ a+b \\ c-d \end{pmatrix}$.

27. $T : \mathcal{M}_{2\times 2} \to \mathbb{R}^4$ defined by $T\begin{pmatrix} a & b \\ c & d \end{pmatrix} = \begin{pmatrix} a \\ b+1 \\ 2b-3c \\ d \end{pmatrix}$.

28. $T : \mathbb{P}_2(\mathbb{R}) \to \mathcal{P}_2(\mathbb{R})$ defined by $T(ax^2 + bx + c) = cx^2 + ax + b$

29. $T : \mathcal{P}_2(\mathbb{R}) \to \mathcal{P}_2(\mathbb{R})$ defined by $T(ax^2 + bx + c) = (a+b)x^2 - b + c$.

30. Let $n \in \mathbb{N}$ and let $T : \mathcal{P}_n(\mathbb{R}) \to \mathcal{P}_n(\mathbb{R})$ be defined by $T(p(x)) = p'(x)$.

31. $T : H_4(\mathbb{R}) \to \mathbb{R}^4$ defined by $T(v) = [v]_Y$, where Y is the basis given in Example 3.4.15.

32. $T : \mathcal{D}(Z_2) \to \mathcal{D}(Z_2)$ defined by $T(x) = x + x$.

33. The transformation of Exercise 12 in Section 4.2 on heat states.

In Exercises 34-38, determine whether each pair of vector spaces is isomorphic. If so, find an isomorphism between them. If not, justify.

34. $\mathbb{R}^3$ and $\mathcal{P}_1(\mathbb{R})$

35. $\mathbb{R}^2$ and $\mathcal{M}_{2\times 3}$

36. $\mathcal{P}_6(\mathbb{R})$ and $\mathcal{M}_{2\times 1}$

37. $\mathcal{D}(Z_2)$ and Z_2^7

38. $\mathcal{D}(Z_2)$ and $\mathcal{P}_2(Z_2)$

Additional Questions

39. Let V and W be finite-dimensional vector spaces. Suppose $T : V \to W$ is linear and $T(U) = S$ for some $U \subseteq V$ and spanning set $S = \text{Span } W$. Prove or disprove that T maps V onto W.

40. Show that the zero transformation $Z : V \to W$, defined by $Z(v) = 0$ for all $v \in V$, can be one-to-one.

41. Give an example to show that the zero transformation can be onto.

42. (The composition of one-to-one maps is one-to-one.) Let $T : U \to V$ and $S : V \to W$ be one-to-one linear maps between vector spaces. Prove that the composition $S \circ T : U \to W$ is one-to-one.

43. (The composition of onto maps is onto.) Let $T : U \to V$ and $S : V \to W$ be onto linear maps between vector spaces. Prove that the composition $S \circ T : U \to W$ is onto.

44. Consider the vector space $\mathcal{D}(Z_2)$ of 7-bar LCD characters and linear transformation $T : \mathcal{D}(Z_2) \to \mathcal{D}(Z_2)$ defined by $T(x) = x + x$. Determine whether T is one-to-one and/or onto.

45. Consider the transformation of Exercise (Fig. 4.31) on heat states. Determine if this transformation is one-to-one and/or onto.

46. Prove that the converse of Lemma 4.7.30 is false.

47. Consider the Radiography/Tomography application that we have been exploring. We want to be able to recover brain images from radiographs. Based on the discussion in this section, why is it important to know whether the radiographic transformation is one-to-one? Discuss what is needed in order for a radiographic setup to be a one-to-one transformation or discuss why this is not possible.

48. Consider the Radiography/Tomography application that we have been exploring. We want to be able to recover brain images from radiographs. Based on the discussion in this section, why is it important to know whether the radiographic transformation is onto? Discuss what is needed in order for a radiographic setup to be an onto transformation or discuss why this is not possible.

49. Prove or disprove the following claim.
   Claim: Suppose $T : V \to W$ is an onto linear transformation. Then $T(V) = W$.

50. Prove or disprove the following claim.
   Claim: Suppose $T : V \to W$ is a linear transformation that maps V onto W and that $\mathcal{B}$ is a basis for V. Then Span $T(\mathcal{B}) = W$.

51. Prove or disprove the following claim.
   Claim: Suppose $T : V \to W$ is a one-to-one linear transformation. Then $\dim T(V) = \dim V$.

52. Determine whether or not a transformation $T : V \to \{0\}$ can be an isomorphism. If so, state the conditions on which this can happen. If not, justify.

53. Prove Theorem 4.7.52(a).

54. Prove Theorem 4.7.52(b).

55. Prove Theorem 4.7.52(c).

56. Prove Theorem 4.7.55.

57. Let V and W be finite-dimensional vector spaces. Prove that if $\dim V = \dim W$ and $T : V \to W$ is an onto linear transformation, then T is an isomorphism.

58. Let V and W be finite-dimensional vector spaces. Prove that if $\dim V = \dim W$ and $T : V \to W$ is a one-to-one linear transformation, then T is an isomorphism.

59. If the results in Exercises 57 and 58 are true, how would this help when creating an isomorphism?

60. If the results in Exercises 57 and 58 are true, how would this help when determining whether a transformation is an isomorphism?

61. Let $S : V \to W$ be a one-to-one linear transformation, $T : V \to W$ be an onto linear transformation. Suppose that $Q : W \to X$ and $R : U \to V$ are arbitrary linear transformations (you have no information whether R or Q are one-to-one or onto.)

   (a) Is $S \circ R$ always one-to-one?
   (b) Is $Q \circ S$ always one-to-one?
   (c) Is $T \circ R$ always onto?
   (d) Is $Q \circ T$ always onto?

Invertibility

5

Our exploration of radiographic transformations has led us to make some key correspondences in our quest to find the best possible brain image reconstructions.

- Every one-to-one transformation has the property that two different radiographs do not correspond to the same brain image.
- Every onto transformation has the property that every radiograph corresponds to at least one brain image.
- Every one-to-one transformation has a left-inverse, which can be used to correctly reconstruct a brain image from any radiograph.

We want to know whether we can invert the radiographic transformation (see Figure 5.1).

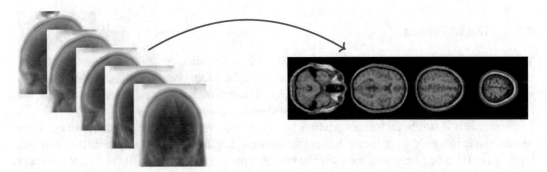

Fig. 5.1 In the radiography exploration, we ask, "Given this radiograph, what does the brain that produced it look like?"

Now, we have also found that invertible transformations are both one-to-one and onto, but radiographic transformations are not expected to have these properties in any practical scenario.

In this chapter, we explore the matrix representations of transformations and the corresponding vector spaces (domain and codomain). We need to understand the properties of these vector spaces (brain image space and radiograph space) in order to fully appreciate the complexity of our tomography goal. This study leads us to understand a variety of properties of invertible transformations and invertible matrices. In the final section, we will perform our first tomographic reconstructions using a left-inverse matrix representation of our radiographic transformation.

© Springer Nature Switzerland AG 2022, corrected publication 2023
H. A. Moon et al., *Application-Inspired Linear Algebra*, Springer Undergraduate Texts
in Mathematics and Technology, https://doi.org/10.1007/978-3-030-86155-1_5

5.1 Transformation Spaces

We have come a long way in understanding the radiography/tomography problem. We understand that brain images and radiographs can be viewed as vectors in their respective vector spaces with associated arithmetic. We have found that these vectors can be described by linear combinations of a smaller subset of vectors. We are able to efficiently define subspaces of vectors using spanning sets or bases. We have gained an understanding of the radiographic transformation that summarizes the physical process of imaging a brain to produce a radiograph. We found that the process, even though described by a linear operation with simple properties, was prone to having both invisible objects and radiographs that are not possible. And then our first attempts at determining which brain image produces a given radiograph have ended in some disappointment. In fact, for a transformation to be invertible, it must be between vector spaces of equal dimension—a condition that we do not expect in any practical situation. But then we noticed that all we really need is for our transformation to have a left inverse. Unfortunately, we found that only one-to-one transformations have left inverses—again a condition which we do not expect. Finally, even if our transformation is onto, this does not guarantee that we can recover the brain image that produced our radiograph.

However, we also made the strong connection between transformations and their matrix representations. Any properties of the linear radiographic transformation T are mimicked in the properties of the corresponding matrix representation $M = [T]_{\mathcal{O}}^{\mathcal{R}}$. Is there information in M that can shed light on more useful properties of T? We are excited to offer the encouraging answer, "Yes, there is useful information in M to shed light on our problem!"

In this section, we consider the radiographic transformation restricted to subspaces to obtain an invertible transformation. The remaining catch is that we still need these two subspaces to be sufficiently rich and descriptive for practical tomography. It will take us several more sections to explore this idea of practicality.

5.1.1 The Nullspace

We begin our exploration by considering objects invisible to the radiographic transformation. Any one-to-one transformation cannot have invisible objects (except the zero vector). That is, whatever domain subspace we choose cannot contain invisible objects. This approach may seem counterproductive, but right now, we need to more deeply understand our transformation.

We saw that if two brain images produce the same radiograph, then the difference image is an invisible brain image. More to the point, from the radiograph itself, there is no way to determine which brain image is the best one—and there may be infinitely many to choose from. If the difference image contains information about a brain abnormality then our choice is critical from a medical standpoint. We will explore these invisible objects in a general vector space setting.

Definition 5.1.1

Let V and W be vector spaces. The **nullspace** of a linear transformation, $T : V \to W$, written Null(T), is the subset of V that maps, through T, to $0 \in W$. That is,

$$\text{Null}(T) = \{v \in V \mid T(v) = 0\}.$$

$\square$

The nullspace of a transformation contains all of the domain vectors that map to the zero vector in the codomain under the transformation. These vectors are invisible to the transformation.

Example 5.1.2 Define $T : \mathbb{R}^3 \to \mathbb{R}^2$ by

$$T\begin{pmatrix} x \\ y \\ z \end{pmatrix} = \begin{pmatrix} x+z \\ y+z \end{pmatrix}.$$

This transformation is linear: if

$$\begin{pmatrix} x \\ y \\ z \end{pmatrix}, \begin{pmatrix} r \\ s \\ t \end{pmatrix} \in \mathbb{R}^3$$

and $\alpha \in \mathbb{R}$, then

$$\begin{aligned}
T\left(\alpha \begin{pmatrix} x \\ y \\ z \end{pmatrix} + \begin{pmatrix} r \\ s \\ t \end{pmatrix} \right) &= T \begin{pmatrix} \alpha x + r \\ \alpha y + s \\ \alpha z + t \end{pmatrix} \\
&= \begin{pmatrix} (\alpha x + r) + (\alpha z + t) \\ (\alpha y + s) + (\alpha z + t) \end{pmatrix} \\
&= \begin{pmatrix} (\alpha x + \alpha z) + (r + t) \\ (\alpha y + \alpha z) + (s + t) \end{pmatrix} \\
&= \begin{pmatrix} \alpha x + \alpha z \\ \alpha y + \alpha z \end{pmatrix} + \begin{pmatrix} r + t \\ s + t \end{pmatrix} \\
&= \alpha \begin{pmatrix} x + z \\ y + z \end{pmatrix} + \begin{pmatrix} r + t \\ s + t \end{pmatrix} \\
&= \alpha T \begin{pmatrix} x \\ y \\ z \end{pmatrix} + T \begin{pmatrix} r \\ s \\ t \end{pmatrix}.
\end{aligned}$$

We now examine the nullspace of T:

$$\begin{aligned}
\mathrm{Null}(T) &= \{ v \in \mathbb{R}^3 \mid T(v) = 0 \} \\
&= \left\{ \begin{pmatrix} x \\ y \\ z \end{pmatrix} \mid x + z = y + z = 0, x, y, z \in \mathbb{R} \right\} \\
&= \left\{ \begin{pmatrix} x \\ y \\ z \end{pmatrix} \mid x = -z, y = -z, z \in \mathbb{R} \right\} \\
&= \left\{ \begin{pmatrix} -z \\ -z \\ z \end{pmatrix} \mid z \in \mathbb{R} \right\} \\
&= \left\{ \alpha \begin{pmatrix} 1 \\ 1 \\ -1 \end{pmatrix} \mid \alpha \in \mathbb{R} \right\} \\
&= \mathrm{Span} \left\{ \begin{pmatrix} 1 \\ 1 \\ -1 \end{pmatrix} \right\}.
\end{aligned}$$

$\square$

The name "nullspace" implies the set is actually a subspace. Indeed, we see that the nullspace in Example 5.1.2 above is a subspace of the domain $\mathbb{R}^3$. We claim this general result in the next theorem.

Theorem 5.1.3
Let V and W be vector spaces and $T : V \to W$ be linear. Then, $\text{Null}(T)$ is a subspace of V.

Proof. Suppose V and W are vector spaces and $T : V \to W$ is linear. We will show that $\text{Null}(T)$ is not empty and that it is closed under vector addition and scalar multiplication.

First, $\text{Null}(T)$ contains $0 \in V$ since T is linear and $T(0) = 0$. Next, consider $u, v \in \text{Null}(T)$ and scalar α. We have $T(\alpha u + v) = \alpha T(u) + T(v) = \alpha(0) + 0 = 0$. So, $\alpha u + v \in \text{Null}(T)$ and by Corollary 2.5.6, $\text{Null}(T)$ is closed under vector addition and scalar multiplication. Therefore, by Theorem 2.5.3 $\text{Null}(T)$ is a subspace of V. $\square$

Definition 5.1.4

Let V and W be vector spaces and $T : V \to W$ linear. The **nullity** of T, written $\text{Nullity}(T)$, is equal to $\dim(\text{Null}(T))$. $\square$

The nullity of a transformation is just the dimension of the nullspace of the transformation. Most English nouns that end in "ity" indicate a quality or degree of something: capacity, purity, clarity. Nullity is no exception. In some sense, it measures the variety present in the nullspace of a transformation. The larger the nullity, the richer the set of null objects. For example, a radiographic transformation with a large nullity will have a greater variety of invisible objects.

Example 5.1.5 Let $V \subseteq \mathcal{P}_2$ be defined by $V = \{ax^2 + (3a - 2b)x + b \mid a, b \in \mathbb{R}\}$. Define the linear transformation $\mathcal{F} : V \to \mathcal{P}_1$ by

$$\mathcal{F}(ax^2 + (3a - 2b)x + b) = 2ax + 3a - 2b.$$

Let us now find the nullspace of $\mathcal{F}$. Using the definition, we have

$$
\begin{aligned}
\text{Null}(\mathcal{F}) &= \{v \in V \mid \mathcal{F}(v) = 0\} \\
&= \{ax^2 + (3a - 2b)x + b \mid 2ax + 3a - 2b = 0\} \\
&= \{ax^2 + (3a - 2b)x + b \mid a = 0, b = 0\} \\
&= \{0\}.
\end{aligned}
$$

In this case, $\text{Nullity}(\mathcal{F}) = 0$. $\square$

When the nullity of a transformation, T, is zero, we say that the transformation has a trivial nullspace, or that the nullspace is the trivial subspace, $\{0\}$.

Example 5.1.6 Define $h : V \to \mathcal{P}_1$, where

$$V = \left\{ \begin{pmatrix} a & b & c \\ 0 & b-c & 2a \end{pmatrix} \;\middle|\; a, b, c \in \mathbb{R} \right\} \subseteq \mathcal{M}_{2\times 3}$$

by

$$h \begin{pmatrix} a & b & c \\ 0 & b-c & 2a \end{pmatrix} = ax + c.$$

Transformation h is linear (which the reader can verify). By definition,

$$
\begin{aligned}
\text{Null}(h) &= \{ v \in V \mid h(v) = 0 \} \\
&= \left\{ \begin{pmatrix} a & b & c \\ 0 & b-c & 2a \end{pmatrix} \;\middle|\; h \begin{pmatrix} a & b & c \\ 0 & b-c & 2a \end{pmatrix} = 0;\; a, b, c \in \mathbb{R} \right\} \\
&= \left\{ \begin{pmatrix} a & b & c \\ 0 & b-c & 2a \end{pmatrix} \;\middle|\; ax + c = 0;\; a, b, c \in \mathbb{R} \right\} \\
&= \left\{ \begin{pmatrix} a & b & c \\ 0 & b-c & 2a \end{pmatrix} \;\middle|\; a = 0, c = 0, b \in \mathbb{R} \right\} \\
&= \left\{ \begin{pmatrix} 0 & b & 0 \\ 0 & b & 0 \end{pmatrix} \;\middle|\; b \in \mathbb{R} \right\} \\
&= \text{Span} \left\{ \begin{pmatrix} 0 & 1 & 0 \\ 0 & 1 & 0 \end{pmatrix} \right\}.
\end{aligned}
$$

In this case Nullity$(h) = 1$ because there is one element in the basis for the nullspace of h. $\qquad\square$

Example 5.1.7 Consider the transformation from the vector space of six-bin histograms to the vector space of three-bin histograms, $T : \mathcal{J}_6(\mathbb{R}) \to \mathcal{J}_3(\mathbb{R})$, defined as the operation of bin pair summation described as follows. Suppose $J \in \mathcal{J}_6(\mathbb{R})$ and $K \in \mathcal{J}_3(\mathbb{R})$. If $K = T(J)$ then the first bin of K has value equal to the sum of the first two bins of J, the second bin of K has value equal to the sum of bins three and four of J, and the third bin of K has value equal to the sum of bins five and six of J. The nullspace of T is set of all $J \in \mathcal{J}_6(\mathbb{R})$, which map to the zero histogram in $\mathcal{J}_3(\mathbb{R})$. Let the ordered bin values of J be $\{b_1, b_2, b_3, b_4, b_5, b_6\}$. Then

$$\text{Null}(T) = \{ J \in \mathcal{J}_6(\mathbb{R}) \mid b_1 + b_2 = b_3 + b_4 = b_5 + b_6 = 0 \}.$$

The nullity of T is 3 because a basis for Null(T) contains three vectors. $\qquad\square$

Example 5.1.8 Consider the linear transformation $T : D(Z_2) \to D(Z_2)$, on the space of 7-bar LCD characters, defined by $T(d) = d + d$. As we have seen previously, T maps every input vector to the zero vector. So, Null$(T) = D(Z_2)$ and

$$\text{nullity}(T) = \dim \text{Null}(T) = \dim D(Z_2) = 7.$$

Any basis for $D(Z_2)$ is also a basis for Null(T). $\qquad\square$

Example 5.1.9 Consider the radiographic scenario of Figure **??**. Recall, the transformation, T is described by

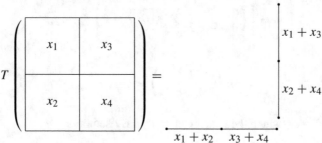

In looking for a nonzero null vector, suppose $x_1 = a \neq 0$. Then we see that $x_2 = -a$, $x_3 = a$, and $x_4 = -a$. Using similar arguments with all object voxels, we see that

$$\text{Null}(T) = \text{Span} \left\{ \begin{array}{|c|c|} \hline 1 & -1 \\ \hline -1 & 1 \\ \hline \end{array} \right\}.$$

We see that $\text{Nullity}(T) = 1$. $\qquad\qquad\qquad\qquad\qquad\qquad\qquad\qquad\qquad\qquad\qquad\qquad\quad\square$

Example 5.1.10 Consider the linear transformation $T : \mathcal{H}_4(\mathbb{R}) \to \mathcal{H}_4(\mathbb{R})$, on heat states of four values, defined as the averaging function of Exercise 12 of Section **??**. T replaces each heat state value (which does not correspond to an endpoint) with the average of its neighbors. The endpoint values are kept at 0. An example of a heat state h and the result of this transformation $T(h)$ is shown in Figure 5.2. The nullspace of T is defined as $\text{Null}(T) = \{h \in \mathcal{H}_4(\mathbb{R}) \mid T(h) = 0\}$. It is straightforward to show that $h = 0$ (the zero vector in $\mathcal{H}_4(\mathbb{R})$), is the only heat state which satisfies the criterion $T(h) = 0$. Thus, $\text{Null}(T) = \{0\}$ and $\text{Nullity}(T) = 0$. $\qquad\qquad\qquad\qquad\qquad\qquad\qquad\qquad\square$

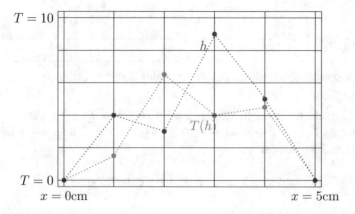

Fig. 5.2 Example of averaging heat state transformation.

5.1.2 Domain and Range Spaces

When considering a transformation, we want to know which vectors can be applied to the transformation. In the case of a radiographic transformation, we wonder what is the shape and size of brain images that the particular radiographic transformation uses. As with most functions, this set is called the **domain**. In linear algebra, we consider only sets that are vector spaces. So, it is often referred to as the **domain space**. If $T : V \to W$, we call V the domain space. There is also an ambient space to which all of the vectors in the domain space map. In the case of a radiographic transformation, this space contains images that satisfy the definition of a radiograph. We say that the **codomain** of a linear transformation, $T : V \to W$, is the ambient vector space W to which domain vectors map. These definitions were first introduced in Section **??**.

In Examples 5.1.5 and 5.1.6, the codomain is $\mathcal{P}_1$. In Example 5.1.9, the codomain is the space of 4-value grayscale radiographs defined in Figure **??**. We have seen that not all transformations map onto their codomain—not all codomain vectors $b \in W$ can be associated with a domain vector $x \in V$ such that $T(x) = b$. In the radiographic sense, this means that not all radiographs are possible. Typically in applications, the set of all possible codomain vectors is more useful than the codomain itself.

Definition 5.1.11

We say that the **range space** of a linear transformation $T : V \to W$, written $\text{Ran}(T)$, is the subset $T(V) \subseteq W$. That is,

$$\text{Ran}(T) = \{T(v) \mid v \in V\}.$$

$\square$

Notice that, by name, we imply that the range space is a vector space and, in fact, it is.

Theorem 5.1.12
Let V and W be vector spaces and let $T : V \to W$ be a linear transformation. Then $\text{Ran}(T)$ is a subspace of W.

The proof is the subject of Exercise 29. See also the proof of Theorem 5.1.3.

As the dimension of the nullspace of a transformation was given a special name, *nullity*, the dimension of the range space is also given the special name, *rank*.

Definition 5.1.13

Let V and W be vector spaces and $T : V \to W$ linear. We define the **rank** of the transformation as the dimension of the range space and write

$$\text{Rank}(T) = \dim \text{Ran}(T).$$

$\square$

In both Examples 5.1.5 and 5.1.6, the range is equal to the codomain, $\mathcal{P}_1(\mathbb{R})$. In both of these examples, the rank of the transformation is $\dim \mathcal{P}_1(\mathbb{R}) = 2$. Let us use the definition of range to verify this for Example 5.1.5. We will leave verification for Example 5.1.6 for the reader. Recall, the definition of $\mathcal{F}$ from Example 5.1.5 gives us that

$$
\begin{aligned}
\mathrm{Ran}(\mathcal{F}) &= \{\mathcal{F}(v) \mid v \in \mathcal{P}_2(\mathbb{R})\} \\
&= \{2ax + 2a - 2b \mid a, b \in \mathbb{R}\} \\
&= \{2a(x+1) - 2b(1) \mid a, b \in \mathbb{R}\} \\
&= \mathrm{Span}\{x+1, 1\} \\
&= \mathcal{P}_2(\mathbb{R}).
\end{aligned}
$$

In general, the range need not equal the codomain. Let's consider several examples.

Example 5.1.14 Define $f : \mathcal{M}_{2\times 2}(\mathbb{R}) \to \mathbb{R}^4$ by

$$
f\begin{pmatrix} a & b \\ c & d \end{pmatrix} = \begin{pmatrix} a \\ b+a \\ b \\ c \end{pmatrix}.
$$

Determine $\mathrm{Ran}(f)$ and $\mathrm{Null}(f)$. First we find the range.

$$
\begin{aligned}
\mathrm{Ran}(f) &= \{f(v) \mid v \in \mathcal{M}_{2\times 2}(\mathbb{R})\} \\
&= \left\{ f\begin{pmatrix} a & b \\ c & d \end{pmatrix} \;\middle|\; a, b, c, d \in \mathbb{R} \right\} \\
&= \left\{ \begin{pmatrix} a \\ b+a \\ b \\ c \end{pmatrix} \;\middle|\; a, b, c \in \mathbb{R} \right\} \\
&= \left\{ a\begin{pmatrix} 1 \\ 1 \\ 0 \\ 0 \end{pmatrix} + b\begin{pmatrix} 0 \\ 1 \\ 1 \\ 0 \end{pmatrix} + c\begin{pmatrix} 0 \\ 0 \\ 0 \\ 1 \end{pmatrix} \;\middle|\; a, b, c \in \mathbb{R} \right\} \\
&= \mathrm{Span}\left\{ \begin{pmatrix} 1 \\ 1 \\ 0 \\ 0 \end{pmatrix}, \begin{pmatrix} 0 \\ 1 \\ 1 \\ 0 \end{pmatrix}, \begin{pmatrix} 0 \\ 0 \\ 0 \\ 1 \end{pmatrix} \right\}.
\end{aligned}
$$

Since $\left\{ \begin{pmatrix} 1 \\ 1 \\ 0 \\ 0 \end{pmatrix}, \begin{pmatrix} 0 \\ 1 \\ 1 \\ 0 \end{pmatrix}, \begin{pmatrix} 0 \\ 0 \\ 0 \\ 1 \end{pmatrix} \right\}$ is a linearly independent spanning set, we have $\mathrm{Rank}(f) = 3$. Now we compute the nullspace.

$$\text{Null}(f) = \{v \in \mathcal{M}_{2\times 2} \mid f(v) = 0\}$$

$$= \left\{ \begin{pmatrix} a & b \\ c & d \end{pmatrix} \;\middle|\; a, b, c, d \in \mathbb{R} \text{ and } f\begin{pmatrix} a & b \\ c & d \end{pmatrix} = 0 \right\}$$

$$= \left\{ \begin{pmatrix} a & b \\ c & d \end{pmatrix} \;\middle|\; a, b, c, d \in \mathbb{R} \text{ and } \begin{pmatrix} a \\ b+a \\ b \\ c \end{pmatrix} = 0 \right\}$$

$$= \left\{ \begin{pmatrix} a & b \\ c & d \end{pmatrix} \;\middle|\; a, b, c = 0, d \in \mathbb{R} \right\}$$

$$= \left\{ \begin{pmatrix} 0 & 0 \\ 0 & d \end{pmatrix} \;\middle|\; d \in \mathbb{R} \right\}$$

$$= \text{Span}\left\{ \begin{pmatrix} 0 & 0 \\ 0 & 1 \end{pmatrix} \right\}.$$

Thus Nullity$(f) = 1$. $\qquad\qquad\square$

In this example, the codomain is $\mathbb{R}^4$ and Ran$(f) \neq \mathbb{R}^4$ that means there are elements in $\mathbb{R}^4$ that are not mapped to through f. That is, f is not onto.

And, there is more than one element in the nullspace. In particular, notice that

$$f\begin{pmatrix} 0 & 0 \\ 0 & 1 \end{pmatrix} = f\begin{pmatrix} 0 & 0 \\ 0 & 2 \end{pmatrix} = \begin{pmatrix} 0 \\ 0 \\ 0 \\ 0 \end{pmatrix}.$$

But,

$$\begin{pmatrix} 0 & 0 \\ 0 & 1 \end{pmatrix} \neq \begin{pmatrix} 0 & 0 \\ 0 & 2 \end{pmatrix}.$$

Thus, f is not one-to-one.

Since the range of a transformation is the set of all vectors in the codomain that can be mapped to by some domain vector, we can express it in terms of a basis of the domain. This can be a very useful tool for finding a spanning set for the range.

Theorem 5.1.15

Let V and W be finite-dimensional vector spaces and $T : V \rightarrow W$ be linear. Suppose $\mathcal{B}$ is a basis for V, then Span $T(\mathcal{B}) = \text{Ran}(T)$.

Proof. Let V and W be finite-dimensional vector spaces and let $T : V \rightarrow W$ be a linear transformation. Suppose $\mathcal{B} = \{v_1, v_2, \ldots, v_n\}$ is a basis for V and $T(v_k) = w_k$ for $k = 1, 2, \ldots, n$. For any vector, $u \in V$ we can write $u = \alpha_1 v_1 + \alpha_2 v_2 + \ldots + \alpha_n v_n$ for some scalars $\alpha_1, \alpha_2, \ldots, \alpha_n$. Then,

$$\text{Ran}(T) = \{T(x) \mid x \in V\}$$
$$= \{T(\alpha_1 v_1 + \alpha_2 v_2 + \ldots + \alpha_n v_n) \mid a_k \in \mathbb{R}, k = 1, 2, \ldots, n\}$$
$$= \{\alpha_1 w_1 + \alpha_2 w_2 + \ldots + \alpha_n w_n \mid a_k \in \mathbb{R}, k = 1, 2, \ldots, n\}$$
$$= \text{Span}\{w_1, w_2, \ldots, w_n\}$$
$$= \text{Span}\ T(\mathcal{B}).$$

$\square$

It is important to note that the spanning set, $T(\mathcal{B})$, in the proof of Theorem 5.1.15 need not be linearly independent.

Example 5.1.16 Consider the histogram transformation T of Example 5.1.7. The range of T is the space of all three-bin histograms, which can be obtained as bin-pair sums of six-bin histograms. We can use Theorem 5.1.15 to find a spanning set for $\text{Ran}(T)$, then use the method of spanning set reduction (see page **??**) to find a basis for $\text{Ran}(T)$.

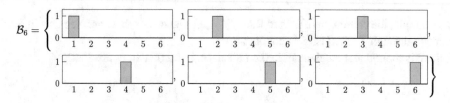

is a basis for $\mathcal{J}_6(\mathbb{R})$. Applying the transformation to each basis vector yields

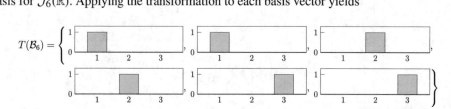

The set $T(\mathcal{B}_6)$ is linearly dependent. We can extract a basis for $(J)_3(\mathbb{R})$ as a subset of $T(\mathcal{B})$. The basis is

$$\mathcal{B}_3 = \left\{ \ldots \right\}.$$

We have $\text{Ran}(T) = \text{Span}\ T(\mathcal{B}_6) = \text{Span}\ \mathcal{B}_3$ and $\text{Rank}(T) = \dim \text{Ran}(T) = 3$. $\square$

Example 5.1.17 Consider the 7-bar LCD transformation T of Example 5.1.8. The range of T is the space of all 7-bar LCD characters that are the result of adding a 7-bar LCD character to itself. In this case, only the zero vector satisfies this requirement. Thus, $\text{Ran}(T) = \{0\}$, $\text{Ran}(T) = \text{Span}\ \emptyset$, and $\text{Rank}(T) = 0$. $\square$

Example 5.1.18 Consider the heat state transformation T of Example 5.1.10. The range of T is the space of all 4-value heat states that are the result of the averaging operation T on a 4-value heat state. We can use Theorem 5.1.15 to find a spanning set for $\text{Ran}(T)$.

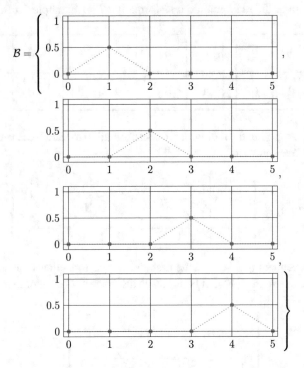

is a basis for $\mathcal{H}_4(\mathbb{R})$. Applying the averaging transformation to each basis vector yields

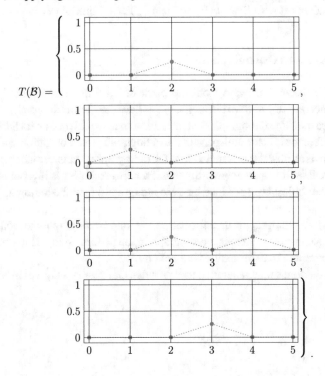

The reader can verify that $T(\mathcal{B})$ is linearly independent and therefore is also a basis for $\mathcal{H}_4(\mathbb{R})$. We have $\mathrm{Ran}(T) = \mathrm{Span}\, T(\mathcal{B}) = \mathcal{H}_4(\mathbb{R})$ and $\mathrm{Rank}(T) = \dim \mathrm{Ran}(T) = 4$. $\square$

Example 5.1.19 Consider the radiographic transformation T of Example 5.1.9.

$$
\mathcal{B} = \left\{ \begin{array}{|c|c|} \hline 1 & 0 \\ \hline 0 & 0 \\ \hline \end{array}, \begin{array}{|c|c|} \hline 0 & 0 \\ \hline 1 & 0 \\ \hline \end{array}, \begin{array}{|c|c|} \hline 0 & 1 \\ \hline 0 & 0 \\ \hline \end{array}, \begin{array}{|c|c|} \hline 0 & 0 \\ \hline 0 & 1 \\ \hline \end{array} \right\}.
$$

is a basis for $\mathcal{I}_{2\times 2}(\mathbb{R})$. Applying the radiographic transformation to each basis vector yields a spanning set for the range of T. We can write

$$
T(\mathcal{B}) = \left\{ \begin{array}{c} 1 \\ 0 \\ \hline 1 \quad 0 \end{array}, \begin{array}{c} 0 \\ 1 \\ \hline 1 \quad 0 \end{array}, \begin{array}{c} 1 \\ 0 \\ \hline 0 \quad 1 \end{array}, \begin{array}{c} 0 \\ 1 \\ \hline 0 \quad 1 \end{array} \right\}.
$$

This set is linearly dependent (the sum of the first and last vectors equals the sum of the second and third vectors). A basis for $\mathrm{Span}\, T(\mathcal{B})$ can be found as a subset of $T(\mathcal{B})$. For example,

$$
\mathcal{C} = \left\{ \begin{array}{c} 1 \\ 0 \\ \hline 1 \quad 0 \end{array}, \begin{array}{c} 0 \\ 1 \\ \hline 1 \quad 0 \end{array}, \begin{array}{c} 1 \\ 0 \\ \hline 0 \quad 1 \end{array} \right\} \subseteq T(\mathcal{B})
$$

is a linearly independent set with $\mathrm{Span}\, T(\mathcal{B}) = \mathrm{Span}\, \mathcal{C}$. Thus, $\mathrm{Rank}(T) = 3$. $\square$

5.1.3 One-to-One and Onto Revisited

We can continue this discussion again from the point of view of radiography. We saw that some transformations are not one-to-one (two different objects have the same radiograph). Also, we found that if two objects produce the same radiograph, that their difference would then be invisible. Another way to say this is that the difference is in the nullspace of the radiographic transformation. Since the nullspace is a vector space, if there is an object that is invisible to the radiographic transformation, any scalar multiple of it will also be invisible. It is also noteworthy that if a nonzero object is invisible (meaning both the zero object and another object both produce the zero radiograph) then the radiographic transformation is not one-to-one.

Recall that, for a given radiographic transformation, we found radiographs that could not be produced from any object. This means that there is a radiograph in the codomain that is not mapped to from the domain. These radiographic transformations are not onto.

We now state the theorems that generalize these results. Our first result gives a statement equivalent to being onto.

> **Theorem 5.1.20**
> Let V and W be vector spaces and let $T : V \to W$ be a linear transformation. T is onto if and only if $\text{Ran}(T) = W$.

Proof. Let V and W be vector spaces and let $T : V \to W$ be linear. Suppose $\text{Ran}(T) = W$ then, by definition of $\text{Ran}(T)$, if $w \in W$, there is a $v \in V$ so that $f(v) = w$. Thus T is onto. Now, if T is onto, then for all $w \in W$ there is a $v \in V$ so that $T(v) = w$. That means that $W \subseteq \text{Ran}(T)$. But, by definition of T and $\text{Ran}(T)$, we already know that $\text{Ran}(T) \subseteq W$. Thus, $\text{Ran}(T) = W$. $\qquad\square$

Next, we give an equivalent statement for one-to-one transformations.

> **Theorem 5.1.21**
> Let V and W be vector spaces. A linear transformation, $T : V \to W$, is one-to-one if and only if $\text{Null}(T) = \{0\}$.

Proof. Let V and W be vector spaces and let $T : V \to W$ be a linear transformation. Suppose T is one-to-one and suppose that $u \in \text{Null}(T)$. Then $T(u) = 0$. But, $T(0) = 0$. So, since T is one-to-one, we know that $u = 0$. Thus, $\text{Null}(T) = \{0\}$. Now, suppose $\text{Null}(T) = \{0\}$. We want to show that T is one-to-one. Notice that if $u, v \in V$ satisfy

$$T(u) = T(v)$$

then

$$T(u) - T(v) = 0.$$

But since T is linear this gives us that

$$T(u - v) = 0.$$

Thus, $u - v \in \text{Null}(T)$. But $\text{Null}(T) = \{0\}$. Thus, $u - v = 0$. That is, $u = v$. So, T is one-to-one. $\quad\square$

> **Corollary 5.1.22**
> A linear transformation, $T : V \to \text{Ran}(T)$ is an isomorphism if and only if $\text{Null}(T) = \{0\}$.

Proof. First, suppose T is an isomorphism. Then, by Theorem 5.1.21, $\text{Null}(T) = \{0\}$. Next, suppose $\text{Null}(T) = \{0\}$. Then, by Theorem 5.1.21, T is one-to-one. And by Theorem 5.1.20, T is onto. Thus, T is an isomorphism. $\qquad\square$

Theorems 5.1.20 and 5.1.21 and Corollary 5.1.22 give us tools to determine whether a transformation is one-to-one and/or onto. Let's consider several examples.

Example 5.1.23 Consider again, the example with $V = \{ax^2 + (3a - 2b)x + b \mid a, b \in \mathbb{R}\} \subseteq \mathcal{P}_2$ and Define $\mathcal{F} : V \to \mathcal{P}_1$, defined by

$$\mathcal{F}(ax^2 + (3a - 2b)x + b) = 2ax + 3a - 2b.$$

We showed in Example 5.1.5 that $\text{Null}(\mathcal{F}) = \{0\}$. Thus $\mathcal{F}$ is one-to-one. We also saw that $\text{Ran}(\mathcal{F}) = \mathcal{P}_1$. Thus, the range and codomain of $\mathcal{F}$ are the same. And, so we know $\mathcal{F}$ is onto. But now we know that $\mathcal{F}$ is an isomorphism. This means that $V \cong \mathcal{P}_1$. And, $\dim V = 2$, $\text{Nullity}(\mathcal{F}) = 0$, and $\text{Rank}(\mathcal{F}) = 2$. $\square$

Example 5.1.24 Define $h : V \to \mathcal{P}_1$, where

$$V = \left\{ \begin{pmatrix} a & b & c \\ 0 & b-c & 2a \end{pmatrix} \;\middle|\; a, b, c \in \mathbb{R} \right\} \subseteq \mathcal{M}_{2 \times 3}$$

by

$$h \begin{pmatrix} a & b & c \\ 0 & b-c & 2a \end{pmatrix} = ax + c.$$

We found that

$$\text{Null}(h) = \text{Span} \left\{ \begin{pmatrix} 0 & 1 & 0 \\ 0 & 1 & 0 \end{pmatrix} \right\}.$$

Thus, h is not one-to-one. But, we also noted (again, be sure you know how to show this) that $\text{Ran}(h) = \mathcal{P}_1$. Thus, h is onto. And, $\dim V = 3$, $\text{Nullity}(h) = 1$, and $\text{Rank}(h) = 2$. $\square$

Example 5.1.25 Define $g : V \to \mathbb{R}^3$, where $V = \mathcal{P}_1$ by

$$g(ax + b) = \begin{pmatrix} a \\ b \\ a+b \end{pmatrix}.$$

Notice that

$$\text{Null}(g) = \{ax + b \mid a, b \in \mathbb{R}, g(ax + b) = 0\}$$

$$= \left\{ ax + b \;\middle|\; a, b \in \mathbb{R}, \begin{pmatrix} a \\ b \\ a+b \end{pmatrix} = \begin{pmatrix} 0 \\ 0 \\ 0 \end{pmatrix} \right\}$$

$$= \{ax + b \mid a = 0, b = 0\}$$

$$= \{0\}.$$

Thus, g is one-to-one. Now we find the range space.

$$\text{Ran}(g) = \{g(ax + b) \mid a, b \in \mathbb{R}\}$$

$$= \left\{ \begin{pmatrix} a \\ b \\ a + b \end{pmatrix} \,\middle|\, a, b \in \mathbb{R} \right\}$$

$$= \text{Span} \left\{ \begin{pmatrix} 1 \\ 0 \\ 1 \end{pmatrix}, \begin{pmatrix} 0 \\ 1 \\ 1 \end{pmatrix} \right\}.$$

Since Rank$(g)=2$ and dim $\mathbb{R}^3=3$, $\mathbb{R}^3 \neq \text{Ran}(g)$ and thus g is not onto. And, dim $V = 2$, Nullity$(g) = 0$, and Rank$(g) = 2$. $\qquad\square$

Example 5.1.26 Consider the histogram rebinning transformation of Examples 5.1.7 and 5.1.16. We found that Nullity$(T) = 3$, Ran$(T) = \mathcal{H}_3(\mathbb{R})$, and Rank$(T) = 3$. Thus, T is not one-to-one but T is onto. $\qquad\square$

Example 5.1.27 Consider the LCD character transformation of Examples 5.1.8 and 5.1.17. We found that Nullity$(T) = 7$, Ran$(T) = \{0\}$, and Rank$(T) = 0$. Thus, T is neither one-to-one nor onto. $\qquad\square$

Example 5.1.28 Consider the Heat State averaging transformation of Examples 5.1.10 and 5.1.18. We found that Nullity$(T) = 0$, Ran$(T) = H_4(\mathbb{R})$, and Rank$(T) = 4$. Thus, T is an isomorphism. $\qquad\square$

Example 5.1.29 Consider the radiographic transformation of Examples 5.1.9 and 5.1.19. We found that Nullity$(T) = 1$. Also, Rank$(T) = 3$ while the dimension of the radiograph space is 4. Thus, T is neither one-to-one nor onto. $\qquad\square$

5.1.4 The Rank-Nullity Theorem

In each of the last examples of the previous section, you will notice a simple relationship: the dimension of the nullspace and the dimension of the range space add up to the dimension of the domain. This is not a coincidence. In fact, it makes sense if we begin putting our theorems together.

> **Theorem 5.1.30 (Rank-Nullity)**
> Let V and W be finite-dimensional vector spaces and let $T : V \to W$ be a linear transformation. Then
> $$\dim V = \text{Rank}(T) + \text{Nullity}(T).$$

Proof. Let V and W be finite-dimensional vector spaces and let $T : V \to W$ be linear. Let $\mathcal{B} = \{v_1, v_2, \ldots, v_n\}$ be a basis for V. We will consider the case when Ran(T) contains only the zero vector and the case when Null(T) contains only the zero vector separately. This is to remove doubt that may arise when considering empty bases for these subspaces.

First, we consider the case when Ran$(T) = \{0\}$. Then, a basis for Ran(T) is the empty set, so rank$(T) = 0$. We also know that if $v \in V$ then $T(v) = 0$. So, $T(\mathcal{B}) = \{0\}$. Thus, $\mathcal{B} \subseteq \text{Null}(T)$ is a basis for the nullspace of T and Nullity$(T) = n$. Thus, rank$(T) + \text{Nullity}(T) = n = \dim V$.

Next, we consider the case when $\text{Null}(T) = \{0\}$. In this case, the basis for $\text{Null}(T)$ is the empty set and $\text{nullity}(T) = 0$. Now, we refer to Theorems 4.7.19, 5.1.15, and 5.1.21. We then know that $\{T(v_1), T(v_2), \ldots, T(v_n)\}$ is linearly independent and we also know that $\text{Span}\{T(v_1), T(v_2), \ldots, T(v_n)\} = \text{Ran}(T)$. Thus, $\{T(v_1), T(v_2), \ldots, T(v_n)\}$ is a basis for $\text{Ran}(T)$ and $\text{Rank}(T) = n$. Thus, $\text{Rank}(T) + \text{Nullity}(T) = n$.

Finally, we consider the case where $\text{Rank}(T) = m \geq 1$ and $\text{Nullity}(T) = k \geq 1$. Let

$$\mathcal{B}_N = \{\tilde{v}_1, \tilde{v}_2, \ldots, \tilde{v}_k\}$$

be a basis for $\text{Null}(T)$. Since $\dim V = n \geq k$ (see Corollary 3.4.22), we can add $n - k$ vectors to $\mathcal{B}_N$ that maintain linear independence and therefore form a basis for V. That is, we can create the set

$$\tilde{\mathcal{B}} = \{\tilde{v}_1, \tilde{v}_2, \ldots, \tilde{v}_k, \tilde{v}_{k+1}, \tilde{v}_{k+2}, \ldots, \tilde{v}_n\}$$

so that $\tilde{\mathcal{B}}$ is a basis for V. Then

$$T(\tilde{\mathcal{B}}) = \{T(\mathcal{B}_N), T(\tilde{v}_{k+1}), T(\tilde{v}_{k+2}), \ldots, T(\tilde{v}_n)\} = \{0, T(\tilde{v}_{k+1}), T(\tilde{v}_{k+2}), \ldots, T(\tilde{v}_n)\}.$$

Let $S = \{T(\tilde{v}_{k+1}), T(\tilde{v}_{k+2}), \ldots, T(\tilde{v}_n)\}$. Notice that S has $n - k$ elements in it and that $\text{Span } S = \text{Span } T(\tilde{\mathcal{B}})$. We will now show that S is a basis for $\text{Ran}(T)$. In doing this, we will have shown that $n - k = \text{rank}(T)$, that is, $n - k = m$.

By Theorem 5.1.15, $\text{Ran}(T) = \text{Span } T(\tilde{\mathcal{B}}) = \text{Span } S$. From Lemma 3.4.23, we know that a basis for $\text{Ran}(T)$ is a minimal spanning set for $\text{Ran}(T)$. So $n - k \geq \text{rank}(T) = m$.

Now, we show that S is linearly independent. Let $\alpha_{k+1}, \alpha_{k+2}, \ldots, \alpha_n$ be scalars so that

$$\alpha_{k+1}T(\tilde{v}_{k+1}) + \alpha_{k+2}T(\tilde{v}_{k+2}) + \ldots + \alpha_n T(\tilde{v}_n) = 0.$$

Then, using linearity of T, we have

$$T(\alpha_{k+1}\tilde{v}_{k+1} + \alpha_{k+2}\tilde{v}_{k+2} + \ldots + \alpha_n\tilde{v}_n) = 0.$$

So, we see that $\alpha_{k+1}\tilde{v}_{k+1} + \alpha_{k+2}\tilde{v}_{k+2} + \ldots + \alpha_n\tilde{v}_n$ is in the nullspace of T. But, $\text{Null}(T) = \text{Span } \mathcal{B}_N$. This means that we can describe $\alpha_{k+1}\tilde{v}_{k+1} + \alpha_{k+2}\tilde{v}_{k+2} + \ldots + \alpha_n\tilde{v}_n$ using a linear combination of the basis elements of $\text{Null}(T)$. That is,

$$\alpha_{k+1}\tilde{v}_{k+1} + \alpha_{k+2}\tilde{v}_{k+2} + \ldots + \alpha_n\tilde{v}_n = \beta_1\tilde{v}_1 + \beta_2\tilde{v}_2 + \ldots + \beta_k\tilde{v}_k$$

for some scalars $\beta_1, \beta_2, \ldots, \beta_k$. Rearranging this equation gives us the following linear dependence relation, for the vectors in $\tilde{\mathcal{B}}$,

$$\beta_1\tilde{v}_1 + \beta_2\tilde{v}_2 + \ldots + \beta_k\tilde{v}_k - \alpha_{k+1}\tilde{v}_{k+1} - \alpha_{k+2}\tilde{v}_{k+2} - \ldots - \alpha_n\tilde{v}_n = 0.$$

Since $\tilde{\mathcal{B}}$ is a basis for V, we know that the above equation is true only when

$$\beta_1 = \beta_2 = \ldots = \beta_k = \alpha_{k+1} = \alpha_{k+2} = \ldots = \alpha_n = 0.$$

So, by definition, S is linearly independent. From Corollary 3.4.22, we know that a basis of $\text{Ran}(T)$ is the largest linearly independent set in $\text{Ran}(T)$. That means that $n - k \leq \text{rank}(T) = m$. Putting our two results together, we find that S is a basis for $\text{Ran}(T)$ and $n - k = m$. Rearranging this equation gives us our result: $\text{rank}(T) + \text{nullity}(T) = \dim V$. $\qquad\square$

For a geometric visualization of Theorem 5.1.30, we encourage the reader to recall Figure **??** and then work through Exercise 30. The proof of the Rank-Nullity Theorem shows us that we can create a basis for V that can be separated into a basis for the nullspace and a set that maps to a basis for the range space.

> **Corollary 5.1.31**
> Let V and W be finite-dimensional vector spaces and let $T : V \to W$ be a linear transformation. Let $\mathcal{B}_N = \{v_1, v_2, \ldots, v_k\}$ be a basis for $\text{Null}(T)$. If $\mathcal{B} = \{v_1, v_2, \ldots, v_n\}$ is a basis for V, then $\{T(v_{k+1}), T(v_{k+2}), \ldots, T(v_n)\}$ is a basis for $\text{Ran}(T)$.

The proof follows directly from the proof of Theorem 5.1.30. Theorem 5.1.30 is useful in determining rank and nullity, along with proving results about subspaces. We are often able to determine some properties of a transformation without knowing any more than the dimensions of the vector spaces from which and to which it maps. Let's look at an example.

Example 5.1.32 Given a linear transformation $T : \mathcal{M}_{2\times 5}(\mathbb{R}) \to \mathcal{P}_4(\mathbb{R})$. We know that T cannot be one-to-one. The Rank-Nullity Theorem says that $\dim \mathcal{M}_{2\times 5}(\mathbb{R}) = \text{rank}(T) + \text{nullity}(T)$. Since $\text{rank}(T) \leq \dim \mathcal{P}_4 = 5$ and $\dim \mathcal{M}_{2\times 5} = 10$, we know that $\text{nullity}(T) \geq 5$. That is, $\text{Null}(T) \neq \{0\}$. So by Theorem 5.1.21, we know that T cannot be one-to-one. $\qquad\square$

Again, we have introduced more terminology and therefore, we need to discuss how to use this language properly.

⋆⋆ **Watch Your Language!** The terminology in this section needs care because some terms are names of vector spaces while others are the dimension of these spaces.

Appropriate use of terms from this section are as follows.

✓ The nullspace of T has basis $\{v_1, v_2, \ldots, v_n\}$.
✓ The nullity of T is n.
✓ The range space has basis $\{w_1, w_2, \ldots, w_m\}$.
✓ The rank of T is m.

Inappropriate use of terms from this section are as follows.

✗ The nullity of T has basis $\{v_1, v_2, \ldots, v_n\}$.
✗ The nullity of T has dimension n.
✗ The rank of T has dimension m.
✗ The rank of the basis $\mathcal{B}$ is m.

5.1.5 Exercises

For Exercises 1–16, state the (a) the domain, V, (b) $\dim(V)$, (c) the codomain, W, (d) $\dim(W)$, (e) Range space, (f) rank, (g) nullspace, (h) nullity, of each of the following transformations. And, then (i) verify the Rank-Nullity Theorem.

1. $T : \mathbb{R}^2 \to \mathbb{R}^3$, where $T\begin{pmatrix} x \\ y \end{pmatrix} = \begin{pmatrix} y \\ x \\ x - y \end{pmatrix}$.

2. $T : \mathbb{R}^3 \to \mathbb{R}^4$, where $T\begin{pmatrix} x \\ y \\ z \end{pmatrix} = \begin{pmatrix} x + y \\ z + y \\ 0 \\ 0 \end{pmatrix}$.

3. Define $\mathcal{F} : V \to \mathcal{P}_1$, where

$$V = \{ax^2 + (3a - 2b)x + b \mid a, b \in \mathbb{R}\} \subseteq \mathcal{P}_2.$$

by

$$\mathcal{F}(ax^2 + (3a - 2b)x + b) = 2ax + 3a - 2b.$$

4. Define $\mathcal{G} : \mathcal{P}_2 \to \mathcal{M}_{2\times2}$ by

$$\mathcal{G}(ax^2 + bx + c) = \begin{pmatrix} a & a - b \\ c - 2a & c + 3a \end{pmatrix}.$$

5. Define $h : V \to \mathcal{P}_1$, where

$$V = \left\{ \begin{pmatrix} a & b & c \\ 0 & b - c & 2a \end{pmatrix} \middle| \; a, b, c \in \mathbb{R} \right\} \subseteq \mathcal{M}_{2\times3}$$

by

$$h\begin{pmatrix} a & b & c \\ 0 & b - c & 2a \end{pmatrix} = ax + c.$$

6. Let

$$\mathcal{I} = \left\{ \begin{pmatrix} 3a \\ -b & \quad & 2a \\ & 0 \\ c & & b \\ & 3c \end{pmatrix} \; a, b, c \in \mathbb{R} \right\}.$$

And define $f : \mathcal{I} \to \mathcal{P}_2$ by

$$f(I) = ax^2 + (b + c)x + (a + c).$$

7. Define $f : \mathcal{M}_{2\times 2} \to \mathbb{R}^4$ by

$$f \begin{pmatrix} a & b \\ c & d \end{pmatrix} = \begin{pmatrix} a \\ b \\ c \\ d \end{pmatrix}.$$

8. Define $f : \mathcal{P}_2 \to \mathbb{R}^2$ by

$$f(ax^2 + bx + c) = \begin{pmatrix} a + b \\ a - c \end{pmatrix}.$$

9. $T : V_{\mathcal{O}} \to V_{\mathcal{R}}$, where $V_{\mathcal{O}}$ is the space of objects with 4 voxels and $V_{\mathcal{R}}$ is the space of radiographs with 4 pixels and

$$T\left(\begin{array}{|c|c|} \hline x_1 & x_3 \\ \hline x_2 & x_4 \\ \hline \end{array}\right) = \begin{array}{|c|} \hline x_1 + x_2 \\ \hline x_3 + x_4 \\ \hline \frac{2}{3}x_1 + x_2 + \frac{2}{3}x_4 \\ \hline \frac{1}{3}x_1 + x_3 + \frac{1}{3}x_4 \\ \hline \end{array}.$$

10. $T : V_{\mathcal{O}} \to \mathbb{R}^4$, where $V_{\mathcal{O}}$ is the space of objects with 4 voxels and

$$T\left(\begin{array}{|c|c|} \hline x_1 & x_3 \\ \hline x_2 & x_4 \\ \hline \end{array}\right) = \left[\begin{array}{|c|c|} \hline x_1 & x_3 \\ \hline x_2 & x_4 \\ \hline \end{array}\right]_{\mathcal{B}},$$

where $\mathcal{B}$ is the standard basis for $V_{\mathcal{O}}$.

11. $T : \mathcal{P}_2(\mathbb{R}) \to \mathcal{P}_2(\mathbb{R})$ defined by $T(ax^2 + bx + c) = cx^2 + ax + b$

12. $T : \mathcal{P}_2(\mathbb{R}) \to \mathcal{P}_2(\mathbb{R})$ defined by $T(ax^2 + bx + c) = (a + b)x^2 - b + c$.

13. $T : \mathcal{P}(\mathbb{R}) \to \mathcal{P}(\mathbb{R})$ defined by $T(p(x)) = p'(x)$, where $\mathcal{P}(\mathbb{R})$ is the space of polynomials with real coefficients. (Note: This is an infinite dimensional problem and therefore more challenging.)

14. $T : H_4(\mathbb{R}) \to \mathbb{R}^4$ defined by $T(v) = [v]_Y$, where Y is the basis given in Example 3.4.15.

15. $T : \mathcal{D}(Z_2) \to \mathcal{D}(Z_2)$ defined by $T(x) = x + x$.

16. The transformation of Exercise 12 in Section **??** on heat states.

For each vector space, V, in Exercises 17–24, create an onto transformation $T : V \to \mathbb{R}^d$ where d is given also. If such a transformation is not possible, justify why not. Determine the range, rank, nullity, and nullspace of the transformation you created.

17. $V = \mathbb{R}^5, d = 3$

18. $V = \mathbb{R}^5, d = 7$

19. $V = \mathcal{P}_2(\mathbb{R}), d = 6$

20. $V = \mathcal{P}_2(\mathbb{R}), d = 3$

21. $V = \mathcal{M}_{2\times 3}(\mathbb{R}), d = 4$

22. $V = H_4(\mathbb{R})$, $d = 6$
23. $V = H_4(\mathbb{R})$, $d = 4$
24. V is the vector space of histograms with 7 bins. $d = 3$.

Consider the three radiographic scenarios in Exercises 25–27, which were previously examined in Section **??**. For each, state the nullspace and range of the radiographic transformation. And discuss how this information helps us understand the transformation? (Figures 5.3, 5.4 and 5.5).

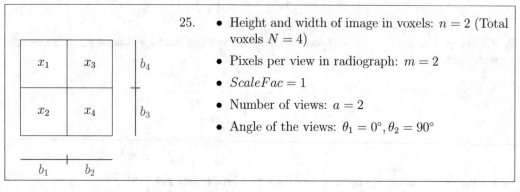

25. • Height and width of image in voxels: $n = 2$ (Total voxels $N = 4$)
     • Pixels per view in radiograph: $m = 2$
     • *ScaleFac* $= 1$
     • Number of views: $a = 2$
     • Angle of the views: $\theta_1 = 0°$, $\theta_2 = 90°$

Fig. 5.3 Tomographic Scenario A. Objects are in the vector space of 2×2 grayscale images. Radiographs are in the vector space of 2 views each with 2 pixels and the geometry as shown.

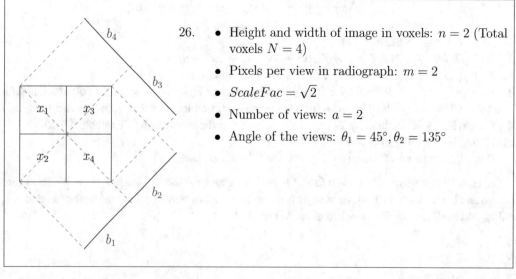

26. • Height and width of image in voxels: $n = 2$ (Total voxels $N = 4$)
     • Pixels per view in radiograph: $m = 2$
     • *ScaleFac* $= \sqrt{2}$
     • Number of views: $a = 2$
     • Angle of the views: $\theta_1 = 45°$, $\theta_2 = 135°$

Fig. 5.4 Tomographic Scenario B. Objects are in the vector space of 2×2 grayscale images. Radiographs are in the vector space of 2 views each with 2 pixels and the geometry as shown.

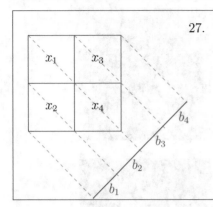

27.
- Height and width of image in voxels: $n = 2$ (Total voxels $N = 4$)
- Pixels per view in radiograph: $m = 4$
- $ScaleFac = \sqrt{2}/2$
- Number of views: $a = 1$
- Angle of the views: $\theta_1 = 45°$

Fig. 5.5 Tomographic Scenario C. Objects are in the vector space of 2×2 grayscale images. Radiographs are in the vector space of 1 view with 4 pixels and the geometry as shown.

Additional Exercises

28. Create a one-to-one transformation that is not onto. Describe the rank, range, nullspace, and nullity of the transformation you create. If no such transformation exists, justify.
29. Prove Theorem 5.1.12.
30. We now reconsider Example 5.1.2.

    (a) Find the matrix associated with this linear transformation.
    (b) Find the nullspace of this matrix and verify that it is the same as the nullspace of the transformation T.
    (c) Recall the superposition principle (Theorem 3.1.30). What does this theorem say about the preimage of the point (x, y) under the transformation T? (The preimage of the point $(0, 0)$ under the transformation T is exactly the nullspace of T.)
    (d) Every point in the range $\text{Ran}(T) = \mathbb{R}^2$ is associated with one of these parallel preimages; and each preimage has dimension 1. The set of these parallel preimages "fills out" the domain $\mathbb{R}^3$. This gives us a nice graphical way to visualize the Rank-Nullity theorem!

5.2 Matrix Spaces and the Invertible Matrix Theorem

As we begin this section, we should revisit our tomographic goal. Suppose we are given radiographs like those on top of Figure 5.6. We know that there is a linear transformation, $T : \mathcal{O} \to \mathcal{R}$, that transforms the brain objects in $\mathcal{O}$ (whose slices can be seen on bottom in Figure 5.6) to these radiographs (vectors in $\mathcal{R}$). Our overall goal is to recover these brain slices. What we know is that, if x is radiographed to get the images, b on top in Figure 5.6, then $T(x) = b$. Our hope is to recover x. Algebraically, this means we want $x = T^{-1}(b)$. We recognized, in Section 5.1, that not all transformations have an inverse. That is, T^{-1} may not even exist. We discussed various properties of a transformation for which an inverse exists. In fact, we have seen several small examples where we have seen a radiographic transformation without an inverse. In these examples, we saw that there are objects that are invisible to the transformation (nullspace vectors) and so the transformation is not one-to-one.

Working with transformations on vector spaces that are less standard (like the vector space of brain images or the vector space of radiographs) can be tedious and sometimes difficult. We saw in Sections ?? and ?? that these n-dimensional spaces are isomorphic (act just like) the Euclidean spaces $\mathbb{R}^n$. In fact, the isomorphism is the transformation that transforms a vector v to its coordinate vector

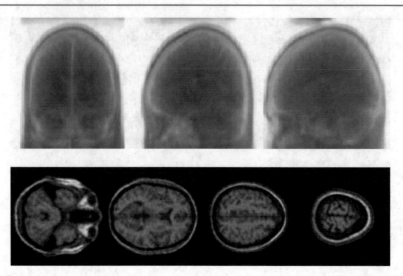

Fig. 5.6 Top: Three examples of radiographs of a human head taken at different orientations. Bottom: Four examples of density maps of slices of a human head.

$[v]_{\mathcal{B}}$ (for some basis $\mathcal{B}$ of our vector space). With this we recognized, in Section **??**, that there is a matrix representation, $M = [T]_{\mathcal{B}_V}^{\mathcal{B}_W}$, of $T : V \to W$ that allows all computations of T to happen in the coordinate spaces instead of directly with the vectors of some obscure or less standard vector space. In this section, we will consider the invertibility of T, but this time we will consider the properties of M that help us recognize when T is invertible or when T has a left inverse. We will discuss spaces related to the matrix M that are analogous to the nullspace and range space of the transformation T.

5.2.1   Matrix Spaces

In Section **??**, we considered an $m \times n$ matrix, M, where left matrix multiplication by M is a linear transformation that maps a vector from $\mathbb{R}^n$ to $\mathbb{R}^m$. Moreover, every linear transformation, that maps an n-dimensional space V to an m-dimensional space W, can be represented by left multiplication by some matrix. We next explore how the transformation spaces (nullspace and range space) are manifested in these matrix representations. Throughout this discussion, we will continue to use the notation $M = [T]_{\mathcal{B}_V}^{\mathcal{B}_W}$. To make the reading clearer, we suggest that whenever you see this notation, you read it with its meaning not its symbols. That is, read it as, "M is the matrix representation of T with respect to the bases $\mathcal{B}_V$ and $\mathcal{B}_W$." At times, we will write this as well.

The Nullspace of a Matrix

Recall that the nullspace of a transformation is the space of all vectors that are invisible to the transformation. Let V and W be vector spaces whose dimensions are n and m, respectively. To find the nullspace, Null(T), of a transformation $T : V \to W$, we look for all vectors $v \in V$ so that $T(v) = 0$. Suppose that M is the matrix representation of T with respect to the bases $\mathcal{B}_V$ and $\mathcal{B}_W$, of V and W, respectively. Then, in the coordinate spaces, we consider the transformation defined by left multiplication by M. We will, for ease of notation, call this transformation T_M. Then, for $v \in \text{Null}(T)$, we should have a corresponding $[v]_{\mathcal{B}_V}$ so that

$$T_M\left([v]_{\mathcal{B}_V}\right) = [T]^{\mathcal{B}_W}_{\mathcal{B}_V}[v]_{\mathcal{B}_V} = M[v]_{\mathcal{B}_V} = [0]_{\mathcal{B}_W}.$$

This result suggests that the matrix representation of a transformation also has a nullspace, a subspace of coordinate (domain) vectors that result in the zero (codomain) coordinate vector upon left multiplication by M.

Definition 5.2.1

Let $M \in \mathcal{M}_{m \times n}(\mathbb{R})$. The **nullspace** of M, denoted $\text{Null}(M)$, is

$$\text{Null}(M) = \left\{ x \in \mathbb{R}^n \mid Mx = 0 \in \mathbb{R}^m \right\}.$$

The **nullity** of M is $\dim \text{Null}(M)$.   □

Definition 5.2.1 does not require a discussion about another transformation T. This allows us to extend the idea of the nullspace of a transformation to matrices in general, without referring to the associated linear transformation.

If V and W are n- and m- dimensional vector spaces, respectively and if $T : V \to W$ has matrix representation $M = [T]^{\mathcal{B}_W}_{\mathcal{B}_V}$, then the nullspace of T is a subspace of V, while the nullspace of M is a subspace of $\mathbb{R}^n$. We now consider the relationship between these two vector spaces. In the following theorem, we show that a basis for the nullspace of a linear transformation can easily be used to find the nullspace for any matrix representation of that transformation.

Theorem 5.2.2

Let V and W be finite-dimensional vector spaces, $T : V \to W$ be a linear transformation, and $\mathcal{B}_V$ and $\mathcal{B}_W$ be bases for V and W, respectively. Also, let $M = [T]^{\mathcal{B}_W}_{\mathcal{B}_V}$ be a matrix representation of T. Suppose $\beta = \{\beta_1, \beta_2, \ldots, \beta_k\}$ is a basis for $\text{Null}(T)$, possibly empty. Then $\mu = \{[\beta_1]_{\mathcal{B}_V}, [\beta_2]_{\mathcal{B}_V}, \ldots, [\beta_k]_{\mathcal{B}_V}\}$ is a basis for $\text{Null}(M)$.

Proof. Suppose V and W are finite-dimensional vector spaces, $T : V \to W$ is a linear transformation, and that $\mathcal{B}_V$ and $\mathcal{B}_W$ are bases for V and W, respectively. Also, suppose that $M = [T]^{\mathcal{B}_W}_{\mathcal{B}_V}$. Let $\beta = \{\beta_1, \beta_2, \ldots, \beta_k\} \subseteq V$ be a basis for $\text{Null}(T)$. If $\dim V = n$, then the corresponding coordinate space is $\mathbb{R}^n$. Now consider the set

$$\mu = \{[\beta_1]_{\mathcal{B}_V}, [\beta_2]_{\mathcal{B}_V}, \ldots, [\beta_k]_{\mathcal{B}_V}\} \subseteq \mathbb{R}^n.$$

If $\beta = \emptyset$, then $\mu = \emptyset$ and $T(0) = 0 \in W$. This means that $M[0]_{\mathcal{B}_V} = [0]_{\mathcal{B}_W}$. Thus, $\text{Null}(M) = \{0\} = \text{Span}(\mu)$. Now, suppose that $\beta \neq \emptyset$. We will show that μ is linearly independent and that $\text{Span}\,\mu = \text{Null}(M)$.

First, let $\alpha_1, \alpha_2, \ldots, \alpha_k$ be scalars that satisfy the linear dependence relation

$$\alpha_1[\beta_1]_{\mathcal{B}_V} + \alpha_2[\beta_2]_{\mathcal{B}_V} + \ldots + \alpha_k[\beta_k]_{\mathcal{B}_V} = 0. \qquad (5.1)$$

In Section **??**, we determined that Equation 5.1 can be rewritten as

$$[\alpha_1\beta_1 + \alpha_2\beta_2 + \ldots + \alpha_k\beta_k]_{\mathcal{B}_V} = [0]_{\mathcal{B}_V}.$$

Because coordinate vectors are unique and because β is a basis, we know that

$$\alpha_1\beta_1 + \alpha_2\beta_2 + \ldots + \alpha_k\beta_k = 0.$$

Because β is linearly independent, $\alpha_1 = \alpha_2 = \ldots = \alpha_k = 0$. Thus, μ is linearly independent.

Next, we will use a set argument to show that Span $\mu = \text{Null}(M)$.

($\subseteq$) Let $x \in$ Span μ be the vector given by

$$x = \alpha_1[\beta_1]_{\mathcal{B}_V} + \alpha_2[\beta_2]_{\mathcal{B}_V} + \ldots + \alpha_k[\beta_k]_{\mathcal{B}_V}.$$

Since $x \in \mathbb{R}^n$, there is a unique $v \in V$ so that $x = [v]_{\mathcal{B}_V}$.

$$\begin{aligned}
Mx &= [T]_{\mathcal{B}_V}^{\mathcal{B}_W} x \\
&= [T]_{\mathcal{B}_V}^{\mathcal{B}_W}(\alpha_1[\beta_1]_{\mathcal{B}_V} + \alpha_2[\beta_2]_{\mathcal{B}_V} + \ldots + \alpha_k[\beta_k]_{\mathcal{B}_V}) \\
&= \alpha_1[T]_{\mathcal{B}_V}^{\mathcal{B}_W}[\beta_1]_{\mathcal{B}_V} + \alpha_2[T]_{\mathcal{B}_V}^{\mathcal{B}_W}[\beta_2]_{\mathcal{B}_V} + \ldots + \alpha_k[T]_{\mathcal{B}_V}^{\mathcal{B}_W}[\beta_k]_{\mathcal{B}_V} \\
&= \alpha_1[T\beta_1]_{\mathcal{B}_W} + \alpha_2[T\beta_2]_{\mathcal{B}_W} + \ldots + \alpha_k[T\beta_k]_{\mathcal{B}_W} \\
&= \alpha_1 0_{\mathcal{B}_W} + \alpha_2 0_{\mathcal{B}_W} + \ldots + \alpha_k 0_{\mathcal{B}_W} \\
&= 0_{\mathcal{B}_W}.
\end{aligned}$$

So, $x \in \text{Null}(M)$ and Span $\mu \subseteq \text{Null}(M)$.

($\supseteq$) Finally, we show that $\text{Null}(M) \subseteq$ Span μ. Let $[x]_{\mathcal{B}_V} \in \text{Null}(M)$. Then $0 = M[x]_{\mathcal{B}_V} = [T]_{\mathcal{B}_V}^{\mathcal{B}_W}[x]_{\mathcal{B}_V} = [Tx]_{\mathcal{B}_W}$. So, $x \in \text{Null}(T)$ and can be written $x = \alpha_1\beta_1 + \alpha_2\beta_2 + \ldots + \alpha_k\beta_k$. Now, $[x]_{\mathcal{B}_V} = \alpha_1[\beta_1]_{\mathcal{B}_V} + \alpha_2[\beta_2]_{\mathcal{B}_V} + \ldots + \alpha_k[\beta_k]_{\mathcal{B}_V}$. Thus, $[x]_{\mathcal{B}_V} \in$ Span μ and $\text{Null}(M) \subseteq$ Span μ.

Therefore, $\text{Null}(M) = $ Span μ. Thus, μ is a basis for $\text{Null}(M)$. $\qquad\qquad\square$

Corollary 5.2.3

Let V and W be finite-dimensional vector spaces, $T : V \to W$ be a linear transformation, and $\mathcal{B}_V$ and $\mathcal{B}_W$ be bases for V and W, respectively. Also let $M = [T]_{\mathcal{B}_V}^{\mathcal{B}_W}$. The transformation $U : \text{Null}(T) \to \text{Null}(M)$ defined by $U(v) = [v]_{\mathcal{B}_V}$ is an isomorphism. Furthermore, $\text{Nullity}(T) = \text{Nullity}(M)$.

The proof of this corollary is the subject of Exercise 9. The main message here is that the nullspace of a matrix M is isomorphic to the nullspace of the corresponding transformation, T.

Example 5.2.4 Suppose $M : \mathbb{R}^n \to \mathbb{R}^m$ is the matrix representation of a linear radiographic transformation $T : V \to W$, relative to some bases. Null(M) is the set of all invisible objects *as represented in the coordinate space* $\mathbb{R}^n$, and Null(T) is the set of all invisible objects in V, which are grayscale images. And, Corollary 5.2.3 tells us that there's a one-to-one and onto transformation that maps from Null(T) to Null(M). $\square$

Example 5.2.5 Given the matrix

$$M = \begin{pmatrix} 1 & 1 & 1 \\ 2 & 1 & -1 \\ -1 & 0 & 2 \end{pmatrix}.$$

We can find Null(M) by solving the matrix equation

$$\begin{pmatrix} 1 & 1 & 1 \\ 2 & 1 & -1 \\ -1 & 0 & 2 \end{pmatrix} \begin{pmatrix} x \\ y \\ z \end{pmatrix} = \begin{pmatrix} 0 \\ 0 \\ 0 \end{pmatrix}.$$

The solution set is

$$\left\{ \begin{pmatrix} 2z \\ -3z \\ z \end{pmatrix} \in \mathbb{R}^3 \,\middle|\, z \in \mathbb{R} \right\}.$$

So, the nullspace of M is given by

$$\text{Null}(M) = \text{Span} \left\{ \begin{pmatrix} 2 \\ -3 \\ 1 \end{pmatrix} \right\}.$$

$\square$

The above example showed no connection to any transformation. We can talk about the nullspace of a matrix without relating it to a transformation.

Example 5.2.6 Let $V = \{ax^3 + bx^2 - ax + c \mid a, b, c \in \mathbb{R}\}$ and $W = \mathcal{M}_{2\times3}(\mathbb{R})$. Now, let us consider the transformation $T : V \to W$ defined by

$$T(ax^3 + bx^2 - ax + c) = \begin{pmatrix} a & a & a \\ a+b & -a & -b \end{pmatrix}.$$

Determine the matrix representation, M, of T and then find Null(M). First, we choose a basis for V.

$$V = \{ax^3 + bx^2 - ax + c \mid a, b, c \in \mathbb{R}\} = \text{Span}\{x^3 - x, x^2, 1\},$$

and $\{x^3 - x, x^2, 1\}$ is linearly independent. Therefore, a basis for V is given by

$$\mathcal{B}_V = \{x^3 - x, x^2, 1\}.$$

Next, we find where each basis element maps to. We will write each result as a coordinate vector using the standard basis for W. Using the definition of T, we get

$$[T(x^3 - x)]_{\mathcal{B}_W} = \left[\begin{pmatrix} 1 & 1 & 1 \\ 1 & -1 & 0 \end{pmatrix}\right]_{\mathcal{B}_W} = \begin{pmatrix} 1 \\ 1 \\ 1 \\ 1 \\ -1 \\ 0 \end{pmatrix},$$

$$[T(x^2)]_{\mathcal{B}_W} = \left[\begin{pmatrix} 0 & 0 & 0 \\ 1 & 0 & -1 \end{pmatrix}\right]_{\mathcal{B}_W} = \begin{pmatrix} 0 \\ 0 \\ 0 \\ 1 \\ 0 \\ -1 \end{pmatrix}, \text{ and}$$

$$[T(1)]_{\mathcal{B}_W} = \left[\begin{pmatrix} 0 & 0 & 0 \\ 0 & 0 & 0 \end{pmatrix}\right]_{\mathcal{B}_W} = \begin{pmatrix} 0 \\ 0 \\ 0 \\ 0 \\ 0 \\ 0 \end{pmatrix}.$$

Recall, these results form the columns of the matrix representation. So,

$$M = \begin{pmatrix} 1 & 0 & 0 \\ 1 & 0 & 0 \\ 1 & 0 & 0 \\ 1 & 1 & 0 \\ -1 & 0 & 0 \\ 0 & -1 & 0 \end{pmatrix}.$$

To find $\text{Null}(M)$, we solve the matrix equation

$$\begin{pmatrix} 1 & 0 & 0 \\ 1 & 0 & 0 \\ 1 & 0 & 0 \\ 1 & 1 & 0 \\ -1 & 0 & 0 \\ 0 & -1 & 0 \end{pmatrix} [v]_{\mathcal{B}_V} = \begin{pmatrix} 0 \\ 0 \\ 0 \\ 0 \\ 0 \\ 0 \end{pmatrix}.$$

The solution set is

$$[v]_{\mathcal{B}_V} = \left\{ \begin{pmatrix} 0 \\ 0 \\ z \end{pmatrix} \,\middle|\, z \in \mathbb{R} \right\}.$$

That is,

$$\text{Null}(M) = \text{Span}\left\{ \begin{pmatrix} 0 \\ 0 \\ 1 \end{pmatrix} \right\}.$$

$\square$

In the last two examples, we did not show the steps to solving the matrix equations. As a review of past materials, we encourage the reader to check these steps.

The Column Space of a Matrix

We just saw how there is a close relationship between the nullspace of a transformation and the nullspace of the corresponding matrix representation. Similarly, we will see here how the range space of a linear transformation is related to the *column space* of the corresponding matrix representation. But first, let us define the column space of a matrix.

Definition 5.2.7

Let M be an $m \times n$ matrix with columns $\alpha_1, \alpha_2, \ldots, \alpha_n \in \mathbb{R}^m$. The **column space** of M, denoted $\text{Col}(M)$ is defined to be

$$\text{Col}(M) = \text{Span}\{\alpha_1, \alpha_2, \ldots, \alpha_n\}.$$

The **rank** of M is $\dim(\text{Col}(M))$. $\square$

The columns of M are not necessarily a basis for $\text{Col}(M)$ (See Exercise 8). But, we know that, because $\text{Col}(M)$ is a subspace of $\mathbb{R}^m$, $\text{Rank}(M) = \dim(\text{Col}(M)) \le m$. Now, since the range space of $T : V \to W$ is a subspace of W, $\text{Rank}(T) \le m$ also. The next two theorems address the relationship between the range of T and $\text{Col}(M)$.

Theorem 5.2.8

Let V and W be vector spaces, $T : V \to W$ be linear, and $M = [T]_{\mathcal{B}_V}^{\mathcal{B}_W}$ where $\mathcal{B}_V$ and $\mathcal{B}_W$ are ordered bases for V and W, respectively. Then,

$$\text{Col}(M) = \left\{ [T(x)]_{\mathcal{B}_W} \mid x \in V \right\}.$$

The proof is the subject of Exercise 10.

Recall that $\text{Ran}(T) = \{T(x) \mid x \in V\}$. So, Theorem 5.2.8 says that the column space of M is the set of all coordinate vectors of the range space of T, relative to the basis $\mathcal{B}_W$. The next theorem makes a connection between the bases of $\text{Ran}(T)$ and $\text{Col}(M)$.

Theorem 5.2.9
Let V and W be finite-dimensional vector spaces, $T : V \to W$ be a linear transformation, and $M = [T]_{\mathcal{B}_V}^{\mathcal{B}_W}$ where $\mathcal{B}_V$ and $\mathcal{B}_W$ are ordered bases for V and W, respectively. Suppose $\beta = \{\beta_1, \beta_2, \ldots, \beta_m\}$ is a basis for $\text{Ran}(T)$. Then $\mu = \{[\beta_1]_{\mathcal{B}_W}, [\beta_2]_{\mathcal{B}_W}, \ldots, [\beta_m]_{\mathcal{B}_W}\}$ is a basis for $\text{Col}(M)$.

The proof is similar to the proof of Theorem 5.2.2 and is the subject of Exercise 11.

Corollary 5.2.10
Let V and W be finite-dimensional vector spaces, $T : V \to W$ be a linear transformation, and $M = [T]_{\mathcal{B}_V}^{\mathcal{B}_W}$ where $\mathcal{B}_V$ and $\mathcal{B}_W$ are ordered bases for V and W, respectively. Then, the transformation $U : \text{Ran}(T) \to \text{Col}(M)$ defined by $U(v) = [v]_{\mathcal{B}_W}$ is an isomorphism. Furthermore, $\text{Rank}(T) = \text{Rank}(M)$.

The proof is the subject of Exercise 12. The main message here is that the column space, $\text{Col}(M)$, of a matrix M is isomorphic to the range space of the corresponding transformation, T.

Example 5.2.11 Suppose $M : \mathbb{R}^n \to \mathbb{R}^m$ is the matrix representation of a linear radiographic transformation $T : V \to W$, relative to some bases. $\text{Col}(M)$ is the set of all possible radiographs *as represented in the coordinate space* $\mathbb{R}^m$. $\text{Ran}(T)$ is the set of all possible radiographs in W, which are grayscale images. Corollary 5.2.10 says that there is a one-to-one and onto transformation that maps $\text{Ran}(T)$ to $\text{Col}(M)$. $\qquad\square$

Next, we consider a method for finding a basis for $\text{Col}(M)$. The following method utilizes matrix reduction to perform the method of spanning set reduction (see page **??**). We want to find all $w \in \mathbb{R}^m$ so that there exists a $v \in \mathbb{R}^n$ such that $Mv = w$.

Example 5.2.12 Let us consider the matrix

$$M = \begin{pmatrix} 1 & 1 & 1 \\ 2 & 1 & -1 \\ -1 & 0 & 2 \end{pmatrix}.$$

To find $\text{Col}(M)$, we find

$$w \in \text{Span} \left\{ \begin{pmatrix} 1 \\ 2 \\ -1 \end{pmatrix}, \begin{pmatrix} 1 \\ 1 \\ 0 \end{pmatrix}, \begin{pmatrix} 1 \\ -1 \\ 2 \end{pmatrix} \right\}.$$

That is,

$$w = x \begin{pmatrix} 1 \\ 2 \\ -1 \end{pmatrix} + y \begin{pmatrix} 1 \\ 1 \\ 0 \end{pmatrix} + z \begin{pmatrix} 1 \\ -1 \\ 2 \end{pmatrix},$$

for some $x, y, z \in \mathbb{R}$. This is equivalent to the matrix equation

$$\begin{pmatrix} 1 & 1 & 1 \\ 2 & 1 & -1 \\ -1 & 0 & 2 \end{pmatrix} \begin{pmatrix} x \\ y \\ z \end{pmatrix} = w.$$

So, we really are looking for all elements of the set

$$\{w \in \mathbb{R}^3 \mid Mv = w \text{ for } v \in \mathbb{R}^3\} = \left\{ \begin{pmatrix} a \\ b \\ c \end{pmatrix} \mid \begin{pmatrix} 1 & 1 & 1 \\ 2 & 1 & -1 \\ -1 & 0 & 2 \end{pmatrix} \begin{pmatrix} x \\ y \\ z \end{pmatrix} = \begin{pmatrix} a \\ b \\ c \end{pmatrix}, \text{ where } x, y, z \in \mathbb{R} \right\}.$$

Next, we formulate the corresponding augmented matrix and examine the nature of the solution set. (Here, the right-handside of the augmented matrix is important to track through the matrix reduction, so we show these steps.)

$$\begin{pmatrix} 1 & 1 & 1 & a \\ 2 & 1 & -1 & b \\ -1 & 0 & 2 & c \end{pmatrix}$$

$$\xrightarrow[R_3=r_1+r_3]{R_2=-2r_1+r_2} \begin{pmatrix} 1 & 1 & 1 & a \\ 0 & -1 & -3 & -2a+b \\ 0 & 1 & 3 & a+c \end{pmatrix}$$

$$\xrightarrow[R_3=r_2+r_3]{R_1=r_2+r_1, R_2=-r_2} \begin{pmatrix} 1 & 0 & -2 & -a+b \\ 0 & 1 & 3 & 2a-b \\ 0 & 0 & 0 & -a+b+c \end{pmatrix}.$$

The last row of the matrix corresponds to the equation

$$0 = -a+b+c.$$

Therefore, as long as $w = \begin{pmatrix} a \\ b \\ c \end{pmatrix}$ with $-a+b+c = 0$ the equation $Mv = w$ has a solution. This means that

$$\text{Col}(M) = \left\{ \begin{pmatrix} a \\ b \\ c \end{pmatrix} \mid -a+b+c = 0 \right\}$$

$$= \left\{ \begin{pmatrix} b+c \\ b \\ c \end{pmatrix} \mid b, c \in \mathbb{R} \right\}$$

$$= \text{Span} \left\{ \begin{pmatrix} 1 \\ 1 \\ 0 \end{pmatrix}, \begin{pmatrix} 1 \\ 0 \\ 1 \end{pmatrix} \right\}.$$

$\square$

The reduced matrix has two leading entries, one in each of the first two columns. This led to a basis containing two vectors. It turns out that there is a simpler method for finding a basis for the column space: Choose as basis vectors the matrix columns corresponding to reduced matrix columns containing leading entries. Now, if $\text{Col}(M) = \text{Span}\{\alpha_1, \alpha_2, \ldots, \alpha_n\}$, then to form a basis, we need to find a maximally linearly independent subset. The following argument shows that the columns corresponding to the leading 1's are such a set.

Choose α_1 to be the first basis element for $\text{Col}(M)$. (If α_1 is all zeros, we just start with the first column that isn't all zeros.) Since α_1 is not the zero vector, $\{\alpha_1\}$ is linearly independent. Now, we check to see if $\{\alpha_1, \alpha_2\}$ is linearly independent. We can do this by solving for a in the equation $a\alpha_1 = \alpha_2$. That is, we reduce the augmented matrix

$$\left(\begin{array}{c|c} | & | \\ \alpha_1 & \alpha_2 \\ | & | \end{array} \right).$$

If the second column has a leading one, then that means there is a row with zeros to the left of the augment and a nonzero on the right of the augment. This would mean that the equation $a\alpha_1 = \alpha_2$ has no solution and they are linearly independent. If there is no leading entry in the second column, then these columns are linearly dependent.

Now, supposing that $\{\alpha_1, \alpha_2\}$ is linearly independent, we check to see if α_3 can be added to the set to form a new linearly independent set, $\{\alpha_1, \alpha_2, \alpha_3\}$. That means we want to solve for a and b in the equation $a\alpha_1 + b\alpha_2 = \alpha_3$. This can be done by reducing the augmented matrix

$$\left(\begin{array}{cc|c} | & | & | \\ \alpha_1 & \alpha_2 & \alpha_3 \\ | & | & | \end{array} \right).$$

If, after reducing, the third column has a leading entry, then $\{\alpha_1, \alpha_3\}$ is linearly independent and $\{\alpha_2, \alpha_3\}$ is also linearly independent. That is, if there is a leading entry to the right of the augment, $\{\alpha_1, \alpha_2, \alpha_3\}$ is linearly independent. If not, then either $\{\alpha_1, \alpha_3\}$ or $\{\alpha_2, \alpha_3\}$ is linearly dependent.

We can continue this process of collecting linearly independent vectors by recognizing that the set of columns corresponding to a leading entry in the reduced matrix is a linearly independent set. So we choose them to be in the basis for $\text{Col}(M)$. All other columns are in the span of these chosen vectors.

Example 5.2.13 Let $V = \{ax^3 + bx^2 - ax + c \mid a, b, c \in \mathbb{R}\}$ and $W = \mathcal{M}_{2\times3}(\mathbb{R})$. Now, let us consider the transformation $T : V \to W$ defined by

$$T(ax^3 + bx^2 - ax + c) = \begin{pmatrix} a & a & a \\ a+b & -a & -b \end{pmatrix}.$$

Recall from Example 5.2.6, that we found a basis

$$\mathcal{B}_V = \{x^3 - x, x^2, 1\}$$

of V and the corresponding matrix representation is

$$M = \begin{pmatrix} 1 & 0 & 0 \\ 1 & 0 & 0 \\ 1 & 0 & 0 \\ 1 & 1 & 0 \\ -1 & 0 & 0 \\ 0 & -1 & 0 \end{pmatrix}.$$

To find $\text{Col}(M)$, we find all $w \in \mathbb{R}^6$ so that

$$w \in \text{Span}\left\{ \begin{pmatrix} 1 \\ 1 \\ 1 \\ 1 \\ -1 \\ 0 \end{pmatrix}, \begin{pmatrix} 0 \\ 0 \\ 0 \\ 1 \\ 0 \\ -1 \end{pmatrix}, \begin{pmatrix} 0 \\ 0 \\ 0 \\ 0 \\ 0 \\ 0 \end{pmatrix} \right\}.$$

In other words, we find all $w \in \mathbb{R}^6$ so that there is a $v \in \mathbb{R}^3$ with $Mv = w$. In this example, it is clear that the last column is not part of a linearly independent set. Also, it is clear that the first two columns are linearly independent (they are not multiples of one another). Thus, a basis for the column space is

$$\left\{ \begin{pmatrix} 1 \\ 1 \\ 1 \\ 1 \\ -1 \\ 0 \end{pmatrix}, \begin{pmatrix} 0 \\ 0 \\ 0 \\ 1 \\ 0 \\ -1 \end{pmatrix} \right\}.$$

□

The Row Space of a Matrix

Now, since an $m \times n$ matrix, M has rows that are vectors in $\mathbb{R}^n$, we can also introduce the subspace of $\mathbb{R}^n$ that relates to the rows of the matrix. The definition of the row space is very similar to Definition 5.2.7.

Definition 5.2.14

Let M be an $m \times n$ matrix with rows $\beta_1^\mathsf{T}, \beta_2^\mathsf{T}, \ldots, \beta_m^\mathsf{T} \in \mathbb{R}^n$. The **row space** of M, denoted $\text{Row}(M)$ is defined to be

$$\text{Row}(M) = \text{Span}\{\beta_1, \beta_2, \ldots, \beta_m\}.$$

□

Recall, for $v \in \mathbb{R}^n$, v^T is the transpose of $v \in \mathbb{R}^n$ (see Definition 2.5.14). Given a matrix M, we find $\text{Row}(M)$ by finding linear combinations of the rows. We can use the method of Spanning Set Reduction (see page **??**) to find a basis for $\text{Row}(M)$. Let us consider how this method can be employed.

Example 5.2.15 Let M be the matrix defined by

$$M = \begin{pmatrix} 1 & 1 & 1 \\ 2 & 1 & -1 \\ -1 & 0 & 2 \end{pmatrix}.$$

Define $r_1^\mathsf{T}, r_2^\mathsf{T}, r_3^\mathsf{T}$ to be the three rows, in order, of M. When performing matrix reduction techniques, we are finding linear combinations of the rows. Therefore, we can consider the reduced echelon form, $\tilde{M}$, of M found in Example 5.2.12,

$$\tilde{M} = \begin{pmatrix} 1 & 0 & -2 \\ 0 & 1 & 3 \\ 0 & 0 & 0 \end{pmatrix}.$$

As, we did in Example 5.2.12, define the rows of $\tilde{M}$ to be $R_1^\mathsf{T}, R_2^\mathsf{T}, R_3^\mathsf{T}$. We learn from this form (and the steps taken to get to this form) that

$$R_1 = -r_1 + r_2$$
$$R_2 = 2r_1 - r_2$$
$$R_3 = -r_1 + r_2 + r_3.$$

From, these equations, we can see that the rows of the reduced echelon form are vectors in Row(M). We also see that, since $R_3 = (0\ 0\ 0)^\mathsf{T}$, the rows of M are linearly dependent. Indeed, $r_3 = r_1 - r_2$. We see also that we have found two vectors, $R_1, R_2 \in \text{Span}\{r_1, r_2, r_3\}$ that are linearly independent. We also know that, since $r_1, r_2, r_3 \in \text{Span}\{R_1, R_2\}$, a basis for Row$(M)$ is given by $\{R_1, R_2\}$. In fact, this process will always lead to a basis for Row(M). □

It is important to note that $\dim(\text{Row}(M))$ is the number of nonzero rows in the reduced echelon form. Putting this together with our process for finding Col(M), we see that the number of columns corresponding to leading entries in the reduced echelon form, $\tilde{M}$, is the same as the number of nonzero rows in $\tilde{M}$. That means that rank(M) is also equal to $\dim(\text{Row}(M))$.

Another method for finding a basis for the row space of a matrix is to consider the transpose, M^T, (see Definition 2.5.14) of the matrix M.

Theorem 5.2.16
Let M be an $m \times n$ matrix. Then $\text{Row}(M) = \text{Col}(M^\mathsf{T})$ and $\text{Rank}(M) = \text{Rank}(M^\mathsf{T})$.

Proof. Let M have rows $r_1^\mathsf{T}, r_2^\mathsf{T}, \ldots, r_m^\mathsf{T} \in \mathbb{R}^n$. By Definitions 5.2.7 and 5.2.14,

$$\text{Row}(M) = \text{Span}\{r_1, r_2, \ldots, r_m\} = \text{Col}(M^\mathsf{T}).$$

This means that $\text{Rank}(M^\mathsf{T}) = \dim(\text{Row}(M))$.

Now, suppose $c_1, c_2, \ldots, c_n \in \mathbb{R}^m$ are the columns of M. We know that $\text{Rank}(M)$ is the number of linearly independent columns of M. Suppose $\text{Rank}(M) = k \leq m$ and suppose, without loss of generality, that $c_1, c_2, \ldots, c_k$ are linearly independent. Then the augmented matrix corresponding to the vector equations

$$\alpha_1 c_1 + \alpha_2 c_2 + \ldots + \alpha_k c_k = c_{k+1}$$
$$\alpha_1 c_1 + \alpha_2 c_2 + \ldots + \alpha_k c_k = c_{k+2}$$
$$\vdots$$
$$\alpha_1 c_1 + \alpha_2 c_2 + \ldots + \alpha_k c_k = c_n$$

is given by

$$\left(\begin{array}{ccccccc} | & | & & | & | & | & & | \\ c_1 & c_2 & \ldots & c_k & c_{k+1} & c_{k+2} & \ldots & c_n \\ | & | & & | & | & | & & | \end{array} \right).$$

This matrix has the same columns as M and has reduced echelon form with no leading entries on the right side of the augment. That is, the reduced echelon form has $m - k$ rows of zeros at the bottom.

In terms of the rows of M, we see that M is the matrix

$$\left(\begin{array}{cccc} | & | & & | \\ c_1 & c_2 & \ldots & c_n \\ | & | & & | \end{array} \right) = \left(\begin{array}{ccc} - & r_1^{\mathsf{T}} & - \\ - & r_2^{\mathsf{T}} & - \\ & \vdots & \\ - & r_m^{\mathsf{T}} & - \end{array} \right)$$

Again, the reduced echelon form has k rows with leading entries and $m - k$ rows of zeros. That is, there are k linearly independent rows in M. Therefore, $\text{Rank}(M^{\mathsf{T}}) = \dim(\text{Row}(M)) = \text{Rank}(M)$. $\square$

5.2.2   The Invertible Matrix Theorem

In this section, we explore the properties of matrix representations of linear transformations that are one-to-one and onto. In particular, we wish to know which matrix properties indicate the invertibility properties of the corresponding linear transformation.

Recall that, if V and W are vector spaces, then we can discuss the invertibility of a linear transformation, $T : V \to W$. We say that T is invertible if and only if there is another linear transformation, $S : W \to V$ so that $T \circ S = S \circ T = id$, where $id : V \to V$ is the identity transformation. We will gather and add to theorems from previous sections to create the Invertible Matrix Theorem.

Before getting deep into theorems, let us first define what we mean by the inverse of a matrix. It is important, as we connect linear transformations to matrices, to consider the connection to an invertible matrix and an invertible linear transformation.

Definition 5.2.17

Let M be a real $n \times n$ matrix. M is **invertible** if there exists a real $n \times n$ matrix, called the **inverse** of M and denoted M^{-1}, such that $MM^{-1} = M^{-1}M = I_n$. $\square$

For a matrix to have an inverse it necessarily has the same number of rows and columns. We can understand this by noticing that the products AB and BA of two matrices A and B only make sense when the number of columns of A matches the number of rows of B and vice versa.

The next example demonstrates a method for finding the inverse of a matrix that will be useful in understanding the theorems that follow.

Example 5.2.18 Let A be the 3×3 matrix given by

$$A = \begin{pmatrix} 1 & 2 & 1 \\ 1 & 1 & 1 \\ 1 & 3 & 3 \end{pmatrix}.$$

If A has an inverse, then there is a 3×3 matrix B so that $AB = BA = I_3$. That is, we can find

$$B = \begin{pmatrix} a & b & c \\ d & e & f \\ g & h & i \end{pmatrix},$$

so that

$$AB = \begin{pmatrix} 1 & 2 & 1 \\ 1 & 1 & 1 \\ 1 & 3 & 3 \end{pmatrix} \begin{pmatrix} a & b & c \\ d & e & f \\ g & h & i \end{pmatrix} = \begin{pmatrix} 1 & 0 & 0 \\ 0 & 1 & 0 \\ 0 & 0 & 1 \end{pmatrix}$$

and

$$BA = \begin{pmatrix} a & b & c \\ d & e & f \\ g & h & i \end{pmatrix} \begin{pmatrix} 1 & 2 & 1 \\ 1 & 1 & 1 \\ 1 & 3 & 3 \end{pmatrix} = \begin{pmatrix} 1 & 0 & 0 \\ 0 & 1 & 0 \\ 0 & 0 & 1 \end{pmatrix}.$$

In Section **??**, we saw that a matrix product can be rewritten as a linear combination of the columns. Suppose $\alpha_1, \alpha_2, \alpha_3$ are the columns of A, then we can rewrite the columns of AB as

$$\begin{aligned} a\alpha_1 + d\alpha_2 + g\alpha_3 &= e_1 \\ b\alpha_1 + e\alpha_2 + h\alpha_3 &= e_2, \\ c\alpha_1 + f\alpha_2 + i\alpha_3 &= e_3 \end{aligned}$$

where e_1, e_2, e_3 are the columns of I_3. Another way to write this is as matrix equations

$$A \begin{pmatrix} a \\ d \\ g \end{pmatrix} = \begin{pmatrix} 1 \\ 0 \\ 0 \end{pmatrix}, \quad A \begin{pmatrix} b \\ e \\ h \end{pmatrix} = \begin{pmatrix} 1 \\ 0 \\ 0 \end{pmatrix}, \quad \text{and} \quad A \begin{pmatrix} c \\ f \\ i \end{pmatrix} = \begin{pmatrix} 0 \\ 0 \\ 1 \end{pmatrix}.$$

We could solve the first of these three equations for the first column vector $\begin{pmatrix} a \\ d \\ g \end{pmatrix}$ of B by row reducing the augmented matrix

$$\left(\begin{array}{ccc|c} 1 & 2 & 1 & 1 \\ 1 & 1 & 1 & 0 \\ 1 & 3 & 3 & 0 \end{array} \right).$$

We could then solve for the other columns of B similarly. However, we would have repeated essentially the same row reduction computation three times. To simultaneously solve these three equations, we can use the method of matrix reduction on the augmented matrix

$$\left(\begin{array}{ccc|ccc} 1 & 2 & 1 & 1 & 0 & 0 \\ 1 & 1 & 1 & 0 & 1 & 0 \\ 1 & 3 & 3 & 0 & 0 & 1 \end{array}\right).$$

If we reduce this and the result is I_3 on the left side of the augment, then we have exactly one solution, B, to the matrix equation $AB = I_3$. This matrix B is the inverse. In fact, we reduce the above and get

$$\left(\begin{array}{ccc|ccc} 1 & 0 & 0 & 0 & \frac{3}{2} & -\frac{1}{2} \\ 0 & 1 & 0 & 1 & -1 & 0 \\ 0 & 0 & 1 & -1 & \frac{1}{2} & \frac{1}{2} \end{array}\right).$$

Therefore, the inverse, B is given by

$$B = \left(\begin{array}{ccc} 0 & \frac{3}{2} & -\frac{1}{2} \\ 1 & -1 & 0 \\ -1 & \frac{1}{2} & \frac{1}{2} \end{array}\right).$$

One can easily check that, indeed, $AB = I_3$ and $BA = I_3$. $\square$

This example suggests that a matrix must be row equivalent to the identity matrix in order for it to be invertible. Another useful result that is suggested by the above example is the relationship between the determinants $\det(M)$ and $\det(M^{-1})$ whenever M^{-1} exists. We present the result in the following theorem.

Theorem 5.2.19

Let M be an $n \times n$ invertible matrix. Then, $\det(M) \det(M^{-1}) = 1$.

The proof of Theorem 5.2.19 follows from the definition of the determinant. In particular, we see in Example 5.2.18, that we can compute the inverse of M by reducing M to I while at the same time, using the same row operations, taking I to M^{-1}. This means that we "undo" the determinant calculation for M to get the determinant calculation for M^{-1}. See Exercise 2 to try an example of this.

In addition, we can list many properties related to when a matrix is invertible. In fact, we have already proven many results (in Sections **??, ??, ??,** and **??**) that characterize invertibility. We combine these theorems to summarize facts about matrix invertibility in the following theorem, known as the Invertible Matrix Theorem. If you completed and proved Theorems 3.1.34, 3.3.20, and 3.4.40 from exercises in previous sections, you will have most of this theorem, but in case you did not, we complete it here.

Theorem 5.2.20 ((Partial) Invertible Matrix Theorem)

Let $M \in \mathcal{M}_{n \times n}(\mathbb{R})$ with columns $c_1, c_2, \ldots, c_n \in \mathbb{R}^n$ and rows $r_1^\mathsf{T}, r_2^\mathsf{T}, \ldots, r_n^\mathsf{T}$. Then the following statements are equivalent.

(a) M is invertible.
(b) The reduced echelon form of M has n leading ones.
(c) The matrix equation $Mx = 0$ has a unique solution.
(d) For every $b \in \mathbb{R}^n$, the matrix equation $Mx = b$ has a unique solution.
(e) The homogeneous system of equations with coefficient matrix M has a unique solution.
(f) The set $\{c_1, c_2, \ldots, c_n\}$ is linearly independent.
(g) The set $\{c_1, c_2, \ldots, c_n\}$ is a basis for $\mathbb{R}^n$.
(h) $\det(M) \neq 0$.
(i) $\mathrm{Rank}(M) = n$.
(j) $\mathrm{Nullity}(M) = 0$.
(k) $\{r_1, r_2, \ldots, r_n\}$ is linearly independent.
(l) $\{r_1, r_2, \ldots, r_n\}$ is a basis for $\mathbb{R}^n$.
(m) M^T is invertible.

To prove Theorem 5.2.20, we will employ results from previous sections.

Proof.   To show these statements are equivalent, we first list the equivalencies we have already established in theorems from previous sections.

By Theorem 2.2.20, (c) $\iff$ (b).
By Corollary 3.1.32, (c) $\iff$ (d).
By Theorem 3.1.22, (c) $\iff$ (e).
By Theorem 3.3.14, (d) $\iff$ (f).
By Corollary 3.4.36, (d) $\iff$ (g).
By Theorem 4.5.20, (c) $\iff$ (h).

The above equivalencies show that statements (b)–(h) are all equivalent.

Next, we show that (i) is equivalent to (g). By Definition 3.4.3, if (g) is true then $\{c_1, c_2, \ldots, c_n\}$ is linearly independent. So, by Definition 5.2.7, $\{c_1, c_2, \ldots, c_n\}$ is a basis for $\mathrm{Col}(M)$ and $\mathrm{Rank}(M) = n$. That is, (i) is true. Now, assume (i), we know that there are n columns of M, and by Definition 5.2.7, we know that these columns span an n dimensional subspace of $\mathbb{R}^n$. That is, the columns span $\mathbb{R}^n$. If the columns form a linearly dependent set, then we would be able to find a spanning set of $\mathbb{R}^n$ with fewer than n elements. Theorem 3.4.20 tells us that this is impossible. So, we know that the columns of M form a basis for $\mathbb{R}^n$. But, since $\mathrm{Span}\{c_1, c_2, \ldots, c_n\} = \mathrm{Col}(M)$ we have that $\mathrm{Col}(M) = \mathbb{R}^n$. Therefore (i) is equivalent to (g).

The Rank-Nullity Theorem tells us that $\mathrm{Nullity}(M) = n - \mathrm{Rank}(M)$, so (i) is equivalent to (j).

Now, to show (a) is equivalent to statements (b)–(j), we show that (a) implies (c) and (b) implies (a). Because (c) is equivalent to (b), we will be able to conclude that (a) is also equivalent to all other statements in the theorem. If M is invertible, then there is a matrix M^{-1} so that $M^{-1}M = I_n$. Therefore, if $Mx = 0$, then

$$x = I_n x = M^{-1}Mx = M^{-1}0 = 0.$$

That is, $Mx = 0$ has only the trivial solution.

Now, suppose that the reduced echelon form of M has n leading entries. Then, when we reduce the augmented matrix $\left(M\,|\,I_n\right)$ to reduced echelon form we get a leading 1 in every column (and therefore every row) to the left of the augment. That is, the reduced echelon form of $\left(M\,|\,I_n\right)$ is a matrix of the form $\left(I_n\,|\,A\right)$, where A is an $n \times n$ matrix whose columns satisfy

$$Ma_1 = e_1$$
$$Ma_2 = e_2$$
$$\vdots$$
$$Ma_n = e_n.$$

That is,

$$MA = \begin{pmatrix} | & | & & | \\ e_1 & e_2 & \dots & e_n \\ | & | & & | \end{pmatrix}.$$

That is, $MA = I_n$. Reversing the matrix reduction steps used to reduce $\left(M\,|\,I_n\right)$ to $\left(I_n\,|\,A\right)$ are the same steps one would take to reduce $\left(A\,|\,I_n\right)$ to the reduced echelon form, $\left(I_n\,|\,M\right)$. This means that a similar argument tells us that $AM = I_n$. That is, $A = M^{-1}$ and M is invertible. Thus, (a) is equivalent to statements (b)–(j).

Finally, Theorem 5.2.16 tells us that $\text{Rank}(M^{\mathsf{T}}) = \text{Rank}(M)$. Therefore, we have (i) is equivalent to (m). Since $r_1, r_2, \ldots, r_n$ are the columns of M^{T}, we have (k) and (l) are both equivalent to (m).

Thus, statements (a)–(m) are all equivalent. $\qquad\square$

This theorem can be used to test a matrix for invertibility or to determine whether any of the equivalent statements are true. For example, we can use it to test whether the columns of a matrix span $\mathbb{R}^n$, or we can use it to determine whether the nullspace of a matrix is $\{0\}$. It tells us that for a square matrix, either all statements are true *or* all statements are false. We will add more equivalent statements to the invertible matrix theorem in later sections.

Now that we have characterized when a matrix is invertible, we want to use this characterization to say something about linear transformations. In the next theorem, we consider the relationship between the invertibility of a linear transformation and the invertibility of its matrix representation.

Theorem 5.2.21
Let V and W be finite-dimensional vector spaces and $T : V \to W$ be linear. Let $\mathcal{B}_V$ and $\mathcal{B}_W$ be ordered bases for V and W, respectively. Also let $M = [T]_{\mathcal{B}_V}^{\mathcal{B}_W}$. Then the following statements are equivalent.

(a) M is invertible.
(b) T is invertible.
(c) $[T^{-1}]_{\mathcal{B}_W}^{\mathcal{B}_V} = M^{-1}$.

Proof. Suppose V and W are vector spaces with dimension n and m, respectively. By Definition 5.2.17 and Theorem 4.7.49, we know that if any of (a)–(c) are true, then $V \cong W$. So we know that $m = n$.

Assume, also, that V and W have ordered bases $\mathcal{B}_V$ and $\mathcal{B}_W$, respectively. Let $T : V \to W$ be a linear transformation with matrix representation M. That is, $[T(v)]_{\mathcal{B}_W} = M[v]_{\mathcal{B}_V}$ Define $T_1 : \mathbb{R}^n \to V$ to be the transformation that maps each vector $v \in V$ to its coordinate vector $[v]_{\mathcal{B}_V} \in \mathbb{R}^n$, $T_2 : \mathbb{R}^n \to \mathbb{R}^n$ to be the transformation defined by multiplying vectors in $\mathbb{R}^n$ by the matrix M, and $T_3 : \mathbb{R}^n \to W$ to be the transformation that maps coordinate vectors $[w]_{\mathcal{B}_W} \in \mathbb{R}^n$ to their corresponding vector $w \in W$. (See Figure **??**.) That is, for $v \in V$, $w \in W$, and $x \in \mathbb{R}^n$,

$$T_1(v) = [v]_{\mathcal{B}_V}, \; T_2(x) = Mx, \; \text{and} \; T_3([w]_{\mathcal{B}_W}) = w.$$

We showed, in Theorems 4.7.43 and 4.7.49, that T_1 and T_3 are isomorphisms and that T_1^{-1} and T_3^{-1} exist. We will show

$$(a) \implies (b) \implies (c) \implies (a).$$

((a) $\implies$ (b)): Suppose M is invertible. Then, by definition, M^{-1} exists. Define $S_2 : \mathbb{R}^n \to \mathbb{R}^n$ (as in Figure **??**) for $y \in \mathbb{R}^n$, by

$$S_2(y) = M^{-1}y.$$

For $x, y \in \mathbb{R}^2$

$$(S_2 \circ T_2)(x) = S_2(T_2(x)) = S_2(Mx) = M^{-1}Mx = I_n x = x$$
$$\text{and}$$
$$(T_2 \circ S_2)(y) = T_2(S_2(y)) = T_2(M^{-1}y) = MM^{-1}y = I_n y = y.$$

Therefore, $S_2 = T_2^{-1}$. Now, $T = T_3 \circ T_2 \circ T_1$. (Here, we have employed Lemma 4.4.4 and Definition 4.4.7.) Define $S = T_1^{-1} \circ T_2^{-1} \circ T_3^{-1}$ and for $v \in V$ and $w \in W$ with $T(v) = w$, we have

$$
\begin{aligned}
(S \circ T)(v) &= (T_1^{-1} \circ T_2^{-1} \circ T_3^{-1}) \circ T(v) \\
&= T_1^{-1}(T_2^{-1}(T_3^{-1}(T(v)))) \\
&= T_1^{-1}(T_2^{-1}(T_3^{-1}(w))) \\
&= T_1^{-1}(T_2^{-1}([w]_{\mathcal{B}_W})) \\
&= T_1^{-1}(M^{-1}[w]_{\mathcal{B}_W}) \\
&= T_1^{-1}([v]_{\mathcal{B}_V}) \\
&= v,
\end{aligned}
$$

$$\text{and}$$

$$
\begin{aligned}
(T \circ S)(w) &= T(S(w)) \\
&= T(T_1^{-1}(T_2^{-1}(T_3^{-1}(w)))) \\
&= T(T_1^{-1}(T_2^{-1}([w]_{\mathcal{B}_W}))) \\
&= T(T_1^{-1}(M^{-1}[w]_{\mathcal{B}_W})) \\
&= T(T_1^{-1}([v]_{\mathcal{B}_V})) \\
&= T(v) \\
&= w.
\end{aligned}
$$

Therefore, $S = T^{-1}$ and so, T is invertible.

((b) $\implies$ (c)): In proving (a) implies (b), we found $T_2^{-1} : \mathbb{R}^n \to \mathbb{R}^n$, defined as multiplication by the matrix M^{-1}. And, if $T(v) = w$ then

$$M[v]_{\mathcal{B}_V} = [w]_{\mathcal{B}_W} \text{ and } T^{-1}(w) = v.$$

So

$$M^{-1}[w]_{\mathcal{B}_W} = [v]_{\mathcal{B}_v} = [T^{-1}(w)]_{\mathcal{B}_V},$$

then, by Lemma 4.4.4 and Definition 4.4.7, M^{-1} is the matrix representation of T^{-1}.

((c) $\implies$ (a)): Finally, suppose that T is invertible. Then, we know that the linear transformation T^{-1} exists. That means that $(T \circ T^{-1})(w) = w$ for all $w \in W$ and $(T^{-1} \circ T)(v) = v$ for all $v \in V$. Suppose that $T(v) = w$. Then $T^{-1}(w) = v$. We also know that there is a matrix A so that

$$[v]_{\mathcal{B}_V} = [T^{-1}(w)]_{\mathcal{B}_V} = A[w]_{\mathcal{B}_W}.$$

We will show that $AM = MA = I_n$. Indeed, we have, for any $x, y \in \mathbb{R}^n$, there is $v \in V$ and $w \in W$ so that $x = [v]_{\mathcal{B}_V}$ and $y = [w]_{\mathcal{B}_W}$. Therefore,

$$
\begin{aligned}
AMx &= AM[v]_{\mathcal{B}_V} \\
&= A[T(v)]_{\mathcal{B}_W} \\
&= A[w]_{\mathcal{B}_W} \\
&= [v]_{\mathcal{B}_V} \\
&= x.
\end{aligned}
$$

We also have for any $y \in \mathbb{R}^n$, there is

$$
\begin{aligned}
MAy &= MA[w]_{\mathcal{B}_W} \\
&= M[T(w)]_{\mathcal{B}_V} \\
&= M[v]_{\mathcal{B}_V} \\
&= [w]_{\mathcal{B}_W} \\
&= y.
\end{aligned}
$$

Therefore $MA = AM = I_n$. That is, M is invertible with $A = M^{-1}$ and $[T^{-1}]_{\mathcal{B}_W}^{\mathcal{B}_V} = M^{-1}$.   □

Theorems 5.2.20 and 5.2.21 tell us other equivalent statements about T such as "T is one-to-one" and "T is onto" (see Exercises 57 and 58 of Section **??**). Theorem 5.2.21 not only tells us that M is invertible if and only if T is invertible but also provides a method for finding the matrix representation of the inverse transformation T^{-1}.

Example 5.2.22 Consider the vector space of functions defined by

$$F = \left\{ f(x) = ce^{-3x} + d \cos 2x + b \sin 2x \mid c, d, b \in \mathbb{R} \right\},$$

over the field $\mathbb{R}$. Let $T : F \to F$ be the linear transformation defined by $T(f(x)) = f'(x)$. Let $\alpha = \{e^{-3x}, \cos 2x, \sin 2x\}$, a basis for F. Theorem 5.2.21 guarantees that the inverse matrix M^{-1} exists if and only if T^{-1} exists. The inverse transformation is defined by

$$T^{-1}(re^{-3x} + s\sin 2x + t\cos 2x) = -\tfrac{1}{3}re^{-3x} + \tfrac{1}{2}t\sin 2x - \tfrac{1}{2}s\cos 2x,$$

for scalars $r, s, t \in \mathbb{R}$. The reader can verify that this definition does provide the inverse transformation by confirming that $T^{-1}(T(\alpha)) = \alpha$ and that $T(T^{-1}(\alpha)) = \alpha$.

So now, we can find this matrix directly from T^{-1}.

$$M^{-1} = [T^{-1}]_\alpha = \begin{pmatrix} -\tfrac{1}{3} & 0 & 0 \\ 0 & 0 & \tfrac{1}{2} \\ 0 & -\tfrac{1}{2} & 0 \end{pmatrix}.$$

The reader can verify that $MM^{-1} = M^{-1}M = I_3$.   □

In this section, we explore the properties of matrix representations of linear transformations that are one-to-one and onto. In particular, we wish to know what matrix properties indicate the invertibility properties of the corresponding linear transformation.

Theorem 5.2.23

Let $T : V \to W$ be linear, dim $V = n$ and dim $W = m$. Let $M = [T]_{\mathcal{B}_V}^{\mathcal{B}_W}$ be the matrix representation of T relative to domain basis $\mathcal{B}_V$ and codomain basis $\mathcal{B}_W$. Then

(a) T is onto if, and only if, the rows of M form a linearly independent set in $\mathbb{R}^n$.
(b) T is one-to-one if, and only if, the columns of M form a linearly independent set in $\mathbb{R}^m$.

Proof. Let V and W be vector spaces with dim $V = n$ and dim $W = m$. Define $T : V \to W$ to be a linear transformation. Let $M = [T]_{\mathcal{B}_V}^{\mathcal{B}_W}$ be the matrix representation of T relative to domain basis $\mathcal{B}_V$ and codomain basis $\mathcal{B}_W$.

We first consider the map $T_2 : \mathbb{R}^n \to \mathbb{R}^m$ defined by $T_2(x) = Mx$ first discussed in Section **??**. Since the coordinate maps $T_1 : V \to \mathbb{R}^n$ and $T_3 : \mathbb{R}^m \to W$ (see Figure **??**) are isomorphisms, then we know that T is onto if and only if T_2 is onto and T is one-to-one if and only if T_2 is one-to-one. Thus, it is sufficient to show that T_2 is onto if and only if the rows of M form a linearly independent set in $\mathbb{R}^n$ and that T_2 is one-to-one if and only if the columns of M form a linearly independent set in $\mathbb{R}^m$.

We know that T_2 is onto if and only if for all $b \in \mathbb{R}^m$, $T_2(x) = Mx = b$ has a solution. But we also know that $Mx = b$ has a solution for all $b \in \mathbb{R}^m$ if and only if the reduced echelon form of M has a leading one in every row. This happens if and only if no row of M can be written as a linear combination of other rows of M. That is, T is onto if and only if the rows of M are linearly independent.

Also, T_2 is one-to-one if and only if $T_2(x) = T_2(y)$ implies $x = y$; i.e., if and only if $T_2(x) = 0$ implies $x = 0$. We know that $T_2(x) = Mx = 0$ has a unique solution (the trivial solution) if and only if the columns of M are linearly independent. Therefore, T is one-to-one if and only if the columns of M are linearly independent.   □

Example 5.2.24 Consider the linear transformation $T: \mathcal{P}_3(\mathbb{R}) \to \mathcal{P}_2(\mathbb{R})$ defined by $T(f(x)) = f'(x)$. Let $\alpha = \{x^3, x^2, x, 1\}$ and $\beta = \{x^2, x, 1\}$. Then

$$M = [T]_\alpha^\beta = \begin{pmatrix} 3 & 0 & 0 & 0 \\ 0 & 2 & 0 & 0 \\ 0 & 0 & 1 & 0 \end{pmatrix}.$$

The rows of M form a linearly independent set, but the columns of M form a linearly dependent set. Theorem 5.2.23 tells us that T must be onto and is not one-to-one. Let's verify these properties.

We know that T is onto if $Tu = w$ has a solution $u \in \mathcal{P}_3(\mathbb{R})$ for every $w \in \mathcal{P}_2(\mathbb{R})$. Let $w = ax^2 + bx + c$ for arbitrary scalars $a, b, c \in \mathbb{R}$. Then $v = \frac{1}{3}ax^3 + \frac{1}{2}bx^2 + cx + d$ is a solution to $Tv = w$ for any $d \in \mathbb{R}$.

We know that T is not one-to-one if there exists $u \in \mathcal{P}_2(\mathbb{R})$, $u \neq 0$, such that $Tu = 0$. Let $u = 0x^2 + 0x + d$ for any nonzero scalar $d \in \mathbb{R}$. Then, $Tu = u'(x) = 0$. Thus, T is not one-to-one.   □

Let us continue connecting invertibility of a linear transformation and invertibility of the corresponding matrix representation.

Corollary 5.2.25
Let $T: V \to W$ be linear, dim $V = n$ and dim $W = m$. Let $M = [T]_{\mathcal{B}_V}^{\mathcal{B}_W}$ be the matrix representation of T relative to domain basis $\mathcal{B}_V$ and codomain basis $\mathcal{B}_W$. Then T is invertible if, and only if, the columns of M form a basis for $\mathbb{R}^m$ and the rows of M form a basis for $\mathbb{R}^n$.

Proof. Let V and W be vector spaces with dim $V = n$ and dim $W = m$. We know that T is invertible if and only if $V \cong W$. That is, $m = n$. We also know, by Theorems 5.2.20 and 5.2.21, that

$$\begin{aligned} T \text{ is invertible} \quad &\Longleftrightarrow \quad M \text{ is invertible} \\ &\Longleftrightarrow \quad \text{The columns of } M \text{ form a basis for } \mathbb{R}^n \\ &\Longleftrightarrow \quad \text{The rows of } M \text{ form a basis for } \mathbb{R}^n. \end{aligned}$$

   □

Example 5.2.26 Recall the transformation given in Example 5.2.22 was invertible. The matrix for the transformation is

$$M = [T]_\alpha = \begin{pmatrix} -3 & 0 & 0 \\ 0 & 0 & -2 \\ 0 & 2 & 0 \end{pmatrix}.$$

The rows of M form a linearly independent set, and the columns of M form a linearly independent set. Corollary 5.2.25 confirms the invertibility of T.   □

** **Watch Your Language!** The common practice of describing both transformations and matrices as "invertible" and having an "inverse." However, it is not appropriate to use the same language for both. Acceptable language for discussing properties of transformations and its corresponding matrix representation is provided here.

✓ The transformation T is one-to-one.
✓ The matrix representation M has linearly independent rows.
✓ The transformation T is onto.
✓ The matrix representation M has linearly independent columns.
✓ The transformation T is an isomorphism.
✓ The matrix representation is row equivalent to the identity matrix.
✓ T is invertible.
✓ M is invertible.
✓ T has an inverse, T^{-1}.
✓ M has an inverse, M^{-1}.

It is *not* accepted practice to say

✗ The transformation T has linearly independent rows.
✗ The matrix representation M is one-to-one.
✗ The transformation T has linearly independent columns.
✗ The matrix representation M is onto.
✗ The transformation T can be reduced to the identity.
✗ The matrix representation is an isomorphism.

☞ **Path to New Applications**
Linear programming is a tool of optimization. Solving systems of equations and matrix equations are necessary tools for finding basic solutions. To determine whether or not there is a basic solution, researchers refer to the Invertible Matrix Theorem (Theorem 7.3.16). See Section **??** to learn more about how to connect linear programming to linear algebra tools.

5.2.3 Exercises

1. For each of the following matrices, find Null(M), Col(M), Rank(M), Nullity(M), size(M), the number of columns without leading entries, and the number of leading entries in the echelon form.

   (a) $M = \begin{pmatrix} 1 & 2 & 3 & -1 \\ 1 & 1 & -1 & -1 \\ 2 & 3 & 2 & -2 \\ 5 & 6 & -1 & -5 \end{pmatrix}$

   (b) $M = \begin{pmatrix} 1 & 2 & 3 & -1 & 0 \\ 1 & 3 & -1 & -1 & 2 \\ 3 & 3 & -1 & -2 & -1 \end{pmatrix}$

   (c) $M = \begin{pmatrix} 1 & 0 & 1 \\ 1 & 1 & -1 \\ 2 & 2 & 2 \\ 3 & 1 & 4 \\ -1 & 0 & 1 \end{pmatrix}$

   (d) $M = \begin{pmatrix} 3 & 0 & 0 & 0 \\ 0 & 2 & 0 & 0 \\ 0 & 1 & 1 & 0 \end{pmatrix}$.

2. Verify by example, using the matrices below, that Theorem 5.2.19 is true. That is, show Theorem 5.2.19 is true by finding M^{-1}, $\det(M)$, and $\det(M^{-1})$ using matrix reduction. This is not a proof, but the proof is very similar.

   (a) $M = \begin{pmatrix} 1 & 1 \\ 2 & 3 \end{pmatrix}$

   (b) $M = \begin{pmatrix} 2 & 7 \\ 3 & 4 \end{pmatrix}$

   (c) $M = \begin{pmatrix} a & b \\ c & d \end{pmatrix}$ (Assume $ad - bc \neq 0$.)

   (d) $M = \begin{pmatrix} 1 & 2 & 3 \\ 1 & 1 & 1 \\ 0 & 1 & 1 \end{pmatrix}$

   (e) $M = \begin{pmatrix} 2 & 0 & 1 \\ -1 & 2 & 1 \\ 1 & 1 & 2 \end{pmatrix}$.

3. How does nullity(M) show up in the echelon form of the matrix reduction?
4. How does rank(M) show up in the echelon form of the matrix reduction?
5. How are dim V and dim W related to M?
6. Use the Rank-Nullity Theorem to make a conjecture that brings together a relationship with all or some of the answers to Exercises 3, 4, and 5.
7. The invertible matrix theorem is an important theorem. Fill in the blanks or circle the correct answer below to complete the statement of the theorem.

   (a) $AX = b$ has a unique solution if _____
      (Choose one: A is invertible or A is not invertible).
   (b) A is invertible if and only if $\det(A)$_____.
   (c) $AX = b$ has a unique solution if $\det(A)$_____.

8. Explain why the columns of M may not form a basis for Col(M) in Definition 5.2.7.
9. Prove Corollary 5.2.3.
10. Prove Theorem 5.2.8.
11. Prove Theorem 5.2.9.
12. Prove Corollary 5.2.10.

Which of the following statements are equivalent to those given in the Invertible Matrix Theorem? Prove or disprove the equivalency of each statement. Use the conditions and notation given in Theorems 5.2.20 and 5.2.21.

13. M is one-to-one.
14. The reduced echelon form of M is the identity matrix I_n.
15. Ran$(T) = \mathbb{R}^n$.
16. $Mx = b$ has at least one solution $x \in \mathbb{R}^n$ for each $b \in \mathbb{R}^n$.
17. M^2 is invertible.
18. Null$(M) = \{0\}$.
19. $[T^{-1}]_{\mathcal{B}_W}^{\mathcal{B}_V} = M^T$.
20. If $\{b_1, b_2, \ldots, b_n\}$ is a basis for $\mathbb{R}^n$ then $\{Mb_1, Mb_2, \ldots, Mb_n\}$ is also a basis for $\mathbb{R}^n$.
21. If $\{b_1, b_2, \ldots, b_n\}$ is a basis for V then $\{Tb_1, Tb_2, \ldots, Tb_n\}$ is a basis for W.
22. Explain why the nullspace of a matrix can be viewed as the solution set of a homogeneous system of equations.

5.3   Exploration: Reconstruction Without an Inverse

In this exploration, we construct and use left-inverse transformations for linear transformations which are one-to-one, but not necessarily an isomorphism.

Suppose we have a radiographic transformation $T : \mathcal{O} \to \mathcal{R}$, where $\mathcal{O}$ is the object space with ordered basis $\mathcal{B}_{\mathcal{O}}$ and $\mathcal{R}$ is the radiograph space with ordered basis $\mathcal{B}_{\mathcal{R}}$. For any object vector $x \in \mathcal{O}$ and corresponding radiograph $b \in \mathcal{R}$, we have the equivalent matrix representation $[T]_{\mathcal{B}_{\mathcal{O}}}^{\mathcal{B}_{\mathcal{R}}} [x]_{\mathcal{B}_{\mathcal{O}}} = [b]_{\mathcal{B}_{\mathcal{R}}}$. We will consider transformations T for which a left-inverse transformation $S : \mathcal{R} \to \mathcal{O}$ exists (see Definition 4.7.53). S has the matrix representation $[S]_{\mathcal{B}_{\mathcal{R}}}^{\mathcal{B}_{\mathcal{O}}}$. The existence of S allows us to recover an object from radiographic data:

$$[S]_{\mathcal{B}_{\mathcal{R}}}^{\mathcal{B}_{\mathcal{O}}} [b]_{\mathcal{B}_{\mathcal{R}}} = [S]_{\mathcal{B}_{\mathcal{R}}}^{\mathcal{B}_{\mathcal{O}}} [T]_{\mathcal{B}_{\mathcal{O}}}^{\mathcal{B}_{\mathcal{R}}} [x]_{\mathcal{B}_{\mathcal{O}}} = [I]_{\mathcal{B}_{\mathcal{O}}}^{\mathcal{B}_{\mathcal{O}}} [x]_{\mathcal{B}_{\mathcal{O}}} = [x]_{\mathcal{B}_{\mathcal{O}}} .$$

In order to simplify the notation for this section, we will write $Tx = b$ and $Sb = S(Tx) = Ix = x$ with the understanding that we are working in the coordinate spaces relative to the bases $\mathcal{B}_{\mathcal{O}}$ and $\mathcal{B}_{\mathcal{R}}$.

5.3.1   Transpose of a Matrix

To begin our study, we recall Definition 2.5.14, the definition of the transpose of a matrix. Here, we restate the definition more concisely. We also provide some properties of the transpose.

Definition 5.3.1

The **transpose** of $n \times m$ matrix A, denoted A^{T}, is the $m \times n$ matrix formed by interchanging the columns and rows. That is, $(A^{\mathsf{T}})_{i,j} = A_{j,i}$ for all $1 \leq i \leq n$ and $1 \leq j \leq m$.   □

Theorem 5.3.2

Properties of the transpose. Let A and B be $m \times n$ matrices and C an $n \times k$ matrix.

1. $(A^{\mathsf{T}})^{\mathsf{T}} = A$.
2. If $D = A^{\mathsf{T}}$, then $D^{\mathsf{T}} = A$.
3. $(AC)^{\mathsf{T}} = C^{\mathsf{T}} A^{\mathsf{T}}$.
4. $(A + B)^{\mathsf{T}} = A^{\mathsf{T}} + B^{\mathsf{T}}$.

Proof. In this proof, let indices i and j be arbitrary over their expected ranges, and let A_{ij} be the entry of A in the ith row and j^{th} column. In each case, we show that two matrices X and Y are equal by showing that arbitrary entries are equal: $X_{ij} = Y_{ij}$.

1. $((A^{\mathsf{T}})^{\mathsf{T}})_{ij} = (A^{\mathsf{T}})_{ji} = A_{ij}$, so $(A^{\mathsf{T}})^{\mathsf{T}} = A$.
2. Suppose $D = A^{\mathsf{T}}$. Then, by (1), $A = (A^{\mathsf{T}})^{\mathsf{T}} = D^{\mathsf{T}}$.
3. Let $\bar{A} = A^{\mathsf{T}}$ and $\bar{C} = C^{\mathsf{T}}$. $((AC)^{\mathsf{T}})_{ij} = (AC)_{ji} = \sum_{\ell=1}^{k} a_{j\ell} c_{\ell i} = \sum_{\ell=1}^{k} \bar{c}_{i\ell} \bar{a}_{\ell j} = (\bar{C}\bar{A})_{ij}$
   $= (C^{\mathsf{T}} A^{\mathsf{T}})_{ij}$. Thus, $(AC)^{\mathsf{T}} = C^{\mathsf{T}} A^{\mathsf{T}}$.

4. $((A + B)^\mathsf{T})_{ij} = (A + B)_{ji} = A_{ji} + B_{ji} = (A^\mathsf{T})_{ij} + (B^\mathsf{T})_{ij} = (A^\mathsf{T} + B^\mathsf{T})_{ij}$. Thus, $(A + B)^\mathsf{T} = A^\mathsf{T} + B^\mathsf{T}$.

$\square$

5.3.2 Invertible Transformation

Now consider the following example: We are given a radiograph with $M = 24$ pixels that was created by applying some radiographic transformation, T to an object with $N = 16$ voxels.

1. Give a scenario for a radiographic transformation T that fits the above example. Don't calculate a T, rather give the following:

    – Size of the object: _____ × _____.
    – Number of pixels per view:
    – Number of views:

2. Suppose we know b and we want to find x. This means that we want

$$x = T^{-1}b.$$

    (a) What properties must the *transformation T* have so that T is invertible?
    (b) What properties must the *transformation T* have so that a left-inverse of T exists?
    (c) What matrix properties must the *matrix representation* of T have so that it is invertible?
    (d) When $N \le M$ (as in the example above), what matrix properties must the *matrix representation* of T have so that it has a left-inverse?

3. For ease of notation, when the bases are understood, we often use the same notation for the matrix and the transformation, that is, we let T represent both the transformation and its associated matrix. Suppose, $N \le M$ and a left-inverse, P of T exists. This means that $x = Pb$. If T were invertible, we would have $P = T^{-1}$. But, in the example above, we know that T is not invertible. Using the following steps, find the left-inverse of T.

    (a) Because $Tx = b$, for any linear transformation (matrix) A, we can write $ATx = Ab$. This is helpful if AT is invertible. Since T is one-to-one, we know that for AT to be invertible, the only vector in $\mathrm{Ran}(T)$ that can be in $\mathrm{Null}(A)$ is the zero vector. What other properties must A have so that AT is invertible?
    (b) Provide a matrix, A so that A has the properties you listed in 3a and so that AT is invertible.
    (c) Solve for x in the matrix equation $ATx = Ab$ using the A you found and provide a representation of the left-inverse of P.

4. Putting this all together now, state the necessary and sufficient condition for T to have a left-inverse?

5.3.3 Application to a Small Example

Consider the following radiographic example.

- Total number of voxels: $N = 16$ ($n = 4$).
- Total number of pixels: $M = 24$
- $ScaleFac = 1$

- Number of views: $a = 6$
- View angles: $\theta_1 = 0°$, $\theta_2 = 20°$, $\theta_3 = 40°$, $\theta_4 = 60°$, $\theta_5 = 80°$, $\theta_6 = 100°$.

5. Use **tomomap.m** to compute T and verify that the left-inverse of T must exist. The function **tomomap** returns a transformation matrix in sparse format. To use and view as a full matrix array use the command T=full(T); after constructing T.

6. Compute the left-inverse P. Use P to find the object that created the following radiograph vector (You should be able to copy and paste this into OCTAVE or MATLAB.

```
b=[0.00000
32.00000
32.00000
0.00000
1.97552
30.02448
30.02448
1.97552
2.71552
29.28448
29.28448
2.71552
2.47520
29.52480
29.52480
2.47520
1.17456
30.82544
30.82544
1.17456
1.17456
30.82544
30.82544
1.17456]
```

5.3.4 Application to Brain Reconstruction

Now, we can reconstruct some brain images from radiographic data. This section will guide you in this process.

7. Collect the following necessary files and place them in your working OCTAVE/MATLAB directory.

   (a) Data File: Lab5radiographs.mat
   (b) Plotting Script: ShowSlices.m
   (c) OCTAVE/MATLAB Code: tomomap.m

8. Choose whether to create a new script file (".m" file) for all of your commands or to work at the OCTAVE/MATLAB prompt.

9. Load the provided radiographic data with the following command.

```
load Lab5radiographs.mat
```

This command loads a variable named B that is a 12960x362 array. Each column is a radiograph vector corresponding to one horizontal slice of the human head. Each radiograph has 12960 total pixels spread across many views. The first 181 columns are noiseless radiographs, and the last 181 columns are corresponding radiographs with a small amount of noise added.

10. Use the familiar function tomomap.m to construct the transformation operator T corresponding to the following scenario (which is the scenario under which the radiographic data was obtained): $n = 108$, $m = 108$, $ScaleFac = 1$, and 120 view angles: the first at $1°$, the last at $179°$, and the rest equally spaced in between (hint: use the linspace command).

11. Some OCTAVE/MATLAB functions do not work with sparse arrays (such as our T). So, just make T a full array with this command:

    ```
    T=full(T);
    ```

12. It is tempting to compute the one-sided inverse P as found in (3c). However, such a large matrix takes time to compute and too much memory for storage. Instead, we can use a more efficient solver provided by OCTAVE/MATLAB. If we seek a solution to $Lx = b$, for invertible matrix L, we find the unique solution by finding L^{-1} and then multiplying it by b. OCTAVE/MATLAB does both operations together in an efficient way (by not actually computing L^{-1}) when we use the command x=L\b.

    As a first test, we will try this with the 50th radiograph in the matrix B. That is, we will reconstruct the slice of the brain that produced the radiograph that is represented in the 50th column of B. We want to solve the equation found in (3c): $ATx = Ab$ using the A matrix which you found in (3b).

    ```
    b=B(:,50);
    x=(A*T)\(A*b);
    ```

    The vector x is the coordinate vector for our first brain slice reconstruction. To view this reconstruction, use the following commands

    ```
    figure;
    x=reshape(x,108,108);
    imagesc(x,[0,255]);
    ```

    The reshape command is necessary above because the result x is a $(108 \cdot 108) \times 1$ vector, but the object is a 108×108 image.

13. Notice also that x and b could be matrices, say X and B. In this case, each column of X is the unique solution (reconstruction) for the corresponding column of B. Use these ideas to reconstruct *all* 362 brain slices using a single OCTAVE/MATLAB command.

    Use the variable name X for the results. Make sure that X is an 11664x362 array. Now, the first 181 columns are reconstructions from noiseless data, and the last 181 columns are reconstructions from noisy data.

14. Run the script file ShowSlices.m that takes the variable X and plots example slice reconstructions. Open ShowSlices.m in an editor and observe the line

    ```
    slices=[50 90 130];
    ```

Any three slices can be chosen to plot by changing the slice numbers. In the figure, the left column of images are reconstructions from noiseless data, the right column of images are reconstructions from the corresponding noisy data. IMPORTANT: Once X is computed it is only necessary to run ShowSlices.m to view different slices; running the other commands described above is time consuming and unnecessary.

Congratulations! You have just performed your first brain scan tomography. Using your new tool, answer the following questions.

15. Are the reconstructions from the noiseless radiographs exact discrete representations of the brain? Based on the concepts you have learned about matrix equations, explain why the reconstructions are exact or why not?

16. Determine a way to find the relative amount of noise in the noisy radiographs. Remember that the noisy radiograph corresponding to the radiograph represented in the 50th column of B is in the $(181 + 50)^{th}$ column of B. There are many ways to answer this question, so be creative. Is there a lot or a little noise in the radiographs?

17. Are the reconstructions from the noisy radiographs exact representations of the brain?

18. Compare the degree of "noisiness" in the noisy radiographs to the degree of "noisiness" in the corresponding reconstructions. Draw some conclusion about what this comparison tells us about the practicality of this method for brain scan tomography.

Diagonalization

6

In studying the behavior of heat states, we used a (linear) heat diffusion transformation E. Given a heat state $u(t)$ at time t, the heat state at a later time $t + \Delta t$ is given by $u(t + \Delta t) = Eu(t)$. But, this works well only if the time step Δt is sufficiently small. So, in order to describe the evolution of heat states over a longer period of time, the heat diffusion transformation must be applied many times. For example, in Figure 6.1, $u(3000\Delta t)$ requires 3000 applications of E. That is,

$$u(3000\Delta t) = E(E(\cdots E(u(t))\cdots)) = E^{3000}u(t).$$

Performing such repeated operations can be tedious, time-consuming, and numerically unstable. We need to find a more robust and practical method for exploring evolutionary transformations that are described by repeated operations of a single transformation.

The key tool in this exploration is the idea that transformations (and vectors) can be represented in terms of any basis we choose for the vector space of interest. We will find that not all representations are useful, but some are surprisingly simplifying. Using our findings, we will be able to study different aspects of evolutionary transformations including basis interpretation, independent state evolution, long-term state behavior, and even new findings on invertibility.

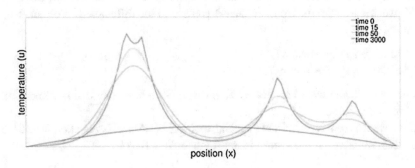

Fig. 6.1 In finding each of the heat states in this figure, we must apply repeated applications of E to the initial heat state, $u(0)$.

© Springer Nature Switzerland AG 2022, corrected publication 2023
H. A. Moon et al., *Application-Inspired Linear Algebra*, Springer Undergraduate Texts
in Mathematics and Technology, https://doi.org/10.1007/978-3-030-86155-1_6

6.1   Exploration: Heat State Evolution

In the exercises of Section 4.3, we explored two ways to compute the heat state along a rod after k time steps. We found the matrix representation, E, for the heat diffusion transformation, relative to the standard basis. We considered raising E to a large power, k, followed by multiplying by the (coordinate vector for the) initial heat state:

$$u(k\Delta t) = E^k u(0).$$

We also considered iteratively multiplying the (coordinate vector for the) current heat state by E:

$$u(\Delta t) = Eu(0), \ u(2\Delta t) = Eu(\Delta t), \ldots, \ u(k\Delta t) = Eu((k-1)\Delta t).$$

In both scenarios, we found that the results became more and more computationally cumbersome. In this section, we will explore heat states for which multiplying by E is not cumbersome.

▶   Throughout the discussion that follows it will be important to pay attention to the basis in which coordinate vectors are expressed.

The following tasks will lead the reader through this exploration.

1. In the following picture, there are 12 different initial heat states (in orange) and their corresponding diffusions (colors varying from orange to dark blue). Group the pictures based on the diffusion behavior. We are not interested in comparing particular snapshots in time (for example, comparing all of the greed curves). Rather, we are interested in features of how each state evolves from orange to blue. Briefly list the criteria you used to group them. As we continue this exploration, a clear description of your chosen criteria will be helpful.

To continue the exploration, the reader needs to recognize the common feature among heat state evolution for heat states 3, 4, and 8. A hint is provided in the first paragraph of the next chapter, but we encourage the readers to attempt to recognize this feature without reading the hint. We will continue this exploration with this particular feature.

2. Write the expression for Eu for the special vectors sharing the features you identified for heat states 3, 4, and 8.
3. Now view the diffusion of linear combinations of these special vectors. Use the MATLAB/OCTAVE code DiffuseLinearCombination.m. What do you see in the diffusion of a vector that is a linear combination of

   (a) Two of these special vectors?
   (b) Three of these special vectors?

4. Write out algebraically what happens in the diffusion of a heat state that is a linear combination of these special vectors.
5. What if we want to find more of these special vectors? What matrix equation would we solve?
6. What do this equation and the invertible matrix theorem tell us?

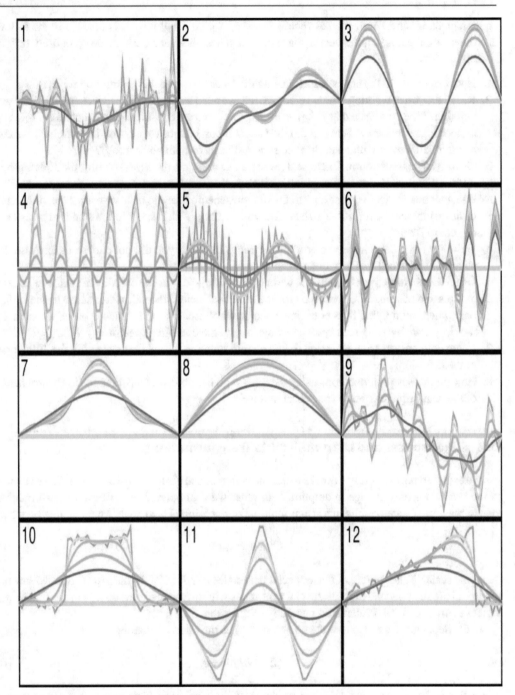

Exercises

Recall that our goal is to find the heat state after k time steps of the heat diffusion. We have observed that some heat states (special vectors) only change in amplitude with time. We see this when considering the matrix representation E and special vectors v_j for which $Ev_j = a_j v_j$. Now we will explore what could happen if some set of special vectors $\beta = \{v_1, v_2, \cdots, v_m\}$ is a basis for $\mathbb{R}^m$. In the following exercises, assume that, for the vector space of heat states, there is a basis made up entirely of special vectors.

1. How can we write the heat state $u(t)$ using the basis β? Write this expression for $u(t)$.
2. Recall that when working within vector spaces, we need to know with respect to which basis we are working. When we write that $\beta = \{v_1, v_2, \ldots, v_m\}$, we are considering the vectors $v_1, v_2, \ldots, v_m$ as coordinate vectors with respect to the standard basis. Is the expression in Exercise 1 written as a coordinate vector with respect to the standard basis or with respect to β?
3. Using your expression in Exercise 1, write an expression for $u(t + \Delta t)$ by multiplying by the diffusion matrix E. Are you working with respect to the standard basis or with respect to β?
4. Use your answer from Question 1 to find a representation for the heat state after the 10th iteration of diffusion, that is, for $u(t + 10\Delta t)$. Are you working with respect to the standard basis or with respect to β?
5. Write $u(t)$ as a coordinate vector with respect to the basis you did not choose in Exercise 2. For simplicity, let's call this new coordinate vector $w(t)$.
6. Using the notation from Question 5, we can write $w(t + \Delta t)$ as the coordinate vector (with respect to the same basis as in Question 5) for the next heat state. Since E makes sense as the diffusion only when multiplying by a coordinate vector with respect to the standard basis, we cannot just multiply $w(t)$ by E. Discuss possible ways to find the coordinate vector $w(t + \Delta t)$.
7. Using your answer from Question 6, find a representation for the heat state after the 10th iteration of the diffusion.
8. How might these computations using the basis β give us information about the long-term behavior of the heat diffusion? State your observations.

6.2   Eigenspaces and Diagonalizable Transformations

As we explored how heat states evolve under the action of a diffusion transformation E, we found that some heat states only change in amplitude. In other words, applying the diffusion transformation to one of these heat states results in a scalar multiple of the original heat state. Mathematically, we write

$$Ev = \lambda v, \tag{6.1}$$

for some scalar λ and for one of these special heat states $v \in \mathcal{H}_m(\mathbb{R})$. In the last section we saw that it is easy to predict how one of these special heat states will diffuse many time steps in the future, which makes them of particular interest in our study of diffusion.

In the Section 6.1 we also rewrote equation (6.3) as the matrix equation

$$(E - \lambda I)v = 0. \tag{6.2}$$

Since this is a homogeneous equation, we know it has a solution. This means that either there is a unique solution (only the trivial solution $v = 0$) or infinitely many solutions. If we begin with the zero heat state (all temperatures are the same everywhere along the rod) then the diffusion is trivial because nothing happens. The special vectors we seek are *nonzero* vectors satisfying the matrix equation (6.2). In order to do this we need to find values of λ so that the matrix $(E - \lambda I)$ has a nontrivial nullspace.

We also observed previously that, since the diffusion transformation is linear, we can readily predict the diffusion of any linear combination of our special heat states. This means that a basis of these special heat states is particularly desirable. If $\mathcal{B} = \{v_1, v_2, \ldots, v_m\}$ is a basis of these special vectors so that $Ev_i = \lambda_i v_i$ and $u(0)$ is our initial heat state, we can write $u(0)$ in coordinates relative to $\mathcal{B}$. That is, we can find scalars $\alpha_1, \alpha_2, \ldots \alpha_m$ so that

$$u(0) = \alpha_1 v_1 + \alpha_2 v_2 + \ldots + \alpha_m v_m.$$

Then, when we apply the diffusion operator to find the heat state, $u(\Delta t)$, one time step (Δt) later, we get

$$\begin{aligned}
u(\Delta t) &= Eu(0) \\
&= E(\alpha_1 v_1 + \alpha_2 v_2 + \ldots + \alpha_m v_m) \\
&= \alpha_1 E v_1 + \alpha_2 E v_2 + \ldots + \alpha_m E v_m \\
&= \alpha_1 \lambda_1 v_1 + \alpha_2 \lambda_2 v_2 + \ldots + \alpha_m \lambda_m v_m.
\end{aligned}$$

Now, using this same idea, we can find $u(k\Delta t)$ for k time steps later (far into the future). We get

$$\begin{aligned}
u(k\Delta t) &= E^k u(0) \\
&= \alpha_1 E^k v_1 + \alpha_2 E^k v_2 + \ldots + \alpha_m E^k v_m \\
&= \alpha_1 \lambda_1^k v_1 + \alpha_2 \lambda_2^k v_2 + \ldots + \alpha_m \lambda_m^k v_m.
\end{aligned}$$

This equation requires *no* matrix multiplication and $u(k\Delta t)$ can be computed directly without the need for computing each intermediate time-step result. Because of this, we can easily predict the long-term behavior of this diffusion in a way that is also computationally efficient.

6.2.1 Eigenvectors and Eigenvalues

Nonzero vectors satisfying equation (6.1) are important in both of the main applications (diffusion welding and radiography/tomography) in this text. Such vectors also arise in important ways in many other linear algebra concepts and applications. We assign the following terminology to these vectors and their corresponding scalars.

Definition 6.2.1

Let V be a vector space and $L : V \to V$ a linear transformation. If for some scalar λ and some nonzero vector $v \in V$, $L(v) = \lambda v$, then we say v is an **eigenvector** of L with **eigenvalue** λ. $\quad\square$

As with heat states, we see that eigenvectors of a linear transformation only change amplitude (and possibly sign) when they are mapped by the transformation. This makes repetitive applications of a linear transformation on its eigenvectors very simple.

▶ **Important Note:** Throughout our discussion of eigenvectors and eigenvalues we will assume that L is a linear transformation on the n-dimensional vector space V. That is, $L : V \to V$ is linear. We will also let M be a matrix representation of L relative to some basis. The reader should be aware of whether we are working directly in the vector space V or whether we have fixed a basis and hence are working in the associated coordinate space.

Example 6.2.2 Consider the linear transformation $L : \mathbb{R}^2 \to \mathbb{R}^2$ defined by $L(x) = Mx$ where $M = \begin{pmatrix} 1 & 1 \\ 3 & -1 \end{pmatrix}$. The vector $v = \begin{pmatrix} 1 \\ 1 \end{pmatrix}$ is an eigenvector of M with eigenvalue 2 because

$$L(v) = Mv = \begin{pmatrix} 1 & 1 \\ 3 & -1 \end{pmatrix} \begin{pmatrix} 1 \\ 1 \end{pmatrix} = \begin{pmatrix} 2 \\ 2 \end{pmatrix} = 2v.$$

The vector $w = \begin{pmatrix} 1 \\ 2 \end{pmatrix}$ is not an eigenvector of L because

$$L(w) = Mw = \begin{pmatrix} 1 & 1 \\ 3 & -1 \end{pmatrix} \begin{pmatrix} 1 \\ 2 \end{pmatrix} = \begin{pmatrix} 3 \\ 1 \end{pmatrix} \neq \lambda \begin{pmatrix} 1 \\ 2 \end{pmatrix} = \lambda w$$

for any scalar λ.   □

Example 6.2.3 Consider the vector space of 7-bar LCD characters, $V = \mathcal{D}(Z_2)$ and the linear transformation T on V whose action is to flip any character upside-down. For example,

We see that any nonzero character x with up-down symmetry is an eigenvector of T with eigenvalue 1 because $T(x) = x = 1x$.   □

In order to find the eigenvalues and eigenvectors of a transformation $L : V \to V$, we need to find λ so that $(L - \lambda I)(v) = 0$ has nontrivial solutions v; equivalently, we can consider the equivalent matrix equation $(M - \lambda I)v = 0$. If we find any eigenvector v with eigenvalue λ, then any (nonzero) scalar multiple of v is also an eigenvector with eigenvalue λ (see Exercise 19). Also, if we add any two eigenvectors v and w with eigenvalue λ, then their (nonzero) sum $v + w$ is an eigenvector with eigenvalue λ (see Exercise 20). This leads to the idea of an eigenspace of L (and of M) for any eigenvalue λ.

It is important to note that eigenvalues are elements of the scalar set over which scalar-vector multiplication is defined. In this text, when the scalar field is $\mathbb{R}$, we consider only eigenvalues $\lambda \in \mathbb{R}$. Here, we define the eigenspace corresponding to an eigenvalue, λ, to be the vector space of all vectors whose eigenvalue is λ.

Definition 6.2.4

Let V be a vector space and $L : V \to V$ a linear transformation. Let M be the matrix of the transformation L with respect to basis $\mathcal{B}$ of V. Suppose that λ is an eigenvalue of L (and hence, also of M). Then

$$\mathcal{E}_\lambda = \{v \in V \mid L(v) = \lambda v\} = \text{Null}(L - \lambda I)$$

is called the **eigenspace** of L (and of M) corresponding to eigenvalue λ.   □

Exercise 33 asks the reader to verify that every eigenspace of a linear transformation L is, indeed, a subspace of V.

We conclude with a brief remark connecting the nullspace of a transformation to its eigenspaces. Suppose v is an eigenvector of the linear transformation L with eigenvalue $\lambda = 0$. Because $L(v) = \lambda v = 0v = 0$, $v \in \text{Null}(L)$. However, there is one vector, namely $w = 0 \in \text{Null}(L)$ that is not an eigenvector of L. We want to be careful to note that eigenspaces contain the zero vector, but *the zero vector is not an eigenvector*.

6.2.2   Computing Eigenvalues and Finding Eigenvectors

In this section, we outline a procedure, called the Factor Algorithm, for determining the eigenvectors and eigenvalues of a matrix. We will give examples showing how to implement this procedure. Finally, we will discuss eigenvalues of diagonal matrices and eigenvalues of the transpose of a matrix. We begin with the algorithm.

▶ **Factor Algorithm for finding Eigenvectors and Eigenvalues** Given an $n \times n$ square matrix M, the algorithm for finding eigenvectors and eigenvalues is as follows.

1. Choose a basis $\mathcal{B} = \{w_1, w_2, \ldots, w_m\}$ for $\mathbb{R}^n$, and select one basis vector $w_\ell \in \mathcal{B}$.
2. Compute the elements of the set $S = \{w_\ell, Mw_\ell, M^2w_\ell, \cdots, M^kw_\ell\}$ where k is the first index for which the set is linearly dependent.
3. Find a linear dependence relation for the elements of S. This linear dependence relation creates a matrix equation $p(M)w_\ell = 0$ for some polynomial $p(M)$ of degree k.
4. Factor the polynomial $p(M)$ to find values of $\lambda \in \mathbb{R}$ and v satisfying $(M - \lambda I)v = 0$. Add any v to the set of eigenvectors and the corresponding λ to the set of eigenvalues.
5. Repeat, from Step 2, using any basis element $w_\ell \in \mathcal{B}$ that has not yet been used and is not in the span of the set of eigenvectors. If all basis vectors have been considered, then the algorithm terminates.

This algorithm is based on a method by W. A. McWorter and L. F. Meyers.[1] It differs from the method that many first courses in linear algebra use. We begin with this approach because we find other calculations less illuminating for our purposes and the method we present in this section produces eigenvectors at the same time as their eigenvalues. For the more standard approach, we refer the reader to Subsection 6.2.3.

We start with a small example of the Factor Algorithm.

Example 6.2.5 Let us consider the matrix M defined by

$$M = \begin{pmatrix} 5 & -3 \\ 6 & -4 \end{pmatrix}.$$

We seek nonzero $v \in \mathbb{R}^2$ and $\lambda \in \mathbb{R}$ so that $(M - \lambda I)v = 0$. This means that $v \in \text{Null}(M - \lambda I)$. Our plan is to write v as a linear combination of a set of linearly *dependent* vectors.

First, we choose the standard basis $\mathcal{B}$ for $\mathbb{R}^2$ and begin the Factor Algorithm with the first basis vector $w = \begin{pmatrix} 1 \\ 0 \end{pmatrix}$. We now create the set S. Notice that

---

[1] William A. McWorter, Jr. and Leroy F. Meyers, "Computing Eigenvalues and Eigenvectors without Determinants." Mathematics Magazine, Vol. 71, No. 1, Taylor & Francis. Feb. 1998.

$$Mw = \begin{pmatrix} 5 & -3 \\ 6 & -4 \end{pmatrix} \begin{pmatrix} 1 \\ 0 \end{pmatrix} = \begin{pmatrix} 5 \\ 6 \end{pmatrix}.$$

Notice also that $\{w, Mw\}$ is a linearly independent set. So, we continue by computing M^2w,

$$M^2w = \begin{pmatrix} 5 & -3 \\ 6 & -4 \end{pmatrix} \begin{pmatrix} 5 \\ 6 \end{pmatrix} = \begin{pmatrix} 7 \\ 6 \end{pmatrix}.$$

Now, we have the linearly dependent set

$$S = \{w, Mw, M^2w\} = \left\{ \begin{pmatrix} 1 \\ 0 \end{pmatrix}, \begin{pmatrix} 5 \\ 6 \end{pmatrix}, \begin{pmatrix} 7 \\ 6 \end{pmatrix} \right\}.$$

Therefore, we can find $\alpha_1, \alpha_2, \alpha_3$, not all zero, that satisfy the linear dependence relation

$$\alpha_1 \begin{pmatrix} 1 \\ 0 \end{pmatrix} + \alpha_2 \begin{pmatrix} 5 \\ 6 \end{pmatrix} + \alpha_3 \begin{pmatrix} 7 \\ 6 \end{pmatrix} = 0.$$

In fact, one linear dependence relation is given by $\alpha_1 = -2$, $\alpha_2 = -1$, and $\alpha_3 = 1$. That is,

$$M^2w - Mw - 2w = 0.$$

Now, we seek a factorization of the left side, of the matrix equation above, in the form $(M - \lambda I)(M + aI)w$ for some scalars λ and a. We will explore two methods to find this factorization. First, we recognize that

$$\begin{aligned} M^2w - Mw - 2w &= (M^2 - M - 2)w \\ &= (M - 2I)(M + I)w \\ &= (M - 2I)(Mw + w) \end{aligned}$$

or

$$= (M + I)(Mw - 2w).$$

From here, we have two eigenvalues, $\lambda_1 = 2$ and $\lambda_2 = -1$, corresponding to the eigenvectors $v_1 = Mw + w$ and $v_2 = Mw - 2w$, respectively.

Alternatively, we can find λ and a so that

$$M^2w - Mw - 2w = (M - \lambda I)(Mw + aw).$$

We have

$$(M - \lambda I)(Mw + aw) = M^2w + (a - \lambda)Mw - a\lambda w \overset{\text{set}}{=} M^2w - Mw - 2w = 0.$$

Matching up like terms and equating their coefficients, we have the system of *nonlinear* equations

$$\begin{aligned} -1 &= a - \lambda, \\ -2 &= -a\lambda. \end{aligned}$$

Eliminating a, we have the equation in λ:

$$\lambda^2 - \lambda - 2 = 0.$$

Again, we find two solutions: $\lambda = -1$, $a = -2$ and $\lambda = 2$, $a = 1$. Consider these two cases separately.

1. When $\lambda = -1$ we have $(M + I)(Mw - 2w) = 0$, from which we see that if we assign $v_1 = Mw - 2w$, we have $(M + I)v_1 = 0$. In other words, $Mv_1 = Iv_1 = v_1$. This means that v_1 (or any multiple of v_1) is an eigenvector with eigenvalue 1. We now compute v_1.

$$v_1 = Mw - 2w = \begin{pmatrix} 3 \\ 6 \end{pmatrix}.$$

Thus,

$$\lambda_1 = -1, \quad \mathcal{E}_{-1} = \operatorname{Span}\{v_1\} = \operatorname{Span}\left\{ \begin{pmatrix} 1 \\ 2 \end{pmatrix} \right\}.$$

2. When $\lambda = 2$ we have $(M - 2I)(Mw - w) = 0$, from which we find

$$v_2 = Mw + w = \begin{pmatrix} 6 \\ 6 \end{pmatrix}.$$

Thus,

$$\lambda_2 = 2, \quad \mathcal{E}_2 = \operatorname{Span}\{v_2\} = \operatorname{Span}\left\{ \begin{pmatrix} 1 \\ 1 \end{pmatrix} \right\}.$$

At this point, we have two eigenvectors v_1 and v_2, and every vector in the starting basis $\mathcal{B}$ is in the span of the set of eigenvectors. So, the algorithm terminates. We have a full description of the eigenvalues and eigenspaces of M because the eigenvectors we have found form a basis for $\mathbb{R}^2$. In other words, every vector in $\mathbb{R}^2$ can be written as a sum of eigenvectors from the eigenspaces, that is, $\mathbb{R}^2 = \mathcal{E}_1 \oplus \mathcal{E}_2$. (See Definition 2.5.27.) $\qquad\square$

In general, if we are given an $n \times n$ matrix M, we can expect at most n eigenvalues for an $n \times n$ matrix. (See Theorem 6.2.16.) The Factor Algorithm procedure was simple in the example above because we obtained two distinct eigenvalues of the 2×2 matrix, M by considering a single set of linearly dependent vectors. Typically, we may need to repeat the steps in the algorithm for finding eigenvalues and eigenvectors. In addition, we may find fewer than n eigenvalues. These two scenarios are demonstrated in the next two examples.

Example 6.2.6 Find the eigenvalues and eigenvectors of the linear transformation $L : \mathcal{P}_2(\mathbb{R}) \to \mathcal{P}_2(\mathbb{R})$ defined by

$$L(a + bx + cx^2) = (a - b + c) + (a + b - c)x + (-a + b + c)x^2.$$

Consider the standard basis $\mathcal{P} = \{1, x, x^2\}$ for $\mathcal{P}_2(\mathbb{R})$. We have the matrix representation of L relative to this basis:

$$M = \begin{pmatrix} 1 & -1 & 1 \\ 1 & 1 & -1 \\ -1 & 1 & 1 \end{pmatrix}.$$

Notice that M has rank 3 and is invertible. We can see this by checking any of the many statements given in Theorem 5.2.20. So, we know that L is an isomorphism. Thus, $\text{Null}(L) = \{0\}$ and $\text{Ran}(L) = \mathcal{P}_2(\mathbb{R})$, and further, 0 is not an eigenvalue of L (or M). We would like to describe the eigenspaces in $\mathbb{R}^3$ relative to M.

We begin the Factor Algorithm by choosing the standard basis $\mathcal{B}$ for $\mathbb{R}^n$ and initial basis vector $w = \begin{pmatrix} 1 \\ 0 \\ 0 \end{pmatrix} \in \mathcal{B}$. We next compute the linearly dependent set of vectors $\{w, Mw, M^2w, \cdots, M^kw\}$ with smallest k. We have

$$S = \left\{ w = \begin{pmatrix} 1 \\ 0 \\ 0 \end{pmatrix}, Mw = \begin{pmatrix} 1 \\ 1 \\ -1 \end{pmatrix}, M^2w = \begin{pmatrix} -1 \\ 3 \\ -1 \end{pmatrix}, M^3w = \begin{pmatrix} -5 \\ 3 \\ 3 \end{pmatrix} \right\}.$$

The first three vectors in the set are linearly independent. We have the linear dependence relation

$$M^3w - 3M^2w + 6Mw - 4w = 0$$

for which we seek scalars λ, α, β that give the factorization

$$(M - \lambda I)(M^2w + \alpha Mw + \beta w) = 0.$$

We have

$$(M - \lambda I)(M^2w + \alpha Mw + \beta w) = M^3w + (\alpha - \lambda)M^2w + (\beta - \alpha\lambda)Mw - \beta\lambda w$$
$$\overset{\text{set}}{=} M^3w - 3M^2w + 6Mw - 4w.$$

Equating coefficients of like terms, we have the nonlinear system of equations:

$$-3 = \alpha - \lambda$$
$$6 = \beta - \alpha\lambda$$
$$-4 = -\beta\lambda.$$

Using substitution, we see this has solutions satisfying

$$0 = \lambda^3 - 3\lambda^2 + 6\lambda - 4 = (\lambda - 1)(\lambda^2 - 2\lambda + 4).$$

This cubic equation has one real root, $\lambda_1 = 1$ and two complex roots, $\lambda_{2,3} = 1 \pm \sqrt{3}i$. (We can arrive at the same result by observing that the linear dependence relation can be written as $(M^3 - 3M^2 + 6M - 4I)w = 0$ which factors to $(M - I)(M^2 - 2M + 4I)w = 0$.) The real root is an eigenvalue of M and the corresponding eigenvector is

$$v_1 = M^2w + \alpha Mw + \beta w = M^2w - 2Mw + 4w = \begin{pmatrix} 1 \\ 1 \\ 1 \end{pmatrix}.$$

We now have a set of eigenvectors consisting of the single vector v_1. Step 5 of the algorithm requires us to choose, if possible, another vector from the basis $\mathcal{B}$ which is not in the span of the set of eigenvectors. Suppose we choose $w = \begin{pmatrix} 0 \\ 1 \\ 0 \end{pmatrix}$. Repeating Steps 2 through 5, we find the following results which the reader is encouraged to check.

$$S = \left\{ w = \begin{pmatrix} 0 \\ 1 \\ 0 \end{pmatrix}, Mw = \begin{pmatrix} -1 \\ 1 \\ 1 \end{pmatrix}, M^2 w = \begin{pmatrix} -1 \\ -1 \\ 3 \end{pmatrix}, M^3 w = \begin{pmatrix} 3 \\ -5 \\ 3 \end{pmatrix} \right\}.$$

$$M^3 w - 3M^2 w + 6Mw - 4w = 0.$$

This is the same linear dependence relation and no new eigenvector information is revealed. We continue by choosing the final basis vector, $w = \begin{pmatrix} 0 \\ 0 \\ 1 \end{pmatrix}$, which also results in the same linear dependence relation. The basis vectors are now exhausted and the algorithm terminates.

We have only

$$\lambda_1 = 1, \quad \mathcal{E}_1 = \text{Span} \left\{ \begin{pmatrix} 1 \\ 1 \\ 1 \end{pmatrix} \right\}.$$

So, any nonzero vector in $\mathcal{E}_1$ is an eigenvector of the matrix M. What are the eigenvectors of the transformation L? We used the standard basis to obtain M from L, so $v_1 = [p_1(x)]_\mathcal{B}$ for some $p_1(x) \in \mathcal{P}_2(\mathbb{R})$. That is

$$[p_1(x)]_\mathcal{P} = \begin{pmatrix} 1 \\ 1 \\ 1 \end{pmatrix}.$$

This tells us that the eigenvector $p_1(x)$ is $1(1) + 1(x) + (1)x^2 = 1 + x + x^2$. Thus, the eigenspace corresponding to eigenvalue $\lambda_1 = 1$ is

$$\mathcal{E}_1 = \text{Span} \left\{ 1 + x + x^2 \right\}.$$

Indeed,

$$\begin{aligned} L(p_1(x)) &= L(1 + x + x^2) \\ &= (1 - 1 + 1) + (1 + 1 - 1)x + (-1 + 1 + 1)x^2 \\ &= 1 + x + x^2 \\ &= 1 \cdot p_1(x) \\ &= \lambda_1 p_1(x). \end{aligned}$$

$\square$

The next example illustrates how to find a full complement of eigenspaces when multiple passes through the algorithm may be needed to find all n potential eigenvalues.

Example 6.2.7 We wish to find the eigenvalues and eigenspaces of the matrix

$$M = \begin{pmatrix} 3 & -4 & 2 \\ 2 & -3 & 2 \\ 0 & 0 & 1 \end{pmatrix}.$$

We proceed, as before, by choosing the standard basis $\mathcal{B}$ for $\mathbb{R}^3$, and selecting as an initial vector $w = \begin{pmatrix} 1 \\ 0 \\ 0 \end{pmatrix}$. We have the linearly dependent set

$$\{w, Mw, M^2w\} = \left\{ \begin{pmatrix} 1 \\ 0 \\ 0 \end{pmatrix}, \begin{pmatrix} 3 \\ 2 \\ 0 \end{pmatrix}, \begin{pmatrix} 1 \\ 0 \\ 0 \end{pmatrix} \right\}.$$

This set has the linear dependence relation $M^2w - w = 0$. This relation factors as $(M - I)(M + I)w = 0$, leading to two different factorizations of the form $(M + \lambda I)v = 0$. First, $(M - I)(Mw + w) = 0$ yields $\lambda_1 = 1$ with $v_1 = Mw + w = \begin{pmatrix} 4 \\ 2 \\ 0 \end{pmatrix}$. Second, $(M + I)(Mw - w) = 0$ yields $\lambda_2 = -1$ with $v_2 = Mw - w = \begin{pmatrix} 2 \\ 2 \\ 0 \end{pmatrix}$.

We now have a set of two linearly independent eigenvectors. We next choose a basis vector that is not in the span of this set. The second basis vector $\begin{pmatrix} 0 \\ 1 \\ 0 \end{pmatrix}$ is a linear combination of v_1 and v_2, so we cannot select it and we need not consider it further. The third basis vector $w = \begin{pmatrix} 0 \\ 0 \\ 1 \end{pmatrix}$ is not in the span, so we proceed with this choice. We have linearly dependent set

$$S = \left\{ w = \begin{pmatrix} 0 \\ 0 \\ 1 \end{pmatrix}, Mw = \begin{pmatrix} 2 \\ 2 \\ 1 \end{pmatrix}, M^2w = \begin{pmatrix} 0 \\ 0 \\ 1 \end{pmatrix} \right\}$$

and corresponding linear dependence relation $M^2w - w = 0$. This relation factors as $(M - I)(M + I)w = 0$. First, $(M - I)(Mw + w) = 0$ yields $\lambda_1 = 1$ with $v_3 = Mw + w = \begin{pmatrix} 2 \\ 2 \\ 2 \end{pmatrix}$.

Second, $(M + I)(Mw - w) = 0$ yields $\lambda_2 = -1$ with $v_2 = Mw - w = \begin{pmatrix} 2 \\ 2 \\ 0 \end{pmatrix}$, an eigenvector that we found previously. We now have a set of three linearly independent eigenvectors v_1, v_2, and v_3 with corresponding eigenvalues $\lambda = 1, -1, 1$. Eigenvalue $\lambda = 1$ is associated with an eigenspace of dimension 2 and eigenvalue $\lambda = -1$ is associated with an eigenspace of dimension 1. We have exhausted the basis $\mathcal{B}$ so the algorithm terminates and our collection is complete.

$$\lambda_1 = 1, \quad \mathcal{E}_1 = \text{Span}\left\{ \begin{pmatrix} 2 \\ 1 \\ 0 \end{pmatrix}, \begin{pmatrix} 1 \\ 1 \\ 1 \end{pmatrix} \right\}.$$

$$\lambda_2 = -1, \quad \mathcal{E}_2 = \text{Span}\left\{ \begin{pmatrix} 1 \\ 1 \\ 0 \end{pmatrix} \right\}.$$

$\square$

Why does the algorithm work?

How do we know that we will find all eigenvalues (and corresponding eigenspaces) using this method? In any one iteration of our algorithm, starting with vector w, the eigenvectors produced will be linear combinations of the vectors $\{w, Mw, \ldots, M^k w\}$. Because we use the minimum k so that this set is linearly dependent, there is essentially (up to constant multiples) only one linear dependence relation, for which there is one overall factorization. Hence after this iteration through the algorithm, we have found all eigenvectors in $\text{Span}\{w, Mw, \ldots, M^k w\}$. As the algorithm allows a search over a complete set of basis vectors, all possible eigenspaces will be found.

Fortunately, the computation of eigenvalues is often simplified by the particular structure of the matrix. Consider the following results.

Lemma 6.2.8

Suppose the kth column of $n \times n$ real matrix M is βe_k, a scalar multiple of the kth standard basis vector for $\mathbb{R}^n$. Then e_k is an eigenvector of M with eigenvalue β.

Proof. Suppose the jth column of $n \times n$ real matrix M is M_j, and $M_k = \beta e_k$. Let $(e_k)_j$ denote the jth entry of the vector e_k. Then, we have

$$\begin{aligned} Me_k &= M_1(e_k)_1 + M_2(e_k)_2 + \cdots + M_n(e_k)_n \\ &= M_1(0) + \ldots + M_k(1) + \ldots + M_n(0) \\ &= \beta e_k. \end{aligned}$$

Therefore, e_k is an eigenvalue of M with eigenvalue β.

$\square$

Example 6.2.9 Consider matrix $M = \begin{pmatrix} 2 & 0 & 0 \\ 0 & 3 & 1 \\ 0 & 0 & 3 \end{pmatrix}$. Because $M_1 = \begin{pmatrix} 2 \\ 0 \\ 0 \end{pmatrix} = 2e_1$, e_1 is an eigenvector of M with eigenvalue 2. A simple matrix-vector multiplication will verify this fact.

$$Me_1 = \begin{pmatrix} 2 & 0 & 0 \\ 0 & 3 & 1 \\ 0 & 0 & 3 \end{pmatrix} \begin{pmatrix} 1 \\ 0 \\ 0 \end{pmatrix} = 1 \begin{pmatrix} 2 \\ 0 \\ 0 \end{pmatrix} + 0 \begin{pmatrix} 0 \\ 3 \\ 0 \end{pmatrix} + 0 \begin{pmatrix} 0 \\ 1 \\ 3 \end{pmatrix} = \begin{pmatrix} 2 \\ 0 \\ 0 \end{pmatrix} = 2e_1.$$

Notice also that e_2 is an eigenvector of M (with eigenvalue 3) because $M_2 = 3e_2$.

$\square$

Theorem 6.2.10
The eigenvalues of an upper (or lower) triangular $n \times n$ matrix M are the diagonal entries, m_{kk}, $k = 1, 2, \cdots, n$.

Proof. Let M be the upper triangular matrix given by

$$M = \begin{pmatrix} m_{11} & m_{12} & \cdots & m_{1n} \\ 0 & m_{22} & & m_{2n} \\ \vdots & & \ddots & \vdots \\ 0 & 0 & \cdots & m_{nn} \end{pmatrix}.$$

Then by Lemma 6.2.8, $v_{m_{11}} = e_1$ is an eigenvector of M with eigenvalue m_{11}. Next, let us assume $m_{22} \neq m_{11}$ (for otherwise, we already have a vector with eigenvalue m_{22}) and consider the vector

$$v_{m_{22}} = \begin{pmatrix} -\dfrac{m_{12}}{m_{11} - m_{22}} \\ 1 \\ 0 \\ \vdots \\ 0 \end{pmatrix}.$$

Then

$$M v_1 = \begin{pmatrix} \dfrac{m_{11}(-m_{12}) + m_{12}(m_{11} - m_{22})}{m_{11} - m_{22}} \\ m_{22} \\ 0 \\ \vdots \\ 0 \end{pmatrix} = \begin{pmatrix} \dfrac{-m_{12} m_{22}}{m_{11} - m_{22}} \\ m_{22} \\ 0 \\ \vdots \\ 0 \end{pmatrix} = m_{22} v_2.$$

Therefore, m_{22} is an eigenvalue of M corresponding to v_2. In the same vein, for $k \leq n$, let us assume that $m_{kk} \neq m_{ii}$ for any $i < k$ and consider the vector

$$v_k = \begin{pmatrix} \alpha_1 \\ \alpha_2 \\ \vdots \\ \alpha_{k-1} \\ \alpha_k \\ 0 \\ \vdots \\ 0 \end{pmatrix},$$

where

$$\alpha_k = 1 \text{ and } \alpha_i = -\frac{1}{m_{ii} - m_{kk}} \sum_{j=i+1}^{k} \alpha_j m_{ij}.$$

Then the ith entry in Mv_k is given by

$$
\begin{aligned}
(Mv_k)_i &= \sum_{j=1}^{n} m_{ij}\alpha_j \\
&= m_{ii}\alpha_i + m_{i(i+1)}\alpha_{i+1} + \ldots + m_{i(k-1)}\alpha_{k-1} + m_{ik} \\
&= -\frac{m_{ii}}{m_{ii} - m_{kk}} \sum_{j=i+1}^{k} \alpha_j m_{ij} + \sum_{j=i+1}^{k} m_{ij}\alpha_j \\
&= \sum_{j=i+1}^{k} \frac{-\alpha_j m_{ii} m_{ij} + m_{ij}\alpha_j(m_{ii} - m_{kk})}{m_{ii} - m_k k} \\
&= -\frac{m_{kk}}{m_{ii} - m_{kk}} \sum_{j=i+1}^{k} m_{ij}\alpha_j.
\end{aligned}
$$

So, $Mv_k = m_{kk}v_k$. Therefore, v_k is an eigenvector of M with eigenvalue m_{kk}. $\square$

It may seem that v_k in the proof above magically appears. In order to produce the vector v_k above, we first did some scratch work. We considered the matrix equation $M - m_{22}I = 0$ for the matrix

$$M = \begin{pmatrix} a & b & c \\ d & e & f \\ g & h & i \end{pmatrix}.$$

We then used matrix reduction to find a solution. We then repeated this for a 4×4 matrix M. We saw a pattern and tried it on an $n \times n$ matrix with diagonal entries m_{ii}, $1 \leq i \leq n$. The important message here is that a proof does not typically get pushed through without behind-the-scenes scratch work. We encourage the reader to do these steps before starting other proofs of this nature.

Another result about eigenvalues that will become important in Section 6.4 is the following theorem.

Theorem 6.2.11
Let M be an $n \times n$ matrix. Then, λ is an eigenvalue of M if and only if λ is an eigenvalue of M^{T}.

Proof. Let M be an $n \times n$ matrix.
($\Rightarrow$) Suppose λ is an eigenvalue of M. Then $M - \lambda I_n$ is not invertible. Then, by Theorem 5.2.20, we know that $(M - \lambda I)^{\mathsf{T}}$ is not invertible. But,

$$M^{\mathsf{T}} - \lambda I = (M - \lambda I)^{\mathsf{T}}. \quad \text{(See Exercise 36)}$$

Therefore, $M^{\mathsf{T}} - \lambda I$ is not invertible. That is, λ is an eigenvalue for M^{T}.

($\Leftarrow$) To show that if λ is an eigenvalue of M^T then it is also an eigenvalue of M, we use the same argument with M^T in place of M and recognize that $(M^\mathsf{T})^\mathsf{T} = M$. $\square$

6.2.3 Using Determinants to Find Eigenvalues

A fairly standard approach to finding eigenvalues involves the use of determinants. Here, we describe the connections between determinants and eigenvalues. First, let us recall the Heat Diffusion exploration in Section 4.3. In the exploration, we found that, for certain heat states, applying the diffusion operator resulted in a change in amplitude only. That is, these heat states satisfy the eigenvector equation

$$Ev = \lambda v, \tag{6.3}$$

where v is an eigenvector with eigenvalue λ. We also found that these heat state vectors satisfy the matrix equation

$$(E - \lambda I)v = 0. \tag{6.4}$$

This is a homogeneous equation, and hence has a solution. This means that either there is a unique solution (only the trivial solution $v = 0$) or infinitely many solutions. If we begin with the zero heat state (all temperatures are the same everywhere along the rod) then the diffusion is trivial because nothing happens. It would be nice to find a nonzero vector satisfying the matrix Equation (6.4) because it gets us closer to the desirable situation of having a basis.

Notice that $\det(M - \lambda I)$ is a polynomial in λ. This polynomial will be very useful in our discussion of eigenvalues and thus it deserves a name.

Definition 6.2.12

The function

$$f(x) = \det(M - xI)$$

is called the **characteristic polynomial** of L (and of M).
The corresponding equation, used to find the eigenvalues,

$$\det(M - \lambda I) = 0$$

is called the **characteristic equation**. $\square$

Theorem 4.5.20 tells us that Equation 6.3 has a nonzero solution as long as λ is a solution to the characteristic equation, that is,

$$\det(E - \lambda I) = 0.$$

This observation gives us an equivalent characterization for eigenvalues that we present in the next theorem.

> **Theorem 6.2.13**
> Let M be an $n \times n$ matrix. Then λ is an eigenvalue of M for some eigenvector $v \in \mathbb{R}^n$ if and only if λ is a solution to
> $$\det(M - \lambda I) = 0.$$

Proof. Let M be and $n \times n$ matrix.

($\Rightarrow$) Suppose λ is an eigenvalue of M. Then $Mv = \lambda v$ for some nonzero vector v and $(M - \lambda I)v = 0$ has nontrivial solutions. Since at least one solution exists, this system of equations is consistent and we must have $\det(M - \lambda I) = 0$. That is, λ is a zero of the characteristic polynomial $\det(M - \lambda I)$.

($\Leftarrow$) Now suppose that λ is a zero of the characteristic polynomial $\det(M - \lambda I)$. Then $\det(L - \lambda I) = 0$ and $(M - \lambda I)v = 0$ has some nontrivial solution v. That is, there exists nonzero v such that $Mv = \lambda v$ and λ is an eigenvalue of L. $\qquad\square$

We now use Theorem 6.2.13 as a tool and revisit the factor algorithm examples in the context of determinants.

Example 6.2.14 Consider the linear transformation of Example 6.2.2 $L : \mathbb{R}^2 \to \mathbb{R}^2$ defined by $Lx = Mx$ where $M = \begin{pmatrix} 1 & 1 \\ 3 & -1 \end{pmatrix}$. We can find the eigenvalues of M by finding the zeros of the characteristic polynomial $\det(M - \lambda I)$ as follows. First we compute the matrix $M - \lambda I$.

$$M - \lambda I = \begin{pmatrix} 1 & 1 \\ 3 & -1 \end{pmatrix} - \begin{pmatrix} \lambda & 0 \\ 0 & \lambda \end{pmatrix}$$
$$= \begin{pmatrix} 1 - \lambda & 1 \\ 3 & -1 - \lambda \end{pmatrix}.$$

Therefore, the characteristic polynomial is

$$\det(M - \lambda I) = \begin{vmatrix} 1 - \lambda & 1 \\ 3 & -1 - \lambda \end{vmatrix} = (1 - \lambda)(-1 - \lambda) - 3.$$

Finally, we find the zeros of the (quadratic) characteristic polynomial. That is, we solve

$$(1 - \lambda)(-1 - \lambda) - 3 = 0.$$

And, we find that $\lambda_1 = -2$ and $\lambda_2 = 2$ are solutions. That is, the eigenvalues of M are $\lambda_1 = -2$ and $\lambda_2 = 2$.

Theorem 6.2.13 gives us a method to find the eigenvectors of M. That is, we now solve the matrix equations $(M - \lambda I)v = 0$ for $\lambda_1 = -2$ and $\lambda_2 = 2$.

For $\lambda_1 = -2$, we have

$$(M - \lambda_1 I)v = 0$$

$$\begin{pmatrix} 1+2 & 1 \\ 3 & -1+2 \end{pmatrix} v = 0$$

$$\begin{pmatrix} 3 & 1 \\ 3 & 1 \end{pmatrix} v = 0.$$

Using matrix reduction, we see that the solutions to $(M - \lambda_1 I)v = 0$ are of the form $v = \begin{pmatrix} \alpha \\ -3\alpha \end{pmatrix}$.
Therefore, the eigenspace corresponding to $\lambda_1 = -2$ is

$$\mathcal{E}_{\lambda_1} = \left\{ \begin{pmatrix} \alpha \\ -3\alpha \end{pmatrix} \mid \alpha \in \mathbb{R} \right\} = \text{Span} \left\{ \begin{pmatrix} 1 \\ -3 \end{pmatrix} \right\}.$$

Similarly, we find that the eigenspace for $\lambda_2 = 2$ is

$$\mathcal{E}_{\lambda_2} = \left\{ \begin{pmatrix} \alpha \\ \alpha \end{pmatrix} \mid \alpha \in \mathbb{R} \right\} = \text{Span} \left\{ \begin{pmatrix} 1 \\ 1 \end{pmatrix} \right\}.$$

$\square$

It is critical to note that eigenvalues are elements of the scalar set over which scalar-vector multiplication is defined. Consider, for example, a possible characteristic polynomial

$$-(\lambda - 4)(\lambda^2 + 1).$$

If the scalar set for the vector space in which we are working is $\mathbb{R}$, then there is only one zero, namely, $\lambda = 4$, of the characteristic polynomial. The complex zeros, $\lambda = \pm i$, are not considered in the context of real-valued vector spaces. Finally, consider the characteristic polynomial $-(\lambda - 4)(\lambda - 1)(\lambda - 1)$. There are two real zeros $\lambda = 1, 4$, but $\lambda = 1$ has a multiplicity of 2 and therefore gets counted twice. This means that we will say that there are three real zeros and will be careful to indicate when these zeros are distinct (no repetition).

We now give one more example where we find the eigenvalues of a matrix. Again, the procedure we will use begins by finding the zeros of the characteristic polynomial. Then, eigenvectors are determined as solutions to Equation 6.4 for the given eigenvalue λ.

Example 6.2.15 Let us find the eigenvalues and eigenvectors of the matrix

$$A = \begin{pmatrix} 5 & -3 \\ 6 & -4 \end{pmatrix}.$$

First, we find the zeros of the characteristic polynomial:

$$\det(A - \lambda I) = 0$$

$$\left| \begin{pmatrix} 5 & -3 \\ 6 & -4 \end{pmatrix} - \lambda \begin{pmatrix} 1 & 0 \\ 0 & 1 \end{pmatrix} \right| = 0$$

$$\begin{vmatrix} 5 - \lambda & -3 \\ 6 & -4 - \lambda \end{vmatrix} = 0$$

$$(5 - \lambda)(-4 - \lambda) + 18 = 0$$

$$\lambda^2 - \lambda - 2 = 0$$

$$(\lambda + 1)(\lambda - 2) = 0$$

$$\lambda = -1, \lambda = 2.$$

So, now we have two eigenvalues $\lambda_1 = -1$ and $\lambda_2 = 2$. (The subscripts on the λ's have nothing to do with the actual eigenvalue. We are just numbering them.) Using these eigenvalues and Equation 6.4, we can find the corresponding eigenvectors. Let's start with $\lambda_1 = -1$. We want to find v so that

$$(A - (-1)I)v = 0.$$

We can set this up as a system of equations:

$$(A + I)v = 0$$

$$\begin{pmatrix} 6 & -3 \\ 6 & -3 \end{pmatrix} \begin{pmatrix} x \\ y \end{pmatrix} = \begin{pmatrix} 0 \\ 0 \end{pmatrix}$$

$$6x + -3y = 0$$
$$6x + -3y = 0.$$

We can see that this system has infinitely many solutions (that's what we expected) and the solution space is

$$\mathcal{E}_1 = \left\{ \begin{pmatrix} x \\ 2x \end{pmatrix} \middle| x \in \mathbb{R} \right\} = \text{Span} \left\{ \begin{pmatrix} 1 \\ 2 \end{pmatrix} \right\}.$$

Using the same process, we can find the eigenvectors corresponding to $\lambda_2 = 2$:

$$(A - 2I)v = 0$$

$$\begin{pmatrix} 3 & -3 \\ 6 & -6 \end{pmatrix} \begin{pmatrix} x \\ y \end{pmatrix} = \begin{pmatrix} 0 \\ 0 \end{pmatrix}.$$

So, the solution space to this system is

$$\mathcal{E}_2 = \left\{ \begin{pmatrix} x \\ x \end{pmatrix} \middle| x \in \mathbb{R} \right\} = \text{Span} \left\{ \begin{pmatrix} 1 \\ 1 \end{pmatrix} \right\}.$$

We can verify that every nonzero vector in $\mathcal{E}_2$ is an eigenvector of A with eigenvalue $\lambda_2 = 2$. Notice,

$$Av = \begin{pmatrix} 5 & -3 \\ 6 & -4 \end{pmatrix} \begin{pmatrix} \alpha \\ \alpha \end{pmatrix} = \begin{pmatrix} 2\alpha \\ 2\alpha \end{pmatrix} = 2v = \lambda_2 v.$$

The reader can verify that any nonzero vector in $\mathcal{E}_1$ is an eigenvector of A with eigenvalue $\lambda = -1$. $\square$

6.2.4 Eigenbases

Now that we can find any eigenvalues and eigenvectors of a linear transformation, we would like to determine if a basis of eigenvectors exists. Computations, such those from our heat state evolution application, would be much simpler when using a basis of eigenvectors. Since a basis consists of linearly independent vectors, the next theorem is a giant step in the right direction.

> **Theorem 6.2.16**
> Suppose V is a vector space. Let $\lambda_1, \lambda_2, \ldots, \lambda_k$ be distinct eigenvalues of the linear transformation $L : V \to V$ with a set of corresponding eigenvectors $v_1, v_2, \ldots, v_k$. Then, $\{v_1, v_2, \ldots, v_k\}$ is linearly independent.

Proof. (By induction on k.) Let $L : V \to V$ be a linear transformation. Suppose $k = 1$, and v_1 is an eigenvector of L. Then $\{v_1\}$ is linearly independent because $v_1 \neq 0$. Now, for $k \geq 1$, let $\lambda_1, \lambda_2, \ldots, \lambda_k$ be distinct eigenvalues of L with eigenvectors $v_1, v_2, \ldots, v_k$. Assume that $\{v_1, v_2, \ldots, v_k\}$ is linearly independent for $k \geq 1$ and for distinct eigenvalues $\lambda_1, \lambda_2, \ldots, \lambda_k$. We show that $\{v_1, v_2, \ldots, v_k, v_{k+1}\}$ is linearly independent when $\lambda_{k+1} \neq \lambda_j$, $1 \leq j \leq k$. Since $v_{k+1} \in \text{Null}(L - \lambda_{k+1}I)$, we need only show that no nonzero vector $w \in \text{Span}\{v_1, v_2, \ldots, v_k\}$ is in $\text{Null}(L - \lambda_{k+1}I)$. Let $w = \alpha_1 v_1 + \alpha_2 v_2 + \ldots + \alpha_k v_k$.

$$\begin{aligned} (L - \lambda_{k+1}I)w &= (L - \lambda_{k+1}I)(\alpha_1 v_1 + \alpha_2 v_2 + \ldots + \alpha_k v_k) \\ &= (\lambda_1 - \lambda_{k+1})\alpha_1 v_1 + (\lambda_2 - \lambda_{k+1})\alpha_2 v_2 + \ldots + (\lambda_k - \lambda_{k+1})\alpha_k v_k. \end{aligned}$$

Since the eigenvalues are distinct and the vectors $v_1, v_2, \ldots, v_k$ are linearly independent, if any $\alpha_k \neq 0$ then $w \notin \text{Null}(L - \lambda_{k+1}I)$. Thus, no nonzero w is in $\text{Null}(L - \lambda_{k+1}I)$ and so we conclude that $v_{k+1} \notin \text{Span}\{v_1, v_2, \ldots, v_k\}$ and $\{v_1, v_2, \ldots, v_k, v_{k+1}\}$ is linearly independent. $\square$

> **Corollary 6.2.17**
> Let V be an n-dimensional vector space and let $L : V \to V$ be linear. If L has n distinct eigenvalues $\lambda_1, \lambda_2, \ldots, \lambda_n$ with corresponding eigenvectors $v_1, v_2, \ldots, v_n$, then $\{v_1, v_2, \ldots, v_n\}$ is a basis for V.

The proof is the subject of Exercise 32.

Up to this point, we have considered only real eigenvalues. If we consider complex eigenvalues, we change the fourth step of the eigenvalue algorithm by looking for *all* roots of the polynomial p. The Fundamental Theorem of Algebra[2] guarantees k (possibly repeated) roots p. Each root provides an eigenvalue. Though we will not prove it, we need the following result that states that the algorithm will produce n eigenvalues (some of which may be repeated or complex).

Theorem 6.2.18

If M is an $n \times n$ matrix, then there are n, not necessarily distinct, eigenvalues of M.

When we are searching for eigenvectors and eigenvalues of an $n \times n$ matrix M, we are considering the linear transformation $L : \mathbb{R}^n \to \mathbb{R}^n$ defined by $L(v) = Mv$ whose eigenvectors are vectors in the domain space that are scaled (by the eigenvalue) when we apply the L to them. That is, v is an eigenvector with corresponding eigenvalue $\lambda \in \mathbb{R}$ if $v \in \mathbb{R}^n$ and $L(v) = \lambda v$. An eigenbasis is just a basis for $\mathbb{R}^n$ made up of eigenvectors.

Similarly when searching for eigenvectors and eigenvalues of a general linear transformation $L : V \to V$, eigenvectors of L are domain vectors which are scaled by a scalar under the action of L. An eigenbasis for V is a basis made up of eigenvectors of L.

Definition 6.2.19

Let V be a vector space and $L : V \to V$ linear. A basis for V consisting of eigenvectors of L is called an **eigenbasis** of L for V. □

Recall Example 6.2.5, where we found eigenvalues and eigenvectors of matrix $A = \begin{pmatrix} 5 & -3 \\ 6 & -4 \end{pmatrix}$. If we create a set consisting of basis elements for each of the eigenspaces $\mathcal{E}_1$ and $\mathcal{E}_2$, we get the set $\mathcal{B} = \left\{ \begin{pmatrix} 1 \\ 2 \end{pmatrix}, \begin{pmatrix} 1 \\ 1 \end{pmatrix} \right\}$, which is a basis for $\mathbb{R}^2$.

In Example 6.2.6, we found that the transformation L had only one eigenvalue. As L is a transformation on a space of dimension 3, Corollary 6.2.17 does not apply, and we do not yet know if we can obtain an eigenbasis.

Example 6.2.20 Consider the matrix

$$A = \begin{pmatrix} 2 & 0 & 0 \\ 0 & 3 & 1 \\ 0 & 0 & 3 \end{pmatrix}.$$

We want to know if there is an eigenbasis for A for $\mathbb{R}^3$. We begin by finding the eigenvectors and eigenvalues. That is, we want to know for which nonzero vectors v and scalars λ does $Av = \lambda v$. We know that the eigenvalues of A are $\lambda = 2, 3$ by Theorem 6.2.10. We also know, by Lemma 6.2.8, that e_1 is an eigenvector with eigenvalue 2 and e_2 is an eigenvector with eigenvalue 3. To obtain the final eigenvector, we choose a vector y so that $\{e_1, e_2, y\}$ is linearly independent and apply the algorithm on page 335. Choosing $y = e_3$ we find linearly dependent set

---

[2] The Fundamental Theorem of Algebra states that any nth degree polynomial with real coefficients has n (not necessarily distinct) zeros. These zeros might be complex or real.

$$\{y, Ay, A^2y\} = \left\{ \begin{pmatrix} 0 \\ 0 \\ 1 \end{pmatrix}, \begin{pmatrix} 0 \\ 1 \\ 3 \end{pmatrix}, \begin{pmatrix} 0 \\ 6 \\ 9 \end{pmatrix} \right\}.$$

The linear dependence relation yields the equation and its factorization

$$0 = A^2y - 6Ay + 9y = (A - 3I)(Ay - 3y).$$

We have the eigenvalue 3 and eigenvector $Ay - 3y = \begin{pmatrix} 0 \\ 1 \\ 0 \end{pmatrix}$, which is the same eigenvector we previously found. Thus, the two eigenspaces are

$$\mathcal{E}_1 = \text{Span} \left\{ \begin{pmatrix} 1 \\ 0 \\ 0 \end{pmatrix} \right\}, \quad \mathcal{E}_2 = \text{Span} \left\{ \begin{pmatrix} 0 \\ 1 \\ 0 \end{pmatrix} \right\}.$$

We cannot form an eigenbasis of A for $\mathbb{R}^3$ because we can collect at most two linearly independent eigenvectors, one from $\mathcal{E}_1$ and one from $\mathcal{E}_2$. $\square$

6.2.5 Diagonalizable Transformations

Our explorations in heat diffusion have shown us that if $\mathcal{B} = \{v_1, v_2, \ldots, v_n\}$ is an eigenbasis for $\mathbb{R}^n$ corresponding to the diffusion matrix E then we can write any initial heat state vector $v \in \mathbb{R}^n$ as

$$v = \alpha_1 v_1 + \alpha_2 v_2 + \ldots + \alpha_n v_n.$$

Suppose these eigenvectors have eigenvalues $\lambda_1, \lambda_2, \ldots, \lambda_n$, respectively. Then with this decomposition into eigenvectors, we can find the heat state at any later time (say k time steps later) by multiplying the initial heat state by E^k. This became an easy computation with the above decomposition because it gives us, using the linearity of matrix multiplication,

$$E^k v = E^k(\alpha_1 v_1 + \alpha_2 v_2 + \ldots + \alpha_n v_n) = \alpha_1 \lambda_1^k v_1 + \alpha_2 \lambda_2^k v_2 + \ldots + \alpha_n \lambda_n^k v_n. \qquad (6.5)$$

We can then apply our knowledge of limits from Calculus here to find the long-term behavior. That is, the long-term behavior is

$$\lim_{k \to \infty} \alpha_1 \lambda_1^k v_1 + \alpha_2 \lambda_2^k v_2 + \ldots + \alpha_n \lambda_n^k v_n.$$

We see that this limit really depends on the size of the eigenvalues. But we have also seen that changing the representation basis to an eigenbasis was convenient for heat state calculations. Let's remind ourselves how we went about that. First, if we want all computations in the eigenbasis, we have to recalculate the diffusion transformation matrix as well. In other words, we want a matrix transformation that does the same thing that $L(v) = Ev$ does, but the new matrix is created using the eigenbasis. Specifically, we want the matrix representation for the linear transformation that takes a coordinate vector $[v]_{\mathcal{E}}$ (where $\mathcal{E} = \{v_1, v_2, \ldots, v_n\}$ is the eigenbasis) and maps it to $[Ev]_{\mathcal{E}}$. Call this matrix E'. What we are saying is that we want $E'[v]_{\mathcal{E}} = [Ev]_{\mathcal{E}}$. As always, the columns of this matrix are the vectors that are the result of applying the transformation to the current basis elements (in $\mathcal{E}$).

Thus, the columns of E' are $[Ev_1]_\mathcal{E}, [Ev_2]_\mathcal{E}, \ldots, [Ev_n]_\mathcal{E}$. But

$$[Ev_1]_\mathcal{E} = [\lambda_1 v_1]_\mathcal{E} = \lambda_1 [v_1]_\mathcal{E} = \lambda_1 \begin{pmatrix} 1 \\ 0 \\ \vdots \\ 0 \end{pmatrix}$$

$$[Ev_2]_\mathcal{E} = [\lambda_2 v_2]_\mathcal{E} = \lambda_2 [v_2]_\mathcal{E} = \lambda_2 \begin{pmatrix} 0 \\ 1 \\ \vdots \\ 0 \end{pmatrix}$$

$$\vdots$$

$$[Ev_n]_\mathcal{E} = [\lambda_n v_n]_\mathcal{E} = \lambda_n [v_n]_\mathcal{E} = \lambda_n \begin{pmatrix} 0 \\ 0 \\ \vdots \\ 1 \end{pmatrix}.$$

So, we have

$$E' = \begin{pmatrix} \lambda_1 & 0 & \ldots & 0 \\ 0 & \lambda_2 & & \\ \ldots & & \ddots & \vdots \\ 0 & \ldots & 0 & \lambda_n \end{pmatrix}.$$

Knowing that a change of basis is a linear transformation (actually, an isomorphism), we can find its matrix representation (usually known as a change of basis matrix). Let's call this matrix Q and see how this works. We know that $Q[v]_\mathcal{E} = [v]_\mathcal{S}$. This means that if we are given a coordinate vector with respect to the basis $\mathcal{E}$, this transformation will output a coordinate vector with respect to the standard basis $\mathcal{S}$. Recall that to get the coordinate vector in the new basis, we solve for the coefficients in

$$v = \alpha_1 v_1 + \alpha_2 v_2 + \ldots + \alpha_n v_n.$$

Then

$$[v]_\mathcal{E} = \begin{pmatrix} \alpha_1 \\ \alpha_2 \\ \vdots \\ \alpha_n \end{pmatrix}.$$

One way to solve this is to set up the matrix equation

$$Q[v]_{\mathcal{E}} = v$$

$$\begin{pmatrix} | & | & & | \\ v_1 & v_2 & \dots & v_n \\ | & | & & | \end{pmatrix} \begin{pmatrix} \alpha_1 \\ \alpha_2 \\ \vdots \\ \alpha_n \end{pmatrix} = \begin{pmatrix} | \\ v \\ | \end{pmatrix}.$$

This is the transformation written in matrix form. The matrix representation that takes a coordinate vector with respect to the basis $\mathcal{E}$ to a coordinate vector with respect to the standard basis is

$$Q = \begin{pmatrix} | & | & & | \\ v_1 & v_2 & \dots & v_n \\ | & | & & | \end{pmatrix}.$$

So the matrix representation for the transformation that changes from the eigenbasis $\mathcal{E}$ to the standard basis is given by Q. The columns of Q are the eigenvectors of E written with respect to standard coordinates. Because $\mathcal{E}$ is a basis of $\mathbb{R}^n$, we can use Theorem 5.2.20 to see that Q^{-1} exists. We use all of this information to rewrite

$$E'[u]_{\mathcal{E}} = [v]_{\mathcal{E}}.$$

That is, $[u]_{\mathcal{E}} = Q^{-1}u$ and $[v]_{\mathcal{E}} = Q^{-1}v$ for some u and v in the standard basis. So, we have

$$Q^{-1}(u(t + \Delta t)) = Q^{-1}(Eu(t)) = E'Q^{-1}u(t),$$

$$u(t + \Delta t) = Eu(t) = QE'Q^{-1}u(t).$$

It is straightforward to show that for time step k:

$$u(t + k\Delta t) = E^k u(t) = Q(E')^k Q^{-1}u(t),$$

$$u(t + k\Delta t) = E^k u(t) = Q \begin{pmatrix} \lambda_1^k & 0 & \dots & 0 \\ 0 & \lambda_2^k & & \\ \vdots & & \ddots & \vdots \\ 0 & \dots & 0 & \lambda_n^k \end{pmatrix} Q^{-1}u(t),$$

$$Q^{-1}u(t + k\Delta t) = \begin{pmatrix} \lambda_1^k & 0 & \dots & 0 \\ 0 & \lambda_2^k & & \\ \vdots & & \ddots & \vdots \\ 0 & \dots & 0 & \lambda_n^k \end{pmatrix} Q^{-1}u(t),$$

$$[u(t + k\Delta t)]_{\mathcal{E}} = \begin{pmatrix} \lambda_1^k & 0 & \dots & 0 \\ 0 & \lambda_2^k & & \\ \vdots & & \ddots & \vdots \\ 0 & \dots & 0 & \lambda_n^k \end{pmatrix} [u(t)]_{\mathcal{E}}.$$

We see that when vectors are represented as coordinate vectors with respect to an eigenbasis, the matrix representation of the transformation is diagonal.

Of course, all of this is dependent on the existence of an eigenbasis for $\mathbb{R}^n$. Exercise 30 gives the necessary tools to show that we indeed have an eigenbasis for the diffusion transformation.

Following the same procedure in a general setting, let us see what this means in any context. That is, we want to know when we can actually decompose a matrix M into a matrix product QDQ^{-1} where Q is invertible and D is diagonal. From above we see that to form the columns of Q we use the eigenvectors of A. This means that as long as we can find an eigenbasis $\{v_1, v_2, \ldots, v_n\}$, then $Q = \begin{pmatrix} | & | & & | \\ v_1 & v_2 & \ldots & v_n \\ | & | & & | \end{pmatrix}$.

The invertibility of Q follows directly from the fact that $\mathrm{Ran}(Q) = \mathrm{Span}\{v_1, v_2, \ldots, v_n\} = \mathbb{R}^n$ and Theorem 5.2.20.

Definition 6.2.21

Given a vector space V, a linear transformation $L : V \to V$ is called a **diagonalizable** transformation if there is an ordered basis $\mathcal{B}$ for V such that $[L]_\mathcal{B}$ is a diagonal matrix. □

As with other descriptors for transformations, we can also talk about diagonalizable matrices. We define that here.

Definition 6.2.22

Given an $n \times n$ matrix M, we say that M is **diagonalizable** if there exist an invertible matrix Q and diagonal matrix D so that $M = QDQ^{-1}$. □

Before we look at some examples, we have the tools to make three very important statements about diagonalizability of linear transformations. The first theorem provides an existence test for diagonalizability, only requiring that one compute and test the set of eigenvalues.

Theorem 6.2.23

Let V be a vector space and let $L : V \to V$ be a linear transformation. If L has n distinct eigenvalues, then L is diagonalizable.

The proof follows from Corollary 6.2.17 and the discussion above. See Exercise 22.

The second theorem provides a somewhat less-desirable test for diagonalizability. It tells us that if an eigenbasis exists for V corresponding to the transformation L then L is diagonalizable.

Theorem 6.2.24

Let V be a vector space and let $L : V \to V$ be a linear transformation. Then L is diagonalizable if and only if L has n linearly independent eigenvectors.

The proof of this theorem is the subject of Exercise 35.

We now consider the connection between the eigenvalues of a matrix and its invertibility.

Theorem 6.2.25
Let M be an $n \times n$ matrix. Then M is invertible if and only if 0 is not an eigenvalue of M.

Proof. Let M be an $n \times n$ matrix. We will show that if 0 is an eigenvalue of M, then M is not invertible. We will then show that if M is not invertible, then 0 is an eigenvalue.

First, assume 0 is an eigenvalue of M. Then there is a nonzero vector v so that $Mv = 0v$. That is, v is a nonzero solution to the matrix equation $Mx = 0$. Therefore, by Theorem 5.2.20, M is not invertible.

Now assume M is not invertible. Then, again, by Theorem 5.2.20, there is a nonzero solution, v, to the matrix equation $Mx = 0$. That is, $Mv = 0 = 0 \cdot v$. That is, v is an eigenvector with eigenvalue 0. $\square$

Theorem 6.2.25 is an additional equivalent statement that can be added to the Invertible Matrix Theorem (Theorem 5.2.20).

If M is an $n \times n$ matrix with n distinct nonzero eigenvalues, $\lambda_1, \lambda_2, \ldots, \lambda_n$, then we have a tool to compute the inverse of M. In fact, because M is diagonalizable, we can write

$$M = QDQ^{-1},$$

where D is the diagonal matrix

$$D = \begin{pmatrix} \lambda_1 & 0 & \ldots & 0 \\ 0 & \lambda_2 & & 0 \\ \vdots & & \ddots & \vdots \\ 0 & \ldots & 0 & \lambda_n \end{pmatrix}.$$

Thus, $M^{-1} = QD^{-1}Q^{-1}$, where

$$D^{-1} = \begin{pmatrix} \frac{1}{\lambda_1} & 0 & \ldots & 0 \\ 0 & \frac{1}{\lambda_2} & & 0 \\ \vdots & & \ddots & \vdots \\ 0 & \ldots & 0 & \frac{1}{\lambda_n} \end{pmatrix}.$$

We leave it to the reader to check that the inverse is, indeed, the matrix above. Exercise 12 is a result of this discussion. We now present a couple of examples about the diagonalizability and invertibility of M.

Example 6.2.26 Let M be the matrix given by

$$M = \begin{pmatrix} 1 & 0 & 1 \\ 2 & 1 & 3 \\ 1 & 0 & 1 \end{pmatrix}.$$

We want to determine whether M is diagonalizable. If we can find 3 linearly independent eigenvectors then M is diagonalizable. By Lemma 6.2.8, $v_1 = \begin{pmatrix} 0 & 1 & 0 \end{pmatrix}$ is an eigenvalue of M with eigenvalue $\lambda_1 = 1$. To find any remaining eigenvalues and eigenvectors, we apply the Factor Algorithm.

Consider vector $y = \begin{pmatrix} 1 & 0 & 0 \end{pmatrix}$ (because $\{v_1, y\}$ is linearly independent). We have linearly dependent set

$$\{y, My, M^2y, M^3y\} = \left\{ \begin{pmatrix} 1 \\ 0 \\ 0 \end{pmatrix}, \begin{pmatrix} 1 \\ 2 \\ 1 \end{pmatrix}, \begin{pmatrix} 2 \\ 7 \\ 2 \end{pmatrix}, \begin{pmatrix} 4 \\ 17 \\ 4 \end{pmatrix} \right\}.$$

The linear dependence relation is $M^3y - 3M^2y + 2My = 0$ with factorizations

$$(M - 0I)(M^2y - 3My + 2y) = 0 \text{ and } (M - 2I)(M^2y - My).$$

Thus, we have the new eigenvalues and eigenvectors:

$$\lambda_2 = 0, \quad v_2 = M^2y - 3My + 2y = \begin{pmatrix} 1 \\ 1 \\ -1 \end{pmatrix},$$

$$\lambda_3 = 2, \quad v_3 = M^2y - My = \begin{pmatrix} 1 \\ 5 \\ 1 \end{pmatrix}.$$

Theorem 6.2.25 tells us that M is not invertible. However, we have the eigenbasis $\{v_1, v_2, v_3\}$ of $n = 3$ vectors, so M is diagonalizable as $M = QDQ^{-1}$ where

$$Q = \begin{pmatrix} 1 & 0 & 1 \\ 1 & 1 & 5 \\ -1 & 0 & 1 \end{pmatrix} \quad \text{and} \quad D = \begin{pmatrix} 0 & 0 & 0 \\ 0 & 1 & 0 \\ 0 & 0 & 2 \end{pmatrix}.$$

$\square$

In Example 6.2.20, we found only two eigenvalues each of which had eigenspace of dimension 1. This means that we cannot form an eigenbasis for $\mathbb{R}^3$. Thus, M is not diagonalizable. This might lead us to think that we can just count the eigenvalues instead of eigenvectors. Let's see an example where this is not the case.

Example 6.2.27 Let M be the matrix given by

$$M = \begin{pmatrix} 3 & 1 & 0 \\ 1 & 3 & 0 \\ 2 & 2 & 2 \end{pmatrix}.$$

We can find the eigenvalues and eigenvectors of M to determine if M is diagonalizable. Using the Factor Algorithm, begin with vector $w = \begin{pmatrix} 1 \\ 0 \\ 0 \end{pmatrix}$. We find the linearly dependent set

$$\{w, Mw, M^2w\} = \left\{ \begin{pmatrix} 1 \\ 0 \\ 0 \end{pmatrix}, \begin{pmatrix} 3 \\ 1 \\ 2 \end{pmatrix}, \begin{pmatrix} 10 \\ 6 \\ 12 \end{pmatrix} \right\},$$

which leads to the linear dependence relation $M^2w - 6Mw + 8w = 0$ and factorizations $(M - 2I)(Mw - 2w) = 0$ and $(M - 4I)(Mw + w)$. Thus, we have some eigenvalues and associated eigenvectors:

$$\lambda_1 = 2, \quad v_1 = Mw - 2w = \begin{pmatrix} 1 \\ 1 \\ 2 \end{pmatrix}.$$

$$\lambda_2 = 4, \quad v_2 = Mw + w = \begin{pmatrix} 4 \\ 1 \\ 2 \end{pmatrix}.$$

Next, we seek a third eigenvector, linearly independent relative to the two we know. Notice that e_3 is an eigenvector of M with eigenvalue 2. Because we have three linearly independent eigenvectors, M is diagonalizable as $M = QDQ^{-1}$ where

$$Q = \begin{pmatrix} 1 & 0 & -1 \\ 1 & 0 & 1 \\ 2 & 1 & 0 \end{pmatrix} \quad \text{and} \quad D = \begin{pmatrix} 4 & 0 & 0 \\ 0 & 2 & 0 \\ 0 & 0 & 2 \end{pmatrix},$$

even though there are only two distinct eigenvalues. Because 0 is not an eigenvalue, Theorem 6.2.25 tells us that M^{-1} exists. In fact,

$$M^{-1} = QD^{-1}Q^{-1} = \begin{pmatrix} 1 & 0 & -1 \\ 1 & 0 & 1 \\ 2 & 1 & 0 \end{pmatrix} \begin{pmatrix} \frac{1}{4} & 0 & 0 \\ 0 & \frac{1}{2} & 0 \\ 0 & 0 & \frac{1}{2} \end{pmatrix} \begin{pmatrix} 1 & 0 & -1 \\ 1 & 0 & 1 \\ 2 & 1 & 0 \end{pmatrix}^{-1}.$$

□

We wrap up this section with a discussion about the appropriate language.

⋆⋆ **Watch Your Language!** Let V be a vector space with ordered basis $\mathcal{B}$. Consider a matrix M and a linear transformation $L : V \to V$. In discussion of eigenvectors, eigenvalues, and diagonalizability, it is important to include the transformation or matrix to which they correspond. We also need to take care in using the terminology only where appropriate. We present the appropriate use of these terms here.

✓ The eigenvalues of M are $\lambda_1, \lambda_2, ..., \lambda_k \in \mathbb{R}$.
✓ The eigenvalues of L are $\lambda_1, \lambda_2, ..., \lambda_k \in \mathbb{R}$.
✓ The eigenvectors corresponding to M are vectors in V.
✓ The eigenvectors of L are the vectors $v_1, v_2, \ldots, v_k$.
✓ L is diagonalizable because the matrix representation $[L]_{\mathcal{B}}$ is a diagonalizable matrix.

It is inappropriate to say the following things:

✗ The eigenvalues are $\lambda_1, \lambda_2, ..., \lambda_k \in \mathbb{R}$.
✗ The eigenvectors are $v_1, v_2, \ldots, v_k$.
✗ L is the diagonalizable matrix $[L]_{\mathcal{B}}$.

☞ **Path to New Applications**

Nonlinear optimization and optimal design problems use approximation techniques that require the tools for finding eigenvalues and eigenvectors of Hessian matrices. See Section 8.3.3 to learn how to connect optimization and optimal design to the tools of linear algebra.

6.2.6 Exercises

For each of the following matrices, use the Factor Algorithm to find eigenvalues and their corresponding eigenspaces.

1. $M = \begin{pmatrix} 2 & -3 \\ 5 & 6 \end{pmatrix}$

2. $M = \begin{pmatrix} 1 & 1 \\ -2 & 3 \end{pmatrix}$

3. $M = \begin{pmatrix} 5 & 10 & -3 \\ 0 & -2 & 0 \\ 0 & 3 & -1 \end{pmatrix}$

4. $M = \begin{pmatrix} 1 & 0 & 0 \\ 1 & 2 & 1 \\ 5 & 4 & 2 \end{pmatrix}$

5. $M = \begin{pmatrix} 3 & -4 & 2 \\ 2 & -3 & 2 \\ 0 & 0 & 1 \end{pmatrix}$

6. $M = \begin{pmatrix} 1 & 0 & 0 & 0 \\ 1 & 2 & 0 & 0 \\ 5 & 4 & 2 & 0 \\ 1 & 1 & 1 & 1 \end{pmatrix}$

For each of the following matrices, find the characteristic polynomial and then find eigenvalues and their corresponding eigenspaces.

7. $M = \begin{pmatrix} 2 & -3 \\ 5 & 6 \end{pmatrix}$

8. $M = \begin{pmatrix} 1 & 1 \\ -2 & 3 \end{pmatrix}$

9. $M = \begin{pmatrix} 5 & 10 & -3 \\ 0 & -2 & 0 \\ 0 & 3 & -1 \end{pmatrix}$

10. $M = \begin{pmatrix} 1 & 0 & 0 \\ 1 & 2 & 1 \\ 5 & 4 & 2 \end{pmatrix}$

11. $M = \begin{pmatrix} 3 & -4 & 2 \\ 2 & -3 & 2 \\ 0 & 0 & 1 \end{pmatrix}$

12. $M = \begin{pmatrix} 1 & 0 & 0 & 0 \\ 1 & 2 & 0 & 0 \\ 5 & 4 & 2 & 0 \\ 1 & 1 & 1 & 1 \end{pmatrix}$

Determine which of the following matrices are diagonalizable. Whenever it is, write out the diagonalization of M.

13. $M = \begin{pmatrix} 2 & -3 \\ 5 & 6 \end{pmatrix}$

14. $M = \begin{pmatrix} 1 & 1 \\ -2 & 3 \end{pmatrix}$

15. $M = \begin{pmatrix} 5 & 10 & -3 \\ 0 & -2 & 0 \\ 0 & 3 & -1 \end{pmatrix}$

16. $M = \begin{pmatrix} 1 & 0 & 0 \\ 1 & 2 & 1 \\ 5 & 4 & 2 \end{pmatrix}$ 　　　　　　　　　　　18. $M = \begin{pmatrix} 1 & 0 & 0 & 0 \\ 1 & 2 & 0 & 0 \\ 5 & 4 & 2 & 0 \\ 1 & 1 & 1 & 1 \end{pmatrix}$

17. $M = \begin{pmatrix} 3 & -4 & 2 \\ 2 & -3 & 2 \\ 0 & 0 & 1 \end{pmatrix}$

Prove or disprove the following statements:

19. If v is an eigenvector of a transformation L, then so is αv for any nonzero scalar α.
20. If v and w are eigenvectors of a transformation L having the same eigenvalue λ, then $v + w$ is also an eigenvector with eigenvalue λ.
21. If v_1 and v_2 are eigenvectors of a transformation L having the same eigenvalue λ, then any linear combination of v_1 and v_2 is also an eigenvalue with the same eigenvalue.
22. If v_1 is an eigenvector of L with eigenvalue λ_1 and v_2 is an eigenvector of L with eigenvalue λ_2 then $v_1 + v_2$ is also an eigenvector of L.
23. If v is an eigenvector of M, then v^T is an eigenvector of M^T.
24. If λ is an eigenvalue of invertible matrix M, then $\frac{1}{\lambda}$ is an eigenvalue of M^{-1}.
25. If λ is an eigenvalue of M then λ^2 is an eigenvalue of M^2.

Additional Exercises.

26. Let a matrix M have eigenspaces $\mathcal{E}_1, \mathcal{E}_2, \ldots \mathcal{E}_k$ with bases $\beta_1, \beta_2, \ldots, \beta_k$. Why does Theorem 6.2.16 tell us that $\beta_1 \cup \beta_2 \cup \ldots \cup \beta_k$ is a linearly independent set? That is, the union of bases of eigenspaces of a matrix forms a linearly independent set.
27. Show that $QDQ^{-1} = A$ in Example 6.2.26.
28. Prove that the eigenvalues of a diagonal matrix are the diagonal entries.
29. Verify that the eigenvalues of the triangular matrix below are its diagonal entries. (This fact is true for upper and lower triangular matrices of any dimension.)

$$\begin{pmatrix} a & 0 & 0 \\ b & c & 0 \\ d & e & f \end{pmatrix}.$$

Hint: if λ is an eigenvalue of a matrix A, what do you know about the matrix $A - \lambda I$?

30. Consider the heat diffusion operator $E : \mathbb{R}^m \to \mathbb{R}^m$ with standard basis matrix representation

$$E = \begin{pmatrix} 1 - 2\delta & \delta & 0 & 0 \\ \delta & 1 - 2\delta & \delta & 0 \ldots \\ 0 & \delta & 1 - 2\delta & \delta \\ & & \vdots & \ddots \end{pmatrix},$$

where $0 < \delta < \frac{1}{4}$. Show that the kth eigenvector v_k ($1 \le k \le m$) is given by

$$v_k = \left(\sin \frac{\pi k}{m+1}, \sin \frac{2\pi k}{m+1}, \sin \frac{3\pi k}{m+1}, \ldots, \sin \frac{(m-1)\pi k}{m+1}, \sin \frac{m\pi k}{m+1} \right)^\mathsf{T}$$

and provide the kth eigenvalue. Discuss the relative size of the eigenvalues. Is the matrix diagonalizable?

31. Complete Example 6.2.3 by finding all eigenvalues and bases for each eigenspace. Is the transformation diagonalizable?

32. Prove Corollary 6.2.17.

33. Prove that every eigenspace $\mathcal{E}_\lambda$ of linear transformation $L : V \to V$ is a subspace of V.

34. Prove Theorem 6.2.23.

35. Prove Theorem 6.2.24.

36. To assist with the proof of Theorem 6.2.11, show that for any square matrix M and diagonal matrix D that $(M + D)^\mathsf{T} = M^\mathsf{T} + D$.

37. Consider an invertible diagonalizable transformation $T = QDQ^{-1}$, where D is diagonal. Show that the inverse transformation $T^{-1} = QD^{-1}Q^{-1}$.

38. Let $\mathcal{F}(\mathbb{R}) = \{f : \mathbb{R} \to \mathbb{R} \mid f$ is differentiable and continuous$\}$ and $L : \mathcal{F}(\mathbb{R}) \to \mathcal{F}(\mathbb{R})$ be defined by $L(f(x)) = f'(x)$. What are the eigenvectors (eigenfunctions) of this transformation?

39. An $n \times n$ matrix A is said to be *similar* to an $n \times n$ matrix B if there exists an invertible matrix Q so that $A = QBQ^{-1}$. Prove that similarity satisfies the following properties[3]. Hence, a matrix is diagonalizable if and only if it is similar to a diagonal matrix.

   (a) A is similar to A.
   (b) If A is similar to B then B is similar to A.
   (c) If A is similar to B and B is similar to C, then A is similar to C.

6.3 Explorations: Long-Term Behavior and Diffusion Welding Process Termination Criterion

A manufacturing company uses a diffusion welding process to adjoin several smaller rods into a single longer rod. These thin Beryllium-Copper alloy parts are used in structures with strict weight limitations but also require strength typical of more common steel alloys. Examples include satellites, spacecraft, non-ferromagnetic tools, miscellaneous small parts, aircraft, and racing bikes.

The diffusion welding process leaves the final rod unevenly heated with hot spots around the weld joints. The rod ends are thermally attached to heat sinks which allow the rod to cool after the bonding process as the heat is drawn out the ends. An example heat state of a rod taken immediately after weld completion is shown in Figure 6.2.

The data points represent the temperature at 100 points along a rod assembly. In this example, there are four weld joints as indicated by the four hot spots (or spikes). We begin this section by considering the heat state several time steps in the future. We then guide the reader through an exploration in determining the criterion for save removal of a welded rod.

6.3.1 Long-Term Behavior in Dynamical Systems

In the example of diffusion welding, we are interested in knowing the heat state of a rod at any time after the welds are completed. We will explore this application in the context of eigenvalues and eigenvectors of the diffusion matrix. Recall, we have the diffusion matrix E so that

$$u(k\Delta t) = E^k u(0).$$

---

[3] These properties define an *equivalence relation* on the set of $n \times n$ matrices.

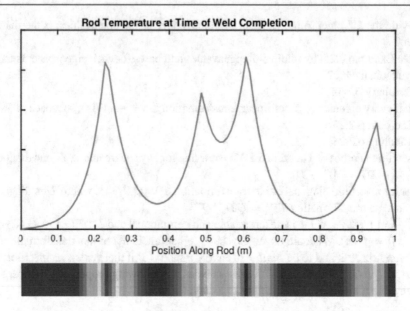

Fig. 6.2 An example heat state right after a weld is finished.

We know that if $v_1, v_2, \ldots, v_m$ are eigenvectors of E with corresponding eigenvalues $\lambda_1, \lambda_2, \ldots, \lambda_m$ and if

$$u(0) = \alpha_1 v_1 + \alpha_2 v_2 + \ldots + \alpha_m v_m$$

then

$$u(k\Delta t) = \alpha_1 \lambda_1^k v_1 + \alpha_2 \lambda_2^k v_2 + \ldots + \alpha_m \lambda_m^k v_m.$$

The remainder of this section is an exploration to discover what these eigenvectors look like, what the corresponding eigenvalues are, and how to describe the behavior of the diffusion based on these. The reader will use the following MATLAB or OCTAVE code found at IMAGEMath.org.

```
HeatEqnClassDemos.m
EigenStuffPlot.m
DiffuseLinearCombination.m
HeatStateLibrary.m
HeatDiffusion.m
EvolutionMatrix.m
```

6.3.2 Using MATLAB/OCTAVE to Calculate Eigenvalues and Eigenvectors

The following exercises include both MATLAB/OCTAVE tasks and discussion points.

1. Watch the graphical demonstration of heat state diffusion by typing the following command:

```
HeatEqnClassDemos(1);
```

What characteristics of heat flow do you observe in this demonstration?

2. Find the eigenvectors and eigenvalues for the heat diffusion transformation. Begin by running the function that creates the diffusion matrix using the commands below:

```
m=5;
E=full(EvolutionMatrix(m))
```

Is E as you expected?

What is the value of δ used by this code? Use the `eig` command to find the eigenvalues and eigenvectors as follows:

```
[V,D]=eig(E);
```

Now, we want to verify that V is the matrix whose columns are eigenvectors and D is the matrix whose diagonal entries are the eigenvalues.

3. Show that the matrix D is actually diagonal by typing

```
D
```

4. Now verify that the first column of V is the eigenvector of E whose eigenvalue is the first diagonal entry of D. Do this by typing

```
E*V(:,1)
D(1,1)*V(:,1)
```

What does the first of these commands do? What does the second do? How do the outputs compare?

5. Using similar syntax, show that the second column of V is an eigenvector of E whose eigenvalue is the second diagonal entry of D. (You may notice that some entries in the eigenvectors may be represented by a very small value $\sim 10^{-16}$. This is a numerical artifact; such small values in relation to other entries should be taken to be zero.

6. Type

```
L=diag(D)
```

This should output a vector. What are the elements of this vector? (Caution: the `diag` command has many uses other than extracting diagonal elements.)

The exercises above have led you through an exploration with $m = 5$. In the following exercises, you will visualize the heat diffusion when $m = 100$; this is the (coordinate) vector space of heat states shown in the demonstration of Exercise 1.

7. Now repeat Exercises 2 and 6 with m=100 to get the new eigenvectors and eigenvalues of E. It is a good idea to suppress outputs by appending commands with a semicolon.

8. Below are commands for viewing 5 eigenvectors with their corresponding eigenvalues. Plot these by typing

```
choices=[80,85,90,95,100];
EigenStuffPlot(V(:,choices),L(choices));
```

How are the individual eigenvectors similar or dissimilar?

9. Make some observations about the relationship between these eigenvectors and eigenvalues.

10. Choose different eigenvectors to view from the list of $m = 100$. Plot these eigenvectors using similar syntax to that given in Exercise 8.

11. Write a list of observations relating eigenvectors and eigenvalues. View more choices of eigenvectors as needed.

In Section 6.1, we wrote an arbitrary heat state as a linear combination of the eigenvectors. In the following exercises, we will explore diffusion on a heat state that is a linear combinations of eigenvectors.

12. In this exercise, we consider a heat state made of a linear combination of two eigenvectors β_{j_1} and β_{j_2} with weights α_{j_1} and α_{j_2}. To view the diffusion of this heat state over $k = 50$ time steps, type the following commands:

    ```
    choices=[60,80];
    alpha=[1,-0.25];
    k=50;
    DiffuseLinearCombination(V(:,choices),L(choices),alpha,k);
    ```

    What linear combination did you plot?
13. Repeat Exercise 12 with a different choice of eigenvectors (`choices`) and scalars (`alpha`). The values in `alpha` should be chosen between -2 and 2 so that the code is numerically stable. You can also change k, but making it larger than 500 can result in a long computation and plotting time. You can also change the pause time between frames by giving a fifth input to the function `DiffuseLinearCombination` which is the pause time in seconds.
14. Make a list of observations about the diffusion of a linear combination of two eigenvectors. Try various linear combinations as needed.
15. Next, consider the evolution of the linear combination of five eigenvectors shown below:

    ```
    choices=[60,70,80,90,100];
    alpha=[1,-1,1,-1,1];
    k=100;
    ```

    What linear combination did you plot?
16. Try various heat states that are linear combinations of 5 eigenvectors and make a list of observations about the diffusion of such heat states.
17. Use the above explorations to make a statement about diffusion details for an arbitrary heat state $u = \alpha_1\beta_1 + \alpha_2\beta_2 + \ldots + \alpha_m\beta_m$, where the β_i are eigenvectors.

6.3.3 Termination Criterion

Let us now consider a company that makes parts such as those described above. The company engineers have determined, through various experiments, that the rod can be removed from the apparatus once the thermal stress drops below a given level. The thermal stress is proportional to the derivative of temperature with respect to position. One way to ensure this condition is to allow the rod temperature to equilibrate to a specified low level, but during this time the machine is unavailable for use.

The Goal

Your goal is to use your knowledge of linear algebra to provide and test a method for determining the earliest possible time at which a rod assembly can be removed safely.

The Safe Removal Criterion

Consider an initial heat state vector $u(0) = (u_1(0), u_2(0), \ldots, u_{100}(0))$ such as the one depicted in Figure 6.2. Consider also an orthonormal eigenbasis $\{v_1, v_2, \cdots, v_{100}\}$ of the heat diffusion operator E.

Let the eigenvectors be ordered by decreasing eigenvalue. Any heat state can be uniquely written in terms of this eigenbasis. In particular, suppose

$$u(t) = \alpha_1(t)v_1 + \alpha_2(t)v_2 + \cdots + \alpha_{100}(t)v_{100}.$$

The rod is safe to remove from the apparatus when $|\alpha_k(t)| < |\alpha_w(t)|$ for all $k > w$, where w is the number of diffusion welds in the rod. In the example above, $w = 4$ and the rod can be safely removed at any time when $|a_k(t)| < |a_4(t)|$ for $k = 5, 6, \cdots, 100$.

Predicting Safe Removal Time

In this task you will predict (based on your intuition and knowledge of heat flow) the order in which several diffusion weld scenarios attain the safe removal criterion.

1. Collect the following files for use in MATLAB/OCTAVE: PlotWeldState.m and WeldStates.mat. The first is a script that makes nice plots of weld states, the second is a data file containing raw temperature heat state vectors with variable names hs1 through hs5.
2. Display an initial heat state by running the MATLAB/OCTAVE commands

    ```
    Load WeldStates.mat
    PlotWeldState(hs)
    ```

    where hs is a vector of heat state temperatures, such as hs5. While not necessarily clear from the heat state plots, each sample heat state corresponds to a diffusion weld process with exactly four weld locations. Some of these welds are close together and difficult to distinguish in the heat state signature.
3. Predict the order in which these example heat states will achieve the safe removal criterion. (Will heat state hs1 reach the safe removal criterion before or after heat state hs2?, etc.) Do not perform any computations. Use your knowledge of heat diffusion and eigenstate evolution. The goal here is to produce sound reasoning and justification, not necessarily the correct answer. We will find the correct answer later in this lab.

Computing the Eigenbasis for E

In this task you will compute the eigenbasis for the diffusion operator E and the initial coefficients $\alpha(0) = (\alpha_1(0), \alpha_2(0), ..., \alpha_{100}(0))$ for the given heat states.

1. Compute the diffusion operator E, a set of eigenvectors as columns of matrix V and corresponding eigenvalues L. Use the methods from Section 6.3.2.
2. Reorder the eigenvectors (and eigenvalues) in order of decreasing eigenvalue. Here is the code to do this.

    ```
    [L,idx]=sort(L,'descend');
    V=V(:,idx);
    ```

3. V is a matrix whose columns are the eigenvectors of E. Examine the matrix $Q = V^{\mathsf{T}}V$. What do the properties of Q tell you about the eigenvectors v_k (columns of V)?
4. How can you compute the eigenbasis coefficients $\alpha = (\alpha_1, \alpha_2, ..., \alpha_{100})$ for any heat state u with a single matrix multiplication? Compute these coefficients for the example heat states (hs1 through hs5) and plot them. Observe that the safe removal criterion is *not* met for any of these initial states.

Determining Safe Removal Times

In this task you will determine the safe removal times for each example heat state by considering the time-dependent coefficients $\alpha(t) = (\alpha_1(t), \alpha_2(t), ..., \alpha_{100}(t))$.

1. Suppose we have a rod with w diffusion welds with initial heat state described by the eigenvector coefficients $\alpha(0) = (\alpha_1(0), \alpha_2(0), ..., \alpha_{100}(0))$ with corresponding eigenvalues $\lambda_1, \lambda_2, ..., \lambda_{100}$. Devise a *computational* procedure for quickly determining how many time steps must pass before the safe removal criterion is met.
2. Apply your procedure to the given five initial heat states, determining the minimum times, say $t_1^*, t_2^*, t_3^*, t_4^*$, and t_5^*, at which the rods can be safely removed.
3. Compare your computed removal time values with the order of safe removal you predicted in Task #1. Explain carefully why your predicted order did, or did not, match your computed order.

6.3.4 Reconstruct Heat State at Removal

In this task you will reconstruct the heat state at the time of safe removal and make some final conclusions.

1. For each of the rods, compute the heat state vector $u(t_j^*)$ where $j = 1, 2, 3, 4, 5$ and t_j^* is the safe removal time for heat state j which you found in Task #3.
2. Discuss the nature of these heat states with particular attention to the original goal of removing the rods when the thermal stress is low.

6.4 Markov Processes and Long-Term Behavior

Our study of the evolution of heat states centered around understanding the effects of multiple repeated applications of the heat diffusion transformation to an initial heat state. We observed the following:

- Heat states that are eigenvectors (we call such heat states "eigenstates") transform by an eigenvalue scaling.
- All eigenvalues for the heat diffusion transformation are between 0 and 1, corresponding to our observation that all heat states diffuse toward the zero heat state.
- Eigenstates are sinusoidal in shape, and eigenstates with higher frequency have lower eigenvalues, corresponding to a faster diffusion.
- The heat diffusion transformation has a basis of eigenvectors, so the evolution of arbitrary heat states can be simply calculated in the coordinate space of an eigenbasis (by scaling each coordinate by the corresponding eigenvalue.)
- In the diffusion of an arbitrary heat state, the components corresponding to higher frequency eigenstates disappear first, while the components corresponding to lower frequency eigenstates persist longer.

In this way, the result of repeated applications is determined by the nature of the eigenvalues. We can make similar observations about linear transformations whose eigenvalues can be greater than one. As with the heat diffusion, the contribution for an eigenvector v_j is diminished over time if it corresponds to the eigenvalue λ_j with $0 < \lambda_j < 1$. But, if some eigenvalue $\lambda_i > 1$ then over time the contribution of the corresponding eigenvector v_i is magnified.

In this section, we seek to understand yet more about the behavior of vector sequences that result from repeated application of a transformation $E : \mathbb{R}^n \to \mathbb{R}^n$. In other words, given a starting vector $u(0) \in \mathbb{R}^n$, we define the sequence $u(0), u(1), u(2), \ldots$ by $u(k) := E^k u(0)$, and we are interested in how $u(k)$ changes as k increases. One important question we might ask is whether the sequence approaches some *limit*, that is, as we repeatedly apply the transformation, is the output getting closer and closer to some vector? In mathematical terms, we are asking whether the limit $u^* = \lim_{k \to \infty} u(k)$ exists, and, if it does, how easily it might be found. This type of analysis appears in a variety of contexts a few of which are mentioned here.

- Consider a diffusion welding process. We have an initial heat state $u(0) \in \mathbb{R}^m$ and a diffusion operator $E : \mathbb{R}^m \to \mathbb{R}^m$. The heat state $u(k) = E^k u(0)$ is the heat state k units of time later than the state $u(0)$. The repeated application of E assisted us in finding a time at which we could safely remove a rod from the welding apparatus. We expect that u^* is the zero heat state because heat is continually lost at the ends of the rod without replacement. We would like to know if this behavior occurs slowly or rapidly and if this property is a general property of transformations.
- Consider the animal populations of a particular habitat as a vector u in $\mathbb{R}^m$: let the jth component, $u_j(k)$ be the population of species j at time k. If future populations can be (linearly) modeled entirely based on current populations, then we have a population dynamics model $u(k) = E^k u(0)$ that predicts the population of all m species at each time step k. We might wonder if any population ever exceeds a critical value or if any species die out. These questions revolve around the limiting behavior.
- Consider a simple weather model that predicts the basic state of the weather (sunny, partly cloudy, rainy) based on the observed weather of recent previous days. The key idea is that the state of the weather tends to change on time scales of more than one day. For example, if it has been sunny for three days, then it is most likely to remain sunny the next day. The state vector contains three values: p_j, the probability that the current day's weather will be of type $j = 1, 2, 3$. This model cannot accurately predict weather because, among other difficulties, it only predicts the probability of the current weather state. Also, the long-term behavior of the model provides information on overall weather type average occurrence.

These examples motivate us to understand when a limit matrix, $\lim_{k \to \infty} E^k$, exists and how to quickly compute it. We will find that limiting behavior and characteristics are determined by the eigenvalues and eigenspaces of the transformation matrix.

Whenever we discuss vectors in this section, it is always in the context of multiplying them by matrices. Therefore, it only makes sense that the definitions herein expect that the vectors lie in a real-valued vector space $\mathbb{R}^m$ such as a coordinate vector space.

6.4.1   Matrix Convergence

We begin with some essential ideas on what it means for a sequence of matrices to converge and some key arithmetic results. In particular, we want convergence to be defined component-wise on matrix elements, and we want to know when matrix multiplication and taking limits commute. These results allow us to perform arithmetic with limit matrices.

Definition 6.4.1

Let $\mathcal{A} = \{A_1, A_2, \ldots\}$ be a sequence of $m \times p$ real matrices. $\mathcal{A}$ is said to **converge** to the $m \times p$ real **limit matrix** L if

$$\lim_{k \to \infty} (A_k)_{ij} = L_{ij}$$

for all $1 \le i \le m$ and $1 \le j \le p$. In the case when $p = 1$, we call the limit a **limit vector**. ☐

This definition simply says that a sequence of matrices (of a fixed size) converges if there is entry-wise convergence.

Notation.

In previous sections of this text, we used the notation A_k to denote a column of the matrix A. In this chapter, we are using the same notation to indicate a term in a sequence of matrices. Be aware of the context as you read this chapter.

The next lemma tells us that left or right multiplication by a fixed matrix to each sequence element does not destroy the existence of a limit matrix.

Lemma 6.4.2

Let $\mathcal{A} = \{A_1, A_2, \ldots\}$ be a sequence of $m \times p$ real matrices that converge to the limit matrix L. Then for any $\ell \times m$ real matrix B, $\lim_{k \to \infty} BA_k$ exists and

$$\lim_{k \to \infty} BA_k = BL.$$

Also, for any $p \times n$ real matrix C, $\lim_{k \to \infty} A_k C$ exists and

$$\lim_{k \to \infty} A_k C = LC.$$

Proof. Suppose $\mathcal{A} = \{A_1, A_2, \ldots\}$ is a sequence of $m \times p$ real matrices that converge to the limit matrix L. Let B be an arbitrary $\ell \times m$ real matrix and let C be an arbitrary $p \times n$ real matrix.

(a) We show that $\lim_{k \to \infty} BA_k$ exists and equals BL by showing that corresponding entries of $\lim_{k \to \infty} BA_k$ and BL are equal.

$$
\begin{aligned}
\left(\lim_{k \to \infty} BA_k \right)_{ij} &= \lim_{k \to \infty} \sum_{r=1}^{m} B_{ir} (A_k)_{rj} \\
&= \sum_{r=1}^{m} B_{ir} \lim_{k \to \infty} (A_k)_{rj} \\
&= \sum_{r=1}^{m} B_{ir} L_{rj} \\
&= (BL)_{ij}
\end{aligned}
$$

(b) We show that $\lim_{k\to\infty} A_k C$ exists and equals LC by showing that corresponding entries of $\lim_{k\to\infty} A_k C$ and LC are equal. [The remainder of the proof is similar to that of part (a) and is the subject of Exercise 6.] □

There are some important consequences associated with Lemma 6.4.2.

- It is possible for $v_k = A_k u$ to exist for every k and for $\lim_{k\to\infty} v_k$ to exist, but for $\lim_{k\to\infty} A_k$ to not exist. For example, if u is the zero vector, then $v_k = A_k u$ is the zero vector. So $\lim_{k\to\infty} v_k = 0$, but A_k may not have a limit.
- Regardless of choice of basis $\mathcal{B}$, if $\lim_{k\to\infty} A_k[v]_{\mathcal{B}}$ exists, the limit will always be the same vector, though expressed as a coordinate vector relative to the chosen basis.
- If we have a sequence of vectors $\{u_k\}$ created by $u_k = A^k u_0$, then (if the limit matrix exists)

$$\lim_{k\to\infty} u_k = \left(\lim_{k\to\infty} A^k \right) u_0.$$

Example 6.4.3 Consider the sequence of matrices $\mathcal{A} = \{A_1, A_2, \cdots\}$ defined by

$$A_n = \begin{pmatrix} 1 + 2^{-n} & 1^n \\ 1 & \frac{1}{n} \end{pmatrix}.$$

That is,

$$A_1 = \begin{pmatrix} \frac{3}{2} & 1 \\ 1 & 1 \end{pmatrix}, \quad A_2 = \begin{pmatrix} \frac{5}{4} & 1 \\ 1 & \frac{1}{2} \end{pmatrix}, \quad A_3 = \begin{pmatrix} \frac{9}{8} & 1 \\ 1 & \frac{1}{3} \end{pmatrix}, \cdots$$

The limit matrix L is defined element-wise:

$$L = \lim_{n\to\infty} A_n = \begin{pmatrix} \lim_{n\to\infty} 1 + 2^{-n} & \lim_{n\to\infty} 1^n \\ \lim_{n\to\infty} 1 & \lim_{n\to\infty} \frac{1}{n} \end{pmatrix} = \begin{pmatrix} 1 & 1 \\ 1 & 0 \end{pmatrix}.$$

Furthermore, for matrix $C = \begin{pmatrix} 1 & -1 & 1 \\ 0 & 2 & -1 \end{pmatrix}$, we have

$$\lim_{n\to\infty} A_n C = \lim_{n\to\infty} \begin{pmatrix} 1 + 2^{-n} & 2(1^n) - 1 - 2^{-n} & -(1^n) + 1 + 2^{-n} \\ 1 & \frac{2}{n} - 1 & 1 - \frac{1}{n} \end{pmatrix} = \begin{pmatrix} 1 & 1 & 0 \\ 1 & -1 & 1 \end{pmatrix} = LC.$$

□

Recall, our goal is to determine the long-term behavior of a process defined by repeated multiplication on the left by a matrix. The next lemma and theorem provide our first important results concerning sequences of matrices given by repeated left matrix multiplication.

Lemma 6.4.4

Let A be a real $n \times n$ matrix and suppose the limit matrix $L = \lim_{k\to\infty} A^k$ exists. Then, for any $n \times n$ invertible real matrix Q, $\lim_{k\to\infty} (QAQ^{-1})^k = QLQ^{-1}$.

Proof. Suppose A is a real $n \times n$ matrix for which the limit matrix $L = \lim_{k\to\infty} A^k$ exists, and let Q be any $n \times n$ invertible real matrix. We show directly that $\lim_{k\to\infty} (QAQ^{-1})^k = QLQ^{-1}$. Using

Lemma 6.4.2, we have

$$\lim_{k \to \infty} \left(QAQ^{-1}\right)^k = \lim_{k \to \infty} \left[\left(QAQ^{-1}\right)\left(QAQ^{-1}\right)\ldots k \text{ times}\ldots\left(QAQ^{-1}\right)\right]$$
$$= \lim_{k \to \infty} QA^k Q^{-1}$$
$$= \left(\lim_{k \to \infty} QA^k\right) Q^{-1}$$
$$= QLQ^{-1}.$$

$\square$

Theorem 6.4.5

Let A be an $n \times n$ real diagonalizable matrix. Every eigenvalue of A satisfies $-1 < \lambda \leq 1$ if and only if $L = \lim_{k \to \infty} A^k$ exists.

Proof. ($\Rightarrow$) Suppose A is an $n \times n$ real diagonalizable matrix and every eigenvalue of A satisfies $-1 < \lambda \leq 1$. Since A is diagonalizable, $A = QDQ^{-1}$, where D is the diagonal matrix of eigenvalues of A and the columns of Q form an eigenbasis for $\mathbb{R}^n$. Notice that if we define D^∞ as $D^\infty \equiv \lim_{k \to \infty} D^k$, then we know D^∞ exists because

$$\lim_{k \to \infty} \lambda^k = \begin{cases} 1 & \text{if } \lambda = 1 \\ 0 & \text{if } 0 < \lambda < 1 \end{cases}$$

Then, using Lemma 6.4.4,

$$\lim_{k \to \infty} A^k = \lim_{k \to \infty} QD_k Q^{-1} = Q\left(\lim_{k \to \infty} D^k\right) Q^{-1} = QD^\infty Q^{-1}.$$

Thus, $\lim_{k \to \infty} A^k$ exists.

($\Leftarrow$) Now, suppose A is an $n \times n$ real diagonalizable matrix for which $L = \lim_{k \to \infty} A^k$ exists. We know that there exist matrices Q and D so that $A = QDQ^{-1}$, where D is the diagonal matrix

$$D = \begin{pmatrix} \lambda_1 & & & \\ & \lambda_2 & & \\ & & \ddots & \\ & & & \lambda_n \end{pmatrix}$$

whose diagonal entries are eigenvalues of A. We also know that, for $k \in \mathbb{N}$, $A^k = QD^kQ^{-1}$, where

$$D^k = \begin{pmatrix} \lambda_1^k & & & \\ & \lambda_2^k & & \\ & & \ddots & \\ & & & \lambda_n^k \end{pmatrix}.$$

Now, $\lim_{k\to\infty} D^k$ exists only when λ_j^k exist for $1 \le j \le n$. That is, $\lim_{k\to\infty} D^k$ exists only when $-1 < \lambda_j \le 1$ for each eigenvalue λ_j. Thus, $\lim_{k\to\infty} A^k$ also exists under the same condition.   □

Example 6.4.6 Consider the matrix

$$A = \begin{pmatrix} 1/2 & 0 \\ -1 & 3/4 \end{pmatrix}.$$

A is diagonalizable as

$$A = QDQ^{-1} = \begin{pmatrix} 1 & 0 \\ 4 & 1 \end{pmatrix} \begin{pmatrix} 1/2 & 0 \\ 0 & 3/4 \end{pmatrix} \begin{pmatrix} 1 & 0 \\ -4 & 1 \end{pmatrix}.$$

Since A has eigenvalues $\lambda_1 = 1/2$ and $\lambda_2 = 3/4$, Theorem 6.4.5 tells us that $\lim_{k\to\infty}$ exists. In this case, $\lim_{k\to\infty} D^k$ is the 2×2 zero matrix. So, $\lim_{k\to\infty} A^k$ is also the 2×2 zero matrix.   □

Example 6.4.7 Consider the matrix

$$A = \begin{pmatrix} -1 & 0 \\ 0 & -1 \end{pmatrix}.$$

A has eigenvalues $\lambda_1 = \lambda_2 = -1$ with corresponding eigenspace

$$\mathcal{E} = \mathrm{Span}\left\{ \begin{pmatrix} 1 \\ 0 \end{pmatrix}, \begin{pmatrix} 0 \\ 1 \end{pmatrix} \right\}.$$

So, A is diagonalizable as

$$A = QDQ^{-1} = \begin{pmatrix} 1 & 0 \\ 0 & 1 \end{pmatrix} \begin{pmatrix} -1 & 0 \\ 0 & -1 \end{pmatrix} \begin{pmatrix} 1 & 0 \\ 0 & 1 \end{pmatrix}.$$

Since the eigenvalues are not in the interval $(-1, 1]$, Theorem 6.4.5 tells us that $\lim_{k\to\infty} A^k$ does not exist. In this case, $A^k = -I_2$ for k odd and $A^k = I_2$ for k even.   □

6.4.2   Long-Term Behavior

Definition 6.4.8

Let $T : \mathbb{R}^n \to \mathbb{R}^n$ be a transformation defined by $T(v) = Mv$ where M is a real $n \times n$ matrix, and let $v_0 \in \mathbb{R}^n$. Then, the **long-term state** of v_0 relative to T is given by $v^* = \lim_{k\to\infty} M^k v_0$ whenever this limit exists.   □

Recall that, in the heat diffusion application, we can describe long-term heat states according to the eigenvalues of the diffusion transformation. In fact, we found that, for initial heat state v_0, the long-term state relative to the diffusion transformation defined by $T(h) = Eh$ was described by

$$\lim_{k\to\infty} E^k v_0 = \lim_{k\to\infty} (\lambda_1^k v_1 + \lambda_2^k v_2 + \ldots + \lambda_n^k v_n,)$$

where $\lambda_1, \lambda_2, \ldots, \lambda_n$ are eigenvalues of E corresponding to eigenvectors $v_1, v_2, \ldots, v_n$. We can extend this idea to any transformation. That is, if a state evolution transformation matrix is diagonalizable with eigenvalues satisfying $-1 < \lambda \le 1$, then we can find a simple expression for the long-term state.

Thus, the long-term state of a vector relative to a diagonalizable state transformation can be simply written in terms of the eigenvectors of M as long as every eigenvalue of M satisfies $-1 < \lambda \leq 1$. The next theorem describes the long-term state in terms of the eigenvalues and their corresponding eigenvectors.

Theorem 6.4.9

Let A be an $n \times n$ real diagonalizable matrix where every eigenvalue of A satisfies $-1 < \lambda_j \leq 1$, with corresponding eigenbasis $\mathcal{E} = \{Q_1, Q_2, \ldots, Q_n\}$. Then the long-term state of any $v \in \mathbb{R}^n$ relative to A is

$$v^* = \sum_{j \in J} \alpha_j Q_j,$$

where $\alpha = [v]_{\mathcal{E}}$ is the corresponding coordinate vector of v and $J = \{j \in \mathbb{N} \mid \lambda_j = 1\}$.

The notation in the theorem allows for repeated eigenvalues. That is, it is possible that $\lambda_j = \lambda_i$ for $i \neq j$. Notice also that J is a set of indices allowing us to keep track of all eigenvalues equal to 1.

Proof. Suppose A is an $n \times n$ real diagonalizable matrix where every eigenvalue of A satisfies $-1 < \lambda_j \leq 1$. Define $\mathcal{E} = \{Q_1, Q_2, \ldots, Q_n\}$ to be an eigenbasis of A. Finally, let $J = \{j \in \mathbb{N} \mid \lambda_j = 1\}$. We know that there are matrices Q and D so that $A = QDQ^{-1}$. In fact, the columns of Q are $Q_1, Q_2, \ldots, Q_n$ and D is diagonal with diagonal entries $\lambda_1, \lambda_2, \ldots, \lambda_n$. Let v be any vector in $\mathbb{R}^n$, and define $\alpha = [v]_{\mathcal{E}}$. Then, $\alpha = Q^{-1}v$.

We also know that, for any $k \in \mathbb{N}$, D^k is the diagonal matrix whose diagonal entries are $\lambda_1^k, \lambda_2^k, \ldots, \lambda_n^k$. Since

$$\lim_{k \to \infty} \lambda_j^k = \begin{cases} 0 & \text{if } j \notin J \\ 1 & \text{if } j \in J \end{cases},$$

we find that the long-term state of v relative to A is

$$v_\infty = \left(\lim_{k \to \infty} A^k \right) v = \lim_{k \to \infty} \left(A^k v \right) = \lim_{k \to \infty} \left((QD^kQ^{-1})v \right)$$

$$= \lim_{k \to \infty} \left(QD^k \alpha \right) = \lim_{k \to \infty} \left(\sum_{j=1}^{n} \lambda_j^k \alpha_j Q_j \right)$$

$$= \sum_{j=1}^{n} \left(\lim_{k \to \infty} \lambda_j^k \right) \alpha_j Q_j$$

$$= \sum_{j \in J} \left(\lim_{k \to \infty} \lambda_j^k \right) \alpha_j Q_j + \sum_{j \notin J} \left(\lim_{k \to \infty} \lambda_j^k \right) \alpha_j Q_j$$

$$= \sum_{j \in J} \alpha_j Q_j.$$

$\square$

Table 6.1 Percent of population described in each column who move to (or remain in) the location described in each row.

	inner city	downtown	suburbs	out-of-area
inner city	93	2	3	0
downtown	1	67	12	3
suburbs	3	30	76	18
out-of-area	3	1	9	79

Example 6.4.10 The heat diffusion operator is an example of a dissipative transformation. The total heat in the heat state is not constant, but decreases over time. This is a direct consequence of the fact that all of the eigenvalues have magnitude less than one: $|\lambda_j| < 1$ for all j. According to Theorem 6.4.9, $v_\infty = 0$, so all heat eventually leaves the system. $\qquad\square$

Example 6.4.11 A recent four-year study of population movement around a large city made the following observations. The study considered four regions: inner city, downtown, suburbs, and out-of-area. Over a four-year period, 93% of inner city people remained living in the inner city, 1% moved to the downtown area, 3% moved to the suburbs, and 3% moved out-of-area. Similar data is reported for each category. A sample population of 10,000 was tracked (including if they moved out-of-area). Participants in the study were, at first, residents of the city or its suburbs.

Table 6.1 shows, for example, that in one year of the study 30% of people who live downtown were reported to have moved to the suburbs within four years. And likewise, 18% of people who were former city residents moved back to the city suburbs. Supposing that the trend indicated in the study continues, we can determine the long-term state of the city's total population of 247,000 which is distributed as follows: 12,500 in the inner city, 4500 downtown, and 230,000 in the suburbs.

This population movement model is not dissipative because the total population does not change (no new residents are counted). The population in any given four-year period is a state vector $v \in \mathbb{R}^4$. We define

$$v = \begin{pmatrix} v_1 \\ v_2 \\ v_3 \\ v_4 \end{pmatrix} = \begin{pmatrix} \text{inner city population} \\ \text{downtown population} \\ \text{suburb population} \\ \text{out-of-area population} \end{pmatrix}.$$

The movement of people over a four-year period is modeled by left matrix multiplication by transformation matrix E defined by

$$E = \begin{pmatrix} 0.93 & 0.02 & 0.03 & 0.00 \\ 0.01 & 0.67 & 0.12 & 0.03 \\ 0.03 & 0.30 & 0.76 & 0.18 \\ 0.03 & 0.01 & 0.09 & 0.79 \end{pmatrix}.$$

If v_k is the population distribution at time period k, then $v_{k+1} = Ev_k$ is the population distribution at time period $k + 1$. We have

$$v_0 = \begin{pmatrix} 12,000 \\ 4,500 \\ 230,000 \\ 0 \end{pmatrix}$$

and

$$v_1 = Ev_0 = \begin{pmatrix} 0.93 & 0.02 & 0.03 & 0.00 \\ 0.01 & 0.67 & 0.12 & 0.03 \\ 0.03 & 0.30 & 0.76 & 0.18 \\ 0.03 & 0.01 & 0.09 & 0.79 \end{pmatrix} \begin{pmatrix} 12{,}000 \\ 4{,}500 \\ 230{,}000 \\ 0 \end{pmatrix} = \begin{pmatrix} 18{,}150 \\ 30{,}735 \\ 176{,}510 \\ 21{,}105 \end{pmatrix},$$

and

$$v_2 = Ev_1 = \begin{pmatrix} 0.93 & 0.02 & 0.03 & 0.00 \\ 0.01 & 0.67 & 0.12 & 0.03 \\ 0.03 & 0.30 & 0.76 & 0.18 \\ 0.03 & 0.01 & 0.09 & 0.79 \end{pmatrix} \begin{pmatrix} 18{,}150 \\ 30{,}735 \\ 176{,}510 \\ 21{,}345 \end{pmatrix} \approx \begin{pmatrix} 22{,}790 \\ 42{,}588 \\ 147{,}712 \\ 33{,}411 \end{pmatrix}.$$

The long-term population state is given by $v^* = \lim_{k \to \infty} E^k v_0$. If E is diagonalizable and $\lim_{k \to \infty} E^k = L$ exists, then we can use the results of this section to find $v^* = QLQ^{-1}v_0$. The eigenvalues of E are $\lambda_1 = 1.0000$, $\lambda_2 \approx 0.9082$, $\lambda_3 \approx 0.7367$, $\lambda_4 \approx 0.5051$. E has four distinct eigenvalues and is diagonalizable as $E = QDQ^{-1}$ where

$$D = \begin{pmatrix} \lambda_1 & 0 & 0 & 0 \\ 0 & \lambda_2 & 0 & 0 \\ 0 & 0 & \lambda_3 & 0 \\ 0 & 0 & 0 & \lambda_4 \end{pmatrix} \quad \text{and} \quad Q \approx \begin{pmatrix} +0.415 & +0.843 & +0.115 & -0.031 \\ +0.322 & -0.219 & -0.439 & -0.543 \\ +0.753 & -0.465 & -0.446 & +0.806 \\ +0.397 & -0.159 & +0.771 & -0.232 \end{pmatrix}.$$

Using the notation of Theorem 6.4.9, we have $J = \{1\}$ and $\alpha_1 \approx 130{,}647$ (the first entry of $Q^{-1}v_0$). So, by Theorem 6.4.9,

$$v^* = (130{,}647)Q_1 = \begin{pmatrix} 54{,}173 \\ 42{,}114 \\ 98{,}328 \\ 51{,}885 \end{pmatrix}.$$

This long-term behavior analysis predicts that in the long term, about 54,000 people will be living in the inner city, about 42,000 downtown, about 98,000 in the suburbs, and about 52,000 will have moved out of the area. □

6.4.3 Markov Processes

The population movement model of Example 6.4.11 is an example of a non-dissipative transformation. The total population at each iteration is constant at 247,000, the initial population. We found that the eigenvalues of the population movement transformation satisfied $-1 < \lambda_i \le 1$, and one of the eigenvalues was exactly 1. The eigenspace corresponding to $\lambda = 1$ contained the long-term state v^*.

On the other hand, the heat diffusion transformation is an example of a dissipative transformation. The total quantity of heat in the initial state v_0 is not preserved under transformation by E. We found that the eigenvalues of this transformation satisfy $-1 < \lambda < 1$, no eigenvalue taking on the exact value 1. The long-term state was $v^* = 0$.

These observations on the difference in behavior of these transformations and their eigenvalues is not a coincidence. While both transformations have matrix entries satisfying $0 \le E_{i,j} \le 1$, they differ in the property of column stochasticity.

Definition 6.4.12

An $n \times n$ real matrix $A = (a_{ij})$ is said to be **column stochastic**, or a **Markov Matrix**, if for all $1 \leq i, j \leq n$, $a_{ij} \geq 0$ and $A^\mathsf{T} \mathbf{1}_n = \mathbf{1}_n$, where $\mathbf{1}_n$ is the vector whose entries are all 1. □

Definition 6.4.12 asserts that a Markov matrix is a square matrix of all nonnegative entries that has columns that sum to 1. The following theorem, along with Theorem 6.4.9, allows us to understand the long-term behavior of any column stochastic matrix.

Theorem 6.2.13

Let A be a column stochastic matrix. Then every eigenvalue λ of A satisfies $|\lambda| \leq 1$ and at least one eigenvalue has value 1.

Proof. Suppose A is a column stochastic matrix. Then, $A^\mathsf{T} \mathbf{1}_n = \mathbf{1}_n$, so $\mathbf{1}_n$ is an eigenvector of A^T with eigenvalue 1. Since, by Theorem 6.2.11, A and A^T have the same eigenvalues, at least one eigenvalue of A has value 1.

Now, consider any eigenvector $v = \begin{pmatrix} v_1 & v_2 & \ldots & v_n \end{pmatrix}^\mathsf{T}$ of A^T with eigenvalue λ. Suppose v_ℓ is the entry in v with largest absolute value and let $a_\ell = \begin{pmatrix} a_{1\ell} & a_{2\ell} & \ldots & a_{n\ell} \end{pmatrix}^\mathsf{T}$ be the ℓth column of A. We have

$$a_\ell^\mathsf{T} v = \lambda v_\ell \text{ and } a_\ell^\mathsf{T} v = \sum_{j=1}^{n} a_{j\ell} v_j.$$

Therefore,

$$|\lambda||v_\ell| = |a_\ell^\mathsf{T} v| \leq \sum_{j=1}^{n} a_{j\ell} |v_\ell| = |v_\ell|.$$

Thus, $|\lambda| \leq 1$. □

It is possible for a Markov matrix, M, to have one or more eigenvalues $\lambda = -1$, in which case Theorem 6.4.5 tells us that $\lim_{k \to \infty} M^k$ does not exist.

Example 6.4.14 The population movement transformation of Example 6.4.11 is a Markov matrix. Both E and E^T are column stochastic. □

Example 6.4.15 State University has initiated a study on the stability of student fields of study. The University accepts students from four major fields of study: Humanities, Physical Sciences, Mathematics, and Engineering. Students are allowed to change their major field at any time. Data collected over the past several years reveals the overall trends in changes of major as follows. Suppose $x_j(k)$ is the number of students in year k studying under major j ($j = 1$, Humanities; $j = 2$, Physical Sciences; $j = 3$, Mathematics; $j = 4$, Engineering), then the change in one year is given by $x(k+1) = Ex(k)$ where

$$E = \begin{pmatrix} 0.94 & 0.02 & 0.01 & 0.00 \\ 0.02 & 0.85 & 0.14 & 0.02 \\ 0.03 & 0.03 & 0.83 & 0.08 \\ 0.01 & 0.10 & 0.02 & 0.90 \end{pmatrix}.$$

For example, the entry 0.14 in the second row and third column indicates that 14% of Mathematics majors will change to become Physical Science majors each year. The University would like to answer two questions:

(Q1) If equal numbers of each major are admitted as first-year students (2500 of each major), what will be the distribution of majors at graduation four years later?

(Q2) What distribution of majors should be admitted so that the same distribution exists at graduation four years later?

To simply answer (Q1), we can compute $x(4) = E^4 x(0)$ where

$$x(0) = (2500 \ 2500 \ 2500 \ 2500)^{\mathsf{T}}.$$

The result is $x(4) \approx (2229 \ 2692 \ 2298 \ 2781)^{\mathsf{T}}$. The students have changed areas of study so that there are more Physical Sciences and Engineering majors than were originally enrolled, and fewer Humanities and Mathematics majors.

In order to answer (Q2), we can use the fact that E is a Markov matrix (be sure and verify this fact). We know that E will have at least one eigenvalue $\lambda = 1$. Any eigenvector in the corresponding eigenspace does not change in time. That is, $x(k + 1) = Ex(k) = \lambda x(k) = x(k)$. So, any initial student distribution that is in the eigenspace of vectors with eigenvalue 1 will remain the same after four years (or any number of years). E has eigenvalues $1, 0.933, 0.793, 0.793$. And the one-dimensional eigenspace associated with $\lambda = 1$ contains the vector $v \approx (0.1342 \ 0.2844 \ 0.2363 \ 0.3451)^{\mathsf{T}}$. This vector v is scaled specifically so that the sum of the entries is 1. So, the entries represent the fraction of students of each type of major that will be stable from year to year.   □

☞ **Path to New Applications**

Researchers modeling dynamical processes with differential (and difference) equations often follow paths similar to the paths discussed in this chapter for the Diffusion Welding example and to solve Markov processes. The path to solving these problems typically require that you find eigenfunction solutions in order to find general solutions (solutions that are linear combinations of the eigenfunctions). See Section 8.3.1 to learn more about connections between these tools and applications.

6.4.4 Exercises

1. Consider a matrix sequence $\mathcal{A} = \{A_1, A_2, \cdots\}$. For each of the following matrix definitions, find the limit matrix $L = \lim_{k \to \infty} A_k$ if it exists.

(a) $A_k = \begin{pmatrix} \frac{1}{k} & 2k \\ 2k & e^{-k} \end{pmatrix}$

(b) $A_k = \begin{pmatrix} 2 & k^{-2} - 1 \\ \left(\frac{1}{3}\right)^k & 1 & 1 \end{pmatrix}$

(c) $A_k = \begin{pmatrix} \cos(2\pi k) & 1 \\ 1 & \sin(2\pi k) \end{pmatrix}$

(d) $A_k = \begin{pmatrix} \cos 1 & \sin 1 \\ -\sin 1 & \cos 1 \end{pmatrix}$

(e) $A_k = \begin{pmatrix} 1 & \frac{\ln k}{k} \\ 1^{-k} & 0 \end{pmatrix}$

(f) $A_k = \begin{pmatrix} 0.45^k & -0.36^k \\ 0 & 0.95^k \end{pmatrix}$

2. For each of the following matrices M, find $L = \lim_{k \to \infty} M^k$ if it exists. If $\lim_{k \to \infty} M^k$ does not exist, then describe the non-convergent behavior.

(a) $M = \begin{pmatrix} 1 & -1 \\ -1 & 1 \end{pmatrix}$

(b) $M = \begin{pmatrix} 1/3 & 1/2 \\ 2/3 & 1/2 \end{pmatrix}$

(c) $M = \begin{pmatrix} 1 & 0 \\ 0 & 0 \end{pmatrix}$

(d) $M = \begin{pmatrix} -1 & 0 & 1 \\ 0 & 1 & 0 \\ 1 & 0 & -1 \end{pmatrix}$

(e) $M = \begin{pmatrix} 0.35 & 0.65 \\ 0.65 & 0.35 \end{pmatrix}$

(f) $M = \begin{pmatrix} 1 & 0 & 0 \\ 0 & 0 & 1 \end{pmatrix}$.

3. For each of the following matrices E, determine if it is column stochastic. If it is column stochastic, then determine $\lim_{k \to \infty} E^k$.

(a) $E = \begin{pmatrix} 0.44 & 0.32 \\ 0.56 & 0.68 \end{pmatrix}$

(b) $E = \begin{pmatrix} 3/7 & 4/7 \\ 3/7 & 5/7 \end{pmatrix}$

(c) $E = \begin{pmatrix} 0.9 & 0.8 & 0.5 \\ 0.1 & 0.2 & 0.5 \end{pmatrix}$

(d) $E = \begin{pmatrix} 1.00 & 0.40 \\ 0.00 & 0.60 \end{pmatrix}$.

4. Construct a 2×2 Markov matrix M with eigenvalues $\lambda = \pm 1$. Show that $\lim_{k \to \infty} M^k$ does not exist. Show that for some $v \in \mathbb{R}^2$, $\lim_{k \to \infty} M^k v$ does exist. Explain your findings in terms of the eigenspaces and eigenvalues of M.

5. Consider a Markov matrix M with eigenvalues $1 = \lambda_1 \geq \lambda_2 \geq \cdots \geq \lambda_{m-1} \geq \lambda_m = -1$. Show that $\lim_{k \to \infty} M^k$ does not exist. For what $v \in \mathbb{R}^m$ does $v^* = \lim_{k \to \infty} M^k v$ exist? Explain your findings in terms of the eigenspaces and eigenvalues of M.

6. Prove part (b.) of Lemma 6.4.2.

7. Consider the population movement study (Example 6.4.11).

(a) Find $[v_0]_Q$, where $Q = \{Q_1, Q_2, Q_3, Q_4\}$, the given eigenbasis for E. Verify that v_0 has components in each of the eigenspaces.

(b) Verify that the given population vectors v_0, v_1, v_2 give the distribution for the same number of people (except for rounding uncertainty).

(c) How can the long-term contribution from the eigenspaces with $\lambda_i < 1$ diminish without v_k loosing population?

8. The Office of Voter Tracking has developed a simple linear model for predicting voting trends based on political party affiliation. They classify voters according to the four categories (in alphabetical order): Democrat, Independent, Libertarian, Republican. Suppose, $x \in \mathbb{R}^4$ is the fraction of voters who voted by the given parties in the last gubernatorial election. In the next election, OVT predicts voting distribution as Ex where

$$E = \begin{pmatrix} 0.81 & 0.07 & 0.04 & 0.01 \\ 0.08 & 0.64 & 0.01 & 0.08 \\ 0.08 & 0.21 & 0.89 & 0.07 \\ 0.03 & 0.08 & 0.06 & 0.84 \end{pmatrix}.$$

Suppose $x = (0.43, 0.08, 0.06, 0.43)^\mathsf{T}$, indicating that 43% voted for the Democratic candidate, 6% voted for the Libertarian candidate, etc. In the next election, OVT predicts the voting to be

$$y = Ex = \begin{pmatrix} 0.81 & 0.07 & 0.04 & 0.01 \\ 0.08 & 0.64 & 0.01 & 0.08 \\ 0.08 & 0.21 & 0.89 & 0.07 \\ 0.03 & 0.08 & 0.06 & 0.84 \end{pmatrix} \begin{pmatrix} 0.43 \\ 0.08 \\ 0.06 \\ 0.43 \end{pmatrix} \approx \begin{pmatrix} 0.361 \\ 0.121 \\ 0.135 \\ 0.384 \end{pmatrix}.$$

(a) Determine whether or not E is Markov. How would you interpret this property in terms of the problem description?

(b) Determine whether or not E^{T} is Markov. How would you interpret this property in terms of the problem description?

(c) Find the long-term behavior of this voter tracking problem. Interpret your result.

9. Consider a two-species ecosystem with a population of x_1 predators and a population of x_2 prey. Each year the population of each group changes according to the combined birth/death rate and the species interaction. Suppose the predatory species suffers from high death rates but flourishes under large prey population. That is, suppose

$$x_1(t + \Delta t) = \frac{1}{2} x_1(t) + \frac{2}{5} x_2(t).$$

Also, suppose that the prey species has a very robust birth rate but looses a large portion of its population to the predator species. In particular,

$$x_2(t + \Delta t) = -\frac{1}{4} x_1(t) + \frac{6}{5} x_2(t).$$

We then have the evolutionary system

$$x(t + \Delta t) = \begin{pmatrix} 0.5 & 0.4 \\ -0.25 & 1.2 \end{pmatrix} x(t).$$

Suppose the year zero population is $x(0) = [100 \ 200]^{\mathsf{T}}$. Then we find some future yearly populations

$$x(1) = \begin{pmatrix} 0.5 & 0.4 \\ -0.25 & 1.2 \end{pmatrix} x(0) = \begin{pmatrix} 130 \\ 215 \end{pmatrix}$$

$$x(2) = \begin{pmatrix} 0.5 & 0.4 \\ -0.25 & 1.2 \end{pmatrix} x(1) \approx \begin{pmatrix} 151 \\ 256 \end{pmatrix}$$

$$x(3) = \begin{pmatrix} 0.5 & 0.4 \\ -0.25 & 1.2 \end{pmatrix} x(2) \approx \begin{pmatrix} 166 \\ 233 \end{pmatrix}.$$

Using eigenvector/eigenvalue analysis, discuss the long-term behavior of the predator-prey problem.

10. In Exercise 9 you investigated a predator-prey model for population growth. In this exercise you will construct what is known as a phase diagram for the same situation that graphically displays the model predictions for various initial populations. In $\mathbb{R}^2$, let the x_1-axis represent the population of prey and the x_2-axis the population of predators.

(a) Draw the eigenspaces and label them with the corresponding eigenvalues. Plot the population trajectory for the first few years: $x(0), x(1), x(2), x(3)$. Using this sketch, predict the population trend over the next several years and the long-term behavior.

(b) The eigenspaces cut the plane into four regions. What will happen if the initial population vector lies in another of the four regions?

(c) How will the population evolve if the initial population lies in an eigenspace?

11. A virus has begun to spread in a large forest and trees are beginning to die. A study by the forest management has revealed some details on how the state of the forest changes from year to year. The state of the forest is given by a vector $x \in \mathbb{R}^4$ indicating what fraction of the acreage is *healthy* (x_1, currently unaffected by the virus), *exposed* (x_2, exposure to the virus, but resistant), *sick* (x_3, actively fighting the virus), and *dead* (x_4). The study revealed the following details:

- 80% of healthy trees will remain healthy the next year—20% will become exposed.
- Among exposed trees, next year 20% will become healthy, 10% will remain exposed, and 70% will become sick.
- Among sick trees, next year 10% will be healthy, 30% will remain sick, and 60% will die.
- 90% of dead acres will remain dead in the next year—10% will have healthy new growth.

(a) If the forest managers take no action, what will be the long-term health state of the forest?

(b) Suppose the managers institute a program which boosts the forest resistance to the virus resulting in only 30% of exposed trees becoming sick the next year, with 60% becoming healthy and 10% remaining exposed. What will be the long-term health state of the forest in this scenario?

The following exercises explore one-dimensional heat diffusion in modified scenarios similar to the diffusion welding application.

12. Consider the one-dimensional heat diffusion operator for the situation in which the rod is thermally insulated from its surroundings. No longer will the heat flow out the ends of the rod. In this situation, heat at the ends of the rod can only flow in one direction and we have the matrix transformation in the standard basis (compare Equation 4.3):

$$
\overline{E} = \begin{pmatrix}
1 - \delta & \delta & 0 & \cdots & & 0 \\
\delta & 1 - 2\delta & \delta & & & \\
0 & \delta & \ddots & \ddots & & \vdots \\
& & & & \delta & 0 \\
\vdots & & \ddots & \ddots & \delta & 0 \\
& & & \delta & 1 - 2\delta & \delta \\
0 & & \cdots & 0 & \delta & 1 - \delta
\end{pmatrix} . \tag{6.6}
$$

In this scenario, temperatures are sampled at locations $x = {}^1\!/_2, {}^3\!/_2, {}^5\!/_2, \cdots, m + {}^1\!/_2$, and $\overline{E}$ is $(m + 1) \times (m + 1)$.

(a) Verify that $\overline{E}$ is a Markov matrix. Is $\overline{E}$ diagonalizable?

(b) Use the theoretical results of this section to characterize the eigenvalues of $\overline{E}$.

(c) Find the eigenvalues and representative eigenvectors for the case $m = 3$. Do your results support your answer of part (12b)?

(d) Verify that the following set is a basis for $\mathbb{R}^{m+1}$ composed of eigenvectors of $\overline{E}$.

$$\overline{\mathcal{B}} = \{w_0, w_1, \cdots, w_m\},$$

where

$$w_k = \left(\cos \frac{k\pi}{2(m+1)}, \cos \frac{3k\pi}{2(m+1)}, \cos \frac{5k\pi}{2(m+1)}, \cdots, \cos \frac{(2m+1)k\pi}{2(m+1)} \right)^{\mathsf{T}}.$$

(e) Compute the eigenvalues of $\overline{E}$.

(f) Describe the long-term heat diffusion behavior for this scenario.

13. Consider the one-dimensional heat diffusion operator for the situation in which some heat from the rod is lost by advection—heat loss to the atmosphere. In this case we have the matrix transformation in the standard basis:

$$\hat{E} = \begin{pmatrix} 1 - 2\delta - \epsilon & \delta & 0 & \cdots & & 0 \\ \delta & 1 - 2\delta - \epsilon & \delta & & & \\ 0 & \delta & \ddots & \ddots & & \vdots \\ \vdots & & \ddots & \ddots & \delta & 0 \\ & & & \delta & 1 - 2\delta - \epsilon & \delta \\ 0 & \cdots & 0 & \delta & 1 - 2\delta - \epsilon \end{pmatrix}, \qquad (6.7)$$

where ϵ is a small positive constant.

(a) Interpret the meaning of the constants δ and ϵ.

(b) Is $\hat{E}$ a Markov matrix? Is $\hat{E}$ diagonalizable?

(c) Use the theoretical results of this section to characterize the eigenvalues of $\hat{E}$.

(d) Find the eigenvalues of $\hat{E}$ and an eigenbasis for $\hat{E}$. Use the observation that $\hat{E} = E - \epsilon I_m$ and your knowledge of the eigenvectors and eigenvalues of E.

(e) Describe the long-term heat diffusion behavior for this scenario.

Inner Product Spaces and Pseudo-Invertibility

7

The tomography problem—finding a brain image reconstruction based on radiographic data—has been challenging for a number of reasons. We have found that the radiographic transformation is, in general, not invertible, and experimental data need not be in its range. This means that if we assume the given linear relationship, then we may either have no solution or infinitely many solutions. We should not be satisfied with such an answer. Instead, we use this result to motivate a deeper study.

Given a non-invertible linear transformation T and co-domain data b, we consider the set of solutions $X = \{x \in \mathbb{R}^n \mid Ax = b\}$, where A is the $m \times n$ matrix representation of T relative to relevant bases for the domain and co-domain of T. There are two cases to consider.

- If X is not empty, then there are infinitely many solutions. Which among them is the best choice?
- If X is empty, then which domain vector is closest to being a solution?

The answers to such questions are very important for any application. If, for example, our brain scan radiographs are not in the range of our transformation, then it is important to provide a "best" or "most likely" brain image *even if it does not satisfy the transformation equation.*

These considerations require us to understand a basic comparative relationship among vectors in a vector space. Questions to consider:

1. What is the *length* of a brain image vector?
2. What would be the *distance* between two radiographs?
3. How *similar* are two brain images? Two radiographs?
4. How might we *project* a vector onto a subspace of pre-determined acceptable solutions?
5. What is the vector in an acceptable set which is *closest* to the set of solutions to a linear transformation?

In this chapter, we explore these key questions and related ideas. In the end, we will be able to find the best possible solutions to *any* linear transformation equation with bounds on the quality of our results. We will reconstruct high-quality brain images from very few moderately noisy radiographs.

7.1 Inner Products, Norms, and Coordinates

We have learned to recognize linearly independent sets and bases for finite-dimensional vector spaces. These ideas have been quite useful in categorizing elements of vector spaces and subspaces. Linear combinations of linearly independent vectors provide a tool for cataloging vectors through coordinate space representations. We can even find matrix representations of linear transformations between

© Springer Nature Switzerland AG 2022, corrected publication 2023
H. A. Moon et al., *Application-Inspired Linear Algebra*, Springer Undergraduate Texts in Mathematics and Technology, https://doi.org/10.1007/978-3-030-86155-1_7

coordinate spaces. In this section, we explore additional questions that bring vector spaces into sharper focus:

- What is the degree to which two vectors are linearly independent?
- Can we develop a way to measure the "length" of a vector? What about the angle between two vectors?

Consider the following three grayscale image vectors in $\mathcal{I}_{4\times4}(\mathbb{R})$:

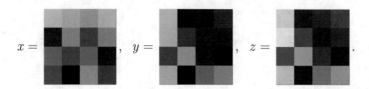

The set $\{x, y, z\}$ is linearly independent and the span of the set is a subspace of dimension three. Also, since the set is linearly independent, any subset is also linearly independent. However, we intuitively see that the set $\{x, y\}$ is somehow "more" linearly independent than the set $\{y, z\}$ because images y and z are very similar, in gray-scale intensity distribution. We might colloquially say that the set $\{y, z\}$ is "nearly linearly dependent" because Span$\{y, z\}$ may not seem to describe as rich a subspace as Span$\{x, y\}$.

Consider a second example of three vectors in $\mathbb{R}^2$ illustrated here:

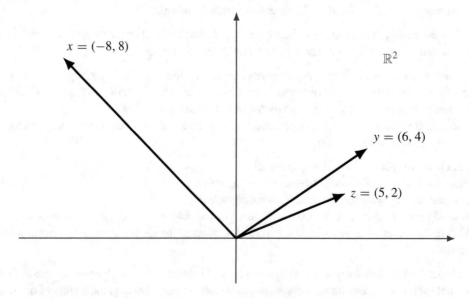

We can make observations analogous to the previous example. We see that $\{y, z\}$ is nearly linearly dependent, whereas we would not be so quick to make this judgment about the set $\{x, y\}$. We say that y and z are nearly linearly dependent because they have a nearly common direction.

In this section, we introduce the *inner product* as a tool for measuring the degree of linear independence between vectors. This tool will allow us to (1) define the distance between vectors, (2), define the length of a vector, and (3) define the angle between two vectors.

7.1.1   Inner Product

Motivating example: the dot product in $\mathbb{R}^n$

Building on the previous example of vectors in $\mathbb{R}^2$, let θ be the positive angle between two vectors x and y in $\mathbb{R}^n$. We expect that vectors that are nearly linearly dependent are separated by angles that are close to zero radians or close to π radians. In these cases, the vectors point nearly along the same line, and $|\cos\theta| \approx 1$. On the other hand, if $\theta \approx \pi/2$ then the two vectors are very much *not* scalar multiples of each other. In fact, they point nearly along perpendicular lines and $\cos\theta \approx 0$.

Angles play a large role in quantifying the "degree of linear dependence," but also, since the zero vector is not part of any linearly independent set, we expect that the length of a vector should also be considered. Recall that the length of a vector $x = (x_1, \ldots, x_n)$ in $\mathbb{R}^n$ is given by the formula

$$\ell(x) = (x_1^2 + \cdots + x_n^2)^{1/2}.$$

It turns out in $\mathbb{R}^2$, and indeed in $\mathbb{R}^n$ in general, the familiar dot product of two vectors encodes information about the degree of linear (in)dependence of the vectors, and hence also about the lengths and the angle between the vectors.

Recall that the dot product is defined as

$$x \cdot y := x_1 y_1 + \cdots + x_n y_n. \tag{7.1}$$

This yields an expression for length of a vector in terms of the dot product of the vector with itself: $x \cdot x = \ell(x)^2$. Moreover, by the law of cosines

$$x \cdot y = \ell(x)\ell(y)\cos\theta. \tag{7.2}$$

Notice that the inner product is symmetric ($x \cdot y = y \cdot x$) and linear in the first argument (see Exercise 18) as we would hope for a function that quantifies the "similarity" between vectors.

Hence the dot product is a useful measure of the "degree of linear dependence"—roughly speaking, larger dot products correspond to larger vectors in similar directions, while smaller dot products correspond to smaller vectors or vectors in very different directions.

Inner product

While this works well in $\mathbb{R}^n$, we would like to extend these ideas to general finite-dimensional vector spaces. Our approach is to define a general similarity function, called an *inner product*, that has the properties we desire.

Definition 7.1.1

Let $(V, +, \cdot)$ be a vector space with scalars in a field $\mathbb{F}$. An **inner product** is a mapping $\langle \cdot, \cdot \rangle : V \times V \to \mathbb{F}$ that satisfies the following three properties. For every $u, v, w \in V$ and $\alpha \in \mathbb{F}$

1. $\langle u, v \rangle = \langle v, u \rangle$ (Symmetric)
2. $\langle \alpha u + v, w \rangle = \alpha \langle u, w \rangle + \langle v, w \rangle$ (Linearity in the first argument)
3. $\langle u, u \rangle \geq 0$ and $\langle u, u \rangle = 0$ if and only if $u = 0$. (Positive definite)

A vector space with an inner product is called an **inner product space**. □

The notation $\langle \cdot, \cdot \rangle$ indicates an inner product that allows inputs where the two dots are located. Also, the above definition of a (real) inner product assumes $\mathbb{F} = \mathbb{R}$ or $\mathbb{F} = Z_2$. Complex inner product spaces are considered in the exercises for the interested reader. This definition does not preclude the possibility that we could define more than one inner product on a vector space.

In Section 7.1.2, we will use inner products to define notions of length and angle between vectors. We now consider several examples of inner products on some familiar vector spaces.

Example 7.1.2 Let us define $\langle \cdot, \cdot \rangle : \mathcal{M}_{3 \times 2}(\mathbb{R}) \times \mathcal{M}_{3 \times 2}(\mathbb{R}) \to \mathbb{R}$ so that $\langle u, v \rangle$ is the sum of the component-wise products of the matrices u and v. That is, if

$$u = \begin{pmatrix} a & b \\ c & d \\ e & f \end{pmatrix} \quad \text{and} \quad v = \begin{pmatrix} g & h \\ j & k \\ \ell & m \end{pmatrix}$$

then,

$$\langle u, v \rangle = ag + bh + cj + dk + e\ell + fm.$$

We will check that the properties of Definition 7.1.1 hold. Notice that because multiplication in $\mathbb{R}$ is commutative, $\langle u, v \rangle = \langle v, u \rangle$. Now, let

$$w = \begin{pmatrix} n & p \\ q & r \\ s & t \end{pmatrix}.$$

Then, because multiplication distributes over addition, we see that

$$\begin{aligned} \langle \alpha u + v, w \rangle &= (\alpha a + g)n + (\alpha b + h)p + (\alpha c + j)q \\ &\quad + (\alpha d + k)r + (\alpha e + \ell)s + (\alpha f + m)t \\ &= \alpha(an + bp + cq + dr + es + ft) \\ &\quad + (gn + hp + jq + kr + \ell s + mt) \\ &= \alpha \langle u, w \rangle + \langle v, w \rangle. \end{aligned}$$

Therefore, the linearity condition holds.

Finally, we compute

$$\langle u, u \rangle = a^2 + b^2 + c^2 + d^2 + e^2 + f^2 \geq 0.$$

And, $\langle u, u \rangle = 0$ if and only if $a = b = c = d = e = f = 0$. That is, u is the 3×2 zero matrix. Thus, this is indeed an inner product on $\mathcal{M}_{2 \times 3}(\mathbb{R})$.　　　□

Example 7.1.3 Let $\langle \cdot, \cdot \rangle : \mathcal{P}_2(\mathbb{R}) \times \mathcal{P}_2(\mathbb{R}) \to \mathbb{R}$ be defined by

$$\langle p_1, p_2 \rangle = \int_0^1 p_1 p_2 \, dx, \quad \text{for } p_1, p_2 \in \mathcal{P}_2(\mathbb{R}).$$

Here, $p_1 p_2$ is the point-wise product: $p_1 p_2(x) = p_1(x) p_2(x)$. For the purposes of this text, we will call this inner product, the standard inner product[1] for $\mathcal{P}_n(\mathbb{R})$. Notice again that because polynomial multiplication is commutative, the symmetric property holds:

$$\langle p_1, p_2 \rangle = \langle p_2, p_1 \rangle.$$

---

[1] This may not be the most useful inner product for various applications. And, in other contexts, you may see more useful inner products on polynomial spaces.

Next, we show that this inner product is linear in the first argument.

$$\langle \alpha p_1 + p_2, p_3 \rangle = \int_0^1 (\alpha p_1 + p_2)p_3 \, dx = \alpha \int_0^1 p_1 p_3 \, dx + \int_0^1 p_2 p_3 \, dx = \alpha \langle p_1, p_3 \rangle + \langle p_2, p_3 \rangle.$$

Finally, if $p \in \mathcal{P}_2(\mathbb{R})$ then $\langle p, p \rangle = \int_0^1 p^2(x) \, dx \geq 0$. If $\langle p, p \rangle = 0$ then $\int_0^1 p^2(x) \, dx = 0$, which means p must be the zero polynomial. So, we now have an inner product on $\mathcal{P}_2(\mathbb{R})$. This example can be readily extended to any polynomial space. □

Example 7.1.4 Let $x, y \in \mathbb{R}^n$, then we define the standard inner product on $\mathbb{R}^n$ as the dot product:

$$\langle x, y \rangle := x_1 y_1 + x_2 y_2 + \ldots + x_n y_n.$$

Consider the vectors from the introduction to this section,

$$x = (-8, 8), \ y = (6, 4), \ z = (5, 2).$$

We have[2]

$$\langle x, y \rangle = (-8)(6) + (8)(4) = -16$$
$$\langle x, z \rangle = (-8)(5) + (8)(2) = -32$$
$$\langle y, z \rangle = (6)(5) + (4)(2) = 38$$
$$\langle x, x \rangle = (-8)(-8) + (8)(8) = 128$$
$$\langle y, y \rangle = (6)(6) + (4)(4) = 52$$
$$\langle z, z \rangle = (5)(5) + (2)(2) = 29.$$ □

The next theorem gives a method for writing the standard inner product on $\mathbb{R}^n$ as a matrix product.

Theorem 7.1.5
Let $x, y \in \mathbb{R}^n$ and let $\langle \cdot, \cdot \rangle$ be the standard inner product on $\mathbb{R}^n$. Then

$$\langle x, y \rangle = x^\mathsf{T} y.$$

Proof. Let $\langle \cdot, \cdot \rangle$ be the standard inner product on $\mathbb{R}^n$ and denote $x, y \in \mathbb{R}^n$ by

$$x = \begin{pmatrix} x_1 \\ x_2 \\ \vdots \\ x_n \end{pmatrix}, y = \begin{pmatrix} y_1 \\ y_2 \\ \vdots \\ y_n \end{pmatrix}.$$

---

[2] Do the following inner product computations align with your geometric intuition?

Then,

$$\langle x, y \rangle = \sum_{k=1}^{n} x_k y_k$$

$$= \begin{pmatrix} x_1 & x_2 & \cdots & x_n \end{pmatrix} \begin{pmatrix} y_1 \\ y_2 \\ \vdots \\ v_n \end{pmatrix}$$

$$= x^{\mathsf{T}} y. \qquad\qquad \square$$

The standard inner product on image spaces is similar to the standard inner product on $\mathbb{R}^n$. For example, suppose images $I, J \in \mathcal{I}_{4 \times 4}(\mathbb{R})$ have ordered entries I_k and J_k for $k = 1, 2, \ldots, 16$. We define the standard inner product $\langle I, J \rangle = \sum_{k=1}^{16} I_k J_k$.

Example 7.1.6 Consider the three images from the introduction to this section,

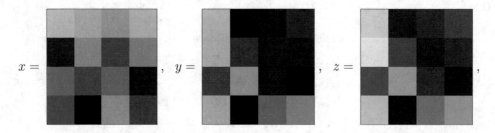

with numerical grayscale entries

$$x = \begin{array}{|c|c|c|c|} \hline 95 & 89 & 82 & 92 \\ \hline 23 & 76 & 44 & 74 \\ \hline 61 & 46 & 62 & 18 \\ \hline 49 & 2 & 79 & 41 \\ \hline \end{array} \,, \quad y = \begin{array}{|c|c|c|c|} \hline 94 & 6 & 14 & 27 \\ \hline 92 & 35 & 20 & 20 \\ \hline 41 & 81 & 20 & 2 \\ \hline 89 & 1 & 60 & 75 \\ \hline \end{array} \,, \quad z = \begin{array}{|c|c|c|c|} \hline 102 & 22 & 30 & 23 \\ \hline 109 & 45 & 21 & 29 \\ \hline 50 & 85 & 33 & 15 \\ \hline 97 & 14 & 67 & 83 \\ \hline \end{array} .$$

Using the standard inner product, we compute

$$\langle x, y \rangle = 39{,}913 \,, \quad \langle x, z \rangle = 47{,}974 \,, \quad \langle y, z \rangle = 51{,}881 \,,$$

$$\langle x, x \rangle = 66{,}183 \,, \quad \langle y, y \rangle = 46{,}079 \,, \quad \langle z, z \rangle = 59{,}527 \,.$$

From these inner products, we can define image vector lengths and angles between vectors according to the ideas behind Equation 7.2. We set $\ell(x) = \sqrt{\langle x, x \rangle}$ and $\cos \theta_{x,y} = \langle x, y \rangle / \ell(x) \ell(y)$

$$\ell(x) \approx 257 \,, \quad \ell(y) \approx 215 \,, \quad \ell(z) \approx 244 \,,$$

$$\cos \theta_{x,y} \approx 0.723 \,, \quad \cos \theta_{x,z} \approx 0.764 \,, \quad \cos \theta_{y,z} \approx 0.991 \,.$$

As we predicted (or because we constructed a good measure of linear dependence) y and z are nearly linearly dependent by the cosine measure. □

7.1.2   Vector Norm

As in Example 7.1.6, we can then use an inner product on a vector space to define vector length and angle between vectors. We want length and angle to match what we would compute in $\mathbb{R}^n$ with the dot product. In $\mathbb{R}^n$, the length of x is

$$\ell(x) = (x \cdot x)^{1/2} \tag{7.3}$$

and the angle θ between vectors x an y is defined to be

$$\theta = \cos^{-1}\left(\frac{x \cdot y}{\ell(x)\ell(y)}\right).$$

In a general inner product space, this yields the following:

Definition 7.1.7

Let V be an inner product space with inner product $\langle \cdot, \cdot \rangle : V \times V \to \mathbb{R}$. We define the **norm** of $u \in V$, denoted $||u||$, by

$$||u|| := (\langle u, u \rangle)^{1/2}.$$

□

Notice that if we consider the standard inner product on $\mathbb{R}^n$, the norm of a vector corresponds to the Euclidean length. If $u = (u_1, u_2, \ldots, u_n)$, then

$$\begin{aligned}
||u|| &= (\langle u, u \rangle)^{1/2} \\
&= (u \cdot u)^{1/2} \\
&= \sqrt{u_1^2 + u_2^2 + \ldots + u_n^2}.
\end{aligned}$$

Definition 7.1.8

A vector v in an inner product space is said to be a **unit vector** if $||v|| = 1$. □

Example 7.1.9 Consider the inner product computations in Example 7.1.4. We find the lengths of each vector:

$$\begin{aligned}
||x|| &= \sqrt{\langle x, x \rangle} = \sqrt{128} = 6\sqrt{2} \approx 11.3 , \\
||y|| &= \sqrt{\langle y, y \rangle} = \sqrt{52} = 2\sqrt{13} \approx 7.21 , \\
||z|| &= \sqrt{\langle z, z \rangle} = \sqrt{29} \approx 5.39 .
\end{aligned}$$

We can also compute the cosine of the angle between pairs of vectors. For example,

$$f(x, y) = \cos\theta_{x,y} = \frac{\langle x, y \rangle}{||x||\,||y||} = \frac{-16}{\left(6\sqrt{2}\right)\left(2\sqrt{13}\right)} \approx -0.261 ,$$

$$f(y, z) = \cos\theta_{y,z} = \frac{\langle y, z \rangle}{\|y\| \, \|z\|} = \frac{38}{\left(2\sqrt{13}\right)\left(\sqrt{29}\right)} \approx 0.979 \, .$$

Notice that $|f(y, z)| > |f(x, y)|$, which indicates that $\{y, z\}$ is closer to being linearly dependent than $\{x, y\}$.   □

Example 7.1.10 Reconsider Example 7.1.3. We can find the "length" of a polynomial $p \in \mathcal{P}_2(\mathbb{R})$. Let $p = ax^2 + bx + c$. Then

$$\begin{aligned}
\|p\| &= \left(\int_0^1 (ax^2 + bx + c)^2 \, dx \right)^{1/2} \\
&= \left(\int_0^1 a^2 x^4 + 2abx^3 + (2ac + b^2)x^2 + 2bcx + c^2 \, dx \right)^{1/2} \\
&= \left(\frac{a^2}{5} + \frac{ab}{2} + \frac{2ac + b^2}{3} + bc + c^2 \right)^{1/2} .
\end{aligned}$$

Notice that $f(x) = \sqrt{5}x^2$ is a unit vector in $\mathcal{P}_2(\mathbb{R})$ with this inner product. For this function, $a = \sqrt{5}$ and $b = c = 0$, so that $\|f\| = 1$.   □

Example 7.1.11 Consider the vector space of 7-bar LCD characters, $\mathcal{D}(Z_2)$. Let $f : \mathcal{D} \times \mathcal{D} \to Z_2$ be defined by

$$f(x, y) = \begin{cases} 1 & \text{if } x = y \neq 0 \\ 0 & \text{otherwise.} \end{cases} \tag{7.4}$$

We can show that f is not an inner product on $\mathcal{D}(Z_2)$. Clearly, f is symmetric and

$$f(ax, y) = \left\{ \begin{array}{ll} 1 & \text{if } a = 1 \text{ and } x = y \neq 0 \\ 0 & \text{otherwise} \end{array} \right\} = af(x, y).$$

We also have positive definiteness: $f(x, x) = 1$ if $x \neq 0$ and $f(x, x) = 0$ if $x = 0$. However, $f(x + y, z) \neq f(x, z) + f(y, z)$ for all $x, y, z \in \mathcal{D}$. This is the subject of Exercise 14.   □

The next example shows that a vector space can have many different inner products and many different associated norms.

Example 7.1.12 Consider the vector space $\mathbb{R}^2$ and define $\langle x, y \rangle = x_1 y_1 + bx_2 y_2$, where $b > 0$ is a scalar. We can show that $\langle \cdot, \cdot \rangle : \mathbb{R}^2 \times \mathbb{R}^2 \to \mathbb{R}$ is an inner product (see Exercise 9). Every positive value of b defines a unique inner product on $\mathbb{R}^2$ and a unique norm. With this inner product, the length of a vector $x = (x_1, x_2)$ is $\|x\| = \sqrt{x_1^2 + bx_2^2}$. The set of unit vectors is the solution set of the ellipse equation $x_1^2 + bx_2^2 = 1$. The sets of unit vectors for $b = 1$ and $b = 4$ are shown in Figure 7.1.   □

Inspired by this last example, we would like to find a generalization of the standard inner product on $\mathbb{R}^n$. We first need to consider a class of matrices with specific properties, which we outline here.

Definition 7.1.13

Let A be an $n \times n$ matrix. We say that A is **symmetric** if and only if $A^\mathsf{T} = A$.   □

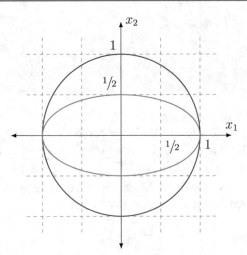

Fig. 7.1 Two example sets of unit vectors in $\mathbb{R}^2$ corresponding to the inner products $\langle x, y \rangle = x_1 y_1 + x_2 y_2$ (circle) and $\langle x, y \rangle = x_1 y_1 + 4 x_2 y_2$ (ellipse).

A symmetric matrix is symmetric with respect to reflection across the main diagonal. That is, if a_{ij} for $1 \le i, j \le n$ is the (i, j)-entry of A, then $a_{ij} = a_{ji}$. For example,

$$M = \begin{pmatrix} 2 & 3 & 5 \\ 3 & -1 & 7 \\ 5 & 7 & 0 \end{pmatrix}$$

is a symmetric matrix, but

$$A = \begin{pmatrix} 2 & 3 & 5 \\ 3 & -1 & 7 \\ 5 & 1 & 0 \end{pmatrix}$$

is not because $a_{3,2} \ne a_{2,3}$.

Definition 7.1.14

We say that a real-valued square matrix A is **positive definite** if and only if A is symmetric and all the eigenvalues of A are positive. □

Example 7.1.15 Let M be the matrix given by

$$M = \begin{pmatrix} 4 & 1 & 1 \\ 1 & 6 & 1 \\ 1 & 1 & 4 \end{pmatrix}.$$

Using ideas from Section 6.2, we find linearly independent eigenvectors

$$v_1 = \begin{pmatrix} 1 \\ 0 \\ -1 \end{pmatrix}, v_2 = \begin{pmatrix} 1 \\ -1 \\ 1 \end{pmatrix}, \text{ and } v_3 = \begin{pmatrix} 1 \\ 2 \\ 1 \end{pmatrix},$$

with corresponding eigenvalues $\lambda_1 = 3$, $\lambda_2 = 4$, and $\lambda_3 = 7$, respectively. Because M is symmetric and has all positive eigenvalues, M is positive definite. □

The next theorem provides a characterization of inner products on $\mathbb{R}^n$.

Theorem 7.1.16
Let $x, y \in \mathbb{R}^n$ and $f : \mathbb{R}^n \times \mathbb{R}^n \to \mathbb{R}^n$ defined by $f(x, y) = x^{\mathsf{T}} A y$ for some $n \times n$ matrix A. Then f is an inner product on $\mathbb{R}^n$ if and only if A is positive definite.

The proof of Theorem 7.1.16 is the subject of Exercise 19.

Theorem 7.1.16 provides a useful generalization of the concept of vector length in $\mathbb{R}^n$. For suitable matrix A, the length of $x \in \mathbb{R}^n$ is given by $\|x\| = \sqrt{x^{\mathsf{T}} A x}$. If A is the $n \times n$ identity matrix, we recover the standard notion of length $\|x\| = \sqrt{x^{\mathsf{T}} I x} = \sqrt{x^{\mathsf{T}} x} = \sqrt{x_1^2 + \ldots + x_n^2}$.

The inner product family of Example 7.1.12 can be written in the form stated in Theorem 7.1.16: $\langle x, y \rangle = x^{\mathsf{T}} A y$ for the symmetric positive definite matrix

$$A = \begin{pmatrix} 1 & 0 \\ 0 & b \end{pmatrix}.$$

(Recall: $b > 0$.) We have

$$x^{\mathsf{T}} A y = \begin{pmatrix} x_1 & x_2 \end{pmatrix} \begin{pmatrix} 1 & 0 \\ 0 & b \end{pmatrix} \begin{pmatrix} y_1 \\ y_2 \end{pmatrix}$$
$$= \begin{pmatrix} x_1 & x_2 \end{pmatrix} \begin{pmatrix} y_1 \\ b y_2 \end{pmatrix}$$
$$= x_1 y_1 + b x_2 y_2.$$

7.1.3   Properties of Inner Product Spaces

The definition of an inner product on vector spaces formalizes the concept of degree of linear dependence. From this idea came a general notion of vector length and angles between vectors. In our study of inner product spaces, we will discover several other useful properties of the inner product and of vector lengths including the familiar triangle inequality and a general concept of perpendicular (orthogonal) vectors. First, we consider several symmetry and uniqueness properties.

Theorem 7.1.17
Let V be a real inner product space. Then for $u, v, w \in V$ and $c \in \mathbb{R}$, the following statements hold.

1. $\langle u, v + w \rangle = \langle u, v \rangle + \langle u, w \rangle$.
2. $\langle u, cv \rangle = \langle cu, v \rangle = c \langle u, v \rangle$.
3. $\langle u, 0 \rangle = \langle 0, u \rangle = 0$.
4. If $\langle u, v \rangle = \langle u, w \rangle$ for all $u \in V$, then $v = w$.

Proof.   Let V be a real inner product space, $u, v, w \in V$ and $c \in \mathbb{R}$. We use the three properties, given in Definition 7.1.1 of real inner product spaces to prove each statement.

1. Using the symmetry of $\langle \cdot, \cdot \rangle$, we have

$$\langle u, v + w \rangle = \langle v + w, u \rangle$$
$$= \langle v, u \rangle + \langle w, u \rangle$$
$$= \langle u, v \rangle + \langle u, w \rangle.$$

2. Again, using the symmetry property, we have

$$\langle u, cv \rangle = \langle cv, u \rangle$$
$$= c \langle v, u \rangle$$
$$= c \langle u, v \rangle$$
$$= \langle cu, v \rangle.$$

3. Let $c = 0$ in statement 2 of the theorem, then, using linearity in the first argument, we have

$$\langle u, 0 \rangle = \langle 0, u \rangle = 0.$$

4. Suppose $\langle u, v \rangle = \langle u, w \rangle$ for all $u \in V$. Then

$$0 = \langle u, v \rangle - \langle u, w \rangle$$
$$= \langle u, v \rangle + \langle u, -w \rangle$$
$$= \langle u, v - w \rangle.$$

Since the above is true for any u, it must be true for $u = v - w$. Therefore, $\langle v - w, v - w \rangle = 0$, which implies that $v - w = 0$. Thus, $v = w$.   □

Theorem 7.1.17 parts (1) and (2) tell us that inner products on real vector spaces are also linear in the second argument. Some of these properties are specific to real inner product spaces. Exercises 25–27 consider the corresponding properties of complex inner product spaces.

Next, we consider properties related to vector length. In particular, we will be satisfied to learn that vector length scales appropriately and the familiar triangle inequality holds. We also find that vector length is non-negative and is only zero when the vector itself is the zero vector of the vector space.

Theorem 7.1.18
Let V be a real inner product space. Then for $u, v \in V$ and $c \in \mathbb{R}$, the following statements hold.

1. $\|cu\| = |c| \|u\|$.
2. $\|u\| \geq 0$ and $\|u\| = 0$ if and only if $u = 0$.
3. $|\langle u, v \rangle| \leq \|u\| \|v\|$. (Cauchy-Schwarz Inequality)
4. $\|u + v\| \leq \|u\| + \|v\|$. (Triangle Inequality)

Proof. Let V be a real inner product space, $u, v \in V$ and $c \in \mathbb{R}$. We use Definition 7.1.1 and properties of a real inner product space (Theorem 7.1.17) to prove each statement.

1. First, we have

$$\|cu\| = \langle cu, cu \rangle^{1/2} = \left(c^2 \langle u, u \rangle \right)^{1/2} = |c| \langle u, u \rangle^{1/2} = |c| \|u\|.$$

2. Now, we know that $\|u\| = \langle u, u \rangle^{1/2}$. Thus, the result follows because V is a real inner product space.

3. If $v = 0$ then the result follows trivially. Otherwise, let $\mu = \langle u, v \rangle / \langle v, v \rangle$. We will show that the inequality $0 \leq \|u - \mu v\|^2$ leads to the desired relation. Using Theorem 7.1.17 and the linearity condition for inner products, we have

$$
\begin{aligned}
0 \leq \|u - \mu v\|^2 \\
= \langle u - \mu v, u - \mu v \rangle \\
= \langle u, u \rangle - 2 \langle u, \mu v \rangle + \langle \mu v, \mu v \rangle \\
= \langle u, u \rangle - 2 \mu \langle u, v \rangle + \mu^2 \langle v, v \rangle \\
= \langle u, u \rangle - 2 \frac{\langle u, v \rangle^2}{\langle v, v \rangle} + \frac{\langle u, v \rangle^2 \langle v, v \rangle}{\langle v, v \rangle^2} \\
= \langle u, u \rangle - \frac{\langle u, v \rangle^2}{\langle v, v \rangle} \\
= \|u\|^2 - \frac{\langle u, v \rangle^2}{\|v\|^2}.
\end{aligned}
$$

Thus, $\|u\| \|v\| \geq |\langle u, v \rangle|$.

4. We now use the Cauchy-Schwarz inequality (Property (3) of this theorem) to establish the triangle inequality.

$$
\begin{aligned}
\|u + v\| = \langle u + v, u + v \rangle^{1/2} \\
= (\langle u, u \rangle + \langle v, v \rangle + 2 \langle u, v \rangle)^{1/2} \\
= \left(\|u\|^2 + \|v\|^2 + 2 \langle u, v \rangle \right)^{1/2} \\
\leq \left(\|u\|^2 + \|v\|^2 + 2 \|u\| \|v\| \right)^{1/2} \\
= \|u\| + \|v\|.
\end{aligned}
$$

Therefore, the triangle inequality holds.   □

Some applications of the triangle inequality and the Cauchy-Schwarz inequality are considered in Exercise 1 and Exercise 15.

Standard inner products, for vector spaces of interest in this text, are compiled in the following table (Table 7.1).

7.1.4   Orthogonality

When the inner product of two nonzero vectors has a value of zero, we understand that these two vectors are "maximally" linearly independent because the angle between them is $\pi/2$. In $\mathbb{R}^2$ or $\mathbb{R}^3$, we know that two such vectors are said to be *perpendicular* to each other. We can generalize this concept to other vector spaces.

Definition 7.1.19

Let V be an inner product space with inner product $\langle \cdot, \cdot \rangle : V \times V \to \mathbb{R}$. Given two vectors $u, v \in V$, we say u and v are **orthogonal** if $\langle u, v \rangle = 0$.   □

Table 7.1 A table of standard inner products.

Vector Space	Standard Inner Product
$(\mathbb{R}^n, +, \cdot)$	Let $u = \begin{pmatrix} u_1 \\ u_2 \\ \vdots \\ u_n \end{pmatrix}$ and $v = \begin{pmatrix} v_1 \\ v_2 \\ \vdots \\ v_n \end{pmatrix}$ then $$\langle u, v \rangle = \sum_{j=1}^{n} u_j v_j \text{ (The dot product)}$$
$(\mathcal{P}_n(\mathbb{R}), +, \cdot)$	$\langle p(x), q(x) \rangle = \displaystyle\int_0^1 p(x) q(x)\, dx$
$(\mathcal{M}_{m \times n}(\mathbb{R}), +, \cdot)$	Let $A = (a_{ij})$ and $B = (b_{ij})$ then $$\langle A, B \rangle = \sum_{i=1}^{m} \sum_{j=1}^{n} a_{ij} b_{ij} \text{ (The Frobenius Inner Product)}$$
$\mathcal{I}_{m \times n}$	Let u be the image with pixel intensities u_{ij} and v be the image with pixel intensities v_{ij}, then $$\langle u, v \rangle = \sum_{i=1}^{m} \sum_{j=1}^{n} u_{ij} v_{ij}.$$

This definition tells us that $0 \in V$ is orthogonal to all vectors in V. The concept of orthogonality can also be extended to sets of more than two vectors.

Definition 7.1.20

We say that a set of nonzero vectors $\{v_1, v_2, \ldots, v_n\}$ is an **orthogonal set** if $\langle v_i, v_j \rangle = 0$ whenever $i \neq j$. If an orthogonal set consists only of unit vectors, then we say that the set is an **orthonormal set**. $\qquad\square$

The first part of Definition 7.1.20 says that the vectors are pairwise orthogonal. The second part of Definition 7.1.20 says the vectors all have unit length. Since orthogonal sets consist of pairwise linearly independent vectors, we might wonder whether orthogonal sets are linearly independent sets.

Theorem 7.1.21
Let $\mathcal{B} = \{v_1, v_2, \ldots, v_n\}$ be an orthogonal set of vectors in an inner product space V. Then $\mathcal{B}$ is linearly independent.

Proof. Let V be an inner product space. Suppose, also, that $\mathcal{B} = \{v_1, v_2, \ldots, v_n\}$ an orthogonal set of vectors in V. If $\mathcal{B} = \{v_1\}$ then it is linearly independent because $v_1 \neq 0$.

Now, suppose $n \geq 2$ and, by way of contradiction, that $\mathcal{B}$ is linearly dependent. That is, we assume without loss of generality that $v_n = a_1 v_1 + a_2 v_2 + \ldots + a_{n-1} v_{n-1}$ for some scalars $a_1, a_2, \ldots, a_{n-1}$. Now, taking the inner product of both the left- and right-hand sides of this equation with v_n, we find

$$\langle v_n, v_n \rangle = \langle a_1 v_1 + a_2 v_2 + \ldots + a_{n-1} v_{n-1}, v_n \rangle$$

$$= a_1 \langle v_1, v_n \rangle + a_2 \langle v_2, v_n \rangle + \ldots + a_{n-1} \langle v_{n-1}, v_n \rangle$$
$$= a_1 0 + a_2 0 + \ldots + a_{n-1} 0$$
$$= 0.$$

Thus, $v_n = 0$, a contradiction. Therefore, $\mathcal{B}$ is linearly independent. □

In the proof of Theorem 7.1.21, there is a suggestion that if we have a linearly independent set we can add a vector, orthogonal to all others, to the set and maintain linear independence. The next corollary formalizes this observation.

Corollary 7.1.22

Let V be an inner product space, $\{u_1, u_2, \ldots, u_k\}$ a linearly independent subset of V and $w \in V$, $w \neq 0$. Then $\{u_1, u_2, \ldots, u_k, w\}$ is linearly independent if $\langle u_j, w \rangle = 0$ for $j = 1, 2, \ldots k$.

Proof. Let V be an inner product space, $S = \{u_1, u_2, \ldots, u_k\}$ a linearly independent subset of V, $w \in V$, $w \neq 0$ and $\langle u_j, w \rangle = 0$ for $j = 1, 2, \ldots k$. We need only show that w cannot be written as a linear combination of the vectors in S.

Suppose, by way of contradiction, that $w = a_1 u_1 + a_2 u_2 + \ldots + a_k u_k$ for some scalars $a_1, a_2, \ldots a_k$. Then the same technique used in the proof of Theorem 7.1.21 gives us

$$\langle w, w \rangle = \langle a_1 u_1 + a_2 u_2 + \ldots + a_k u_k, w \rangle$$
$$= a_1 \langle u_1, w \rangle + a_2 \langle u_2, w \rangle + \ldots + \langle u_k, w \rangle$$
$$= 0.$$

However, $\langle w, w \rangle = 0$ only if $w = 0$, a contradiction. So, w cannot be written as a linear combination of $u_1, u_2, \ldots, u_k$. Thus, $\{u_1, u_2, \ldots, u_k, w\}$ is linearly independent. □

In the next corollary, we recognize that Theorem 7.1.21 leads us to a "maximally" linear independent basis for an inner product space. That is, a sufficiently large orthogonal set in an inner product space is a basis. We will see how to find such bases in the next section.

Corollary 7.1.23

Let V be an n-dimensional inner product space and let $\mathcal{B} = \{v_1, v_2, \ldots, v_n\}$ be an orthogonal set of vectors in V. Then $\mathcal{B}$ is a basis for V.

Proof. By Theorem 7.1.21, we have that $\mathcal{B}$ is a linearly independent set of n vectors in V. Thus, by Theorem 3.4.24, $\mathcal{B}$ is a basis for V. □

Example 7.1.24 Let

$$\mathcal{B} = \left\{ \begin{pmatrix} 1 \\ 1 \\ 1 \end{pmatrix}, \begin{pmatrix} -1 \\ 0 \\ 1 \end{pmatrix}, \begin{pmatrix} -1 \\ 2 \\ -1 \end{pmatrix} \right\} \subseteq \mathbb{R}^3.$$

We can show that $\mathcal{B}$ is an orthogonal set in $\mathbb{R}^3$, but is not orthonormal. Indeed,

$$\left\| \begin{pmatrix} 1 \\ 1 \\ 1 \end{pmatrix} \right\| = \sqrt{3}.$$

But,

$$\left\langle \begin{pmatrix} 1 \\ 1 \\ 1 \end{pmatrix}, \begin{pmatrix} -1 \\ 0 \\ 1 \end{pmatrix} \right\rangle = -1 + 0 + 1 = 0$$

$$\left\langle \begin{pmatrix} 1 \\ 1 \\ 1 \end{pmatrix}, \begin{pmatrix} -1 \\ 2 \\ -1 \end{pmatrix} \right\rangle = -1 + 2 - 1 = 0$$

$$\left\langle \begin{pmatrix} -1 \\ 0 \\ 1 \end{pmatrix}, \begin{pmatrix} -1 \\ 2 \\ -1 \end{pmatrix} \right\rangle = 1 + 0 - 1 = 0.$$

Therefore, $\mathcal{B}$ is an orthogonal set. Furthermore, Corollary 7.1.23 tells us that $\mathcal{B}$ is a basis for $\mathbb{R}^3$.   □

Example 7.1.25 Let $\mathcal{C}([0, 1])$ be the set of continuous functions defined on $[0, 1]$. Define the inner product on $\mathcal{C}([0, 1])$ as in Example 7.1.3. The set $S = \left\{ \sqrt{2} \cos(n\pi x) \mid n \in \mathbb{N} \right\}$ is an orthonormal set.

We can show this by considering $n, k \in \mathbb{N}$, $n \neq k$. Then

$$\langle \sqrt{2} \cos(n\pi x), \sqrt{2} \cos(k\pi x) \rangle = \int_0^1 2 \cos(n\pi x) \cos(k\pi x) \, dx$$

$$= \int_0^1 \cos((n+k)\pi x) + \cos((n-k)\pi x) \, dx$$

$$= \frac{\sin((n+k)\pi)}{(n+k)\pi} + \frac{\sin((n-k)\pi)}{(n-k)\pi}$$

$$= 0.$$

So, S is an orthogonal set. Now, we show that the vectors in S are unit vectors.

$$\left\| \sqrt{2} \cos(n\pi x) \right\| = \int_0^1 2 \cos^2(n\pi x) \, dx$$

$$= \int_0^1 1 + \cos(2n\pi x) \, dx$$

$$= 1 + \frac{\sin(2n\pi)}{2n\pi}$$

$$= 1.$$

Thus, S is an orthonormal set.   □

We can use the inner product to find vectors orthogonal to a given vector. This ability will be useful in the next section for constructing orthogonal bases. Here we give a few examples.

Example 7.1.26 Suppose $u \in \mathbb{R}^3$ is defined by $u = (1 \ 1 \ 1)^{\mathsf{T}}$. Using the standard inner product for $\mathbb{R}^3$, we see that, for an arbitrary $v = (a \ b \ c)^{\mathsf{T}}$,

$$\langle u, v \rangle = a + b + c.$$

To find v orthogonal to u, we set $\langle u, v \rangle$ to zero. If $a = 1, b = 1$, and $c = -2$, then v is orthogonal to u. There are infinitely many such vectors.   □

Example 7.1.27 Suppose we have the vector $p(x) = 1 + x + x^2 \in \mathcal{P}_2(\mathbb{R})$. An orthogonal vector $q(x) = a + bx + cx^2$ must satisfy $\langle p(x), q(x) \rangle = 0$. If we use the inner product of Example 7.1.3, we find

$$
\begin{aligned}
\langle p(x), q(x) \rangle &= \int_0^1 p(x)q(x) \, dx \\
&= \int_0^1 \left(1 + x + x^2\right)\left(a + bx + cx^2\right) dx \\
&= \int_0^1 \left(a + (a+b)x + (a+b+c)x^2 + (b+c)x^3 + cx^4\right) dx \\
&= a + \frac{1}{2}(a+b) + \frac{1}{3}(a+b+c) + \frac{1}{4}(b+c) + \frac{1}{5}c \\
&= \frac{11}{6}a + \frac{13}{12}b + \frac{47}{60}c.
\end{aligned}
$$

For $p(x)$ and $q(x)$ to be orthogonal, $\langle p(x), q(x) \rangle = 0$. There are many such orthogonal pairs. For example, we can choose $a = -\frac{47}{110}, b = 0$ and $c = 1$. That is, $q(x) = -(47/110) + x^2$ is orthogonal to $p(x) = 1 + x + x^2$.   □

Example 7.1.28 Find a vector in $\mathbb{R}^3$ orthogonal to both $u = \begin{pmatrix} 1 \\ 1 \\ -1 \end{pmatrix}$ and $v = \begin{pmatrix} 0 \\ 2 \\ 1 \end{pmatrix}$. We seek a vector $w = \begin{pmatrix} a \\ b \\ c \end{pmatrix}$ that satisfies $\langle u, w \rangle = 0$ and $\langle v, w \rangle = 0$. These conditions lead to the system of linear equations:

$$
\begin{aligned}
\langle u, w \rangle &= a + b - c = 0, \\
\langle v, w \rangle &= 2b + c = 0.
\end{aligned}
$$

This system of linear equations has solution set $\mathrm{Span}\left\{\begin{pmatrix} 3 \\ -1 \\ 2 \end{pmatrix}\right\}$. If we let $w = \begin{pmatrix} 3 \\ -1 \\ 2 \end{pmatrix}$, then by Corollary 7.1.22, $\mathcal{B} = \{u, v, w\}$ is linearly independent. Also, because $\dim \mathbb{R}^3 = 3$, $\mathcal{B}$ is a basis for $\mathbb{R}^3$.   □

In each of the above examples, we found the set of orthogonal vectors by solving a homogeneous system of equations.

7.1.5 Inner Product and Coordinates

The inner product turns out to be very useful in expressing coordinate representations of vectors. Because the inner product measures a degree of linear dependence, scaled by vector lengths, we might expect it to quantify the unique basis decomposition built into our idea of coordinates.

Consider an example in $\mathbb{R}^2$. Suppose we wish to find the coordinate vector of $v = (2, 1)$ relative to the basis $\mathcal{B} = \left\{ u_1 = \begin{pmatrix} 3 \\ 4 \end{pmatrix}, u_2 = \begin{pmatrix} -2 \\ 0 \end{pmatrix} \right\}$. We know that v can be written as a unique linear combination of u_1 and u_2 with coefficients a_1 and a_2. That is,

$$v = a_1 u_1 + a_2 u_2.$$

Part 4 of Theorem 7.1.17 and the linearity of the inner product guarantees that

$$\begin{aligned} \langle u_1, v \rangle &= \langle u_1, a_1 u_1 + a_2 u_2 \rangle \\ &= a_1 \langle u_1, u_1 \rangle + a_2 \langle u_1, u_2 \rangle, \end{aligned}$$

and

$$\langle u_2, v \rangle = a_1 \langle u_2, u_1 \rangle + a_2 \langle u_2, u_2 \rangle.$$

These two linear equations form the matrix equation

$$\begin{pmatrix} \langle u_1, u_1 \rangle & \langle u_1, u_2 \rangle \\ \langle u_2, u_1 \rangle & \langle u_2, u_2 \rangle \end{pmatrix} \begin{pmatrix} a_1 \\ a_2 \end{pmatrix} = \begin{pmatrix} \langle u_1, v \rangle \\ \langle u_2, v \rangle \end{pmatrix}.$$

For the particular vectors of interest, we have

$$\begin{pmatrix} 25 & -6 \\ -6 & 4 \end{pmatrix} \begin{pmatrix} a_1 \\ a_2 \end{pmatrix} = \begin{pmatrix} 10 \\ -4 \end{pmatrix}.$$

The solution is

$$[v]_\mathcal{B} = \begin{pmatrix} a_1 \\ a_2 \end{pmatrix} = \begin{pmatrix} 1/4 \\ -5/8 \end{pmatrix}.$$

The reader can verify that $v = \frac{1}{4} u_1 - \frac{5}{8} u_2$.

We can readily generalize this approach.

Theorem 7.1.29

Suppose $\mathcal{B} = \{u_1, u_2, \ldots, u_n\}$ is a basis for n-dimensional inner product space V and $v \in V$. The coordinate vector $a = [v]_\mathcal{B}$ is the solution to

$$\begin{pmatrix} \langle u_1, u_1 \rangle & \langle u_1, u_2 \rangle & \ldots & \langle u_1, u_n \rangle \\ \langle u_2, u_1 \rangle & \langle u_2, u_2 \rangle & \ldots & \langle u_2, u_n \rangle \\ \vdots & \vdots & \ddots & \vdots \\ \langle u_n, u_1 \rangle & \langle u_n, u_2 \rangle & \ldots & \langle u_n, u_n \rangle \end{pmatrix} \begin{pmatrix} a_1 \\ a_2 \\ \vdots \\ a_n \end{pmatrix} = \begin{pmatrix} \langle u_1, v \rangle \\ \langle u_2, v \rangle \\ \vdots \\ \langle u_n, v \rangle \end{pmatrix}.$$

The proof is directly analogous to the previous discussion.

The following two corollaries show that coordinate computations are sometimes significantly simplified with orthogonal or orthonormal bases.

Corollary 7.1.30

Suppose $\mathcal{B} = \{u_1, u_2, \ldots, u_n\}$ is an orthogonal basis for n-dimensional inner product space V and $v \in V$. The coordinate vector $a = [v]_\mathcal{B}$ has entries

$$a_k = \frac{\langle u_k, v \rangle}{\|u_k\|^2} = \frac{\langle u_k, v \rangle}{\langle u_k, u_k \rangle}, \quad \text{for } k = 1, 2, \ldots, n.$$

Proof. Suppose $\mathcal{B} = \{u_1, u_2, \ldots, u_n\}$ is an orthogonal basis for n-dimensional inner product space V and $v \in V$. Note that

$$\langle u_i, u_j \rangle = \begin{cases} 0 & \text{if } i \neq j \\ \|u_i\|^2 & \text{if } i = j \end{cases}$$

Thus, the matrix equation in Theorem 7.1.29 is

$$\begin{pmatrix} \|u_1\|^2 & 0 & \ldots & 0 \\ 0 & \|u_2\|^2 & \ldots & 0 \\ \vdots & \vdots & \ddots & \vdots \\ 0 & 0 & \ldots & \|u_n\|^2 \end{pmatrix} \begin{pmatrix} a_1 \\ a_2 \\ \vdots \\ a_n \end{pmatrix} = \begin{pmatrix} \langle u_1, v \rangle \\ \langle u_2, v \rangle \\ \vdots \\ \langle u_n, v \rangle \end{pmatrix}.$$

That is, we find each coordinate a_k by solving

$$\|u_k\|^2 a_k = \langle u_k, v \rangle \text{ for all } k = 1, 2, \ldots, n.$$

Therefore,

$$a_k = \frac{\langle u_k, v \rangle}{\|u_k\|^2} \text{ for all } k = 1, 2, \ldots, n.$$

$\square$

Corollary 7.1.31

Suppose $\mathcal{B} = \{u_1, u_2, \ldots, u_n\}$ is an orthonormal basis for n-dimensional inner product space V and $v \in V$. The coordinate vector $a = [v]_\mathcal{B}$ has entries

$$a_k = \langle u_k, v \rangle, \quad \text{for } k = 1, 2, \ldots, n.$$

Proof. Suppose $\mathcal{B} = \{u_1, u_2, \ldots, u_n\}$ is an orthonormal basis for n-dimensional inner product space V and $v \in V$. By Corollary 7.1.30, for $k = 1, 2, \ldots, n$:

$$a_k = \frac{\langle u_k, v \rangle}{\|u_k\|^2}.$$

However, $\|u_k\| = 1$ for all $k = 1, 2, \ldots, n$. Thus $a_k = \langle u_k, v \rangle$.

$\square$

Example 7.1.32 Consider the basis $\mathcal{B} = \{1, x, x^2\}$ for inner product space $P_2(\mathbb{R})$ with the standard inner product. We use Theorem 7.1.29 to verify that for $p(x) = 2 + 3x + 4x^2$, $[p]_{\mathcal{B}} = (2, 3, 4) \in \mathbb{R}^3$. First, we compute the necessary inner products.

$$\langle 1, 1 \rangle = \int_0^1 dx = 1$$

$$\langle 1, x \rangle = \int_0^1 x\, dx = \tfrac{1}{2}$$

$$\langle 1, x^2 \rangle = \int_0^1 x^2\, dx = \tfrac{1}{3}$$

$$\langle x, x \rangle = \int_0^1 x^2\, dx = \tfrac{1}{3}$$

$$\langle x, x^2 \rangle = \int_0^1 x^3\, dx = \tfrac{1}{4}$$

$$\langle x^2, x^2 \rangle = \int_0^1 x^4\, dx = \tfrac{1}{5}$$

$$\langle 1, 2 + 3x + 4x^2 \rangle = 2\langle 1, 1 \rangle + 3\langle 1, x \rangle + 4\langle 1, x^2 \rangle = \tfrac{29}{6}$$

$$\langle x, 2 + 3x + 4x^2 \rangle = 2\langle x, 1 \rangle + 3\langle x, x \rangle + 4\langle x, x^2 \rangle = 3$$

$$\langle x^2, 2 + 3x + 4x^2 \rangle = 2\langle x^2, 1 \rangle + 3\langle x^2, x \rangle + 4\langle x^2, x^2 \rangle = \tfrac{133}{60}.$$

The matrix equation is

$$\begin{pmatrix} 1 & 1/2 & 1/3 \\ 1/2 & 1/3 & 1/4 \\ 1/3 & 1/4 & 1/5 \end{pmatrix} [p]_{\mathcal{B}} = \begin{pmatrix} 29/6 \\ 3 \\ 133/60 \end{pmatrix}$$

with unique solution $[p]_{\mathcal{B}} = \begin{pmatrix} 2 & 3 & 4 \end{pmatrix}^{\mathsf{T}}$. $\qquad\square$

Example 7.1.33 Consider the orthogonal (ordered) basis for $\mathbb{R}^3$:

$$\mathcal{B} = \{\begin{pmatrix} 1 & -4 & -7 \end{pmatrix}^{\mathsf{T}}, \begin{pmatrix} 1 & 2 & -1 \end{pmatrix}^{\mathsf{T}}, \begin{pmatrix} 3 & -1 & 1 \end{pmatrix}^{\mathsf{T}}\}.$$

Using Corollary 7.1.30, we find $[v]_{\mathcal{B}}$ for $v = \begin{pmatrix} 6 & -1 & -8 \end{pmatrix}^{\mathsf{T}}$. Let $a = [v]_{\mathcal{B}}$ and $\mathcal{B} = \{b_1, b_2, b_3\}$ then we have

$$a_1 = \frac{\langle b_1, v \rangle}{\langle b_1, b_1 \rangle} = \frac{6 + 4 + 56}{1 + 16 + 49} = 1$$

$$a_2 = \frac{\langle b_2, v \rangle}{\langle b_2, b_2 \rangle} = \frac{6 - 2 + 8}{1 + 4 + 1} = 2$$

$$a_3 = \frac{\langle b_3, v \rangle}{\langle b_3, b_3 \rangle} = \frac{18 + 1 - 8}{9 + 1 + 1} = 1.$$

Thus,

$$a = [v]_{\mathcal{B}} = \begin{pmatrix} 1 & 2 & 1 \end{pmatrix}^{\mathsf{T}}.$$

The reader should verify that $\mathcal{B}$ is indeed orthogonal and that $v = b_1 + 2b_2 + b_3$.   □

Example 7.1.34 The heat diffusion transformation in $H_3(\mathbb{R})$ in the standard basis, S, has representation

$$[E]_S = \begin{pmatrix} 1/2 & 1/4 & 0 \\ 1/4 & 1/2 & 1/4 \\ 0 & 1/4 & 1/2 \end{pmatrix}.$$

This matrix transformation has orthonormal eigenbasis

$$\mathcal{B} = \{b_1, b_2, b_3\} = \left\{ \begin{pmatrix} 1/2 \\ 1/\sqrt{2} \\ 1/2 \end{pmatrix}, \begin{pmatrix} 1/\sqrt{2} \\ 0 \\ -1/\sqrt{2} \end{pmatrix}, \begin{pmatrix} 1/2 \\ -1/\sqrt{2} \\ 1/2 \end{pmatrix} \right\}.$$

We now find the heat state $[h]_\mathcal{B}$ for $[h]_S = \begin{pmatrix} 1 & 2 & 3 \end{pmatrix}^\mathsf{T}$.

We seek coefficients $[h]_\mathcal{B} = a = \begin{pmatrix} a_1 & a_2 & a_3 \end{pmatrix}^\mathsf{T}$ so that $[h]_S = a_1 b_1 + a_2 b_2 + a_3 b_3$. Using Corollary 7.1.31, we have

$$a_1 = \left\langle \begin{pmatrix} 1/2 \\ 1/\sqrt{2} \\ 1/2 \end{pmatrix}, \begin{pmatrix} 1 \\ 2 \\ 3 \end{pmatrix} \right\rangle = 2 + \sqrt{2}$$

$$a_2 = \left\langle \begin{pmatrix} 1/\sqrt{2} \\ 0 \\ -1/\sqrt{2} \end{pmatrix}, \begin{pmatrix} 1 \\ 2 \\ 3 \end{pmatrix} \right\rangle = -\sqrt{2}$$

$$a_2 = \left\langle \begin{pmatrix} 1/2 \\ -1/\sqrt{2} \\ 1/2 \end{pmatrix}, \begin{pmatrix} 1 \\ 2 \\ 3 \end{pmatrix} \right\rangle = 2 - \sqrt{2}$$

Thus, $[h]_\mathcal{B} = \begin{pmatrix} 2 + \sqrt{2}. \\ -\sqrt{2} \\ 2 - \sqrt{2} \end{pmatrix}$. The reader should verify that $\mathcal{B}$ is indeed an orthonormal eigenbasis for

$[E]_S$ and that $[h]_S = (2 + \sqrt{2})b_1 + (-\sqrt{2})b_2 + (2 - \sqrt{2})b_3$.   □

☞ **Path to New Applications**

Clustering and support vector machines are used to classify data. Classification is based on measuring similarity between data vectors. Inner products provide one method for measuring similarity. See Section 8.3.2 to read more about connections between linear algebra tools and these techniques along with other machine learning techniques. In Exercise 30, we ask you to begin exploring linear algebra techniques used for data classification.

7.1.6   Exercises

1. For each set of given vectors in $\mathbb{R}^n$, compute $\langle u, v \rangle$, $\|u\|$, $\|v\|$, $\|u + v\|$ and show that both the Cauchy-Schwarz and Triangle Inequalities hold. Use the standard inner product on $\mathbb{R}^n$.

(a) $u = \begin{pmatrix} 1 & 2 \end{pmatrix}^\mathsf{T}, v = \begin{pmatrix} 2 & 1 \end{pmatrix}^\mathsf{T}$.

(b) $u = \begin{pmatrix} -3 & 2 \end{pmatrix}^\mathsf{T}, v = \begin{pmatrix} 7 & 6 \end{pmatrix}^\mathsf{T}$.

(c) $u = \begin{pmatrix} 0 & 0 \end{pmatrix}^\mathsf{T}, v = \begin{pmatrix} 3 & 4 \end{pmatrix}^\mathsf{T}$.

(d) $u = \begin{pmatrix} 1 & 2 & 3 \end{pmatrix}^\mathsf{T}, v = \begin{pmatrix} -1 & -2 & -3 \end{pmatrix}^\mathsf{T}$.

(e) $u = \begin{pmatrix} -1 & 3 & -1 \end{pmatrix}^\mathsf{T}, v = \begin{pmatrix} 2 & 2 & 4 \end{pmatrix}^\mathsf{T}$.

(f) $u = \begin{pmatrix} 2 & 1 & 7 \end{pmatrix}^\mathsf{T}, v = \begin{pmatrix} 1 & 2 & 7 \end{pmatrix}^\mathsf{T}$.

2. Let $b_k \in \mathbb{R}^4$ represent the annual revenues of business k over four consecutive fiscal years 2018–2021. The revenue vectors of five businesses are given here:

$$b_1 = \begin{pmatrix} 2 & 3 & 4 & 5 \end{pmatrix}^\mathsf{T},$$
$$b_2 = \begin{pmatrix} 3 & 3 & 1 & 6 \end{pmatrix}^\mathsf{T},$$
$$b_3 = \begin{pmatrix} 0 & 0 & 5 & 5 \end{pmatrix}^\mathsf{T},$$
$$b_4 = \begin{pmatrix} 2 & 3 & 2 & 4 \end{pmatrix}^\mathsf{T},$$
$$b_5 = \begin{pmatrix} 2 & 3 & 4 & 4 \end{pmatrix}^\mathsf{T}.$$

Which of the five businesses has a revenue vector, which is farthest from the vector of median values? Answer this question using two different vector norms using the standard inner product in $\mathbb{R}^n$:

(a) The Euclidean norm: $\|x\| = (\langle x, x \rangle)^{1/2}$.

(b) The weighted norm: $\|x\| = (\langle x, Ax \rangle)^{1/2}$ where $A = \begin{pmatrix} \frac{1}{8} & 0 & 0 & 0 \\ 0 & \frac{1}{8} & 0 & 0 \\ 0 & 0 & \frac{1}{4} & 0 \\ 0 & 0 & 0 & \frac{1}{2} \end{pmatrix}$.

3. Consider the vector space $\mathcal{M}_{2\times3}(\mathbb{R})$ with Frobenious inner product. Determine the subset of matrices $M \subseteq \mathcal{M}_{2\times3}(\mathbb{R})$ containing all matrices orthogonal to both

$$A = \begin{pmatrix} 1 & 2 & 0 \\ 0 & 0 & 0 \end{pmatrix} \quad \text{and} \quad B = \begin{pmatrix} 0 & 0 & 0 \\ 0 & 3 & 1 \end{pmatrix},$$

show that M is a subspace of $\mathcal{M}_{2\times3}(\mathbb{R})$ and find a basis for M.

4. Consider the three images of Example 7.1.6. Suppose each image was normalized (scaled to unit length using the standard image inner product). Describe any differences in the grayscale representation of these new images.

5. Consider the vector space of continuous functions $f : [1, 2] \to \mathbb{R}$ and inner product $\langle f, g \rangle = \int_1^2 f(x)g(x)dx$. Argue that the vector space norm $\|f\| = \langle f, f \rangle^{1/2}$ is an intuitive choice for measuring vector length. Use this norm to find a function $g \in \mathcal{P}_1(\mathbb{R})$ closest in norm to $f(x) = 1/x$, on the interval $[1, 2]$.

6. Using the ideas in Exercise 5, find the polynomial $f(x) \in \mathcal{P}_2(\mathbb{R})$, which most closely approximates the following functions on the given intervals. Plot the functions and discuss your results.

(a) $g(x) = \sin x$ on $[0, \pi]$.

(b) $h(x) = \sqrt{x}$ on $[0, 1]$.

(c) $r(x) = |x|$ on $[-1, 1]$.

(d) $t(x) = 1/x^2$ on $[1, 2]$.

7. A scientist recorded five biochemical properties of the seeds of several varieties of pumpkin. Later it was determined that the pumpkin fruit that was the most successful, in terms of size and consumer acceptance, came from the seed variety with data vector

$$u = \begin{pmatrix} 6 & 13 & 4 & 17 & 10 \end{pmatrix}^{\mathsf{T}}.$$

Before the following growing season, the same biochemical signatures were obtained for three new varieties:

$$x_1 = \begin{pmatrix} 8 & 10 & 4 & 16 & 10 \end{pmatrix}^{\mathsf{T}},$$
$$x_2 = \begin{pmatrix} 10 & 17 & 4 & 13 & 6 \end{pmatrix}^{\mathsf{T}},$$
$$x_3 = \begin{pmatrix} 7 & 11 & 5 & 15 & 9 \end{pmatrix}^{\mathsf{T}}.$$

Describe and justify a reasonable (linear algebra) method for determining which of the three new varieties is most likely to be successful. Test your method.

8. Prove or disprove the following. Claim: Suppose $\langle \cdot, \cdot \rangle_1$ and $\langle \cdot, \cdot \rangle_2$ are both inner products on vector space V. Then, $\langle \cdot, \cdot \rangle = \langle \cdot, \cdot \rangle_1 + \langle \cdot, \cdot \rangle_2$ is an inner product on V.

9. This exercise shows that a vector space can have many different inner products and vector norms. Consider the vector space $\mathbb{R}^2$ and define $\langle x, y \rangle = ax_1y_1 + x_2y_2$, where $a > 0$ is a scalar. Show that $\langle \cdot, \cdot \rangle : \mathbb{R}^2 \times \mathbb{R}^2 \to \mathbb{R}$ is an inner product.

10. Let $\langle \cdot, \cdot \rangle : \mathcal{P}_2(\mathbb{R}) \times \mathcal{P}_2(\mathbb{R}) \to \mathbb{R}$ be defined by

$$\langle p_1, p_2 \rangle = \int_0^1 p_1 p_2 \, dx, \quad \text{for } p_1, p_2 \in \mathcal{P}_2(\mathbb{R}).$$

For what values of c is the vector $p(x) = x^2 + x + c$ a unit vector?

11. Let $\langle \cdot, \cdot \rangle : \mathcal{P}_2(\mathbb{R}) \times \mathcal{P}_2(\mathbb{R}) \to \mathbb{R}$ be defined by

$$\langle p_1(x), p_2(x) \rangle = \int_0^1 p_1(x) p_2'(x) \, dx, \quad \text{for } p_1, p_2 \in \mathcal{P}_2(\mathbb{R}).$$

Is $\langle p_1(x), p_2(x) \rangle$ an inner product on $\mathcal{P}_2(\mathbb{R})$?

12. Consider the function $f : \mathbb{R}^3 \times \mathbb{R}^3 \to \mathbb{R}$ defined by $f(x, y) = |x_1y_1| + |x_2y_2| + |x_3y_3|$. Is f an inner product on $\mathbb{R}^3$?

13. Consider the standard inner product $\langle \cdot, \cdot \rangle : \mathbb{R}^n \times \mathbb{R}^n \to \mathbb{R}$. For what scalars p is $\langle \cdot, \cdot \rangle^p : \mathbb{R}^n \times \mathbb{R}^n \to \mathbb{R}$ also an inner product?

14. Show that the function $f(x, y)$ defined in Example 7.1.11 does not satisfy the inner product linearity condition $f(u + v, w) = f(u, w) + f(v, w)$ and is therefore not an inner product on $\mathcal{D}(Z_2)$.

15. Show that $\|u - v\|^2 = \|u\|^2 + \|v\|^2 - 2\langle u, v \rangle$. Relate this result to the law of cosines in $\mathbb{R}^2$.

16. Consider an orthogonal set of vectors $U = \{u_1, u_2, \cdots, u_n\}$ in vector space V with inner product $\langle \cdot, \cdot \rangle$. Suppose $v \in \text{Span } U$. Write $\|v\|$ in terms of $\|u_k\|$, $k = 1, 2, \cdots, n$.

17. Consider the given functions $f : \mathbb{R}^2 \times \mathbb{R}^2 \to \mathbb{R}$. For each, show that f is an inner product on $\mathbb{R}^2$ and sketch the set of normal vectors.

    (a) $f(x, y) = 4x_1y_1 + 4x_2y_2$.

(b) $f(x, y) = x_1 y_1 + 9x_2 y_2$.

(c) $f(x, y) = 2x_1 y_1 + x_1 y_2 + x_2 y_1 + 2x_2 y_2$.

18. Let $\ell(x)$ be the Euclidean length of any vector $x \in \mathbb{R}^2$, and $f(x, y)$ be the cosine of the angle between any two vectors $x, y \in \mathbb{R}^2$. Prove that the function $\langle \cdot, \cdot \rangle : \mathbb{R}^2 \times \mathbb{R}^2 \to \mathbb{R}$, defined by $\langle x, y \rangle = \ell(x)\ell(y) f(x, y)$, is linear in the first argument.

19. Prove Theorem 7.1.16.

20. Provide a direct proof of Theorem 7.1.21.

21. Consider the vector space of 7-bar LCD digits, $D(Z_2)$. Let u_k be the value of the kth bar for $u \in D(Z_2)$. Determine whether or not each of the given $f : D(Z_2) \times D(Z_2) \to Z_2$ is an inner product on $D(Z_2)$.

(a) $f(u, v) = \displaystyle\sum_{k=1}^{7} u_k v_k$

(b) $f(u, v) = \max_{k}\{u_k v_k\}$

(c) $f(u, v) = \begin{cases} 0 & \text{if } u = 0 \text{ or } v = 0 \\ 1 & \text{otherwise} \end{cases}$

22. Find the coordinate vector $[v]_B$ for the vector v in inner product space V for the given basis B. Use the standard inner product and methods of Section 7.1.5. Some of the given bases are orthogonal, others are not.

(a) $V = \mathbb{R}^2, v = (3\ 3)^\mathsf{T}, B = \{(1\ 1)^\mathsf{T}, (1\ 2)^\mathsf{T}\}$.

(b) $V = \mathbb{R}^2, v = (3\ 3)^\mathsf{T}, B = \{(1\ 1)^\mathsf{T}, (1\ -1)^\mathsf{T}\}$.

(c) $V = \mathbb{R}^3, v = (1\ 2\ -1)^\mathsf{T}, B = \{(1\ 1\ 0)^\mathsf{T}, (0\ 1\ 2)^\mathsf{T}, (1\ 0\ 1)^\mathsf{T}\}$.

(d) $V = \mathbb{R}^3, v = (1\ 2\ -1)^\mathsf{T}, B = \{(1\ 1\ 0)^\mathsf{T}, (-1\ 1\ 0)^\mathsf{T}, (0\ 0\ 1)^\mathsf{T}\}$.

(e) $V = \mathbb{R}^4, v = (1\ 2\ 3\ 4)^\mathsf{T}$,

$$B = \left\{ \left(\tfrac{1}{2}\ 0\ \tfrac{-\sqrt{3}}{2}\ 0\right)^\mathsf{T}, \left(0\ \tfrac{\sqrt{2}}{2}\ 0\ \tfrac{-\sqrt{2}}{2}\right)^\mathsf{T}, \left(\tfrac{\sqrt{3}}{2}\ 0\ \tfrac{1}{2}\ 0\right)^\mathsf{T}, \left(0\ \tfrac{\sqrt{2}}{2}\ 0\ \tfrac{\sqrt{2}}{2}\right)^\mathsf{T} \right\}.$$

(f) $V = P_2(\mathbb{R}), v = x + 3, B = \{1 + x, 2 - x, x^2\}$.

(g) $V = P_2(\mathbb{R}), v = x + 3, B = \{1, 2x - 1, 6x^2 - 6x + 1\}$.

(h) $V = P_2(\mathbb{R}), v = x + 3, B = \{1, \sqrt{12}x - \sqrt{3}, \sqrt{180}x^2 - \sqrt{180}x + \sqrt{5}\}$.

(i) $V = \mathcal{F}(\mathbb{R}), v = \sin(\pi x), B = \left\{\sqrt{2}\cos n\pi x \mid n \in \mathbb{N}\right\}$. (See Example 7.1.25.)

The following exercises explore inner product spaces over the field of complex numbers $\mathbb{C}$.

23. Let $u, v \in \mathbb{C}^n$. Show that in order to preserve the concept of vector length $\|u\| = \sqrt{\langle u, u \rangle}$, the standard inner product should be defined as

$$\langle u, v \rangle = u_1 \bar{v}_1 + u_2 \bar{v}_2 + \ldots + u_n \bar{v}_n,$$

where $\bar{x}$ indicates the complex conjugate of x.

24. Using the standard inner product of Exercise 23, show that the symmetry property no longer holds in Definition 7.1.1. Show that

$$\overline{\langle u, v \rangle} = \langle v, u \rangle$$

holds for both real and complex inner product spaces.

25. Generalize Theorem 7.1.16 for complex inner product spaces and provide a proof.

26. Generalize Theorem 7.1.17 for complex inner product spaces and provide a proof.

27. Show that the zero vector is the only vector orthogonal to nonzero vector $z = a + bi \in \mathbb{C}^1$.

28. Find the set X of all vectors orthogonal to $\begin{pmatrix} 1+i \\ 2-i \end{pmatrix} \in \mathbb{C}^2$. Show that X is a one-dimensional subspace of $\mathbb{C}^2$ by finding a basis for X.

29. Given any nonzero vector v in $\mathbb{R}^n$, prove that its orthogonal complement $v^\perp$ is an $(n-1)$-dimensional subspace.

30. (**Data Classification.**) Any $(n-1)$-dimensional subspace (or more generally, any $(n-1)$-*dimensional hyperplane*[3]) divides $\mathbb{R}^n$ into two regions. (See Exercise 11 of Section 3.5.) This fact is very useful in data classification: if we are interested in developing a classifier for points, we look for subspaces that separate data points with different characteristics. The following exercises point us in this direction.

   (a) Graph the following two sets of points in $\mathbb{R}^2$:

$$A = \{(3,4), (-5,1), (-3,-3), (-6,-4)\},$$

$$B = \{(8,-3), (0,-4), (1,-6), (-2,-6)\}.$$

   (b) Which of the following subspaces separate the sets in the previous part? Sketch the subspaces on your graph from part (a).
      i. $\mathrm{Span}\{(0,1)^\mathsf{T}\}$
      ii. $\mathrm{Span}\{(1,1)^\mathsf{T}\}$
      iii. $\mathrm{Span}\{(6,7)^\mathsf{T}\}$
      iv. $\mathrm{Span}\{(-1,1)^\mathsf{T}\}$
      v. $\mathrm{Span}\{(3,5)^\mathsf{T}\}$

   (c) If the sets in part (a) are representative of two populations, which subspace from part (b) would be the best distinguisher between the populations? Why?

   (d) Reconsider the subspace that you selected in part (c). Can you shift the subspaces by a nonzero vector to get a hyperplane that better separates the populations? What shift(s) would be best?

   (e) Now let's use the language of inner products to make your observations more formal. For the subspace from part (c) and the hyperplane from part (d), do the following. Compute the distance from each point in set A to the subspace/hyperplane and take the minimum. Do the same for distances from points in set B to the subspace/hyperplane. What do you notice about the shifted hyperplane you selected in part (d) compared to the original subspace?

   (f) Based on your responses, make a conjecture about what properties you would like a separating hyperplane to have in order for it to separate two classes of data vectors. These ideas lead to the development of a data analytics tool called a *support vector machine*; see Section 8.3.2 for more information.

7.2 Projections

In Section 7.1, we discussed a notion of how close two vectors are to being linearly dependent. The inner product of two vectors provides a scalar value indicative of how close two vectors are to being scalar multiples of one another. Nonzero orthogonal vectors have no components that are scalar multiples of

---

[3] A k-dimensional hyperplane is a translation, possibly by zero, of a k-dimensional subspace

one another. In this section, we want to take this further. We would like to understand (a) how close a vector is to a given subspace and (b) how to express a vector as a sum of two vectors: one in a subspace and the other orthogonal to the subspace. Consider several examples:

1. Suppose a subspace of brain images is known to be associated with a particular condition. We might wish to compare a given image with this subspace of images. Is our image in the subspace? If not, is it close?
2. Suppose the heat stress on a rod in a welding apparatus is closely linked to a particular subspace of heat states. If a heat state has a large component in this subspace, it may be indicative of high stress, a smaller component may be indicative of lower stress.
3. Suppose we wish to know if an object in an imaging system has a large nullspace component, so that we can better understand our imaging system. We may wish to have some idea of how much of our object information can be recovered from an inverse (object reconstruction) process.
4. Suppose we have two brain images that have been reconstructed from two different radiographs. We may wish to know if these two images differ by something significant or by something of little importance. Again, this can be measured using a known subspace of significant (or insignificant) features.
5. Suppose we have a polynomial function that describes a production strategy over time. We may also know a subspace of functions that are known to have high marketing value. How close is our strategy to this subspace of desirable strategies?

The key idea in this section is the partitioning of vector spaces into direct sums of subspaces that are of particular interest in an application, and the description of vectors in terms of these subspaces. Along the way, we will need to understand the bases for these subspaces. We will also understand that distances are best measured in terms of orthogonal subspaces and orthogonal bases, so we will need to understand how to construct them.

Let us consider a visual example in $\mathbb{R}^3$. Suppose $v \in \mathbb{R}^3$ and W is a two-dimensional subspace of $\mathbb{R}^3$. We want to find $u \in \mathbb{R}^3$ and $w \in W$ so that $v = u + w$ and u is orthogonal to every vector in W. Geometrically, we see that W is a plane in $\mathbb{R}^3$ that passes through the origin, u is orthogonal (perpendicular) to the plane, and w lies entirely within the plane. See Figure 7.2. We say that w is the closest vector to v within W. We can say that $\|w\|/\|v\|$ is a measure of how close v is to being within W.

Next, consider the example of digital images in $\mathcal{I}_{4 \times 4}(\mathbb{R})$. In Section 2.1, we found that

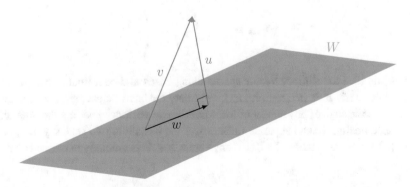

Fig. 7.2 The vector $v \in \mathbb{R}^3$ is decomposed into a vector u, orthogonal to all vectors in W—a subspace of dimension 2—and a vector $w \in W$.

$$v = \quad\quad\quad \in \mathcal{I}_{4\times4}(\mathbb{R}), \text{ but}$$

Image 4

$$v \notin \operatorname{Span}\left\{ \quad , \quad , \quad \right\} \subseteq \mathcal{I}_{4\times4}(\mathbb{R}).$$

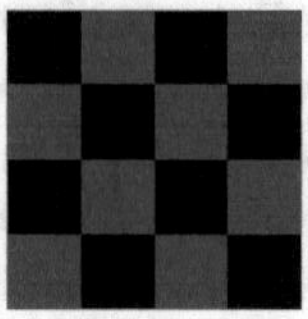

Image A   Image B   Image C

Is there an image in the subspace spanned by Images A, B, and C that is "closest," to Image 4? If so, we will call this image, the *projection* of image v onto the subspace spanned by Images A, B, and C. While Image 4 cannot be written as a linear combination of Images A, B, and C, it may be that the projection, which can be written as such a linear combination, is close to Image 4.

In this section, we will formally define projections as functions from a vector space into a subspace of the vector space satisfying certain properties[4]. First, we will define coordinate projections, which don't require any notion of orthogonality or inner product. Then we will define a subclass of coordinate projections called orthogonal projections. We will discuss how orthogonal projections can be used to find the vector closest to another vector in a subspace. Finally, since orthogonal subspaces and orthogonal bases are needed for computing these projections, we will describe a technique for finding an orthogonal basis called the Gram-Schmidt process.

7.2.1 Coordinate Projection

Our first attempt at defining a projection mapping is based on coordinates. Consider the case of $\mathbb{R}^3$ with the standard basis for motivation. Suppose we want a projection from the whole space to the xy-plane. It would be natural to map any set of vectors to its "shadow" in the xy-plane, as shown in Figure 7.3. Algebraically, this map is just the map T that sets the z coordinate to zero:

$$T : \mathbb{R}^3 \to \mathbb{R}^3, \quad T\begin{pmatrix} x \\ y \\ z \end{pmatrix} := \begin{pmatrix} x \\ y \\ 0 \end{pmatrix}.$$

We can extend this idea to arbitrary vector spaces using bases and coordinates. Suppose we have a vector space V with basis $\beta = \{u_1, u_2, \ldots, u_n\}$. We know the basis determines coordinates for V, so we can project onto the span of any subset of these coordinate vectors by setting the other coordinates to zero. More specifically, we can separate (arbitrarily) the basis into two sets, say, $\{u_1, \ldots, u_k\}$ and $\{u_{k+1}, \ldots, u_n\}$. Then by properties of a basis, any vector $v \in V$ is uniquely expressed as $v = v_1 + v_2$

---

[4] In the larger mathematical context, a projection is any function from a set into a subset that maps any point in the subset to itself, but the projections we will introduce satisfy additional properties related to linear algebra.

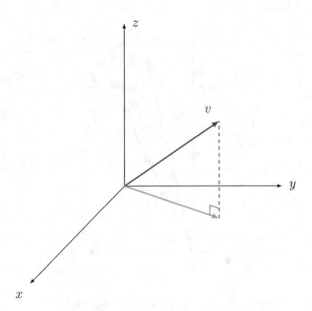

Fig. 7.3 Projection of $v \in \mathbb{R}^3$ onto the xy-plane results in the "shadow" seen in gray.

where $v_1 \in W_1 = \text{Span}\{u_1, \ldots, u_k\}$ and $v_2 \in W_2 = \text{Span}\{\{u_{k+1}, \ldots, u_n\}\}$. Then $V = W_1 \oplus W_2$, and we can define a coordinate projection onto W_1 (or W_2) in the following way.

Definition 7.2.1

Let V be an inner product space over $\mathbb{R}$ and W_1 and W_2 subspaces of V with bases $\beta_1 = \{u_1, \ldots, u_k\}$ and $\beta_2 = \{u_{k+1}, \ldots, u_n\}$, respectively, such that $V = W_1 \oplus W_2$. Let $v = a_1 u_1 + a_2 u_2 + \ldots + a_n u_n$. Then, $v_1 = a_1 u_1 + \ldots + a_k u_k$ is said to be the **coordinate projection of v onto W_1**, $v_2 = a_{k+1} u_{k+1} + \ldots + a_n u_n$ is said to be the **coordinate projection of v onto W_2** and we write $v_1 = \text{cproj}_{W_1} v$ and $v_2 = \text{cproj}_{W_2} v$. $\square$

The idea of coordinate projections generalizes naturally to an arbitrary number of subspaces $W_1, W_2, \ldots, W_j, \ldots, W_m$, provided that $V = W_1 \oplus \ldots \oplus W_m$. Each coordinate projection v_j depends upon *all* subspaces. That is, we cannot compute any v_j without knowing the entire set of basis vectors and all coordinate values. The following three examples illustrate this.

Example 7.2.2 Consider the inner product space $\mathbb{R}^2$ with basis

$$\mathcal{B} = \{u_1, u_2\} = \left\{ \begin{pmatrix} 1 \\ 2 \end{pmatrix}, \begin{pmatrix} -1 \\ 0 \end{pmatrix} \right\}.$$

Let

$$W_1 = \text{Span}\left\{ \begin{pmatrix} 1 \\ 2 \end{pmatrix} \right\} \text{ and } W_2 = \text{Span}\left\{ \begin{pmatrix} -1 \\ 0 \end{pmatrix} \right\}.$$

Consider $v = (1, 4)^\mathsf{T}$. We find the coordinate projections of v onto W_1.

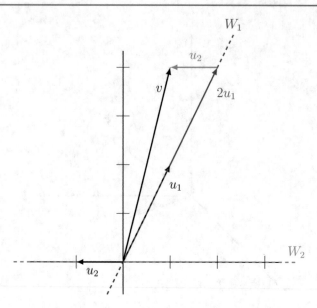

Fig. 7.4 Graphical depiction of Example 7.2.2, $v = 2u_1 + u_2$, so $\mathrm{cproj}_{W_1} v = 2u_1$.

First, note that $\mathbb{R}^2 = W_1 \oplus W_2$ and we can write v as the linear combination

$$v = \begin{pmatrix} 1 \\ 4 \end{pmatrix} = 2 \begin{pmatrix} 1 \\ 2 \end{pmatrix} + \begin{pmatrix} -1 \\ 0 \end{pmatrix}.$$

Thus,

$$v_1 = \mathrm{cproj}_{W_1} v = \begin{pmatrix} 2 \\ 4 \end{pmatrix}.$$

Figure 7.4 graphically shows the coordinate projection onto W_1. Exercise 2 asks you to calculate the coordinate projection onto W_2.   □

Example 7.2.3 Consider Example 7.2.2, replacing the second basis vector by $u_2 = \begin{pmatrix} 1 \\ 1 \end{pmatrix}$, so that

$$\mathcal{B} = \{u_1, u_2\} = \left\{ \begin{pmatrix} 1 \\ 2 \end{pmatrix}, \begin{pmatrix} 1 \\ 1 \end{pmatrix} \right\}.$$

Then

$$W_1 = \mathrm{Span}\left\{ \begin{pmatrix} 1 \\ 2 \end{pmatrix} \right\} \text{ and } W_2 = \mathrm{Span}\left\{ \begin{pmatrix} 1 \\ 1 \end{pmatrix} \right\}.$$

Now we write v as a linear combination using the new basis

$$v = 3 \begin{pmatrix} 1 \\ 2 \end{pmatrix} - 2 \begin{pmatrix} 1 \\ 1 \end{pmatrix}.$$

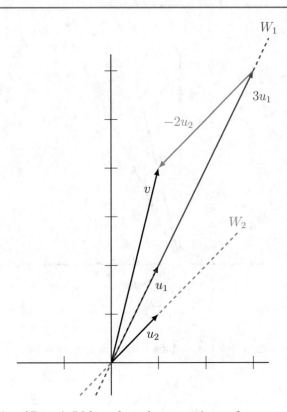

Fig. 7.5 Graphical depiction of Example 7.2.3 $v = 3u_1 - 2u_2$, so $\text{cproj}_{W_1} v = 3u_1$.

Hence,

$$v_1 = \text{cproj}_{W_1} v = \begin{pmatrix} 3 \\ 6 \end{pmatrix}.$$

Figure 7.5 graphically shows the coordinate projection onto W_1. Exercise 2 asks you to calculate the coordinate projection onto W_2. $\square$

Example 7.2.4 Consider once again Example 7.2.2, this time replacing the second basis vector by $u_2 = \begin{pmatrix} 1 \\ -1/2 \end{pmatrix}$, so that

$$\mathcal{B} = \{u_1, u_2\} = \left\{ \begin{pmatrix} 1 \\ 2 \end{pmatrix}, \begin{pmatrix} 1 \\ -1/2 \end{pmatrix} \right\}.$$

Then

$$W_1 = \text{Span}\left\{ \begin{pmatrix} 1 \\ 2 \end{pmatrix} \right\} \text{ and } W_2 = \text{Span}\left\{ \begin{pmatrix} 1 \\ -1/2 \end{pmatrix} \right\}.$$

With some behind-the-scenes trigonometry (try it yourself!) we write v as

$$v = 1.8 \begin{pmatrix} 1 \\ 2 \end{pmatrix} - 0.8 \begin{pmatrix} 1 \\ -1/2 \end{pmatrix}.$$

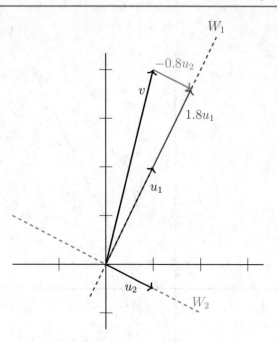

Fig. 7.6 Graphical depiction of Example 7.2.4, $v = 1.8u_1 - 0.8u_2$, so $\mathrm{cproj}_{W_1} v = 1.8u_1$.

Hence,

$$v_1 = \mathrm{cproj}_{W_1} v = \begin{pmatrix} 1.8 \\ 3.6 \end{pmatrix}.$$

Figure 7.6 graphically shows the coordinate projection onto W_1. Exercise 2 asks you to calculate the coordinate projection onto W_2.   □

In each of these three examples, the subspace W_1, the basis for W_1, and the vector v were the same, but with different choices of the basis for W_2 we found different coordinate projections onto W_1. Our desire for measures of closeness and similarity will guide us in making advantageous basis choices. One particularly useful aspect of our definition of a coordinate projection is that it can be applied to vector spaces other than $\mathbb{R}^n$:

Example 7.2.5 Consider the inner product space $P_2(\mathbb{R})$ with basis

$$\mathcal{B} = \{u_1 = 1 + x, u_2 = x, u_3 = 1 + x^2\}.$$

Let $W_1 = \mathrm{Span}\{u_1, u_2\}$ and $W_2 = \mathrm{Span}\{u_3\}$. Consider an arbitrary polynomial $p(x) = a + bx + cx^2 \in P_2(\mathbb{R})$. Then, $p(x)$ is uniquely written as

$$p(x) = (a - c)u_1 + (-a + b + c)u_2 + (c)u_3.$$

So, we have

$$p_1(x) = \mathrm{cproj}_{W_1} p(x) = (a - c)u_1 + (-a + b + c)u_2 = a - c + bx$$

and

$$p_2(x) = \mathrm{cproj}_{W_2} p(x) = cu_3 = c + cx^2.$$

Notice that $p(x) = p_1(x) + p_2(x)$ for any $a, b, c \in \mathbb{R}$. $\square$

Example 7.2.6 Consider the vector space of 7-bar LCD characters, $D(Z_2)$ with basis

$$\mathcal{B} = \left\{ \text{} \right\}.$$

let d be the "4" digit. From Example 3.5.8, we know that

$$d = \text{}.$$

If we consider $D(Z_2) = W_1 \oplus W_2$ where

$$W_1 = \mathrm{Span}\left\{ \text{} \right\},$$

$$W_2 = \mathrm{Span}\left\{ \text{} \right\}.$$

Then we have $d = d_1 + d_2$, where

$$d_1 = \mathrm{cproj}_{W_1} d = \text{} \quad \text{and} \quad d_2 = \mathrm{cproj}_{W_2} d = \text{}.$$

$\square$

7.2.2 Orthogonal Projection

Coordinate projections are useful for describing vector space decomposition into subspaces whose direct sum is the parent vector space. However, we may not know, or may not be interested in, all of the possible subspace decompositions. For example, suppose we have identified a subspace of brain images that indicate a certain brain condition. Understanding how close a brain image is to be in this subspace might give us information about whether the brain has the condition.

Suppose $B_1 = \{u_1, u_2, \ldots, u_k\}$ and $X = \text{Span } B_1$ is a subspace of brain images for which a certain medical condition is indicated. Given a brain image v, we want to know how close v is to being in X. One approach would be to use a coordinate projection $v = v_1 + v_2$ where v_1 is in the subspace X and then measure $\|v_1\|/\|v\|$ as an indication of closeness. However, we have seen (for instance, in Examples 7.2.2, 7.2.3, and 7.2.4) that v_1 depends upon our choice of basis for describing the remainder of the vector space *not* in X. Notice that among Examples 7.2.2, 7.2.3, and 7.2.4, the orthogonal basis in Example 7.2.4 yields the shortest vector v_2 (in Figures 7.4, 7.5, and 7.6). This leads us to conclude that the most natural choice of remaining basis vectors $B_2 = \{u_{k+1}, \ldots, u_n\}$ is the one whose elements are all orthogonal to the elements of B_1. That is, we wish v_2 to measure how much of the vector v is *not* in X.

Toward this end, we define the orthogonal complement of a set of vectors, which we will use to identify appropriate bases for projection.

Definition 7.2.7

Let V be a vector space and $W \subseteq V$. The **orthogonal complement of W**, denoted $W^\perp$ and read "W perp", is the set

$$W^\perp := \{v \in V \mid \langle v, y \rangle = 0 \text{ for all } y \in W\}.$$

$\square$

In words, $W^\perp$ is just the set of vectors orthogonal to all vectors in W.

Example 7.2.8 Consider $y = (1, -2, 3)^\mathsf{T} \in \mathbb{R}^3$. We find $\{y\}^\perp$ by finding all vectors $v \in \mathbb{R}^3$ such that $\langle v, y \rangle = 0$. So, we have

$$\{y\}^\perp = \left\{ \begin{pmatrix} a \\ b \\ c \end{pmatrix} \in \mathbb{R}^3 \,\middle|\, \left\langle \begin{pmatrix} 1 \\ -2 \\ 3 \end{pmatrix}, \begin{pmatrix} a \\ b \\ c \end{pmatrix} \right\rangle = 0 \right\}$$

$$= \left\{ \begin{pmatrix} a \\ b \\ c \end{pmatrix} \in \mathbb{R}^3 \,\middle|\, a - 2b + 3c = 0 \right\}$$

$$= \left\{ \begin{pmatrix} 2 \\ 1 \\ 0 \end{pmatrix} b + \begin{pmatrix} -3 \\ 0 \\ 1 \end{pmatrix} c \in \mathbb{R}^3 \,\middle|\, b, c \in \mathbb{R} \right\}$$

$$= \text{Span} \left\{ \begin{pmatrix} 2 \\ 1 \\ 0 \end{pmatrix}, \begin{pmatrix} -3 \\ 0 \\ 1 \end{pmatrix} \right\}.$$

We see $y^\perp$ in Figure 7.7.

$\square$

Example 7.2.9 Consider the set

$$W = \left\{ \begin{pmatrix} 1 \\ 1 \\ 1 \end{pmatrix}, \begin{pmatrix} 0 \\ 3 \\ -1 \end{pmatrix} \right\} \subset \mathbb{R}^3.$$

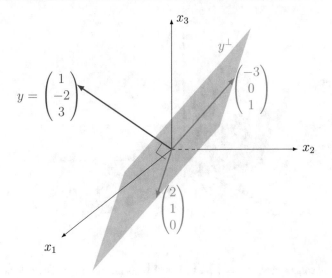

Fig. 7.7 The orthogonal complement, $y^\perp = \left\{ \begin{pmatrix} 1 \\ -2 \\ 3 \end{pmatrix} \right\}^\perp$ of Example 7.2.8 is the plane perpendicular to y and is spanned

by $\begin{pmatrix} 2 \\ 1 \\ 0 \end{pmatrix}$ and $\begin{pmatrix} -3 \\ 0 \\ 1 \end{pmatrix}$.

We find $W^\perp$ by finding all vectors $v \in \mathbb{R}^3$ such that both

$$\left\langle \begin{pmatrix} 1 \\ 1 \\ 1 \end{pmatrix}, v \right\rangle = 0$$

and

$$\left\langle \begin{pmatrix} 0 \\ 3 \\ -1 \end{pmatrix}, v \right\rangle = 0.$$

Then we have

$$W^\perp = \left\{ \begin{pmatrix} a \\ b \\ c \end{pmatrix} \in \mathbb{R}^3 \ \Big| \ \left\langle \begin{pmatrix} 1 \\ 1 \\ 1 \end{pmatrix}, \begin{pmatrix} a \\ b \\ c \end{pmatrix} \right\rangle = 0, \left\langle \begin{pmatrix} 0 \\ 3 \\ -1 \end{pmatrix}, \begin{pmatrix} a \\ b \\ c \end{pmatrix} \right\rangle = 0 \right\}$$

$$= \left\{ \begin{pmatrix} a \\ b \\ c \end{pmatrix} \in \mathbb{R}^3 \ \Big| \ a + b + c = 0, \ 3b - c = 0 \right\}$$

$$= \left\{ \begin{pmatrix} -4 \\ 1 \\ 3 \end{pmatrix} b \ \Big| \ b \in \mathbb{R} \right\}$$

$$= \text{Span} \left\{ \begin{pmatrix} -4 \\ 1 \\ 3 \end{pmatrix} \right\}.$$

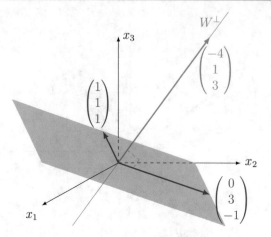

Fig. 7.8 The orthogonal complement, $W^\perp = \left\{ \begin{pmatrix} 1 \\ 1 \\ 1 \end{pmatrix}, \begin{pmatrix} 0 \\ 3 \\ -1 \end{pmatrix} \right\}^\perp$ is the line perpendicular to both vectors in W. (See Example 7.2.9).

We see $W^\perp$ in Figure 7.8. $\square$

Example 7.2.10 Consider $\mathbb{R}^3$ as a subspace of itself. $(\mathbb{R}^3)^\perp$ is the set of all vectors in $\mathbb{R}^3$ which are orthogonal to every vector in $\mathbb{R}^3$. The only vector satisfying this very restrictive requirement is the zero vector. That is, $(\mathbb{R}^3)^\perp = \{0\}$. $\square$

The orthogonal complements in the examples above are all written as a span. Therefore, each is a subspace. In fact, it is always the case that the orthogonal complement is a subspace whether or not the set W is a subspace. We see this result in the next lemma.

Lemma 7.2.11

Let V be an inner product space and $W \subseteq V$. Then $W^\perp$ is a subspace of V.

Proof. Suppose V is a vector space and $W \subseteq V$. We will show that (a) the zero vector of V, 0_V, is in $W^\perp$ and (b) for every $u, v \in W^\perp$ and scalar α, $\alpha u + v \in W^\perp$.

(a) First, notice that by Theorem 7.1.17, $\langle 0_V, y \rangle = 0$ for all $y \in W$, so $0_V \in W^\perp$.
(b) Now, let $u, v \in W^\perp$ and α be a scalar. Then, for all $y \in W$,

$$\langle \alpha u + v, y \rangle = \alpha \langle u, y \rangle + \langle v, y \rangle = \alpha \cdot 0 + 0 = 0.$$

Thus, $\alpha u + v \in W^\perp$.

Therefore, by Theorem 2.5.3 and Corollary 2.5.6, $W^\perp$ is a subspace of V. $\square$

If, in addition, W is a subspace, then we create a natural decomposition of V as the following direct sum.

> **Theorem 7.2.12**
> Let W be a subspace of the inner product space V. Then $V = W \oplus W^{\perp}$.

Proof. Let V be an inner product space and let $W \subseteq V$ be a subspace. We need to show that if $v \in W \cap W^{\perp}$, then v must necessarily be the zero vector of V. Then, we need to show that v can be uniquely written as $v = v_1 + v_2$ where $v_1 \in W$ and $v_2 \in W^{\perp}$. We leave the details of this argument to Exercise 12. □

> **Corollary 7.2.13**
> Let V be an inner product space and $W \subseteq V$. Then
>
> $$V = \text{Span}(W) \oplus W^{\perp}.$$

Proof. Suppose V is an inner product space and $W \subseteq V$. $\text{Span}(W)$ is a subspace of V and the result follows directly from Theorem 7.2.12 and Exercise 16. □

We now have a natural decomposition of a vector space into the direct product of a subspace W and the orthogonal complement subspace $W^{\perp}$. Every vector $v \in V$ can be written $v = w + n$ where $w \in W$ and $n \in W^{\perp}$. Furthermore, $\langle w, n \rangle = 0$.

Now we can use $r = \|w\|^2/\|v\|^2$ as a measure of how close v is to being in W. If $v \in W$ then $w = v$ and $n = 0$. In this case, $r = 1$. If $v \in W^{\perp}$ then $w = 0$ and $n = v$. In this case, $r = 0$. In fact, because w and n are always orthogonal,

$$0 \le r = \frac{\|w\|^2}{\|v\|^2} = \frac{\|w\|^2}{\|w+n\|^2} = \frac{\|w\|^2}{\|w\|^2 + \|n\|^2} \le 1.$$

Definition 7.2.14

Let W be a subspace of inner product space V. Let $v \in V$, then there are unique vectors $w \in W$ and $n \in W^{\perp}$ so that $v = w + n$. We say that w is the **orthogonal projection**, or equivalently, the **projection** of v onto W and write $w = \text{proj}_W(v)$. □

In Definition 7.2.14, we say that $w \in W$ and $n \in W^{\perp}$ are unique. Suppose v can be decomposed into $v = w_1 + n_1$ and $v = w_2 + n_2$. Then

$$
\begin{aligned}
0 &= \langle v - v, v - v \rangle \\
&= \langle (w_1 - w_2) + (n_1 - n_2), (w_1 - w_2) + (n_1 - n_2) \rangle \\
&= \langle w_1 - w_2, w_1 - w_2 \rangle + 2\langle w_1 - w_2, n_1 - n_2 \rangle + \langle n_1 - n_2, n_1 - n_2 \rangle \\
&= \langle w_1 - w_2, w_1 - w_2 \rangle + \langle n_1 - n_2, n_1 - n_2 \rangle.
\end{aligned}
$$

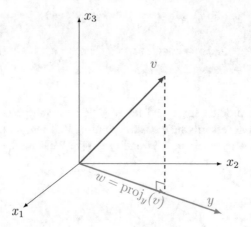

Fig. 7.9 Orthogonal projection onto a vector (or a line).

Now, since $\langle w_1 - w_2, w_1 - w_2 \rangle$ and $\langle n_1 - n_2, n_1 - n_2 \rangle$ are nonnegative, they must both be zero. Therefore, $w_1 = w_2$ and $n_1 = n_2$. This shows the uniqueness assumed in Definition 7.2.14.

For clarity in notation, it is common to write $v_\perp$ instead of n, to indicate the component of v orthogonal to W, and $v_\|$ instead of w, to indicate the component of v that is within the space W.

Both orthogonal projections and coordinate projections result from decomposition of the inner product space V into a direct sum $W_1 \oplus W_2$. The projections agree if and only if $W_2 = W_1^\perp$. For example, see Exercise 18.

Next, suppose we have a vector $v \in \mathbb{R}^3$ and we want to project v onto a vector y in $\mathbb{R}^n$. We are looking for a vector on the line made of all scalar multiples of y. That is, we project onto the subspace

$$L = \{\alpha y : \alpha \in \mathbb{R}\}.$$

Then we are looking for the vector $w = \text{proj}_y(v)$ shown in Figure 7.9. We now consider projecting a vector onto any other vector along the same line.

Projection onto a Line

We begin with $W = \text{Span}\{u_1\}$, the span of only one vector. This means, we want the projection of v onto a line (also depicted in Figure 7.9). First, we recognize that we are looking for $w \in W$ so that $v = w + n$ where $n \in W^\perp$. Then notice, $w \in W$ means $w = \alpha u_1$ for some scalar α. Next, $n \in W^\perp$ says that $\langle n, u_1 \rangle = 0$. Putting these together with $v = w + n$, we see that

$$\begin{aligned}
\langle v, u_1 \rangle &= \langle \alpha u_1 + n, u_1 \rangle \\
&= \alpha \langle u_1, u_1 \rangle + \langle n, u_1 \rangle \\
&= \alpha \langle u_1, u_1 \rangle + 0 \\
&= \alpha \langle u_1, u_1 \rangle.
\end{aligned}$$

Thus,

$$\alpha = \frac{\langle v, u_1 \rangle}{\langle u_1, u_1 \rangle} \text{ and, therefore, } w = \alpha u_1 = \frac{\langle v, u_1 \rangle}{\langle u_1, u_1 \rangle} u_1.$$

Let us, now, consider two examples.

Example 7.2.15 Consider $\mathbb{R}^3$ with the standard inner product. Let $W = \text{Span}\{(2, 2, 1)^T\}$. We can find the orthogonal projection of $v = (1, 3, -1) \in \mathbb{R}^3$ onto W.

$$\text{proj}_W v = \frac{\langle (1, 3, -1)^T, (2, 2, 1)^T \rangle}{\langle (2, 2, 1)^T, (2, 2, 1)^T \rangle} u_1 = \tfrac{7}{9} u_1 = \left(\tfrac{14}{9}, \tfrac{14}{9}, \tfrac{7}{9} \right).$$

$\square$

Example 7.2.16 Find the orthogonal projection of

$$f(x) = 2 - x + x^2 \in P_2(\mathbb{R})$$

onto $W = \text{Span}\{1 - x\}$. We will use the standard inner product on $P_2(\mathbb{R})$.

$$\begin{aligned}
\text{proj}_W f(x) &= \frac{\langle 2 - x + x^2, 1 - x \rangle}{\langle 1 - x, 1 - x \rangle}(1 - x) \\
&= \frac{\int_0^1 (2 - x + x^2)(1 - x)dx}{\int_0^1 (1 - x)^2 dx}(1 - x) \\
&= \frac{^{11}/_{12}}{^1/_3}(1 - x) \\
&= \tfrac{11}{4}(1 - x).
\end{aligned}$$

$\square$

In these examples, we found the scalar multiple of a vector that is closest to another given vector. We now consider projection onto a subspace. That is, we find a vector in a subspace that is closest to a given vector.

Projection onto a Subspace

The scenario is illustrated in Figure 7.10 for a two-dimensional subspace. Suppose we wish to project a vector v onto a k-dimensional subspace W with basis $\mathcal{B}_1 = \{u_1, u_2, \ldots, u_k\}$. We know that every vector in $W^\perp$ is orthogonal to every vector in W. Thus, given a basis for $W^\perp$, $\mathcal{B}_2 = \{u_{k+1}, \ldots, u_n\}$, we know that $\langle u_i, u_j \rangle = 0$ for all $u_i \in \mathcal{B}_1$ and $u_j \in \mathcal{B}_2$.

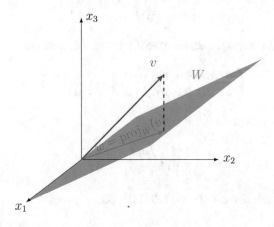

Fig. 7.10 Orthogonal projection onto a subspace (in this case, a plane).

Now, the projection is given by the coordinates $(a_1, a_2, \ldots, a_k) = [v]_{\mathcal{B}_1}$. From Theorem 7.1.29, we have

$$
\begin{pmatrix}
\langle u_1, u_1 \rangle \cdots \langle u_1, u_k \rangle & & \\
\vdots \quad \ddots \quad \vdots & 0 & \\
\langle u_k, u_1 \rangle \cdots \langle u_k, u_k \rangle & & \\
& \langle u_{k+1}, u_{k+1} \rangle \cdots \langle u_{k+1}, u_n \rangle \\
0 & \vdots \quad \ddots \quad \vdots \\
& \langle u_n, u_{k+1} \rangle \cdots \langle u_n, u_n \rangle
\end{pmatrix}
\begin{pmatrix}
a_1 \\ \vdots \\ a_k \\ a_{k+1} \\ \vdots \\ a_n
\end{pmatrix}
=
\begin{pmatrix}
\langle u_1, v \rangle \\ \vdots \\ \langle u_k, v \rangle \\ \langle u_{k+1}, v \rangle \\ \vdots \\ \langle u_n, v \rangle
\end{pmatrix}.
$$

Notice that this matrix equation can be separated into two matrix equations because of its block matrix structure. That is, we can find $a_1, a_2, \ldots, a_k$ by solving the matrix equation

$$
\begin{pmatrix}
\langle u_1, u_1 \rangle \cdots \langle u_1, u_k \rangle \\
\vdots \quad \ddots \quad \vdots \\
\langle u_k, u_1 \rangle \cdots \langle u_k, u_k \rangle
\end{pmatrix}
\begin{pmatrix}
a_1 \\ \vdots \\ a_k
\end{pmatrix}
=
\begin{pmatrix}
\langle u_1, v \rangle \\ \vdots \\ \langle u_k, v \rangle
\end{pmatrix}.
\tag{7.5}
$$

And, we can find $a_{k+1}, a_{k+2}, \ldots, a_n$ by solving the matrix equation

$$
\begin{pmatrix}
\langle u_{k+1}, u_{k+1} \rangle \cdots \langle u_{k+1}, u_n \rangle \\
\vdots \quad \ddots \quad \vdots \\
\langle u_n, u_{k+1} \rangle \cdots \langle u_n, u_n \rangle
\end{pmatrix}
\begin{pmatrix}
a_{k+1} \\ \vdots \\ a_n
\end{pmatrix}
=
\begin{pmatrix}
\langle u_{k+1}, v \rangle \\ \vdots \\ \langle u_n, v \rangle
\end{pmatrix}.
$$

Notice that the solution to Equation 7.5 gives us the projection onto the subspace spanned by the vectors in $\mathcal{B}_1$.

Example 7.2.17 In this example, consider $v = (1, 0, 3)^{\mathsf{T}} \in \mathbb{R}^3$. We will find the orthogonal projection of v onto the subspace

$$
W = \mathrm{Span} \left\{ \begin{pmatrix} 1 \\ 1 \\ 0 \end{pmatrix}, \begin{pmatrix} 2 \\ -1 \\ 1 \end{pmatrix}, \begin{pmatrix} 3 \\ 0 \\ 1 \end{pmatrix} \right\}.
$$

A basis for W is

$$
\mathcal{B}_1 = \left\{ u_1 = \begin{pmatrix} 1 \\ 1 \\ 0 \end{pmatrix}, u_2 = \begin{pmatrix} 3 \\ 0 \\ 1 \end{pmatrix} \right\}.
$$

If we write $\mathrm{proj}_W v = a_1 u_1 + a_2 u_2$, then we have the system of equations

$$
\begin{pmatrix}
\langle u_1, u_1 \rangle & \langle u_1, u_2 \rangle \\
\langle u_2, u_1 \rangle & \langle u_2, u_2 \rangle
\end{pmatrix}
\begin{pmatrix} a_1 \\ a_2 \end{pmatrix}
=
\begin{pmatrix}
\langle u_1, v \rangle \\
\langle u_2, v \rangle
\end{pmatrix}.
$$

Computing the inner products gives

$$
\begin{pmatrix} 2 & 3 \\ 3 & 10 \end{pmatrix}
\begin{pmatrix} a_1 \\ a_2 \end{pmatrix}
=
\begin{pmatrix} 1 \\ 6 \end{pmatrix}.
$$

The matrix equation has solution $a = (-8/11, 9/11)^{\mathsf{T}}$. So,

$$\text{proj}_W v = a_1 u_1 + a_2 u_2 = -\frac{8}{11} \begin{pmatrix} 1 \\ 1 \\ 0 \end{pmatrix} + \frac{9}{11} \begin{pmatrix} 3 \\ 0 \\ 1 \end{pmatrix} = \begin{pmatrix} 19/11 \\ -8/11 \\ 9/11 \end{pmatrix}. \qquad \square$$

To find the projection of a vector v onto any subspace $W = \text{Span}\{u_1, u_2, \ldots, u_k\}$, we make use of the fact that W and $W^\perp$ are orthogonal sets. We can solve for either $\text{proj}_W(v)$ or $\text{proj}_{W^\perp}(v)$, whichever is simpler, and then use the fact that $v = \text{proj}_W(v) + \text{proj}_{W^\perp}(v)$ to solve for the other.

In Example 7.2.17, we could have, instead, found $\text{proj}_W v$ by first projecting v onto $W^\perp$ and then subtracting from v. To see this, we first compute $W^\perp$,

$$W^\perp = \left\{ u \in \mathbb{R}^3 \mid \langle u, u_1 \rangle = 0 \text{ and } \langle u, u_2 \rangle = 0 \right\}$$

$$= \left\{ \begin{pmatrix} a \\ b \\ c \end{pmatrix} \mid a + b = 0, 3a + c = 0 \right\}$$

$$= \left\{ \begin{pmatrix} 1 \\ -1 \\ -3 \end{pmatrix} \alpha \mid \alpha \in \mathbb{R} \right\}.$$

Let $w' = \begin{pmatrix} 1 \\ -1 \\ -3 \end{pmatrix}$, then we know that

$$\text{proj}_{W^\perp}(v) = \text{proj}_{w'}(v) = \frac{\langle v, w' \rangle}{\langle w', w' \rangle} w' = \frac{-8}{11} \begin{pmatrix} 1 \\ -1 \\ -3 \end{pmatrix}.$$

Now find $\text{proj}_W(v)$ as described above.

$$\text{proj}_W(v) = v - \text{proj}_{W^\perp}(v) = \begin{pmatrix} 1 \\ 0 \\ 3 \end{pmatrix} - \frac{-8}{11} \begin{pmatrix} 1 \\ -1 \\ -3 \end{pmatrix} = \begin{pmatrix} 19/11 \\ -8/11 \\ 9/11 \end{pmatrix}$$

which is the same as our result in Example 7.2.17.

Example 7.2.18 Consider the three images of Example 7.1.6 (also shown below). Suppose images y and z are two known examples of traffic density distributions in the inner city that correlate with significant increase in evening commute times. We wish to understand if another density distribution, say x, exhibits similar density properties. To answer this question, we begin by computing the projection of x onto the subspace $W = \text{Span}\{y, z\}$.

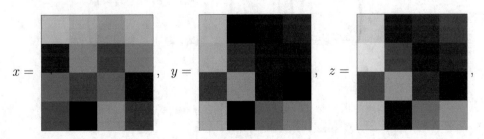

with numerical grayscale entries

$$
x =
\begin{pmatrix}
95 & 89 & 82 & 92 \\
23 & 76 & 44 & 74 \\
61 & 46 & 62 & 18 \\
49 & 2 & 79 & 41
\end{pmatrix},
\quad
y =
\begin{pmatrix}
94 & 6 & 14 & 27 \\
92 & 35 & 20 & 20 \\
41 & 81 & 20 & 2 \\
89 & 1 & 60 & 75
\end{pmatrix},
\quad
z =
\begin{pmatrix}
102 & 22 & 30 & 23 \\
109 & 45 & 21 & 29 \\
50 & 85 & 33 & 15 \\
97 & 14 & 67 & 83
\end{pmatrix}.
$$

If we write $\operatorname{proj}_W x = a_1 y + a_2 z$, then we have the system of equations

$$
\begin{pmatrix}
\langle y, y \rangle & \langle y, z \rangle \\
\langle z, y \rangle & \langle z, z \rangle
\end{pmatrix}
\begin{pmatrix}
a_1 \\
a_2
\end{pmatrix}
=
\begin{pmatrix}
\langle y, x \rangle \\
\langle z, x \rangle
\end{pmatrix}.
$$

Computing the inner products (see Example 7.1.6) gives

$$
\begin{pmatrix}
46{,}079 & 51{,}881 \\
51{,}881 & 59{,}527
\end{pmatrix}
\begin{pmatrix}
a_1 \\
a_2
\end{pmatrix}
=
\begin{pmatrix}
39{,}913 \\
497{,}974
\end{pmatrix}
$$

with solution $(a_1, a_2)^{\mathsf{T}} \approx (-2.203, +2.726)^{\mathsf{T}}$. So, $x = x_{\parallel} + x_{\perp}$, where

$$
x_{\parallel} = \operatorname{proj}_W x \approx
\begin{pmatrix}
71 & 47 & 51 & 3 \\
94 & 46 & 13 & 35 \\
46 & 53 & 46 & 36 \\
68 & 36 & 50 & 61
\end{pmatrix},
\quad
x_{\perp} = \operatorname{proj}_{W^{\perp}}(x) \approx
\begin{pmatrix}
24 & 42 & 31 & 89 \\
-71 & 30 & 31 & 39 \\
15 & -7 & 16 & -18 \\
-19 & -34 & 29 & -20
\end{pmatrix}.
$$

We have $\|\operatorname{proj}_W x\| \approx 198$ and $\|x_{\perp}\| \approx 121$. We also have $\dfrac{\|\operatorname{proj}_W x\|}{\|x\|} \approx 0.853$. So we see that x has a non-trivial projection onto $\operatorname{Span}\{y, z\}$, but perhaps not significant from a traffic congestion standpoint.   □

Matrix Representations of Orthogonal Projection Transformations

Orthogonal projection of a vector $v \in \mathbb{R}^n$ onto any subspace W is determined by the solution to Equation 7.5. We need only have a basis, $\mathcal{B} = \{u_1, u_2, \ldots, u_k\}$, for the subspace and an inner product $\langle \cdot, \cdot \rangle$. Let U be the matrix whose columns are the basis vectors. Notice that

$$
U^{\mathsf{T}} U =
\begin{pmatrix}
- u_1 - \\
\vdots \\
- u_k -
\end{pmatrix}
\begin{pmatrix}
| & & | \\
u_1 & \cdots & u_k \\
| & & |
\end{pmatrix}
=
\begin{pmatrix}
\langle u_1, u_1 \rangle & \cdots & \langle u_1, u_k \rangle \\
\vdots & \ddots & \vdots \\
\langle u_k, u_1 \rangle & \cdots & \langle u_k, u_k \rangle
\end{pmatrix}
$$

and

$$
U^{\mathsf{T}} v =
\begin{pmatrix}
- u_1 - \\
\vdots \\
- u_k -
\end{pmatrix}
\begin{pmatrix}
| \\
v \\
|
\end{pmatrix}
=
\begin{pmatrix}
\langle u_1, v \rangle \\
\vdots \\
\langle u_k, v \rangle
\end{pmatrix}.
$$

Combining this with Equation 7.5, we see that the projection coordinates, $a_1, a_2, \ldots, a_k$ satisfy $U^\mathsf{T} U a = U^\mathsf{T} v$. The following theorem shows how one can obtain a projection vector as a series of left matrix multiplication operations, a method that is especially useful for computational tasks on a computer.

Theorem 7.2.19

Let W be a k-dimensional subspace of $\mathbb{R}^n$ and $v \in \mathbb{R}^n$. Let $\{u_1, u_2, \ldots, u_k\}$ be a basis for W and U the matrix whose columns are $u_1, u_2, \ldots, u_k$. Then

$$\mathrm{proj}_W v = U(U^\mathsf{T} U)^{-1} U^\mathsf{T} v.$$

Proof. Let W be a subspace of $\mathbb{R}^n$ and let $v \in \mathbb{R}^n$. Define $\mathcal{B} = \{u_1, u_2, \ldots, u_k\}$ to be a basis for W. Suppose U is the matrix whose columns are $u_1, u_2, \ldots, u_k$. Then $\mathrm{Rank}(U) = k$. Therefore, $\mathrm{Row}(U) = \mathbb{R}^k$. Now, by the Rank-Nullity Theorem (Theorem 5.1.30), we know that $\mathrm{Nullity}(U) = n - k$. Let $\{u_{k+1}, u_{k+2}, \ldots, u_n\}$ be a basis for $\mathrm{Null}(U)$. By Theorem 5.2.23, we know that the transformation, $T : \mathbb{R}^k \to \mathbb{R}^n$, defined by left multiplication by U is one-to-one and $S : \mathbb{R}^n \to \mathbb{R}^k$ defined by left multiplication by U^T is onto. We now show that $S \circ T : \mathbb{R}^k \to \mathbb{R}^k$ is onto and therefore invertible. Notice that

$$\mathrm{Ran}(S) = \mathrm{Col}(U^\mathsf{T}) = \mathrm{Row}(U) = \mathbb{R}^k.$$

That is, if $u \in \mathbb{R}^k$, then there is $w \in \mathbb{R}^n$ so that $S(w) = u$ and

$$w = \alpha_1 u_1 + \ldots + \alpha_k u_k + \alpha_{k+1} u_{k+1} + \ldots + \alpha_n u_n.$$

Therefore, $S(\tilde{w}) = u$, for $\tilde{w} = \alpha_1 u_1 + \ldots + \alpha_k u_k \in \mathrm{Col}(U)$. Notice, also,

$$\mathrm{Ran}(T) = \mathrm{Col}(U).$$

Therefore, there is a $\tilde{u} \in \mathbb{R}^k$ so that $T(\tilde{u}) = \tilde{w}$. That is,

$$(S \circ T)(\tilde{u}) = S(T(\tilde{u})) = S(\tilde{w}) = u$$

and $S \circ T$ is onto. That is, $\mathrm{Ran}(S \circ T) = \mathbb{R}^k$. Thus, $S \circ T$ is invertible. Therefore, by Theorems 4.4.14 and 5.2.21, $U^\mathsf{T} U$ is invertible. We know that $U^\mathsf{T} U a = U^\mathsf{T} v$, so $a = (U^\mathsf{T} U)^{-1} U^\mathsf{T} v$. Now $[\mathrm{proj}_W v]_\mathcal{B} = a$, so

$$\mathrm{proj}_W v = U a = U(U^\mathsf{T} U)^{-1} U^\mathsf{T} v.$$

$\square$

Consider again the matrix representation of the projection transformation. Suppose we want to find an orthogonal projection of a vector $v \in \mathbb{R}^n$ onto a k-dimensional subspace W. Notice that if we have an orthogonal basis $\mathcal{B} = \{u_1, u_2, \ldots, u_k\}$ for W, then

$$U^\mathsf{T} U = \begin{pmatrix} \langle u_1, u_1 \rangle & & \dots & \langle u_1, u_k \rangle \\ \langle u_2, u_1 \rangle & \langle u_2, u_2 \rangle & & \langle u_2, u_k \rangle \\ \vdots & & \ddots & \vdots \\ \langle u_k, u_1 \rangle & & \dots & \langle u_k, u_k \rangle \end{pmatrix} = \begin{pmatrix} \|u_1\| & 0 & \dots & 0 \\ 0 & \|u_2\| & & 0 \\ \vdots & & \ddots & \vdots \\ 0 & \dots & 0 & \|u_k\| \end{pmatrix}.$$

Therefore, since u_j is not the zero vector for any $1 \le j \le k$, we have

$$(U^\mathsf{T} U)^{-1} = \begin{pmatrix} 1/\|u_1\| & 0 & \dots & 0 \\ 0 & 1/\|u_2\| & & 0 \\ \vdots & & \ddots & \vdots \\ 0 & & \dots & 0 & 1/\|u_k\| \end{pmatrix}.$$

The next two corollaries tell us how the projection transformation is simplified when the subspace basis is orthogonal or orthonormal.

Corollary 7.2.20
Let W be a k-dimensional subspace of $\mathbb{R}^n$ and $v \in \mathbb{R}^n$. Let $\{u_1, u_2, \dots, u_k\}$ be an orthogonal basis for W and U the matrix whose columns are $u_1, u_2, \dots, u_k$. Then $\operatorname{proj}_W v = \hat{U}\hat{U}^\mathsf{T} v$, where $\hat{U}$ is the $n \times k$ matrix whose columns are the normalized basis vectors.

The proof is the subject of Exercise 13.

Corollary 7.2.21
Let W be a k-dimensional subspace of $\mathbb{R}^n$ and $v \in \mathbb{R}^n$. Let $\{u_1, u_2, \dots, u_k\}$ be an orthonormal basis for W and U the matrix whose columns are $u_1, u_2, \dots, u_k$. Then $\operatorname{proj}_W v = UU^\mathsf{T} v$.

Proof. Suppose W is a subspace of $\mathbb{R}^n$ and $v \in \mathbb{R}^n$, $\{u_1, u_2, \dots, u_k\}$ is an orthonormal basis for W and U is the matrix whose columns are $u_1, u_2, \dots, u_k$. Then, the result follows from Corollary 7.2.20 by noting that each basis vector is a unit vector. □

Example 7.2.22 Let $v = (1, 2, 3)^\mathsf{T} \in \mathbb{R}^3$ and define

$$W = \operatorname{Span}\left\{ \begin{pmatrix} 1 \\ 0 \\ 1 \end{pmatrix}, \begin{pmatrix} 2 \\ 1 \\ 2 \end{pmatrix} \right\}.$$

We can find $\operatorname{proj}_W v$ using Theorem 7.2.19. Note that B is a linearly independent set, but not orthogonal, so we define

$$U = \begin{pmatrix} 1 & 2 \\ 0 & 1 \\ 1 & 2 \end{pmatrix}.$$

Then

$$U^\mathsf{T} U = \begin{pmatrix} 2 & 4 \\ 4 & 9 \end{pmatrix} \text{ and } (U^\mathsf{T} U)^{-1} = \begin{pmatrix} 9/2 & -2 \\ -2 & 1 \end{pmatrix}.$$

So, we have

$$U\left(U^\mathsf{T} U\right)^{-1} U^\mathsf{T} = \frac{1}{2} \begin{pmatrix} 1 & 0 & 1 \\ 0 & 2 & 0 \\ 1 & 0 & 1 \end{pmatrix}.$$

Therefore,

$$\mathrm{proj}_W v = \frac{1}{2} \begin{pmatrix} 1 & 0 & 1 \\ 0 & 2 & 0 \\ 1 & 0 & 1 \end{pmatrix} \begin{pmatrix} 1 \\ 2 \\ 3 \end{pmatrix} = \frac{1}{2} \begin{pmatrix} 4 \\ 4 \\ 4 \end{pmatrix} = \begin{pmatrix} 2 \\ 2 \\ 2 \end{pmatrix}.$$

$\square$

Example 7.2.23 Now, let $v = (1, 2, 3)^\mathsf{T} \in \mathbb{R}^3$ and

$$W = \mathrm{Span}\left\{ \begin{pmatrix} 1 \\ 0 \\ 1 \end{pmatrix}, \begin{pmatrix} 2 \\ 1 \\ -2 \end{pmatrix} \right\}.$$

We can find $\mathrm{proj}_W v$ using Corollary 7.2.20 because the basis for W is an orthogonal basis. We define

$$U = \begin{pmatrix} 1 & 2 \\ 0 & 1 \\ 1 & -2 \end{pmatrix}.$$

So,

$$\hat{U} = \begin{pmatrix} 1/\sqrt{2} & 2/3 \\ 0 & 1/3 \\ 1/\sqrt{2} & -2/3 \end{pmatrix}.$$

Then,

$$
\begin{aligned}
\mathrm{proj}_W v &= \hat{U}\hat{U}^\mathsf{T} v \\
&= \begin{pmatrix} 1/\sqrt{2} & 2/3 \\ 0 & 1/3 \\ 1/\sqrt{2} & -2/3 \end{pmatrix} \begin{pmatrix} 1/\sqrt{2} & 0 & 1/\sqrt{2} \\ 2/3 & 1/3 & -2/3 \end{pmatrix} \begin{pmatrix} 1 \\ 2 \\ 3 \end{pmatrix} \\
&= \frac{1}{9} \begin{pmatrix} 14 \\ -2 \\ 22 \end{pmatrix}.
\end{aligned}
$$

$\square$

If, in Example 7.2.23, we had the orthonormal basis

$$\mathcal{B} = \left\{ \begin{pmatrix} 1/\sqrt{2} \\ 0 \\ 1/\sqrt{2} \end{pmatrix}, \begin{pmatrix} 2/3 \\ 1/3 \\ -2/3 \end{pmatrix} \right\},$$

Corollary 7.2.21, tells us that we can directly use U as its columns are already normalized.

7.2.3   Gram-Schmidt Process

We have seen that orthonormal bases for vector spaces are useful in quantifying how close a vector is to being within a subspace of interest. We have also seen that projections, which help us quantify "closeness," are most easily computed and understood in terms of orthonormal bases. In this section, we explore the Gram-Schmidt process for constructing an orthonormal basis from an arbitrary basis.

First, consider the following series of pictures showing how this process works in $\mathbb{R}^3$. Suppose we have a basis $\mathcal{C} = \{v_1, v_2, v_3\}$ (Figure 7.11) and we would like an orthonormal basis $\mathcal{B} = \{u_1, u_2, u_3\}$, with one vector pointing in the direction of v_1. We find an orthogonal basis and then normalize the vectors later.

We begin building the orthogonal basis by starting with the set $\mathcal{B} = \{v_1\}$. This set will get larger as we add orthogonal vectors to it. Notice that, currently, $\mathcal{B}$ is an orthogonal (and, therefore, linearly independent) set (Figure 7.12), but, $\mathcal{B}$ is not a basis for $\mathbb{R}^3$. We rename v_1 as u_1 to make clear that we are creating a new basis.

We use v_2 to obtain a vector orthogonal to u_1 by subtracting, from v_2, the component of v_2 that is parallel to u_1. (Recall, the vector we subtract is $\text{proj}_{u_1} v_2$.) When we do this, all that will be left is the component of v_2 orthogonal to u_1. (Figure 7.13)

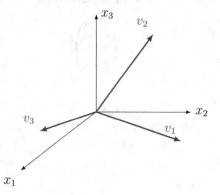

Fig. 7.11   The set $\mathcal{C} = \{v_1, v_2, v_3\}$ is a basis for $\mathbb{R}^3$.

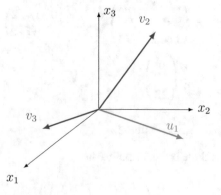

Fig. 7.12   Step 1: Set $u_1 = v_1$.

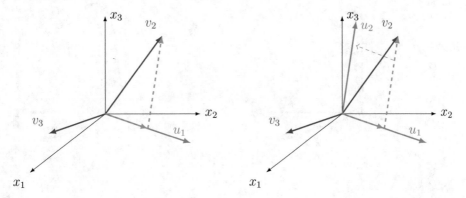

Fig. 7.13 Step 2: Set u_2 equal to the component of v_2 that is orthogonal to u_1 (recall, vectors make more sense drawn with tail at the origin).

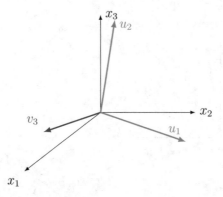

Fig. 7.14 The vectors u_1 and u_2 of the orthogonal basis $\mathcal{B}$.

We label this left over vector u_2 and include it in the set $\mathcal{B}$. So now

$$\mathcal{B} = \{u_1, u_2\}$$

is an orthogonal set, but still not a basis for $\mathbb{R}^3$. We can see the current status in Figure 7.14.

To find u_3, we repeat the process with v_3, but removing the components parallel to each of u_1 and u_2 as in Figures 7.15 and 7.16. That is, we find

$$u_3 = v_3 - \text{proj}_{u_1} v_3 - \text{proj}_{u_2} v_3.$$

We put u_3 into $\mathcal{B}$ to obtain the orthogonal basis

$$\mathcal{B} = \{u_1, u_2, u_3\}$$

for $\mathbb{R}^3$ see in Figure 7.17.

Now, to normalized each, we divide by the respective length. The resulting set

$$\mathcal{B} = \left\{ \frac{u_1}{\|u_1\|}, \frac{u_2}{\|u_2\|}, \frac{u_3}{\|u_3\|} \right\}$$

is an orthonormal basis for $\mathbb{R}^3$. We now give an example.

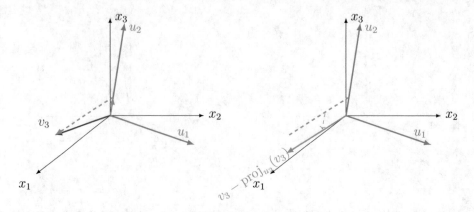

Fig. 7.15 We subtract $\text{proj}_{u_2}(v_3)$ from v_3 to get a vector orthogonal to u_2.

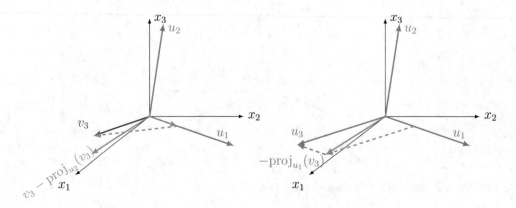

Fig. 7.16 Step 3: We subtract $\text{proj}_{u_1}(v_3)$ from $v_3 - \text{proj}_{u_2}(v_3)$ to get a vector orthogonal to both u_2 and u_1. Set u_3 to be this orthogonal vector.

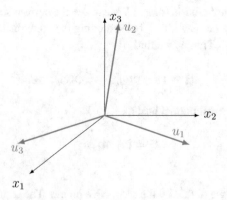

Fig. 7.17 The vectors u_1, u_2 and u_3 of the orthogonal basis $\mathcal{B}$.

Example 7.2.24 Given a basis

$$\left\{ v_1 = \begin{pmatrix} 1 \\ 2 \\ -2 \end{pmatrix}, v_2 = \begin{pmatrix} 4 \\ 4 \\ -3 \end{pmatrix}, v_3 = \begin{pmatrix} -1 \\ 7 \\ 2 \end{pmatrix} \right\}$$

for $\mathbb{R}^3$, find an orthonormal basis containing one vector in the same direction as v_1. As above, we keep v_1 and call it u_1:

$$u_1 = \begin{pmatrix} 1 \\ 2 \\ -2 \end{pmatrix}.$$

Next, we find u_2 using the process described above with v_2. That is,

$$u_2 = v_2 - \text{proj}_{u_1}(v_2) = v_2 - \frac{\langle v_2, u_1 \rangle}{\langle u_1, u_1 \rangle} u_1$$

$$= \begin{pmatrix} 4 \\ 4 \\ -3 \end{pmatrix} - \frac{18}{9} \begin{pmatrix} 1 \\ 2 \\ -2 \end{pmatrix} = \begin{pmatrix} 2 \\ 0 \\ 1 \end{pmatrix}.$$

Finally, we find u_3 as suggested above.

$$u_3 = v_3 - \text{proj}_{u_1}(v_3) - \text{proj}_{u_2}(v_3)$$

$$= v_3 - \frac{\langle v_3, u_1 \rangle}{\langle u_1, u_1 \rangle} u_1 - \frac{\langle v_3, u_2 \rangle}{\langle u_2, u_2 \rangle} u_2$$

$$= \begin{pmatrix} -1 \\ 7 \\ 2 \end{pmatrix} - \frac{9}{9} \begin{pmatrix} 1 \\ 2 \\ -2 \end{pmatrix} - \frac{0}{5} \begin{pmatrix} 2 \\ 0 \\ 1 \end{pmatrix}$$

$$= \begin{pmatrix} -2 \\ 5 \\ 4 \end{pmatrix}.$$

Therefore, an orthonormal basis with one vector in the direction of v_1 is

$$\mathcal{B} = \left\{ \frac{u_1}{\|u_1\|}, \frac{u_2}{\|u_2\|}, \frac{u_3}{\|u_3\|} \right\}$$

$$= \left\{ \begin{pmatrix} 1/3 \\ 2/3 \\ -2/3 \end{pmatrix}, \begin{pmatrix} 2/\sqrt{5} \\ 0 \\ 1/\sqrt{5} \end{pmatrix}, \begin{pmatrix} -2/\sqrt{45} \\ 5/\sqrt{45} \\ 4/\sqrt{45} \end{pmatrix} \right\}.$$

□

The next theorem formalizes this procedure, called the Gram-Schmidt process.

Theorem 7.2.25 [Gram-Schmidt]
Let V be an inner product space with basis $\{v_1, v_2, \ldots, v_n\}$. Set

$$u_1 = v_1$$

and

$$u_k = v_k - \sum_{j=1}^{k-1} \frac{\langle u_j, v_k \rangle}{\langle u_j, u_j \rangle} u_j, \quad \text{for } k = 2, 3, \ldots, n.$$

Then $\{u_1, u_2, \ldots, u_n\}$ is an orthogonal basis for V.

Proof (by induction). Suppose V is an inner product space with basis $\{v_1, v_2, \ldots, v_n\}$. Suppose, also, that

$$u_k = v_k - \sum_{j=1}^{k-1} \frac{\langle u_j, v_k \rangle}{\langle u_j, u_j \rangle} u_j, \quad \text{for } k = 2, 3, \ldots, n.$$

Define $\mathcal{B} = \{u_1, u_2, \ldots, u_n\}$. We will show that $\mathcal{B}$ is an orthogonal set, and so by Corollary 7.1.23, $\mathcal{B}$ is an orthogonal basis for V.

Let $\mathcal{B}_k = \{u_1, u_2, \ldots, u_k\}$. Notice that $\mathcal{B}_1 = \{u_1\}$ has only one element and is, therefore, an orthogonal set. Now, suppose $\mathcal{B}_k$ is an orthogonal set. We show that $\mathcal{B}_{k+1}$ is also an orthogonal set. Let W_k be the set spanned by $\mathcal{B}_k$. Notice that

$$W_k = \text{Span}\{u_1, u_1, \ldots, u_k\}$$
$$= \text{Span}\{v_1, v_2, \ldots, v_k\}.$$

Since $v_{k+1} \notin W_k$, we know that $u_{k+1} \neq 0$. Also, by properties of $\langle \cdot, \cdot \rangle$ and the assumption that $\mathcal{B}_k$ is orthogonal, we have, for $1 \leq \ell \leq k$, that

$$\langle u_{k+1}, v_\ell \rangle = \left\langle v_{k+1} - \sum_{j=1}^{k} \frac{\langle u_j, v_{k+1} \rangle}{\langle u_j, u_j \rangle} u_j, v_\ell \right\rangle$$

$$= \langle v_{k+1}, v_\ell \rangle - \sum_{j=1}^{k} \frac{\langle u_j, v_{k+1} \rangle}{\langle u_j, u_j \rangle} \langle u_j, v_\ell \rangle$$

$$= \langle v_{k+1}, v_\ell \rangle - \frac{\langle u_\ell, v_{k+1} \rangle}{\langle u_\ell, u_\ell \rangle} \langle u_\ell, v_\ell \rangle$$

$$= 0.$$

Therefore, u_{k+1} is orthogonal to each vector $u_1, u_2, \ldots, u_k$. Thus, $\mathcal{B}_{k+1}$ is an orthogonal set. By the principle of mathematical induction, $\mathcal{B}$ is orthogonal. $\qquad\square$

Corollary 7.2.26

Let $\{v_1, v_2, \ldots, v_n\}$ be a basis for $\mathbb{R}^n$. Define

$$u_1 = v_1$$

and

$$u_k = v_k - \hat{U}_{k-1}\hat{U}_{k-1}^\mathsf{T}v_k, \quad \text{for } k = 2, 3, \ldots, n,$$

where $\hat{U}_j$ is the matrix with columns $u_1, u_2, \ldots, u_j$. Then $\{u_1, u_2, \ldots, u_n\}$ is an orthogonal basis for V.

The proof is the subject of Exercise 14.

Example 7.2.27 Let $T : \mathbb{R}^3 \to \mathbb{R}^3$ be the linear transformation defined by left multiplication by matrix

$$M = \begin{pmatrix} 1 & -2 & 2 \\ -2 & 4 & -4 \\ 2 & -4 & 4 \end{pmatrix}.$$

We will find an orthonormal basis for $\text{Null}(T) = \text{Null}(M)$. We have

$$\text{Null}(M) = \left\{ u = (a, b, c) \in \mathbb{R}^3 \mid Mu = 0 \right\}$$

$$= \left\{ u = \begin{pmatrix} a \\ b \\ c \end{pmatrix} \in \mathbb{R}^3 \mid a - 2b + 2c = 0 \right\}$$

$$= \left\{ u = \begin{pmatrix} -2 \\ 1 \\ 0 \end{pmatrix} b + \begin{pmatrix} 2 \\ 0 \\ 1 \end{pmatrix} c \in \mathbb{R}^3 \mid b, c \in \mathbb{R} \right\}$$

$$= \text{Span} \left\{ \begin{pmatrix} -2 \\ 1 \\ 0 \end{pmatrix}, \begin{pmatrix} 2 \\ 0 \\ 1 \end{pmatrix} \right\},$$

so that a basis for $\text{Null}(M)$ is $\left\{ \begin{pmatrix} -2 \\ 1 \\ 0 \end{pmatrix}, \begin{pmatrix} 2 \\ 0 \\ 1 \end{pmatrix} \right\}$. We begin by formulating an orthogonal basis. We use the first vector, $u_1 = (-2, 1, 0)^\mathsf{T}$, as the first vector in the orthogonal basis. The second vector is found using the Gram-Schmidt Procedure.

$$u_2 = \begin{pmatrix} 2 \\ 0 \\ 1 \end{pmatrix} - \frac{\left\langle \begin{pmatrix} -2 \\ 1 \\ 0 \end{pmatrix}, \begin{pmatrix} 2 \\ 0 \\ 1 \end{pmatrix} \right\rangle}{\left\langle \begin{pmatrix} -2 \\ 1 \\ 0 \end{pmatrix}, \begin{pmatrix} -2 \\ 1 \\ 0 \end{pmatrix} \right\rangle} \begin{pmatrix} -2 \\ 1 \\ 0 \end{pmatrix}$$

$$= \begin{pmatrix} 2 \\ 0 \\ 1 \end{pmatrix} - \frac{-4}{5} \begin{pmatrix} -2 \\ 1 \\ 0 \end{pmatrix}$$

$$= \frac{1}{5} \begin{pmatrix} 2 \\ 4 \\ 5 \end{pmatrix}.$$

So, an orthogonal basis is

$$\left\{ \begin{pmatrix} -2 \\ 1 \\ 0 \end{pmatrix}, \begin{pmatrix} 2 \\ 4 \\ 5 \end{pmatrix} \right\}.$$

Therefore, after scaling each vector to be a unit vector, we have the orthonormal basis

$$\left\{ \frac{1}{\sqrt{5}} \begin{pmatrix} -2 \\ 1 \\ 0 \end{pmatrix}, \frac{1}{3\sqrt{5}} \begin{pmatrix} 2 \\ 4 \\ 5 \end{pmatrix} \right\}.$$

□

We conclude with two examples from the inner product space of polynomials on $[0, 1]$.

Example 7.2.28 We can find an orthogonal basis for $P_2([0, 1])$ by the Gram-Schmidt process using the standard basis $\{v_1 = 1, v_2 = x, v_3 = x^2\}$.

$$u_1 = v_1 = 1.$$

$$u_2 = v_2 - \frac{\langle u_1, v_2 \rangle}{\langle u_1, u_1 \rangle} u_1 = x - \frac{\int_0^1 (1)(x)dx}{\int_0^1 (1)(1)dx} 1 = x - \tfrac{1}{2}$$

$$u_3 = v_3 - \frac{\langle u_1, v_3 \rangle}{\langle u_1, u_1 \rangle} u_1 - \frac{\langle u_2, v_3 \rangle}{\langle u_2, u_2 \rangle} u_2$$

$$= x^2 - \frac{\int_0^1 (1)(x^2)dx}{\int_0^1 (1)(1)dx} 1 - \frac{\int_0^1 \left(x - \tfrac{1}{2} \right)(x^2)dx}{\int_0^1 \left(x - \tfrac{1}{2} \right)^2 dx} \left(x - \tfrac{1}{2} \right)$$

$$= x^2 - \tfrac{1}{3} - \tfrac{1}{2}(2x - 1)$$

$$= x^2 - x + \tfrac{1}{6}.$$

Thus, we have the orthogonal basis $B = \{1, 2x - 1, 6x^2 - 6x + 1\}$, where each vector was scaled to eliminate fractions. We can check orthogonality by computing the pairwise inner products.

$$\langle 1, 2x - 1 \rangle = \int_0^1 (2x - 1)dx = (x^2 - x) \Big|_0^1 = 0.$$

$$\left\langle 1, 6x^2 - 6x + 1 \right\rangle = \int_0^1 (6x^2 - 6x + 1)dx = (2x^3 - 3x^2 + 1) \Big|_0^1 = 0.$$

$$\left\langle 2x - 1, 6x^2 - 6x + 1 \right\rangle = \int_0^1 (12x^3 - 18x^2 + 8x - 1)dx$$

$$= \left(3x^4 - 6x^3 + 4x^2 - x\right)\Big|_0^1 = 0.$$

$\square$

Example 7.2.29 We can find an orthonormal basis for $P_2([0, 1])$ from the orthogonal basis found in Example 7.2.28: $B = \{u_1 = 1, u_2 = 2x - 1, u_3 = 6x^2 - 6x + 1\}$. We only need to normalize each basis vector. Let $v_1 = a_1 u_1$ where $a_1 \in \mathbb{R}$ and $\|v_1\| = 1$. That is, v_1 is the normalized basis vector which is a scalar multiple of u_1.

$$1 = \|v_1\| = a_1 \|u_1\| = a_1 \left(\int_0^1 (1)^2 dx\right)^{1/2} = a_1.$$

So, $a_1 = 1$. Similarly,

$$1 = \|v_2\| = a_2 \|u_2\| = a_2 \left(\int_0^1 (2x - 1)^2 dx\right)^{1/2} = \frac{a_2}{\sqrt{3}}.$$

So, $a_2 = \sqrt{3}$.

$$1 = \|v_3\| = a_3 \|u_3\| = a_3 \left(\int_0^1 (6x^2 - 6x + 1)^2 dx\right)^{1/2} = \frac{a_3}{\sqrt{5}}.$$

So, $a_3 = \sqrt{5}$. We have the orthonormal basis

$$\left\{1, \sqrt{3}(2x - 1), \sqrt{5}(6x^2 - 6x + 1)\right\}.$$

$\square$

☞ **Path to New Applications**
Researchers use Fourier analysis to find patterns in data and images. This method uses an orthonormal basis of sinusoidal functions to rewrite data vectors, using the Fourier transform, as a linear combination of these functions. The coefficients in the linear combination allow researchers to view data vectors as coordinate vectors in the Fourier space (the corresponding coordinate space). See Section 8.3.2 to read more about connections between linear algebra techniques and applications.

7.2.4 Exercises

1. In each of the following exercises, the set B is a basis of the vector space V, the subspace X is the span of some of the vectors in B, and w is a vector in V. Find the coordinate projection $\text{cproj}_X w$.

   (a) $V = \mathbb{R}^2$, $B = \left\{\begin{pmatrix} 1 \\ 5 \end{pmatrix}, \begin{pmatrix} 2 \\ 1 \end{pmatrix}\right\} \subseteq \mathbb{R}^2$, $X = \text{Span}\left\{\begin{pmatrix} 2 \\ 1 \end{pmatrix}\right\}$, $w = \begin{pmatrix} 1 \\ 1 \end{pmatrix}$.

(b) $V = \mathbb{R}^3$, $B = \left\{ \begin{pmatrix} 1 \\ 1 \\ 0 \end{pmatrix}, \begin{pmatrix} 1 \\ 0 \\ 1 \end{pmatrix}, \begin{pmatrix} 0 \\ 1 \\ 1 \end{pmatrix} \right\} \subseteq \mathbb{R}^3$, $X = \text{Span} \left\{ \begin{pmatrix} 1 \\ 0 \\ 1 \end{pmatrix}, \begin{pmatrix} 0 \\ 1 \\ 1 \end{pmatrix} \right\}$, $w = \begin{pmatrix} 1 \\ 0 \\ 0 \end{pmatrix}$.

(c) $V = P_2(\mathbb{R})$, $\quad B = \{1, 1 + x, 1 + x + x^2\} \subseteq P_2(\mathbb{R})$, $\quad X = \text{Span}\{1, 1 + x + x^2\}$, $\quad w = 2 - 3x - x^2$.

2. (a) Calculate the coordinate projection of $\text{cproj}_{W_2} v$ of v onto W_2 from Example 7.2.2. Repeat for Example 7.2.3 and Example 7.2.4.

   (b) Verify that $v = \text{cproj}_{W_1} v + \text{cproj}_{W_2} v$ in all three cases, even though in all three cases the coordinate projections were different.

3. For each of the following exercises, find the orthogonal projection $\text{proj}_X w$.

   (a) $X = \mathbb{R}^3$, $w = (-1, 2, 0)^\mathsf{T} \in \mathbb{R}^3$.

   (b) $X = \text{Span}\{(1, 0, 1)^\mathsf{T}, (0, 2, 0)^\mathsf{T}\} \subseteq \mathbb{R}^3$, $w = (1, 2, 3)^\mathsf{T} \in \mathbb{R}^3$.

   (c) $X = \text{Span}\{1 + x^2, x - 3\} \subseteq P_2(\mathbb{R})$, $w = 2 + 3x + 4x^2 \in P_2(\mathbb{R})$.

   (d) $X = \text{Span}\{1, x, x^2, x^3\} \subseteq \mathcal{C}([0, \pi])$, $w = \sin x \in \mathcal{C}([0, \pi])$, where $\mathcal{C}([0, \pi])$ is the vector space of continuous functions on $[0, \pi]$ and $\langle f(x), g(x) \rangle = \int_0^\pi f(x)g(x)dx$.

   (e)

$$X = \text{Span} \left\{ \text{Image A}, \text{Image B}, \text{Image C} \right\} \subseteq \mathcal{I}_{4 \times 4}(\mathbb{R}),$$

$$w = \text{Image 4} \in \mathcal{I}_{4 \times 4}(\mathbb{R}).$$

   (f)

$$X = \text{Span} \left\{ \quad , \quad , \quad \right\} \subseteq D(Z_2),$$

$$w = \quad \in D(Z_2).$$

Use the inner product $\langle u, v \rangle = 1$ if u and v have at least one lit bar in common, and $\langle u, v \rangle = 0$ otherwise.

4. Let $B = \{u_1, u_2\}$ be an orthogonal basis of $\mathbb{R}^2$, and let $W = \text{Span}\{u_1\}$. For $v \in \mathbb{R}^2$, what is the difference between the coordinate projection $\text{cproj}_W(v)$ and the orthogonal projection $\text{proj}_W(v)$? [Hint: what must be true about u_1 in order to have $\text{cproj}_W(v) = \text{proj}_W(v)$?]

5. Consider the inner product space V and subspace $W = \{0\}$. Describe the sets $V^\perp$ and $W^\perp$.

6. For each subspace, find orthonormal bases for W and for $W^\perp$.

(a) $W = \text{Span}\left\{ \begin{pmatrix} -1 \\ 1 \\ 0 \end{pmatrix}, \begin{pmatrix} 0 \\ -1 \\ 0 \end{pmatrix}, \begin{pmatrix} 1 \\ -1 \\ 1 \end{pmatrix} \right\} \subseteq \mathbb{R}^3$

(b) $W = \left\{ \begin{pmatrix} a \\ b \\ c \end{pmatrix} \middle| a + b + c = 0 \right\} \subseteq \mathbb{R}^3.$

(c) $W = \left\{ \begin{pmatrix} a \\ b \\ c \\ d \end{pmatrix} \middle| a + 2c = 0 \right\} \subseteq \mathbb{R}^4.$

(d) $W = \{a + bx + cx^2 \mid a = b\} \subseteq P_2(\mathbb{R}).$

7. Let $\mathbf{1}_n$ be the vector in $\mathbb{R}^n$ whose entries are all 1, and let L be the subspace of all linear combinations of $\mathbf{1}_n$. Show that for any vector u in $\mathbb{R}^n$, $\text{proj}_L u = \bar{u}\mathbf{1}$, where $\bar{u}$ is the mean value of the entries in u.

8. Given the basis $B = \{1, x, x^2, x^3\}$ for $P_3(\mathbb{R})$, and the inner product (note the limits of integration)

$$\langle p_1(x), p_2(x) \rangle = \int_{-1}^{1} p_1(x)p_2(x)dx.$$

Find an orthonormal basis for $P_3(\mathbb{R})$ which includes a scalar multiple of the first basis vector in B. Compare your result with the first four Legendre Polynomials.

9. Find the projections of the vector $x = (1, 2, 0, 1) \in \mathbb{R}^4$ onto each of the eigenspaces of the matrix operator

$$M = \begin{pmatrix} -4 & 1 & 1 & 2 \\ -9 & 2 & 1 & 6 \\ 0 & 0 & 1 & 0 \\ -3 & 1 & 1 & 1 \end{pmatrix}.$$

10. Formulate an inner product on $P_1(\mathbb{R})$ in which the standard basis $\{1, x\}$ is orthonormal.

11. Formulate an inner product on $P_2(\mathbb{R})$ in which the standard basis $\{1, x, x^2\}$ is orthonormal.

12. Prove Theorem 7.2.12.

13. Prove Corollary 7.2.20.

14. Prove Corollary 7.2.26.

15. Let V be an inner product space and $W \subseteq V$ be a subspace. Prove that $\dim W + \dim W^\perp = \dim V$.

16. Let V be an inner product space and suppose that $W \subseteq V$. Prove that $(\text{Span } W)^\perp = W^\perp$. In other words, adding linear combinations of vectors in W to W does not change the orthogonal complement.

17. Let V be an inner product space and $W \subseteq V$ be a subspace. Prove that $(W^\perp)^\perp = W$.

18. Let $V = \mathbb{R}^2$, $W = \left\{ \begin{pmatrix} 1 \\ 2 \end{pmatrix} \right\}$, and $v = \begin{pmatrix} 1 \\ 4 \end{pmatrix}$. Compute the orthogonal projection of v onto W. Compare with Example 7.2.4.

19. Let V be an inner product space and $W_1 \subseteq W_2 \subseteq V$. What can you say about $W_1^\perp$ and $W_2^\perp$? Explain.

20. Prove that if the columns of $n \times k$ matrix M form a linearly independent set, then $\text{rank}(MM^\mathsf{T}) = k$.

21. Modify the Gram-Schmidt process to find an orthogonal basis directly from a finite spanning set.

22. Consider $\mathbb{R}^2$ with the basis $B = \left\{ \begin{pmatrix} 1 \\ 1 \end{pmatrix}, \begin{pmatrix} a \\ b \end{pmatrix} \right\}$, and let $W = \mathrm{Span} \left\{ \begin{pmatrix} 1 \\ 1 \end{pmatrix} \right\}$. For any $v \in \mathbb{R}^2$, denote $v_1 = \mathrm{cproj}_W(v)$. In this exercise, we will explore the quantity $\|v_1\|/\|v\|$. In the introduction, we claimed that this quantity gives a measure of closeness of a vector to the subspace W. This exercise will demonstrate why geometrically it is preferable to use orthogonal projections rather than coordinate projections for this purpose.

(a) Set the basis vector $\begin{pmatrix} a \\ b \end{pmatrix} = \begin{pmatrix} 0 \\ 1 \end{pmatrix}$.

   i. Sketch the subspaces $W = \mathrm{Span} \left\{ \begin{pmatrix} 1 \\ 1 \end{pmatrix} \right\}$ and $X = \mathrm{Span} \left\{ \begin{pmatrix} a \\ b \end{pmatrix} \right\}$ on the same axes.

   ii. To develop some intuition for the next part of this problem, calculate $\|v_1\|/\|v\|$ for $v = \begin{pmatrix} 1 \\ 2 \end{pmatrix}$ and $v = \begin{pmatrix} 1 \\ 0 \end{pmatrix}$.

   iii. Find all vectors in $\mathbb{R}^2$ with the property that $\|v_1\|/\|v\| = 1$.

(b) Repeat part (a) with the basis vector $\begin{pmatrix} a \\ b \end{pmatrix} = \begin{pmatrix} 1 \\ -1 \end{pmatrix}$.

(c) Based on your observations, does the quantity $\|v_1\|/\|v\|$ give a better measurement of closeness of v to W in part (a) or in part (b)? Explain.

Exercises 23 through 32 further explore projections as transformations. Let V be an n-dimensional inner product space, W a subspace of V, $X = W^\perp$ and $v \in V$. Let $P_W : V \to V$ be the transformation defined by $P_W v = \mathrm{proj}_W v$.

23. Show that P is linear.
24. Show that $P_W^2 v = P_W v$.
25. Show that $\|P_W v\| \leq \|v\|$.
26. Show that $v = P_W v + P_X v$.
27. Show that $\mathrm{Null}(P_W) = W^\perp$.
28. Show that if $\dim W > 0$ then $\lambda = 1$ is an eigenvalue of P_W.
29. Show that if $\dim W < n$ then $\lambda = 0$ is an eigenvalue of P_W.
30. Show that $\lambda = 0$ and $\lambda = 1$ are the only eigenvalues of P_W.
31. Show that P_W is diagonalizable.
32. Show that if $\dim W < n$ then P_W is neither injective nor surjective.

7.3   Orthogonal Transformations

In our study of the heat diffusion transformation, we found that diagonalizing the matrix representation was a significant simplifying step in analyzing long-term behavior. Recall that, because the diffusion matrix E is diagonalizable, we were able to, for some initial heat state h_0, write the heat state k time steps later as

$$h_k = E^k h_0 = Q D^k Q^{-1} h_0,$$

where Q is the matrix of whose columns are eigenvectors of E and D is the diagonal matrix of the corresponding eigenvalues. The computation was much simpler because computing D^k consists only of raising the diagonal elements (eigenvalues) to the kth power, resulting in far fewer computations than those involving matrix multiplication.

   We also saw that the eigenvalues and eigenspaces of the transformation were much easier to visualize and interpret than the linear combinations that made up particular heat states. This approach led to a clearer understanding of the time evolution of heat signatures. We also noted that the diffusion matrix, Equation 4.3,

$$E = \begin{pmatrix} 1-2\delta & \delta & 0 & \cdots & & & 0 \\ \delta & 1-2\delta & \delta & & & & \\ 0 & \delta & \ddots & \ddots & & & \vdots \\ & & & & & \delta & 0 \\ \vdots & & & \ddots & \ddots & \delta & 0 \\ & & & & \delta & 1-2\delta & \delta \\ 0 & & \cdots & & 0 & \delta & 1-2\delta \end{pmatrix}, \qquad (7.6)$$

(where $0 < \delta < \frac{1}{4}$), is symmetric. This symmetry reflects the fact that heat flows in all directions equally (in our one-dimensional case, both left and right) scaled only by the magnitude of the heat gradient. In Exercise 18 of Section 6.2, we found that the normalized eigenvectors of E are

$$u_k = \sqrt{\frac{2}{m+1}} \left(\sin\frac{\pi k}{m+1}, \sin\frac{2\pi k}{m+1}, \cdots, \sin\frac{m\pi k}{m+1} \right)^{\mathsf{T}}, \quad k = 1, 2, \cdots, m \qquad (7.7)$$

and the corresponding eigenvalues are

$$\lambda_k = 1 - 2\delta \left(1 - \cos\frac{\pi k}{m+1} \right). \qquad (7.8)$$

We know that $\mathcal{U} = \{u_1, u_2, \cdots, u_m\}$ is an eigenbasis for $\mathbb{R}^m$. However, using topics from Sections 7.1 and 7.2, we can now show that $\mathcal{U}$ is an *orthonormal* basis for $\mathbb{R}^m$ (see Exercise 19).

In this section, we will focus on the relationship between the diagonalizability of a matrix and the symmetry of the corresponding transformation. We begin by examining the properties of matrices that have orthonormal columns.

7.3.1 Orthogonal Matrices

In Section 7.2, we employed orthogonal projections to create orthogonal and orthonormal bases using the Gram-Schmidt process. Now, suppose we have such a basis $\mathcal{U}$, and define Q to be the coordinate transformation matrix from the orthonormal basis $\mathcal{U}$ to the standard basis $\mathcal{S}$. That is, $Q[x]_{\mathcal{U}} = [x]_{\mathcal{S}}$ for any $x \in \mathbb{R}^n$.

Notice that if $\mathcal{U} = \{u_1, u_2, \ldots u_n\}$, then for $1 \le k \le n$, $Q[u_k]_{\mathcal{U}} = u_k$. But, $[u_k]_{\mathcal{U}} = e_k \in \mathbb{R}^n$. Therefore, the columns of Q must be the orthonormal basis vectors, $u_1, u_2, \ldots, u_k$ (written in standard coordinates). Next, we give a name to matrices whose columns are made up of orthonormal basis elements.

Definition 7.3.1

The $n \times n$ matrix Q with columns $u_1, u_2, \ldots, u_n$ is said to be **orthogonal** if $\mathcal{U} = \{u_1, u_2, \ldots, u_n\}$ is an orthonormal basis for $\mathbb{R}^n$. $\qquad\square$

The definition of an *orthogonal* matrix requires that the columns form an *orthonormal* basis, not just an *orthogonal* basis. For orthogonal matrix Q, we can write

$$Q = \begin{pmatrix} | & | & & | \\ u_1 & u_2 & \cdots & u_n \\ | & | & & | \end{pmatrix}, \quad \text{where} \quad \langle u_j, u_k \rangle = \begin{cases} 1 & \text{if } j = k \\ 0 & \text{if } j \neq k \end{cases}.$$

Lemma 7.3.2

Let Q be an $n \times n$ orthogonal matrix. Then $Q^{\mathsf{T}} Q = I_n$.

Proof. Suppose Q is an $n \times n$ orthogonal matrix with columns $u_1, u_2, \cdots, u_n$. We compute the jk-entry of $Q^{\mathsf{T}} Q$ to be

$$(Q^{\mathsf{T}} Q)_{jk} = \langle u_j, u_k \rangle = \begin{cases} 1 & \text{if } j = k \\ 0 & \text{if } j \neq k \end{cases}.$$

Thus, $Q^{\mathsf{T}} Q = I_n$. $\square$

Orthogonal matrices have several interesting properties with important interpretations as coordinate transformations. Recall that u_i, the ith column of Q, represents the image of e_i, the ith standard basis vector. Hence, as a transformation, $T_Q : \mathbb{R}^n \to \mathbb{R}^n$, defined by $T_Q(v) = Qv$, maps the orthonormal set $\{e_1, e_2 \ldots, e_n\}$ to the orthonormal set $\{u_1, u_2, \ldots, u_n\}$ as follows

$$e_1 \mapsto u_1,$$
$$e_2 \mapsto u_2,$$
$$\vdots$$
$$e_n \mapsto u_n.$$

Since T_Q preserves lengths of basis vectors and angles between them, it is reasonable to expect that it will preserve lengths of all vectors and angles between any two vectors. We formalize this result along with other algebraic properties in the following theorem.

Theorem 7.3.3

Let Q be an $n \times n$ real orthogonal matrix, x, y arbitrary vectors in $\mathbb{R}^n$ and $\theta_{x,y}$ the angle between x and y. Then

1. $Q^{-1} = Q^{\mathsf{T}}$,
2. Q^{T} is orthogonal,
3. $\|Qx\| = \|x\|$ (Q preserves vector lengths),
4. $\langle Qx, Qy \rangle = \langle x, y \rangle$,
5. $\cos \theta_{Qx, Qy} = \cos \theta_{x,y}$ (Q preserves angles between vectors), and
6. If λ is a (real) eigenvalue of Q, then $\lambda = \pm 1$.

Proof. Suppose Q is an $n \times n$ real orthogonal matrix and

$$x = \begin{pmatrix} x_1 \\ x_2 \\ \vdots \\ x_n \end{pmatrix}, y = \begin{pmatrix} y_1 \\ y_2 \\ \vdots \\ y_n \end{pmatrix} \in \mathbb{R}^n.$$

We write

$$Q = \begin{pmatrix} | & | & & | \\ u_1 & u_2 & \dots & u_n \\ | & | & & | \end{pmatrix},$$

where $\{u_1, u_2, \dots, u_n\}$ forms an orthonormal set in $\mathbb{R}^n$.

1. By Lemma 7.3.2, $Q^{\mathsf{T}} Q = I_n$. Thus, by the invertible matrix theorem (Theorem 5.2.20), $Q^{\mathsf{T}} = Q^{-1}$.
2. The proof of part 2 above is the subject of Exercise 16.
3. Using the result of part (1) and Theorem 7.1.5, we have

$$\begin{aligned} \|Qx\| &= \langle Qx, Qx \rangle^{1/2} \\ &= \left((Qx)^{\mathsf{T}} (Qx) \right)^{1/2} \\ &= \left(x^{\mathsf{T}} Q^{\mathsf{T}} Qx \right)^{1/2} \\ &= \left(x^{\mathsf{T}} I_n x \right)^{1/2} \\ &= \left(x^{\mathsf{T}} x \right)^{1/2} \\ &= \langle x, x \rangle^{1/2} \\ &= \|x\|. \end{aligned}$$

4. Again, using the result of part (1) and Theorem 7.1.5, we have

$$\begin{aligned} \langle Qx, Qy \rangle &= (Qx)^{\mathsf{T}} (Qy) \\ &= x^{\mathsf{T}} Q^{\mathsf{T}} Qy \\ &= x^{\mathsf{T}} y \\ &= \langle x, y \rangle. \end{aligned}$$

5. Using the results of parts (3) and (4), we have

$$\begin{aligned} \cos \theta_{Qx,Qy} &= \frac{\langle Qx, Qy \rangle}{\|Qx\| \|Qy\|} \\ &= \frac{\langle x, y \rangle}{\|x\| \|y\|} \\ &= \cos \theta_{x,y}. \end{aligned}$$

6. Suppose v is an eigenvector of Q with eigenvalue λ. Then, using the result of part (2), $\|v\| = \|Qv\| = \|\lambda v\| = |\lambda| \|v\|$. Thus $|\lambda| = 1$. As $\lambda \in \mathbb{R}$, $\lambda = \pm 1$.

$\square$

This theorem not only shows that vector norms and angles between vectors are preserved under left multiplication by Q but also we get the very nice result that the inverse of an orthogonal matrix is simply its transpose. This has very useful implications for the heat diffusion scenario. Since we found the diagonalization of the heat diffusion matrix as $E = QDQ^{-1}$ and we found Q to be an orthogonal matrix we apply Theorem 7.3.3 part 1 to write

$$E = QDQ^{-1} = QDQ^{\mathsf{T}}.$$

Example 7.3.4 Consider the orthonormal basis for $\mathbb{R}^2$

$$\mathcal{U} = \left\{ u_1 = \begin{pmatrix} 1/2 \\ \sqrt{3}/2 \end{pmatrix}, u_2 = \begin{pmatrix} -\sqrt{3}/2 \\ 1/2 \end{pmatrix} \right\}$$

and the orthogonal matrix

$$Q = \begin{pmatrix} 1/2 & -\sqrt{3}/2 \\ \sqrt{3}/2 & 1/2 \end{pmatrix}.$$

The reader should verify that Q in indeed orthogonal and that $Q^{\mathsf{T}} = Q^{-1}$. For arbitrary vector $x = \begin{pmatrix} a \\ b \end{pmatrix} \in \mathbb{R}^2$,

$$\begin{aligned}
\|Qx\| &= \left\| \begin{pmatrix} 1/2 & -\sqrt{3}/2 \\ \sqrt{3}/2 & 1/2 \end{pmatrix} \begin{pmatrix} a \\ b \end{pmatrix} \right\| \\
&= \left\| \frac{1}{2} \begin{pmatrix} a - \sqrt{3}b \\ \sqrt{3}a + b \end{pmatrix} \right\| = \left(a^2 + b^2 \right)^{1/2} = \|x\|.
\end{aligned}$$

Consider also the arbitrary vector $y = \begin{pmatrix} c \\ d \end{pmatrix} \in \mathbb{R}^2$. We have

$$\begin{aligned}
\langle Qx, Qy \rangle &= (Qx)^{\mathsf{T}}(Qy) \\
&= \frac{1}{4} \left(a - \sqrt{3}b \quad \sqrt{3}a + b \right) \begin{pmatrix} c - \sqrt{3}d \\ \sqrt{3}c + d \end{pmatrix} \\
&= \frac{1}{4} \left[\left(a - \sqrt{3}b \right) \left(c - \sqrt{3}d \right) + \left(\sqrt{3}a + b \right) \left(\sqrt{3}c + d \right) \right] \\
&= \frac{1}{4} \left[4ac + 4bd \right] \\
&= ac + bd \\
&= \langle x, y \rangle.
\end{aligned}$$

The matrix Q has no real eigenvalues. Indeed, if we choose as starting vector u_1 and employ the Factor Method for finding eigenvalues, we find

$$\{u_1, Qu_1, Q^2u_1\} = \left\{ \begin{pmatrix} 1/2 \\ \sqrt{3}/2 \end{pmatrix}, \begin{pmatrix} -1/2 \\ \sqrt{3}/2 \end{pmatrix}, \begin{pmatrix} -1 \\ 0 \end{pmatrix} \right\}.$$

We know that this set is linearly dependent because it has three vectors in $\mathbb{R}^2$. In fact, we have the linear dependence relation

$$0 = Q^2 u_1 + Q u_1 + u_1 = (Q^2 + Q + I) u_1.$$

Notice that if we try to factor $2Q^2 + Q + 1$, we find that it does not factor into something of the form $(aQ + bI)(cQ + dI)$ for any real numbers a, b, c, and d. Therefore, Q does not have any real eigenvalues. In fact, we have

$$Q^2 u_1 + Q u_1 + u_1 = \left(Q - \left(-\frac{1}{2} + \frac{i\sqrt{3}}{2} \right) \right) \left(Q - \left(-\frac{1}{2} - \frac{i\sqrt{3}}{2} \right) \right) u_1.$$

This means that there are two complex eigenvalues $\lambda = \frac{1}{2} \pm \frac{3}{2}i$. Notice that this situation does not violate the eigenvalue property stated in Theorem 7.3.3. However, the absence of two real eigenvalues does mean that Q is *not* diagonalizable.

As a coordinate transformation, $[x]_U = Q^\mathsf{T} x$, we observe that left multiplication by Q^T performs a rotation about the coordinate origin in $\mathbb{R}^2$ by an angle of $-\pi/3$. The reader should check this by plotting a vector in $\mathbb{R}^2$ and the result after multiplying by Q. □

Example 7.3.5 In this example, we construct a family of 2×2 orthogonal matrices M_θ for which left multiplication by M_θ of a vector in $\mathbb{R}^2$ results in a counterclockwise rotation of that vector through an angle θ. We build the transformation matrix column-by-column by transforming a set of basis vectors. Let $T_\theta : \mathbb{R}^2 \to \mathbb{R}^2$ be the transformation defined by $T_\theta(x) = M_\theta x$. If we work in the standard basis for $\mathbb{R}^2$, we have

$$M_\theta = \left(T_\theta \begin{pmatrix} 1 \\ 0 \end{pmatrix} \ \ T_\theta \begin{pmatrix} 0 \\ 1 \end{pmatrix} \right) = \begin{pmatrix} \cos\theta & -\sin\theta \\ \sin\theta & \cos\theta \end{pmatrix}.$$

If $\theta = \frac{\pi}{3}$ then M_θ is the matrix Q from Example 7.3.4. Notice that if $v = (1, 2)^\mathsf{T}$, we have

$$T_{\pi/3}(v) = \begin{pmatrix} \cos \pi/3 & -\sin \pi/3 \\ \sin \pi/3 & \cos \pi/3 \end{pmatrix} \begin{pmatrix} 1 \\ 2 \end{pmatrix} = \begin{pmatrix} 1/2 & -\sqrt{3}/2 \\ \sqrt{3}/2 & 1/2 \end{pmatrix} \begin{pmatrix} 1 \\ 2 \end{pmatrix} = \begin{pmatrix} 1/2 - \sqrt{3} \\ 1 + \sqrt{3}/2 \end{pmatrix}.$$

In Figure 7.18, we see that T transforms v by rotating it counterclockwise by an angle of $\frac{\pi}{3}$. □

Example 7.3.6 Consider the orthonormal basis for $\mathbb{R}^2$

$$\mathcal{U} = \left\{ u_1 = \begin{pmatrix} 1/2 \\ \sqrt{3}/2 \end{pmatrix}, u_2 = \begin{pmatrix} \sqrt{3}/2 \\ -1/2 \end{pmatrix} \right\}$$

and the orthogonal matrix

$$Q = \begin{pmatrix} 1/2 & \sqrt{3}/2 \\ \sqrt{3}/2 & -1/2 \end{pmatrix}.$$

The reader should verify the properties of Theorem 7.3.3. Even though Q looks quite similar to the matrix in Example 7.3.4, we will see that this orthogonal matrix does not represent a rotation. The eigenvalues of Q are $\lambda_1 = 1$ and $\lambda_2 = -1$. The eigenspaces associated with each eigenvalue are

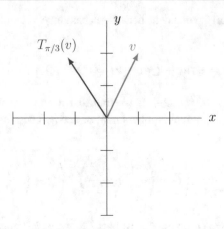

Fig. 7.18 $T_{\pi/3}$ transforms v by rotating it by $\frac{\pi}{3}$ about the origin.

$$\mathcal{E}_1 = \left\{ \alpha \begin{pmatrix} \sqrt{3} \\ 1 \end{pmatrix} \ \middle| \ \alpha \in \mathbb{R} \right\},$$

$$\mathcal{E}_2 = \left\{ \alpha \begin{pmatrix} -1 \\ \sqrt{3} \end{pmatrix} \ \middle| \ \alpha \in \mathbb{R} \right\}.$$

Notice that $\mathcal{E}_1$ and $\mathcal{E}_2$ are orthogonal subspaces and $\mathbb{R}^2 = \mathcal{E}_1 \oplus \mathcal{E}_2$. So, every $v \in \mathbb{R}^2$ can be uniquely expressed as $v = v_1 + v_2$ where $v_1 \in \mathcal{E}_1$ and $v_2 \in \mathcal{E}_2$. Notice also that $Qv = Qv_1 + Qv_2 = v_1 - v_2$. That is, Q represents a reflection transformation about $\mathcal{E}_1$. Exercise 5 requests that the reader creates a visualization of this result.   □

Example 7.3.7 We can construct a family of 2×2 orthogonal matrices M_u for which left multiplication of M_u by a vector in $\mathbb{R}^2$ results in a reflection of that vector about the line $L_u = \mathrm{Span}\{u\} = \{\alpha u \mid \alpha \in \mathbb{R}\}$. Let u be the unit vector $u = (a, b)^\mathsf{T}$. Then, we have (see Exercise 15)

$$M_u = \begin{pmatrix} a^2 - b^2 & 2ab \\ 2ab & b^2 - a^2 \end{pmatrix}.$$

                                                                     □

Next, we explore the diagonalizability of rotation transformations in $\mathbb{R}^2$.

Example 7.3.8 Consider the family of rotation matrices for vectors in $\mathbb{R}^2$

$$M_\theta = \begin{pmatrix} \cos\theta & -\sin\theta \\ \sin\theta & \cos\theta \end{pmatrix}.$$

By considering when $M_\theta - \lambda I$ is not invertible, we can see that any eigenvalues λ of M_θ are zeros of $C(\lambda) = (\cos\theta - \lambda)^2 + \sin^2\theta$. So, we have $\lambda = \cos\theta \pm \sqrt{\cos^2\theta - 1}$, with real-valued solutions when $\theta = k\pi$, $k \in \mathbb{Z}$, of $\lambda = \cos k\pi = (-1)^k$. The eigenspaces are of dimension 2. However, if θ is not an integer multiple of π, then real eigenvalues do not exist. Thus, by the results of Section 6.2.5, M_θ is diagonalizable, if and only if, $\theta = k\pi$, for some $k \in \mathbb{Z}$.   □

7.3.2 Orthogonal Diagonalization

We have considered transformation matrices whose columns form an orthonormal basis of $\mathbb{R}^n$. In this section, we explore matrices whose eigenvectors form an orthonormal basis. Transformations that are diagonalizable with a set of orthogonal eigenvectors have special properties and deserve a special name.

Definition 7.3.9

We say that matrix M is **orthogonally diagonalizable** if

$$M = QDQ^{\mathsf{T}},$$

where Q is orthogonal and D is diagonal. $\qquad\square$

Orthogonally diagonalizable transformations are decomposed into three successive transformations $M = QDQ^{\mathsf{T}}$: an orthogonal coordinate transformation (Q^{T}), followed by a coordinate scaling (D), and ending with the inverse coordinate transformation (Q). The heat diffusion transformation is an example of an orthogonally diagonalizable transformation. We will explore this more in Section 7.3.4.

For now, because the heat diffusion transformation is also symmetric, we want to answer a key question: What is the relationship between symmetric matrices and orthogonally diagonalizable transformations? It turns out that they are the same. In fact, a necessary and sufficient condition[5] for a matrix to be orthogonally diagonalizable is that it be symmetric. We will prove this beautiful and surprising fact, known as the Spectral Theorem, using the following two lemmas.

Lemma 7.3.10
Let M be a real $n \times n$ symmetric matrix and $u, v \in \mathbb{R}^n$. Then $\langle Mu, v \rangle = \langle u, Mv \rangle$.

Proof. Suppose M is a real symmetric matrix and let $u, v \in \mathbb{R}^n$. Then, $M^{\mathsf{T}} = M$. So, we have $\langle Mu, v \rangle = (Mu)^{\mathsf{T}} v = u^{\mathsf{T}} M^{\mathsf{T}} v = u^{\mathsf{T}} Mv = \langle u, Mv \rangle$. $\qquad\square$

In fact, the converse of this theorem is also true: Given a real-valued $n \times n$ matrix M, if for all u, v in $\mathbb{R}^n$, $\langle Mu, v \rangle = \langle u, Mv \rangle$, then M is symmetric. The proof is the subject of Exercise 21.

Lemma 7.3.11
Let M be a real $n \times n$ symmetric matrix. Then M has n real eigenvalues, not necessarily distinct.

Proof. Suppose M is a real $n \times n$ symmetric matrix. Suppose v is a nonzero vector such that $Mv = \lambda v$ for some, possibly complex, λ. By Theorem 6.2.18, M has n, possibly complex and not necessarily

---

[5] A mathematician will use the words "necessary and sufficient condition" for a result to indicate that the condition is equivalent to the result. That is, the result is true if and only if the condition is true. Condition A is called **necessary** for result R if R implies A. Condition A is called **sufficient** for result R if A implies R.

distinct, eigenvalues. We need only show that all n eigenvalues are real. Using the result of Exercise 24 of Section 7.1, we have

$$\|Mv\|^2 = \langle Mv, Mv \rangle = \overline{(Mv)}^\mathsf{T} Mv = \overline{v}^\mathsf{T} \overline{M}^\mathsf{T} Mv = \overline{v}^\mathsf{T} M^2 v = \lambda^2 \overline{v}^\mathsf{T} v = \lambda^2 \|v\|^2.$$

So, $\lambda^2 = \|Mv\|^2 / \|v\|^2 \in \mathbb{R}$ and $\lambda^2 \geq 0$. Thus, $\lambda \in \mathbb{R}$. That is, M has n real eigenvalues. $\square$

Theorem 7.3.12 [Spectral Theorem]
A real matrix M is orthogonally diagonalizable if and only if M is symmetric.

Proof ($\Rightarrow$). Let M be a real matrix that is orthogonally diagonalizable. Then, there exists orthogonal matrix Q and diagonal matrix D so that $M = QDQ^\mathsf{T}$. Then,

$$M^\mathsf{T} = (QDQ^\mathsf{T})^\mathsf{T} = (Q^\mathsf{T})^\mathsf{T} D^\mathsf{T} Q^\mathsf{T} = QDQ^\mathsf{T} = M.$$

Therefore M is symmetric.

($\Leftarrow$) Now, let M be a real $n \times n$ symmetric matrix. We will show, by induction on n, that M is orthogonally diagonalizable. First observe that every 1×1 matrix M is orthogonally diagonalizable. It remains to show that every real $n \times n$ symmetric matrix is orthogonally diagonalizable if every real $(n - 1) \times (n - 1)$ symmetric matrix is orthogonally diagonalizable.

Suppose every real $(n - 1) \times (n - 1)$ symmetric matrix is orthogonally diagonalizable. Let M be a real $n \times n$ symmetric matrix. By Lemma 7.3.11, M has n real eigenvalues. Let λ be an eigenvalue of M with corresponding unit eigenvector v. Let $W = \text{Span}\{v\}$. Suppose $w \in W^\perp$, then, using Lemma 7.3.10,

$$\langle v, Mw \rangle = \langle Mv, w \rangle = \langle \lambda v, w \rangle = \lambda \langle v, w \rangle.$$

So, $\langle v, Mw \rangle = 0$ whenever $\langle v, w \rangle = 0$, and $Mw \in W^\perp$ whenever $w \in W^\perp$. Let $u, w \in W^\perp$ and consider transformation $T : W^\perp \to W^\perp$ defined by $T(u) = Mu$ for all $u \in W^\perp$.

$$\langle u, T(w) \rangle = \langle u, Mw \rangle = \langle Mu, w \rangle = \langle T(u), w \rangle.$$

As a transformation from $W^\perp$ to $W^\perp$, we can fix a basis for $W^\perp$ and represent T by a real $(n - 1) \times (n - 1)$ dimensional matrix. By Problem 21, this matrix is symmetric, so by the induction hypothesis it is orthogonally diagonalizable. So, there exists an orthonormal basis β of eigenvectors for $W^\perp$. Thus, an orthonormal basis of eigenvectors of M for $\mathbb{R}^n$ is $\beta \cup \{v\}$. $\square$

The Spectral Theorem 7.3.12 tells us that symmetry is a necessary and sufficient condition for a real matrix to be orthogonally diagonalizable. In particular, we have the following two facts.

1. If a real matrix M is symmetric, then M is orthogonally diagonalizable. That is, symmetry is a **sufficient condition** for a real matrix to be orthogonally diagonalizable.
2. If a real matrix M is orthogonally diagonalizable, then M is symmetric. That is, symmetry is a **necessary condition** for a real matrix to be orthogonally diagonalizable.

> **Corollary 7.3.13**
>
> Suppose M is a real $n \times n$ symmetric matrix. Then, M has k distinct eigenvalues with associated orthogonal eigenspaces $\mathcal{E}_1, \ldots, \mathcal{E}_k$ for which $\mathbb{R}^n = \mathcal{E}_1 \oplus \ldots \oplus \mathcal{E}_k$.

The proof is the subject of Exercise 17.

Example 7.3.14 Consider the symmetric matrix

$$M = \begin{pmatrix} -2 & 1 & 1 \\ 1 & -2 & 1 \\ 1 & 1 & -2 \end{pmatrix}.$$

We can show that M has $n = 3$ real eigenvalues and is orthogonally diagonalizable. The distinct eigenvalues of M are $\lambda_1 = 0$ and $\lambda_2 = -3$. Because $\lambda_1 = 0$, $\mathcal{E}_1 = \text{Null}(M)$. A basis for $\mathcal{E}_1$ is $\{v_1 = (1, 1, 1)^\mathsf{T}\}$. A basis for $\mathcal{E}_2$ is $\{v_2 = (-1, 1, 0)^\mathsf{T}, v_3 = (-1, 0, 1)^\mathsf{T}\}$. Notice that v_1 is orthogonal to both v_2 and v_3 as expected by Corollary 7.3.13. However, v_2 and v_3 are not orthogonal. We use the Gram-Schmidt process to find an orthonormal basis for $\mathbb{R}^3$ from the eigenvectors of M. We find the orthogonal eigenbasis $\{(1, 1, 1)^\mathsf{T}, (-1, 1, 0)^\mathsf{T}, (-1, -1, 2)^\mathsf{T}\}$. Notice that the first vector is in $\mathcal{E}_1$ and the second and third vectors are in $\mathcal{E}_2$. Normalizing each vector in the basis yields an orthonormal basis, which we express as columns in the (orthogonal) matrix Q:

$$Q = \begin{pmatrix} \frac{1}{\sqrt{3}} & \frac{-1}{\sqrt{2}} & \frac{-1}{\sqrt{6}} \\ \frac{1}{\sqrt{3}} & \frac{1}{\sqrt{2}} & \frac{-1}{\sqrt{6}} \\ \frac{1}{\sqrt{3}} & 0 & \frac{2}{\sqrt{6}} \end{pmatrix}.$$

With diagonal matrix

$$D = \begin{pmatrix} 0 & 0 & 0 \\ 0 & -3 & 0 \\ 0 & 0 & -3 \end{pmatrix},$$

we have that M is orthogonally diagonalizable as $M = QDQ^\mathsf{T}$. $\qquad\square$

Example 7.3.15 Consider the transformation $T : \mathcal{P}_2(\mathbb{R}) \to \mathcal{P}_2(\mathbb{R})$ defined by $T(f(x)) = \frac{d}{dx}(xf(x))$. Let $\beta = \{x^2 + 1, x^2 - 1, x\}$, a basis for $\mathcal{P}_2(\mathbb{R})$. We have

$$T(\beta_1) = 3x^2 + 1 = 2\beta_1 + \beta_2,$$
$$T(\beta_2) = 3x^2 - 1 = \beta_1 + 2\beta_2,$$
$$T(\beta_3) = x = \beta_3,$$

$$M = [T]_\beta = \begin{pmatrix} 2 & 1 & 0 \\ 1 & 2 & 0 \\ 0 & 0 & 1 \end{pmatrix}.$$

The transformation is symmetric, so there exists an orthogonal diagonalization of M. For this example, $M = QDQ^\mathsf{T}$ where Q is the orthogonal matrix

$$Q = \begin{pmatrix} \frac{1}{\sqrt{2}} & -\frac{1}{\sqrt{2}} & 0 \\ \frac{1}{\sqrt{2}} & \frac{1}{\sqrt{2}} & 0 \\ 0 & 0 & 1 \end{pmatrix},$$

and D is the diagonal matrix

$$D = \begin{pmatrix} 3 & 0 & 0 \\ 0 & 1 & 0 \\ 0 & 0 & 1 \end{pmatrix}.$$

Notice that Q can also be expressed as

$$Q = \begin{pmatrix} \cos\theta & -\sin\theta & 0 \\ \sin\theta & \cos\theta & 0 \\ 0 & 0 & 1 \end{pmatrix}, \text{ for } \theta = \pi/4,$$

which is a rotation in the first two coordinates (see Example 7.3.5) and the identity transformation in the third coordinate. The transformation T, expressed as the orthogonal diagonalization in basis β, is the result of an orthogonal (rotation) transformation (Q^T) followed by a coordinate scaling (D) and then by the inverse rotation (Q).   □

7.3.3   Completing the Invertible Matrix Theorem

Recall that one of our main goals in this text stems from the following scenario. We are studying a process that can be represented by a linear transformation (e.g., radiographic processes). Suppose we have an output of this process and we seek the particular input that gave this output. Then, we are essentially solving a matrix equation $Mx = b$. If M is invertible, we find the input as $x = M^{-1}b$.

One of the important consequences of the Spectral Theorem 7.3.12 is another characterization of the invertibility of a matrix. Theorem 7.3.16, below, collects all of the characterizations that we have been collecting throughout this text; part (o) is the new statement for which the equivalence with part (n) can be proven using the Spectral Theorem.

Theorem 7.3.16 [Invertible Matrix Theorem]
Let M be an $n \times n$ real matrix with columns $c_1, c_2, \ldots, c_n \in \mathbb{R}^n$ and rows $r_1^\mathsf{T}, r_2^\mathsf{T}, \ldots, r_n^\mathsf{T}$. Then the following statements are equivalent.

    (a) M is invertible.
    (b) The reduced echelon form of M has n leading ones.
    (c) The matrix equation $Mx = 0$ has a unique solution.
    (d) For every $b \in \mathbb{R}^n$, the matrix equation $Mx = b$ has a unique solution.
    (e) The homogeneous system of equations with coefficient matrix M has a unique solution.

(f) The set $\{c_1, c_2, \ldots, c_n\}$ is linearly independent.

(g) The set $\{c_1, c_2, \ldots, c_n\}$ is a basis for $\mathbb{R}^n$.

(h) $\det(M) \neq 0$.

(i) $\text{Rank}(M) = n$.

(j) $\text{Nullity}(M) = 0$.

(k) $\{r_1, r_2, \ldots, r_n\}$ is linearly independent.

(l) $\{r_1, r_2, \ldots, r_n\}$ is a basis for $\mathbb{R}^n$.

(m) M^{T} is invertible.

(n) All eigenvalues of M are nonzero.

(o) $M^{\mathsf{T}}M$ has n (not necessarily distinct) positive eigenvalues.

The proof of Theorem 7.3.16 is a result of Theorems 5.2.20 and 6.2.25 and the result of Exercise 12. Now, we return to the heat diffusion example.

7.3.4 Symmetric Diffusion Transformation

The heat diffusion matrix, Equation (7.6) is symmetric. It is also orthogonally diagonalizable, as asserted by the Spectral Theorem 7.3.12 and verified in Exercise 19. We investigate many of the results in this section in the context of a specific example heat state with $m = 3$. For this numerical example, we let $\delta = 1/5$ so the diffusion matrix is

$$E = \begin{pmatrix} 3/5 & 1/5 & 0 \\ 1/5 & 3/5 & 1/5 \\ 0 & 1/5 & 3/5 \end{pmatrix}.$$

The eigenvalues are $\lambda_k = 1 - \frac{2}{5}\left(1 - \cos\frac{\pi k}{4}\right) = \frac{3}{5} + \frac{2}{5}\cos\frac{\pi k}{4}$. We have

$$\lambda_1 = \frac{3 + \sqrt{2}}{5} \approx 0.883\,,$$

$$\lambda_2 = \frac{3}{5} = 0.600\,,$$

$$\lambda_3 = \frac{3 - \sqrt{2}}{5} \approx 0.317\,.$$

The eigenvalues are real (see Lemma 7.3.11) and, in this instance, distinct. The corresponding normalized eigenvectors are

$$u_1 = \frac{1}{\sqrt{2}}\begin{pmatrix} \sin\frac{\pi}{4} \\ \sin\frac{\pi}{2} \\ \sin\frac{3\pi}{4} \end{pmatrix} = \begin{pmatrix} 1/2 \\ 1/\sqrt{2} \\ 1/2 \end{pmatrix},$$

$$u_2 = \frac{1}{\sqrt{2}}\begin{pmatrix} \sin\frac{2\pi}{4} \\ \sin\frac{4\pi}{4} \\ \sin\frac{6\pi}{4} \end{pmatrix} = \begin{pmatrix} 1/\sqrt{2} \\ 0 \\ -1/\sqrt{2} \end{pmatrix},$$

$$u_3 = \frac{1}{\sqrt{2}} \begin{pmatrix} \sin\frac{3\pi}{4} \\ \sin\frac{6\pi}{4} \\ \sin\frac{9\pi}{4} \end{pmatrix} = \begin{pmatrix} 1/2 \\ -1/\sqrt{2} \\ 1/2 \end{pmatrix}.$$

The reader can verify that these vectors form an orthonormal basis for $\mathbb{R}^3$ *and* that they are eigenvectors of E. The orthogonal diagonalization is $E = QDQ^\mathsf{T}$ where

$$D = \begin{pmatrix} \lambda_1 & 0 & 0 \\ 0 & \lambda_2 & 0 \\ 0 & 0 & \lambda_3 \end{pmatrix} = \begin{pmatrix} \frac{3+\sqrt{2}}{5} & 0 & 0 \\ 0 & \frac{3}{5} & 0 \\ 0 & 0 & \frac{3-\sqrt{2}}{5} \end{pmatrix}$$

and

$$Q = \begin{pmatrix} | & | & | \\ u_1 & u_2 & u_3 \\ | & | & | \end{pmatrix} = \begin{pmatrix} 1/2 & 1/\sqrt{2} & 1/2 \\ 1/\sqrt{2} & 0 & -1/\sqrt{2} \\ 1/2 & -1/\sqrt{2} & 1/2 \end{pmatrix}.$$

Next, consider the particular heat state, represented in the standard basis,

$$v = \begin{pmatrix} 1 \\ 1 \\ 2 \end{pmatrix}.$$

The heat state, w, resulting from a single time step in the heat diffusion process is given by left multiplication by E.

$$w = Ev = \begin{pmatrix} 4/5 \\ 6/5 \\ 7/5 \end{pmatrix}.$$

One can also view this transformation as a sequence of more elementary operations. First, the heat state is written as a linear combination of the eigenvectors of E. The coordinates in this basis are computed as $a = [v]_U$ where $a_k = \langle u_k, v \rangle$ (see Corollary 7.1.31).

$$[v]_U = Q^\mathsf{T}v = \begin{pmatrix} \langle u_1, v \rangle \\ \langle u_2, v \rangle \\ \langle u_3, v \rangle \end{pmatrix} = \begin{pmatrix} \frac{3+\sqrt{2}}{2} \\ \frac{-1}{\sqrt{2}} \\ \frac{3-\sqrt{2}}{2} \end{pmatrix}.$$

This orthogonal change in coordinates should preserve the vector norm. We can compute and compare $\|v\|$ and $\|[v]_U\|$, or equivalently $\langle v, v \rangle$ and $\langle [v]_U, [v]_U \rangle$:

$$\langle v, v \rangle = 1^2 + 1^2 + 2^2 = 6,$$

$$\langle [v]_U, [v]_U \rangle = \left(\tfrac{3+\sqrt{2}}{2}\right)^2 + \left(\tfrac{-1}{\sqrt{2}}\right)^2 + \left(\tfrac{3-\sqrt{2}}{2}\right)^2 = \tfrac{11}{4} + \sqrt{2} + \tfrac{1}{2} + \tfrac{11}{4} - \sqrt{2} = \tfrac{24}{4} = 6.$$

Second, the coordinates of $[v]_U$ are scaled through left multiplication by D.

$$D[v]_U = \begin{pmatrix} \frac{3+\sqrt{2}}{5} \cdot \frac{3+\sqrt{2}}{2} \\ \frac{3}{5} \cdot \frac{-1}{\sqrt{2}} \\ \frac{3-\sqrt{2}}{5} \cdot \frac{3-\sqrt{2}}{2} \end{pmatrix} = \begin{pmatrix} \frac{11+6\sqrt{2}}{10} \\ \frac{-3}{5\sqrt{2}} \\ \frac{11-6\sqrt{2}}{10} \end{pmatrix}.$$

Finally, this vector is written in terms of the original coordinates by left multiplication by $\left(Q^{\mathsf{T}}\right)^{-1} = Q$.

$$\begin{aligned} w &= [D[v]_U]_S \\ &= Q\left(D[v]_U\right) \\ &= \tfrac{11+6\sqrt{2}}{10}u_1 + \tfrac{-3}{5\sqrt{2}}u_2 + \tfrac{11-6\sqrt{2}}{10}u_3 \\ &= \begin{pmatrix} 4/5 \\ 6/5 \\ 7/5 \end{pmatrix}. \end{aligned}$$

The reader should verify that this final orthogonal transformation preserves the vector norm (that is, $\|w\| = \|D[v]_U\|$).

As a final observation, consider the fact that $\|w\| < \|v\|$. The action of the heat diffusion transformation is decomposed into three successive transformations, two of which preserve vector norm, and a third that performs a coordinate-wise vector scaling by the eigenvalues of the transformation. As the eigenvalues of E all satisfy $0 < \lambda_k < 1$, $\|Ev\| < \|v\|$ for any nonzero v.

7.3.5   Exercises

1. For each of the following matrices, determine whether or not A is an orthogonal matrix. If not, find a matrix whose column space is the same as the column space of A, but is orthogonal.

   (a) $A = \begin{pmatrix} 1/2 & -1/2 \\ 1/2 & 1/2 \end{pmatrix}$

   (b) $A = \begin{pmatrix} 1 & 1 & 1 \\ 2 & -2 & 0 \\ 1 & 0 & -1 \end{pmatrix}$

   (c) $A = \begin{pmatrix} 0 & 1 & 0 \\ 1 & 0 & 0 \\ 0 & 0 & 1 \end{pmatrix}$

   (d) $A = \begin{pmatrix} \frac{1}{2} & \frac{1}{\sqrt{2}} & -\frac{1}{2} \\ \frac{1}{\sqrt{2}} & 0 & \frac{1}{\sqrt{2}} \\ -\frac{1}{2} & \frac{1}{\sqrt{2}} & \frac{1}{2} \end{pmatrix}.$

2. Suppose Q and R are $n \times n$ orthogonal matrices. Is QR an orthogonal matrix? Justify.

3. Show that the matrix M of Example 7.3.14 is not invertible.

4. Let $A_\theta = \begin{pmatrix} \sin\theta & \cos\theta \\ \cos\theta & -\sin\theta \end{pmatrix}.$

   (a) For what values of θ is A_θ orthogonal?

   (b) Draw the vector $u = \begin{pmatrix} 1 \\ 1 \end{pmatrix}$ and its images under the transformation $T(u) = A_\theta u$ for $\theta = 45°, 90°, 30°, 240°$.

   (c) Use your results to state what multiplication by A_θ is doing geometrically.

5. Draw a figure that shows that the transformation defined in Example 7.3.6 does not represent a rotation, but is in fact the reflection stated in the example. (Hint: To where do e_1 and e_2 map?)

6. Provide an orthogonal diagonalization of the following symmetric matrices.

(a) $M = \begin{pmatrix} 2 & 1 \\ 1 & -1 \end{pmatrix}$.

(b) $B = \begin{pmatrix} 0 & 2 \\ 2 & 0 \end{pmatrix}$.

(c) $C = \begin{pmatrix} \sqrt{3} & 1 & 0 \\ 1 & -\sqrt{3} & 0 \\ 0 & 0 & 1 \end{pmatrix}$.

(d) $R = \begin{pmatrix} 0 & 1 & 2 \\ 1 & -2 & 1 \\ 2 & 1 & 0 \end{pmatrix}$.

7. Is the set of all 3×3 orthogonal matrices a subspace of $\mathcal{M}_{3\times 3}(\mathbb{R})$? Justify.

8. Use geometric arguments to show that the counterclockwise rotation transformation in Example 7.3.8 is diagonalizable only if $\theta = n\pi$, $n \in \mathbb{Z}$.

9. Show that the reflection transformation (about a line) in $\mathbb{R}^2$ has two distinct eigenvalues λ_1 and λ_2 for which the dimension of each of the corresponding eigenspaces has dimension 1.

10. Use the family of rotation matrices of Example 7.3.5 and Corollary 4.2.27 to derive the angle sum formulae and triple angle formulae for sine and cosine:

$$\cos(\theta + \phi) = \cos\theta \cos\phi - \sin\theta \sin\phi,$$

$$\sin(\theta + \phi) = \cos\theta \sin\phi + \cos\phi \sin\theta,$$

$$\cos 3\theta = 4\cos^3\theta - 3\cos\theta,$$

$$\sin 3\theta = 3\sin\theta - 4\sin^3\theta.$$

11. Show that the transformation in Example 7.3.15 preserves angles between vectors but does not preserve vector norm. Explain this result.

12. Show that for every real matrix A, both AA^T and $A^\mathsf{T}A$ are orthogonally diagonalizable. What are the necessary and sufficient conditions on A for AA^T to be invertible? For $A^\mathsf{T}A$ to be invertible?

13. Let Q be an $n \times n$ orthogonal matrix. Prove that $QQ^\mathsf{T} = I_n$. (Notice the difference between this claim and Lemma 7.3.2.)

14. Is the converse of Lemma 7.3.2 true? That is, if $Q^\mathsf{T}Q = I_n$ is Q necessarily an orthogonal matrix? This question is asking whether or not the condition $Q^\mathsf{T}Q = I$ is a **necessary and sufficient condition** for a Q to be orthogonal. Prove or give a counter example.

15. Show that left multiplication by the matrix M_u given in Example 7.3.7 performs the indicated reflection transformation.

16. Complete the proof of Theorem 7.3.3.

17. Prove Corollary 7.3.13. Notice that because every symmetric matrix is orthogonally diagonalizable, $0 \le k \le n$.

18. Show that the nullspace of the heat diffusion transformation is the trivial subspace.

19. Verify that the eigenvectors of the heat diffusion transformation, given in Equation 7.7, form an orthonormal set. In particular, show that

$$\langle u_j, u_k \rangle = \begin{cases} 0 & \text{if } j \ne k \\ 1 & \text{if } j = k \end{cases}.$$

20. Verify that the eigenvalues of the heat diffusion transformation are those given in Equation 7.8.
21. Prove the converse of Lemma 7.3.10.

The following exercises explore properties of transformations on complex-valued vector spaces.

22. Consider transformation $T : \mathbb{C}^n \to \mathbb{C}^n$ defined by left multiplication by matrix $M \in \mathcal{M}_{n \times n}(\mathbb{C})$. The **adjoint** of M, written M^*, is $M^* = \overline{M}^\mathsf{T}$. A matrix equal to its adjoint, $M = M^*$, is said to be **self-adjoint** or **Hermitian**. Show that every real symmetric matrix is self-adjoint.
23. Prove that every eigenvalue of a self-adjoint transformation is real.
24. Consider transformation $T : \mathbb{C}^n \to \mathbb{C}^n$ and define $T_1 = \frac{1}{2}(T + T^*)$ and $T_2 = \frac{1}{2i}(T - T^*)$. Notice that $T = T_1 + iT_2$. Prove that T_1 and T_2 are self-adjoint.
25. A matrix $M \in \mathcal{M}_{n \times n}(\mathbb{C})$ is said to be **normal** if $MM^* = M^*M$. Show that every real symmetric matrix is normal. Show, by specific example, that not every normal matrix is symmetric.
26. A matrix $Q \in \mathcal{M}_{n \times n}(\mathbb{C})$ is said to be **unitary** if $Q^{-1} = Q^*$. Prove that if Q is a unitary matrix, then $\|Qx\| = \|x\|$ for all $x \in \mathbb{C}^n$.
27. A matrix $M \in \mathcal{M}_{n \times n}(\mathbb{C})$ is said to be **unitarily diagonalizable** if $M = QDQ^*$, where Q is unitary and $D \in \mathcal{M}_{n \times n}(\mathbb{C})$ is diagonal. Prove that matrix M is normal if, and only if, M is unitarily diagonalizable.

7.4 Exploration: Pseudo-Inverting the Non-invertible

Our study of the heat diffusion process leads us to the discovery of a variety of important linear algebra topics. We found that for some linear transformations $T : V \to V$, there are entire subspaces of eigenvectors that are transformed only by multiplication by a scalar (eigenvalue). Whenever we are able to find a basis (for V) of such vectors, transformation computations become almost trivial because we represent vectors as coordinate vectors relative to that basis. The key step is to recognize that such transformations are diagonalizable—relative to an eigenbasis, the matrix representation is diagonal. This led to our ability to perform repeated application of the transformation quickly and allows us to study long-term behavior. We also noticed that the heat diffusion process is represented as a symmetric matrix and our investigation in Section 7.3 led us to see that *every* real symmetric matrix is diagonalizable.

Meanwhile, we are still investigating how we might "invert" the radiographic transformation T, a transformation that is typically not one-to-one. The matrix representation M—relative to any basis—is typically non-square and therefore not invertible. However, in our study of transformation spaces in Section 5.1, we came very close to discovering an isomorphism between the column space and row space of a matrix. In this section, we will revisit this idea as the Maximal Isomorphism Theorem, then learn how this theorem provides a way to recover any non-invisible object from the corresponding radiograph.

7.4.1 Maximal Isomorphism Theorem

The transformation that allows us to solve $T(x) = b$ when T is not invertible is known as the pseudo-inverse and is defined as follows.

Definition 7.4.1

Let V and W be vector spaces and $T : V \to W$ linear. The transformation $P : W \to V$ is called the **pseudo-inverse** of T if and only if $P(T(x)) = x$ for all $x \in \text{Null}(T)^{\perp}$. □

The pseudo-inverse differs from the left inverse (see Definition 4.7.53) in that if S is a left-inverse of T, $S(T(x)) = x$ for all $x \in V$ instead of only for $x \in \text{Null}(T)^{\perp}$.

We are now ready to state a key invertibility theorem. We see that a linear transformation T is invertible from the range to a restricted subspace of the domain.

Theorem 7.4.2

Let V and W be vector spaces of dimension n and m, respectively, and let $T : V \to W$ be a linear transformation. Then $T : \text{Null}(T)^{\perp} \to \text{Ran}(T)$ is an isomorphism.

Proof. Let V and W be vector spaces and suppose $T : V \to W$ is a linear transformation. We know that the only vector in $\text{Null}(T)^{\perp}$ that maps to 0_W is 0_V. Therefore, by Theorem 5.1.21, $T : \text{Null}(T)^{\perp} \to \text{Ran}(T)$ is one-to-one. By Theorem 5.1.20, $T : \text{Null}(T)^{\perp} \to \text{Ran}(T)$ is an onto transformation. Therefore, $T : \text{Null}(T)^{\perp} \to \text{Ran}(T)$ is an isomorphism. □

Theorem 7.4.2 tells us that the set of nonzero possible radiographs is in one-to-one correspondence with the set of non-invisible objects that have no nullspace components. And furthermore, the isomorphism between them is the radiographic transformation itself. To help us find the inverse transformation, we again consider the corresponding matrix representation and spaces.

Theorem 7.4.3 [Maximal Isomorphism Theorem]

Let $U : \mathbb{R}^n \to \mathbb{R}^m$ be the linear transformation defined by $U(x) = Mx$ where M is an $m \times n$ matrix. Then, $\widetilde{U} : \text{Row}(M) \to \text{Col}(M)$, defined by $\widetilde{U}(x) = U(x) = Mx$ for all $x \in \text{Row}(M)$, is an isomorphism.

Proof. Let U and M be as above. Also, suppose $r_1^{\mathsf{T}}, r_2^{\mathsf{T}}, \ldots, r_m^{\mathsf{T}}$ are the rows of M. For $v \in \text{Null}(U)$,

$$
0 = Mv = \begin{pmatrix} r_1^{\mathsf{T}} v \\ r_2^{\mathsf{T}} v \\ \vdots \\ r_m^{\mathsf{T}} v \end{pmatrix} = \begin{pmatrix} \langle r_1, v \rangle \\ \langle r_2, v \rangle \\ \vdots \\ \langle r_m, v \rangle \end{pmatrix}.
$$

Therefore, if $r_k \in \text{Null}(U)$ for some $1 \leq k \leq m$, then $\langle r_k, r_k \rangle = 0$. That is, $r_k = 0$. So, we know that $r_k \in \text{Null}(U)^{\perp}$ for all $1 \leq k \leq m$. That is, $\text{Row}(M) \subseteq \text{Null}(U)^{\perp}$. Now, we know that $\mathbb{R}^n = \text{Null}(U) \oplus \text{Null}(U)^{\perp}$. Therefore, $n = \dim(\mathbb{R}^n) = \text{Nullity}(U) + \dim(\text{Null}(U)^{\perp})$. We also know, by the Rank-Nullity Theorem (Theorem 5.1.30), that $n = \text{Nullity}(U) + \text{Rank}(U)$. Therefore, $\dim(\text{Row}(M)) = \text{Rank}(U) = \dim(\text{Null}(U)^{\perp})$. Therefore, $\text{Row}(M) = \text{Null}(U)^{\perp}$.

Now, by Theorem 7.4.2, $U : \text{Row}(M) \to \text{Col}(M)$ is an isomorphism. □

Theorem 7.4.3 tells us that we may restrict the transformation U to a particular subspace of its domain (the row space of the associated matrix) and the resulting map will be an isomorphism. The theorem is called the Maximal Isomorphism Theorem because one cannot restrict U to any larger subspace and still have an isomorphism.

Corollary 7.4.4

Let M be an $m \times n$ matrix and $U : \mathbb{R}^n \to \mathbb{R}^m$ be the transformation defined by $U(x) = Mx$. Suppose $V \subseteq \mathbb{R}^n$ and $W \subseteq \mathbb{R}^m$ are subspaces of dimension k, and $\tilde{U} : V \to W$ given by $\tilde{U}(x) = U(x) = Mx$ is an isomorphism. Then $k \leq \text{Rank}(M)$.

We leave the proof as Exercise 15 in the Additional Exercises section.

The Maximal Isomorphism Theorem is a powerful step toward tomographic reconstruction of radiographs formed by the radiographic transformation T. In the matrix representation $M = [T]_{\mathcal{O}}^{\mathcal{R}}$, it is possible to reconstruct any object represented as a linear combination of the rows of M. These objects are in one-to-one correspondence with the radiographs represented as linear combinations of the columns of M. This notion of invertibility does not rely on any specific conditions on the rank or nullity of M, nor does it require the transformation to be one-to-one, nor does it require T to be onto. The inverse of the maximal isomorphism is the pseudo-inverse given in Definition 7.4.1. However, we do not yet know exactly how to find such a maximal isomorphism. Until we can, we are also not able to find the pseudo-inverse. The rest of this section is an exploration that will guide the reader toward a better understanding of the pseudo-inverse.

7.4.2 Exploring the Nature of the Data Compression Transformation

In this section, we will explore the conditions under which transformations that are not one-to-one can be partially invertible. For this exercise, suppose we have a vector space $\mathcal{I}$ of 2×3 pixel images with the geometry:

$$v = \begin{array}{|c|c|c|} \hline x_1 & x_3 & x_5 \\ \hline x_2 & x_4 & x_6 \\ \hline \end{array}.$$

Coordinate vectors for vectors in $\mathcal{I}$ are in $\mathbb{R}^6$. We would like to see if we can store the image data as a vector in $\mathbb{R}^5$ as a simple data compression scheme. Suppose we store only the column and row sums as a vector in $\mathbb{R}^5$. That is, we let $f : \mathcal{I} \to \mathbb{R}^5$ be the transformation that takes an image and outputs the compressed data vector b defined as

$$b = \begin{pmatrix} b_1 \\ b_2 \\ b_3 \\ b_4 \\ b_5 \end{pmatrix} = \begin{pmatrix} x_1 + x_2 \\ x_3 + x_4 \\ x_5 + x_6 \\ x_1 + x_3 + x_5 \\ x_2 + x_4 + x_6 \end{pmatrix} = f \left(\begin{array}{|c|c|c|} \hline x_1 & x_3 & x_5 \\ \hline x_2 & x_4 & x_6 \\ \hline \end{array} \right) = f(v).$$

1. Using the standard ordered bases for $\mathcal{I}$ and for $\mathbb{R}^5$, find $[u]$, $[f(u)]$ and $M = [f]$, where

$$
u = \begin{array}{|c|c|c|}
\hline
0 & 3 & 6 \\
\hline
3 & 4 & 5 \\
\hline
\end{array}.
$$

   (To simplify notation, we use $[v]$ to mean the coordinate vector and $[f]$ to mean the matrix representation of f according to the standard bases.)
2. By inspection (and trial and error), find two linearly independent null vectors of f.
3. Determine $\mathrm{Col}(M)$, $\mathrm{Null}(M)$, $\mathrm{Row}(M)$, $\mathrm{Rank}(M)$, and $\mathrm{Nullity}(M)$.
   (It is advisable to use MATLAB/OCTAVE to find these. It might be useful to know that, in MATLAB/OCTAVE, M^{T} is found using transpose(M). The command M' computes the complex-conjugate transpose of M, so for real matrices, these two methods are equivalent.)
4. Is f one-to-one? Onto? Invertible?
5. For a general linear transformation, A, relate $\dim((\mathrm{Null}(A))^{\perp})$ to $\mathrm{Rank}(A)$. Justify your conclusion.

A Pseudo-Inverse

Now, consider the vectors

$$
v_1 = \begin{array}{|c|c|c|}
\hline
1 & 1 & 1 \\
\hline
1 & 1 & 1 \\
\hline
\end{array}, \quad
v_2 = \begin{array}{|c|c|c|}
\hline
1 & 5 & 4 \\
\hline
3 & 2 & 6 \\
\hline
\end{array}, \quad
v_3 = \begin{array}{|c|c|c|}
\hline
2 & 4 & 6 \\
\hline
3 & 5 & 7 \\
\hline
\end{array} \in \mathcal{I}
$$

and the matrix transformation $P : \mathbb{R}^5 \to \mathbb{R}^6$ defined by (note the prefactor $1/30$):

$$
P(v) = \frac{1}{30} \begin{pmatrix}
12 & -3 & -3 & 8 & -2 \\
12 & -3 & -3 & -2 & 8 \\
-3 & 12 & -3 & 8 & -2 \\
-3 & 12 & -3 & -2 & 8 \\
-3 & -3 & 12 & 8 & -2 \\
-3 & -3 & 12 & -2 & 8
\end{pmatrix} v.
$$

6. Use MATLAB/OCTAVE to compute $P(Mv_k)$ for $k = 1, 2, 3$. (You should find that $P(Mv_k) = v_k$ for some of the test vectors and $P(Mv_k) \neq v_k$ for others. You can also complete this task by direct hand computation.)
7. Draw some conclusions about the possible nature of the matrix transformation P. Use the results of Questions 5 and 6 and recall Corollary 4.7.50 and Theorem 4.7.37.
8. Based on your work thus far, how would you assess the effectiveness of recovering images compressed using M and recovered using P? How is this idea similar to, or distinct from, the radiography/tomography problem?

Conjecturing about a General Pseudo-Inverse

9. You will recall that if S is an $n \times n$ symmetric matrix of rank n, then S is orthogonally diagonalizable. So, we can write

$$
S = QDQ^{\mathsf{T}} \text{ and } S^{-1} = QD^{-1}Q^{\mathsf{T}},
$$

where Q is an $n \times n$ orthogonal matrix satisfying $Q^{\mathsf{T}} = Q^{-1}$ and D is a diagonal matrix whose entries are the (nonzero) eigenvalues of S. Now, we know, for S with rank $r < n$, that D has r nonzero entries. Verify that a pseudo-inverse of S is

$$P = \widetilde{Q}\widetilde{D}^{-1}\widetilde{Q}^{\mathsf{T}},$$

where $\widetilde{D}$ is the $r \times r$ diagonal matrix whose entries are the nonzero eigenvalues of S, and $\widetilde{Q}$ is the $n \times r$ matrix composed of the columns of Q corresponding to the same eigenvalues.

10. Use the interpretation of the orthogonal diagonalization on Page 439 to interpret the fact that if $Sx = b$ and $S = QDQ^{\mathsf{T}}$, then $D(Q^{\mathsf{T}}x) = (Q^{\mathsf{T}}b)$.

11. We wish to employ similar methods for creating a pseudo-inverse of a non-square matrix M. What type of decomposition of M would lend itself to such a construction? (Be careful to consider the properties of Q and D for a square matrix. Which of these properties do we need when M is $m \times n$ instead?) Applying your conjecture, construct the pseudo-inverse of M.

7.4.3 Additional Exercises

12. Let S be an $n \times n$ symmetric matrix of rank $r < n$. Show that S can be exactly reconstructed from r eigenvectors and the r nonzero eigenvalues.

13. Let S be an $n \times n$ symmetric matrix with nonnegative eigenvalues. Find the "square root" matrix A where $A^2 = S$.

14. Let V and W be vector spaces. Suppose $T : V \to W$ is a one-to-one linear transformation. Show that the left inverse of T is a pseudo-inverse.

15. Prove Corollary 7.4.4.

7.5 Singular Value Decomposition

The tomography problem—recovering an object description from radiographic data—has, thus far, remained stubbornly intractable. We have found that if the radiographic transformation is one-to-one, then we can utilize the left inverse transformation. However, this would require an unreasonably large number of radiographs. Now, we understand that at least some transformations T that are *not* one-to-one *do* have a pseudo-inverse P. If this is the case for the tomography problem, then we can reconstruct brain images from, possibly few, radiographs.

The big concept of this section is that, in fact, *every* matrix has a (generalized) diagonalization and a pseudo-inverse. This diagonal form is called the singular value decomposition, or SVD.

The singular value decomposition of a matrix is important and useful in a variety of contexts. Consider a few representative applications.

Inverse Problems. The singular value decomposition can be used to construct pseudo-inverse transformations. Our key example is in brain scan tomography. Figure 7.19 shows example brain image reconstructions obtained from non-invertible radiographic transformations using varying number of views. More views result in better reconstruction detail. This example uses very-low noise radiographic data.

Data Compression. The singular value decomposition can be used to approximate data sets stored as arrays, such as images. Approximate representations are stored with relatively small memory footprint as a small collection of vectors. The image can be approximately recovered from these vectors. Figure 7.20 shows an example of the quality of recovery.

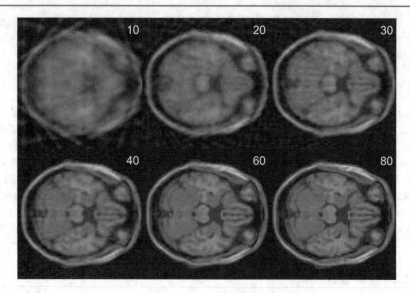

Fig. 7.19 Brain scan pseudo-inverse reconstructions of slice 50 using varying number of radiographic views (indicated in each subfigure).

Fig. 7.20 Example of data compression using singular value decomposition. At left is the original RGB image of Indian Paintbrush. At right is a reconstruction obtained from vectors comprising 30% of the memory footprint as the original image.

Principal Components. The singular value decomposition can be used to determine the principal features in a data set in which each instance is a vector in $\mathbb{R}^n$. Sunspots observed on the surface of the sun wax and wane over a cycle of roughly 9–13 years. Figure 7.21 shows five representative sunspot abundance curves as colored lines[6]. Times are normalized by percent of time through a given cycle, and sunspot abundance is reported according to a standard formula as sunspot number (the Wolf number). The five curves illustrate the diversity of sunspot evolution from cycle to cycle. When the sunspot data are represented as a data matrix, we will see that singular value decomposition provides a principal (unit) vector v and a vector of amplitudes a. We will also see that the principal vector is the sunspot number evolution curve that explains most of the trend. The individual curves are approximated by v scaled by the corresponding amplitude from a. The principal vector is the gray curve in the figure (scaled for visual clarity). Because of the scaling, **it is important to recognize the trend (shape and**

---

[6] Monthly sunspot data was obtained from the SILSO, Royal Observatory of Belgium, Brussels (http://www.sidc.be/silso/datafiles).

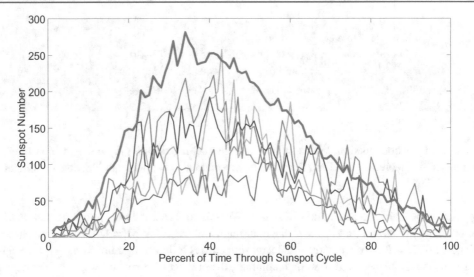

Fig. 7.21 Five representative sunspot number cycle curves (colors) and the principal vector (gray) that explains most of the trend. The principal vector is arbitrarily scaled for visual clarity.

location of various features) in the gray curve and to not focus on the amplitude. Particular features that may be of interest to researchers are where the maximum occurs, how long it takes to reach half the maximum, how fast the curve rises, what is the typical variation around the maximum, etc. This principal vector can also be used to predict the evolution of a partially complete sunspot cycle.

7.5.1 The Singular Value Decomposition

We have seen that every symmetric matrix has an eigenvalue diagonalization. In particular, every symmetric matrix can be represented as an orthogonal change to eigenbasis coordinates, followed by a diagonal transformation stretching each eigenvector by its eigenvalue, followed by a change back to the original coordinates through an orthogonal transformation. We will show that, in an analogous way, every matrix has a (generalized) diagonal representation. That is, there exist orthogonal coordinates for the row space and column space so that the transformation is diagonal. First, we need to generalize the definition of diagonal matrix.

Definition 7.5.1

Let M be an $m \times n$ matrix with (i, j)-entry m_{ij}. We say that M is a **diagonal matrix** if and only if $m_{ij} = 0$ whenever $i \neq j$. $\qquad\square$

An example of a 3×4 diagonal matrix is D, defined by $D = \begin{pmatrix} 1 & 0 & 0 & 0 \\ 0 & 4 & 0 & 0 \\ 0 & 0 & 1 & 0 \end{pmatrix}$.

Now, we present the main result of this section.

Theorem 7.5.2 [Singular Value Decomposition]
Let M be an $m \times n$ real matrix of rank r. There exist $m \times m$ orthogonal matrix U, $n \times n$ orthogonal matrix V, and $m \times n$ diagonal matrix Σ such that $M = U\Sigma V^{\mathsf{T}}$, where the diagonal entries of Σ are $\sigma_1 \geq \sigma_2 \geq \ldots \geq \sigma_{min(n,m)} \geq 0$ and $\sigma_k = 0$, for $k > r$.

The proof is by construction. We show how to construct matrices U, Σ, and V, which satisfy the conditions of the theorem. Therefore, the proof is important also in understanding how to compute the SVD.

Proof. Let M be an $m \times n$ real matrix of rank r. We will construct the $m \times m$ orthogonal matrix U, $n \times n$ orthogonal matrix V, and $m \times n$ diagonal matrix Σ such that $M = U\Sigma V^{\mathsf{T}}$. By construction, the diagonal of Σ will have r positive nonincreasing entries followed by $\min(m, n) - r$ zero entries. (This means that if M has more rows than columns, the number of zero entries along the diagonal is $n - r$, but if M has more columns than rows, there will be $m - r$ zeros along the diagonal.)

We know that the $n \times n$ symmetric matrix $M^{\mathsf{T}}M$ and the $m \times m$ symmetric matrix MM^{T} both have rank r. Therefore, by Theorem 7.3.12, both are orthogonally diagonalizable. So, we can write

$$M^{\mathsf{T}}M = VDV^{\mathsf{T}},$$

where V is orthogonal with columns $v_1, \cdots, v_n$ and D is diagonal with entries $d_1, \cdots, d_n$. Suppose, for notational convenience, that the nonzero eigenvalues are $d_1, \cdots, d_r$, and $d_{r+1} = \cdots = d_n = 0$. Similarly, we write

$$MM^{\mathsf{T}} = UCU^{\mathsf{T}},$$

where U is orthogonal with columns $u_1, \cdots, u_m$ and C is diagonal with entries $c_1, \cdots, c_m$.

Now, consider an arbitrary eigenvector v_k for which the corresponding eigenvalue d_k is nonzero. Because v_k is an eigenvector of $M^{\mathsf{T}}M$ with eigenvalue d_k, we see that

$$(MM^{\mathsf{T}})(Mv_k) = M(M^{\mathsf{T}}Mv_k) = M(d_kv_k) = d_k(Mv_k).$$

Therefore, Mv_k and u_k are eigenvectors of MM^{T} with eigenvalue d_k. Thus, as we have freedom to choose orthogonal eigenvectors as a basis for any eigenspace, each eigenvector Mv_k of eigenvalue d_k can be expressed as a scalar multiple of such a chosen eigenvector u_j of eigenvalue $c_j = d_k$. We can list the vectors $u_1, \cdots, u_m$, in a different order, so that $Mv_k = \sigma_k u_k$ for suitable nonzero constants σ_k and for all $1 \leq k \leq r$. Now, define $\sigma_{r+1} = \sigma_{r+2} = \ldots = \sigma_{\min\{m,n\}} = 0$ and define the Σ to be the $m \times n$ diagonal matrix with diagonal entries $\sigma_1, \cdots, \sigma_{\min\{m,n\}}$.

We now claim that the matrix decomposition with these properties is realized by $M = U\Sigma V^{\mathsf{T}}$. Since $\{v_1, v_2, \ldots, v_n\}$ is a basis for $\mathbb{R}^n$, we can show this is true, by showing that $Mv_k = U\Sigma V^{\mathsf{T}}v_k$ for $1 \leq k \leq n$. Indeed, if we let e_k be the vector whose kth entry is 1 and all others are 0, we have

$$U\Sigma V^{\mathsf{T}}v_k = U\Sigma \begin{pmatrix} - v_1^{\mathsf{T}} - \\ - v_2^{\mathsf{T}} - \\ \vdots \\ - v_n^{\mathsf{T}} - \end{pmatrix} \begin{pmatrix} | \\ v_k \\ | \end{pmatrix}$$

$$= U\Sigma \begin{pmatrix} \langle v_1, v_k \rangle \\ \langle v_2, v_k \rangle \\ \vdots \\ \langle v_n, v_k \rangle \end{pmatrix}$$

$$= U\Sigma e_k$$

$$= \begin{pmatrix} | & | & & | \\ u_1 & u_2 & \dots & u_m \\ | & | & & | \end{pmatrix} (\sigma_k e_k)$$

$$= \sigma_k u_k$$

$$= M v_k.$$

Thus, $M = U\Sigma V^\mathsf{T}$.

Now, we show that $\sigma_k \geq 0$ for $1 \leq k \leq \min(n, m)$. This is done by showing that $D = \Sigma^\mathsf{T}\Sigma$ and $C = \Sigma\Sigma^\mathsf{T}$ in which case, we have that σ_k^2 are the eigenvalues of $M^\mathsf{T}M$ and MM^T. We know that

$$VDV^\mathsf{T} = M^\mathsf{T}M$$

$$= (U\Sigma V^\mathsf{T})^\mathsf{T}(U\Sigma V^\mathsf{T})$$

$$= V\Sigma^\mathsf{T}U^\mathsf{T}U\Sigma V^\mathsf{T}$$

$$= V\Sigma^\mathsf{T}\Sigma V^\mathsf{T}.$$

Since V and V^T are invertible, $D = \Sigma^\mathsf{T}\Sigma$. Similarly, $C = \Sigma\Sigma^\mathsf{T}$. Now, $\Sigma^\mathsf{T}\Sigma$ is an $n \times n$ diagonal matrix with $\sigma_1^2, \sigma_2^2, \ldots, \sigma_n^2$ along the diagonal. Thus, the real eigenvalues d_k satisfy $d_k = \sigma_k^2$, for $1 \leq k \leq n$. Similarly, the real eigenvalues, c_k satisfy $c_k = \sigma_k^2$ for $1 \leq k \leq m$. We choose $\sigma_k \geq 0$ for $1 \leq k \leq \min(n, m)$ and appropriately adjust the signs on the corresponding column vectors in U (or V), if necessary.

The decompositions of $M^\mathsf{T}M$ and MM^T are not unique because the ordering of the columns of V and U gives different decompositions, we can choose to order the columns so that the corresponding σ_k are in order so that $\sigma_1 \geq \sigma_1 \geq \ldots$. Finally, by the rank-nullity theorem, whenever $k > r$, $\sigma_k = 0$. $\square$

The fact that *every* real matrix admits a singular value decomposition is very enlightening. Every linear transformation, $T : V \to W$, can be expressed as a matrix relative to bases for the domain and codomain. This matrix representation has a singular value decomposition $M = U\Sigma V^\mathsf{T}$. Consider two equivalent interpretations.

1. Left multiplication by M is equivalent to a sequence of three matrix products. First, there is an orthogonal transformation (a generalized rotation preserving distances and angles). Second, there is a scaling along each new coordinate direction, expanding if $\sigma_k > 1$ and contracting if $\sigma_k < 1$. Third, there is another orthogonal transformation to some codomain coordinates. Every linear transformation can be viewed in terms of this sequence of three operations.

2. Consider the equation $Mx = b$ with equivalent representation $U\Sigma V^\mathsf{T}x = b$. Because the transpose of an orthogonal matrix is also its inverse, we have $\Sigma\left(V^\mathsf{T}x\right) = \left(U^\mathsf{T}b\right)$. With this view, we see that left multiplication by U^T and V^T represents codomain and domain coordinate transformations, respectively, for which the overall transformation is diagonal. That is, if we use the columns of U as a basis $\mathcal{B}_c$ for the codomain and the columns of V as a basis $\mathcal{B}_d$ for the domain, we have $[M]_{\mathcal{B}_d}^{\mathcal{B}_c}[x]_{\mathcal{B}_d} = [b]_{\mathcal{B}_c}$, where $[M]_{\mathcal{B}_c}^{\mathcal{B}_d} = \Sigma$ is diagonal.

The proof of Theorem 7.5.2 demonstrates one method for finding the singular value decomposition of an arbitrary real matrix M. According to the proof, we can decompose any matrix M by finding the eigenvectors and eigenvalues of $M^{\mathsf{T}}M$ and MM^{T}. The **right singular vectors** (the columns of V) are eigenvectors of symmetric matrix $M^{\mathsf{T}}M$. The **singular values** are square roots of the eigenvalues of $M^{\mathsf{T}}M$. The **left singular vectors** (columns of U) can then be determined by $Mv_k = \sigma_k u_k$. Alternately, we can find the left singular vectors as the eigenvectors of MM^{T}. We form the matrix Σ using the singular values. The conventional ordering for columns of U and V and corresponding diagonal entries of Σ are in order of non-increasing singular values, σ_k. Throughout this section, we will keep to this convention. The following examples illustrate these ideas on small matrices.

Example 7.5.3 Consider the matrix

$$M = \begin{pmatrix} -1 & 2 \\ -3 & 3 \end{pmatrix},$$

which has no real eigenvalues. We can construct a singular value decomposition for M. The first step is to orthogonally diagonalize $M^{\mathsf{T}}M$ as VDV^{T}. We have

$$M^{\mathsf{T}}M = VDV^{\mathsf{T}}, \text{ where } V \approx \begin{pmatrix} -0.6576 & -0.7534 \\ +0.7534 & -0.6576 \end{pmatrix} \text{ and } D \approx \begin{pmatrix} +22.602 & +0.0000 \\ +0.0000 & +0.3982 \end{pmatrix}.$$

We have ordered the eigenvectors so that the eigenvalues are in non-increasing order. The singular values of M are $\sigma_1 = \sqrt{d_1} \approx +4.7541$ and $\sigma_2 = \sqrt{d_2} \approx +0.6310$. The left singular vectors are

$$u_1 = \frac{Mv_1}{\sigma_1} \approx \begin{pmatrix} +0.4553 \\ +0.8904 \end{pmatrix},$$

$$u_2 = \frac{Mv_2}{\sigma_2} \approx \begin{pmatrix} -0.8904 \\ +0.4553 \end{pmatrix}.$$

Alternately, we can compute u_1 and u_2 by diagonalizing the matrix MM^{T} as

$$MM^{\mathsf{T}} = UDU^{\mathsf{T}} = \begin{pmatrix} +0.4553 & -0.8904 \\ +0.8904 & +0.4553 \end{pmatrix} \begin{pmatrix} +22.602 & +0.0000 \\ +0.0000 & +0.3982 \end{pmatrix} \begin{pmatrix} +0.4553 & +0.8904 \\ -0.8904 & +0.4553 \end{pmatrix}.$$

We find u_1 and u_2 as the columns of U. Thus, we have the singular value decomposition of M

$$M = U\Sigma V^{\mathsf{T}} \approx \begin{pmatrix} +0.4553 & -0.8904 \\ +0.8904 & +0.4553 \end{pmatrix} \begin{pmatrix} +4.7541 & +0.0000 \\ +0.0000 & +0.6310 \end{pmatrix} \begin{pmatrix} -0.6576 & +0.7534 \\ -0.7534 & -0.6576 \end{pmatrix}.$$

$\square$

In the above example, we can find the eigenvectors and eigenvalues of a matrix using mathematical software. In fact, the above were found using the eig command in OCTAVE.

We know that every real symmetric matrix is orthogonally diagonalizable. Such a diagonalization is *not* necessarily equivalent to the singular value decomposition because the matrix may have negative eigenvalues. However, the singular values are all nonnegative. This is possible because the left and right singular vectors need not be the same ($U \neq V$ in general).

Example 7.5.4 Consider the real symmetric matrix

$$M = \begin{pmatrix} 1 & 2 \\ 2 & 1 \end{pmatrix},$$

which has orthogonal diagonalization

$$M = QDQ^{\mathsf{T}} = \begin{pmatrix} -\sqrt{2}/2 & \sqrt{2}/2 \\ \sqrt{2}/2 & \sqrt{2}/2 \end{pmatrix} \begin{pmatrix} -1 & 0 \\ 0 & 3 \end{pmatrix} \begin{pmatrix} -\sqrt{2}/2 & \sqrt{2}/2 \\ \sqrt{2}/2 & \sqrt{2}/2 \end{pmatrix}.$$

We can readily create the SVD in two steps. First, we change the sign of the first column of the left Q matrix and the sign of the corresponding eigenvalue in D:

$$M = \begin{pmatrix} \sqrt{2}/2 & \sqrt{2}/2 \\ -\sqrt{2}/2 & \sqrt{2}/2 \end{pmatrix} \begin{pmatrix} 1 & 0 \\ 0 & 3 \end{pmatrix} \begin{pmatrix} -\sqrt{2}/2 & \sqrt{2}/2 \\ \sqrt{2}/2 & \sqrt{2}/2 \end{pmatrix}.$$

Then, we reorder the entries of the diagonal matrix (to non-increasing order) and the corresponding columns of the orthogonal matrices.

$$M = U\Sigma V^{\mathsf{T}} = \begin{pmatrix} \sqrt{2}/2 & \sqrt{2}/2 \\ \sqrt{2}/2 & -\sqrt{2}/2 \end{pmatrix} \begin{pmatrix} 3 & 0 \\ 0 & 1 \end{pmatrix} \begin{pmatrix} \sqrt{2}/2 & \sqrt{2}/2 \\ -\sqrt{2}/2 & \sqrt{2}/2 \end{pmatrix}.$$

□

The previous example suggests a general method for finding the SVD of any symmetric matrix. Suppose $n \times n$ symmetric matrix M has orthogonal diagonalization QDQ^{T} where the eigenvalues λ_k (and corresponding eigenvectors, q_k) are ordered so that $|\lambda_1| \geq |\lambda_2| \geq \cdots \geq |\lambda_n| \geq 0$. Then the SVD is given by $M = U\Sigma V^{\mathsf{T}}$, where $\sigma_k = |\lambda_k|$, $V = Q$, and the columns of U are $\mathrm{sgn}(\lambda_k)q_k$, where

$$\mathrm{sgn}(x) = \begin{cases} -1 & \text{if } x < 0 \\ 0 & \text{if } x = 0 \\ 1 & \text{if } x > 0 \end{cases}.$$

The next example demonstrates the construction of the singular value decomposition of an $m \times n$ matrix with $m < n$.

Example 7.5.5 Consider the nonsquare matrix

$$M = \begin{pmatrix} 1 & 2 & 1 \\ 0 & 1 & 1 \end{pmatrix}.$$

In order to compute the SVD, we begin by finding the orthogonal diagonalization $M^{\mathsf{T}}M = VDV^{\mathsf{T}}$ where D is diagonal with entries $d_1 \approx 7.606$, $d_2 \approx 0.394$, $d_3 = 0$; and

$$V \approx \begin{pmatrix} -0.3197 & -0.7513 & +0.5774 \\ -0.8105 & -0.0988 & -0.5774 \\ -0.4908 & +0.6525 & +0.5774 \end{pmatrix}.$$

We have the singular values of M: $\sigma_1 = \sqrt{d_1} \approx 2.758$ and $\sigma_2 = \sqrt{d_2} \approx 0.6280$. We also have the right singular vectors $v_k = q_k$. The left singular vectors are computed as $u_k = Mv_k/\sigma_k$ for all k such that $\sigma_k > 0$. We have

$$u_1 = \frac{Mv_1}{\sigma_1} \approx \begin{pmatrix} +0.8817 \\ +0.4719 \end{pmatrix},$$

$$u_2 = \frac{Mv_2}{\sigma_2} \approx \begin{pmatrix} -0.4719 \\ +0.8817 \end{pmatrix}.$$

The singular value decomposition is

$$M = U\Sigma V^\mathsf{T} \approx \begin{pmatrix} +0.8817 & -0.4719 \\ +0.4719 & +0.8817 \end{pmatrix} \begin{pmatrix} 2.758 & 0 & 0 \\ 0 & 0.628 & 0 \end{pmatrix} \begin{pmatrix} -0.3197 & -0.8105 & -0.4908 \\ -0.7513 & -0.0988 & +0.6525 \\ +0.5774 & -0.5774 & +0.5774 \end{pmatrix}.$$

□

The matrix spaces associated with a transformation are simply defined in terms of the columns of the orthogonal matrices U and V.

Corollary 7.5.6

Let M be an $m \times n$ real matrix of rank r with singular value decomposition $M = U\Sigma V^\mathsf{T}$. Then

(a) $\{v_{r+1}, \cdots, v_n\}$ is an orthonormal basis for Null(M).
(b) $\{v_1, \cdots, v_r\}$ is an orthonormal basis for Null(M)$^\perp$ = Row(M)
(c) $\{u_1, \cdots, u_r\}$ is an orthonormal basis for Ran(M) = Col(M).

Proof. Suppose M is an $m \times n$ real matrix of rank r with singular value decomposition $M = U\Sigma V^\mathsf{T}$. Let $\sigma_1, \sigma_2, \ldots, \sigma_{\min\{m,n\}}$ be the diagonal entries of Σ. Notice that $Mv = 0$ for any vector in $S = \{v_{r+1}, v_{r+2}, \cdots, v_n\}$ because $\sigma_k = 0$ for $k > r$. Therefore, $S \subset$ Null(M). By the rank-nullity theorem, the nullspace of M has dimension $n - r$, and S is a linearly independent set of $n - r$ null vectors of M (see Equation 7.9) and so is also a basis for Null(M). The dimension of Null(M)$^\perp = r$. Now, $\{v_1, \cdots, v_r\}$ is a linearly independent set of r vectors in $\mathbb{R}^n$ that are not null vectors and so the set is a basis for Null(M)$^\perp$ = Row(M). Finally, by Equation 7.9, Ran(M) = Span$\{u_1, u_2, \cdots, u_r\}$. And as $\{u_1, u_2, \cdots, u_r\}$ is a set of r linearly independent vectors in $\mathbb{R}^m$, it is also basis for Ran(M).   □

Example 7.5.7 Consider the matrix representation, T, of the radiographic transformation of the following scenario, and interpret the singular value decomposition $T = U\Sigma V^\mathsf{T}$.

Suppose we have an object space of $N = 4$ voxel values and $M = 6$ radiograph pixels in a three-view arrangement at angles $-45°$, $0°$, and $+45°$. Suppose each view has two pixels and the pixels are scaled (ScaleFac$= \sqrt{2}$) to include a full view of the object space. The reader should take the time to verify the following results. First, the radiographic transformation is given by

$$T = \begin{pmatrix} 1 & {}^1/_2 & {}^1/_2 & 0 \\ 0 & {}^1/_2 & {}^1/_2 & 1 \\ 1 & 1 & 0 & 0 \\ 0 & 0 & 1 & 1 \\ {}^1/_2 & 1 & 0 & {}^1/_2 \\ {}^1/_2 & 0 & 1 & {}^1/_2 \end{pmatrix}.$$

The symmetric matrix $T^\mathsf{T}T$ has orthogonal diagonalization $V S V^\mathsf{T}$ where the columns of V are

$$v_1 = \begin{pmatrix} +^1\!/_2 \\ +^1\!/_2 \\ +^1\!/_2 \\ +^1\!/_2 \end{pmatrix}, \quad v_2 = \begin{pmatrix} -^1\!/_2 \\ -^1\!/_2 \\ +^1\!/_2 \\ +^1\!/_2 \end{pmatrix}, \quad v_3 = \begin{pmatrix} -^1\!/_2 \\ +^1\!/_2 \\ -^1\!/_2 \\ +^1\!/_2 \end{pmatrix}, \quad v_4 = \begin{pmatrix} -^1\!/_2 \\ +^1\!/_2 \\ +^1\!/_2 \\ -^1\!/_2 \end{pmatrix},$$

and the corresponding eigenvalues (diagonal elements of D) are $d_1 = 6$, $d_2 = 3$, $d_3 = 1$ and $d_4 = 0$. The singular values of T (diagonal elements of Σ) are $\sigma_k = \sqrt{d_k}$: $\sigma_1 = \sqrt{6}$, $\sigma_2 = \sqrt{3}$, $\sigma_3 = 1$ and $\sigma_4 = 0$. The transformation has rank $r = 3$ (the number of nonzero singular values) and the first three columns of U are given by $u_k = T v_k / \sigma_k$.

$$u_1 = \begin{pmatrix} +^1\!/\!\sqrt{6} \\ +^1\!/\!\sqrt{6} \\ +^1\!/\!\sqrt{6} \\ +^1\!/\!\sqrt{6} \\ +^1\!/\!\sqrt{6} \\ +^1\!/\!\sqrt{6} \end{pmatrix}, \quad u_2 = \begin{pmatrix} -^1\!/\!\sqrt{12} \\ +^1\!/\!\sqrt{12} \\ -^1\!/\!\sqrt{3} \\ +^1\!/\!\sqrt{3} \\ -^1\!/\!\sqrt{12} \\ +^1\!/\!\sqrt{12} \end{pmatrix}, \quad u_3 = \begin{pmatrix} -^1\!/_2 \\ +^1\!/_2 \\ 0 \\ 0 \\ +^1\!/_2 \\ -^1\!/_2 \end{pmatrix}.$$

The remaining three columns of U consist of a set of orthogonal unit vectors which are also orthogonal to $\mathrm{Span}\{u_1, u_2, u_3\}$. The Gram-Schmidt process is one method for finding such vectors $\{u_4, u_5, u_6\}$. Consider the following observations.

1. The nullspace of T is $\mathrm{Null}(T) = \mathrm{Span}\{v_{r+1}, \cdots, v_n\} = \mathrm{Span}\{v_4\}$. Notice that v_4 is the coordinate vector for the object

$$x_4 = \begin{array}{|c|c|} \hline -^1\!/_2 & ^1\!/_2 \\ \hline ^1\!/_2 & -^1\!/_2 \\ \hline \end{array} ,$$

for which $T v_4 = 0$. This result follows from the fact that v_4 is a singular vector with singular value of zero. It is also apparent by computing the transformation geometrically. The nullspace of T is the set of all invisible objects.

2. The range of T (the column space of T) is $\mathrm{Ran}(T) = \mathrm{Span}\{u_1, \cdots, u_r\} = \mathrm{Span}\{u_1, u_2, u_3\}$. The range of T is the set of all possible radiographs—radiographs which correspond to some object through the transformation.

3. The row space of T is $\mathrm{Row}(T) = \mathrm{Span}\{v_1, \cdots, v_r\} = \mathrm{Span}\{v_1, v_2, v_3\}$ is the set of all objects that are orthogonal to the set of invisible objects. These are also the objects that are fully recoverable from radiographs.

$\square$

The singular value decomposition is a powerful tool for understanding transformations, and the next few sections will make extensive use of its properties. We will need to solidify some notation to help us understand key ideas and formulate related transformations—such as pseudo-inverse transformations.

▶ **Notation** Let M be an $m \times n$ real matrix of rank r with singular value decomposition $M = U\Sigma V^\mathsf{T}$.

1. We denote the columns of U by u_k for $k \in \{1, 2, \cdots, m\}$ and the columns of V by v_k for $k \in \{1, 2, \cdots, n\}$.

2. We define $\widetilde{U}_s$ as the $m \times s$ matrix whose columns are the first s columns of U and $\widehat{U}_s$ as the $n \times (n-s)$ matrix whose columns are the last $n-s$ columns of U:

$$\widetilde{U}_s = \begin{pmatrix} | & | & & | \\ u_1 & u_2 & \cdots & u_s \\ | & | & & | \end{pmatrix}, \widehat{U}_s = \begin{pmatrix} | & | & & | \\ u_{s+1} & u_{s+2} & \cdots & u_n \\ | & | & & | \end{pmatrix}.$$

3. We define $\widetilde{V}_s$ as the $n \times s$ matrix whose columns are the first s columns of V and $\widehat{V}_s$ as the $n \times (n-s)$ matrix whose columns are the last $n-s$ columns of V:

$$\widetilde{V}_s = \begin{pmatrix} | & | & & | \\ v_1 & v_2 & \cdots & v_s \\ | & | & & | \end{pmatrix}, \widehat{V}_s = \begin{pmatrix} | & | & & | \\ v_{s+1} & v_{s+2} & \cdots & v_n \\ | & | & & | \end{pmatrix}.$$

4. We denote the ordered singular values of M, the diagonal elements of Σ, as $\sigma_1 \geq \sigma_2 \geq \cdots \geq \sigma_r > 0$, and $\sigma_{r+1} = \cdots = \sigma_{\min\{m,n\}} = 0$.

5. We define the $s \times s$ diagonal matrix $\widetilde{\Sigma}_s$, for $s \leq r$ as

$$\left(\widetilde{\Sigma}_s\right)_{ij} = \begin{cases} \sigma_i & \text{if } i = j \\ 0 & \text{otherwise} \end{cases}.$$

We can use this new notation to simply express the result of mapping an arbitrary vector $x \in \mathbb{R}^n$ to $b \in \mathbb{R}^m$ using the matrices $\widetilde{U}$, $\widetilde{V}$, and $\widetilde{\Sigma}$. In the following construction, we assume $r \leq m \leq n$. The other case, $r \leq n \leq m$, is similarly demonstrated with the same result. (See Exercise 8.)

$$Mx = U\Sigma V^{\mathsf{T}} x$$

$$= \begin{pmatrix} | & | & & | \\ u_1 & u_2 & \cdots & u_m \\ | & | & & | \end{pmatrix} \begin{pmatrix} \sigma_1 & & & & 0 \ldots 0 \\ & \sigma_2 & & & \\ & & \ddots & & \vdots \quad \vdots \\ & & & \sigma_m & 0 \ldots 0 \end{pmatrix} \begin{pmatrix} \underline{\quad} & v_1^{\mathsf{T}} & \underline{\quad} \\ \underline{\quad} & v_2^{\mathsf{T}} & \underline{\quad} \\ & \vdots & \\ \underline{\quad} & v_n^{\mathsf{T}} & \underline{\quad} \end{pmatrix} x$$

$$= \begin{pmatrix} | & | & & | \\ u_1 & u_2 & \cdots & u_m \\ | & | & & | \end{pmatrix} \begin{pmatrix} \sigma_1 & & & & 0 \ldots 0 \\ & \sigma_2 & & & \\ & & \ddots & & \vdots \quad \vdots \\ & & & \sigma_m & 0 \ldots 0 \end{pmatrix} \begin{pmatrix} \langle v_1, x \rangle \\ \langle v_2, x \rangle \\ \vdots \\ \langle v_n, x \rangle \end{pmatrix}$$

$$= \begin{pmatrix} | & | & & | \\ u_1 & u_2 & \cdots & u_m \\ | & | & & | \end{pmatrix} \begin{pmatrix} \sigma_1 \langle v_1, x \rangle \\ \sigma_2 \langle v_2, x \rangle \\ \vdots \\ \sigma_m \langle v_m, x \rangle \end{pmatrix}$$

$$= \sigma_1 \langle v_1, x \rangle u_1 + \sigma_2 \langle v_2, x \rangle u_2 + \cdots + \sigma_m \langle v_m, x \rangle u_m$$

$$= \sum_{k=1}^{m} \sigma_k \langle v_k, x \rangle u_k$$

$$= \sum_{k=1}^{r} \sigma_k \langle v_k, x \rangle u_k$$

$$= \tilde{U}_r \tilde{\Sigma}_r \tilde{V}_r^{\mathsf{T}} x.$$

We summarize this general result for a matrix transformation of rank r:

$$Mx = U\Sigma V^{\mathsf{T}} x = \sum_{k=1}^{r} \sigma_k \langle v_k, x \rangle u_k = \tilde{U}_r \tilde{\Sigma}_r \tilde{V}_r^{\mathsf{T}} x. \qquad (7.9)$$

Example 7.5.8 Returning to Example 7.5.7. We can write the transformation $T = \tilde{U}_3 \tilde{\Sigma}_3 \tilde{V}_3^{\mathsf{T}}$ using the following matrices

$$\tilde{U}_3 = \begin{pmatrix} +^1/\sqrt{6} & -^1/\sqrt{12} & -^1/2 \\ +^1/\sqrt{6} & +^1/\sqrt{12} & +^1/2 \\ +^1/\sqrt{6} & -^1/\sqrt{3} & 0 \\ +^1/\sqrt{6} & +^1/\sqrt{3} & 0 \\ +^1/\sqrt{6} & -^1/\sqrt{12} & +^1/2 \\ +^1/\sqrt{6} & +^1/\sqrt{12} & -^1/2 \end{pmatrix}, \tilde{\Sigma}_3 = \begin{pmatrix} \sqrt{6} & 0 & 0 \\ 0 & \sqrt{3} & 0 \\ 0 & 0 & 1 \end{pmatrix} \text{ and } \tilde{V}_3 = \begin{pmatrix} +^1/2 & -^1/2 & -^1/2 \\ +^1/2 & -^1/2 & +^1/2 \\ +^1/2 & +^1/2 & -^1/2 \\ +^1/2 & +^1/2 & +^1/2 \end{pmatrix}.$$

$\square$

The singular value decomposition has properties that lead to important results in constructing pseudo-inverse transformations and low-rank approximations of matrices. Before proceeding, we define the trace of a matrix in order to ease notation and then define a way of measuring the "size" of a matrix.

Definition 7.5.9

Let A be an $n \times n$ matrix with entries a_{ij}. The **trace** of A, denoted $\mathrm{tr}(A)$, is the sum of the diagonal entries of A: $\mathrm{tr}(A) = \sum_{k=1}^{n} a_{kk}$. $\square$

The trace of a matrix is indeed as simple as it sounds. For example, if we have the matrix

$$M = \begin{pmatrix} 3 & 4 & 1 \\ -2 & 2 & 1 \\ 4 & 2 & 5 \end{pmatrix},$$

then $\mathrm{tr}(M) = 3 + 2 + 5 = 10$.

Our measurement of the size of a matrix is the Frobenius norm.

Definition 7.5.10

Let A be an $m \times n$ matrix. We define the **Frobenius norm** of A as

$$\|A\|_F = \sqrt{tr(A^T A)}.$$

$\square$

The Frobenius norm is indeed a norm, derived from the Frobenius Inner Product, found in Table 7.1, and satisfies the properties outlined in Theorem 7.1.18. We leave it to the reader to show this result.

Now, if A is the 2×3 matrix given by

$$A = \begin{pmatrix} 1 & 2 & 0 \\ 2 & 1 & 1 \end{pmatrix},$$

then

$$A^T A = \begin{pmatrix} 1 & 2 \\ 2 & 1 \\ 0 & 1 \end{pmatrix} \begin{pmatrix} 1 & 2 & 0 \\ 2 & 1 & 1 \end{pmatrix} = \begin{pmatrix} 5 & 4 & 2 \\ 4 & 5 & 1 \\ 2 & 1 & 1 \end{pmatrix}.$$

So,

$$\|A\|_F = \sqrt{\mathrm{tr}(A^T A)} = \sqrt{11}.$$

Another way to compute $\|A\|_F$ is

$$\|A\|_F = \sqrt{\sum_{i=1}^{m} \sum_{j=1}^{n} a_{i,j}^2}.$$

Now, given an $m \times n$ matrix, M, we present a method, using the singular value decomposition, for creating a rank s approximation of M. The approximation will be close to M as measured by the Frobenius norm.

Theorem 7.5.11

Let M be an $m \times n$ real matrix of rank r with singular value decomposition $M = U\Sigma V^T$. Let $M_s = \sum_{k=1}^{s} \sigma_k u_k v_k^T$ for all $1 \leq s \leq r$. Then

(a) $M = M_r$,

(b) $\|M - M_s\|_F \leq \sum_{s+1}^{r} \sigma_k$, and

(c) M_s has rank s.

Proof. (a) Suppose M is a real $m \times n$ matrix of rank r with singular value decomposition $M = U\Sigma V^T$. Let $v_1, v_2, \ldots, v_n$ be the columns of V, $u_1, u_2, \ldots, u_m$ be the columns of U, and $\sigma_1, \sigma_2, \ldots, \sigma_r$ be the nonzero diagonal entries of Σ. Then $\{v_1, v_2, \cdots, v_n\}$ is an orthonormal basis for $\mathbb{R}^n$. We know that the two matrices M and M_r are equal, if their corresponding linear transformations are the same. Using Corollary 4.2.27, we verify that the transformations are equivalent by showing that $Mv_j = U\Sigma V^T v_j = \left(\sum_{k=1}^{r} \sigma_k u_k v_k^T\right) v_j$ for all j. We have been using block matrices (See pages 182 and 416.)

$$Mv_j = U\Sigma V^\mathsf{T} v_j$$

$$= (\widetilde{U}_r\ \widehat{U}_r)\begin{pmatrix}\widetilde{\Sigma}_r & 0 \\ 0 & 0\end{pmatrix}\begin{pmatrix}\widetilde{V}_r^\mathsf{T} \\ \widehat{V}_r^\mathsf{T}\end{pmatrix} v_j$$

$$= \widetilde{U}_r \widetilde{\Sigma}_r \widetilde{V}_r^\mathsf{T} v_j$$

$$= \begin{pmatrix} | & | & & | \\ u_1 & u_2 & \dots & u_r \\ | & | & & | \end{pmatrix}\begin{pmatrix}\sigma_1 & & \\ 0 & \ddots & 0 \\ & & \sigma_r\end{pmatrix}\begin{pmatrix}\underline{\quad v_1^\mathsf{T} \quad} \\ \underline{\quad v_2^\mathsf{T} \quad} \\ \vdots \\ \underline{\quad v_r^\mathsf{T} \quad}\end{pmatrix}\begin{pmatrix}| \\ v_j \\ |\end{pmatrix}$$

$$= \begin{pmatrix} | & | & & | \\ \sigma_1 u_1 & \sigma_2 u_2 & \dots & \sigma_r u_r \\ | & | & & | \end{pmatrix}\begin{pmatrix} v_1^\mathsf{T} v_j \\ \vdots \\ v_r^\mathsf{T} v_j \end{pmatrix}$$

$$= \sum_{k=1}^{r} \sigma_k u_k (v_k^\mathsf{T} v_j)$$

$$= \left(\sum_{k=1}^{r} \sigma_k u_k v_k^\mathsf{T}\right) v_j.$$

(b) Next, let $M_s = \sum_{k=1}^{s} \sigma_k u_k v_k^\mathsf{T}$ for $1 \le s \le r$. Then we have

$$\|M - M_s\|_F = \left\|\sum_{k=1}^{r} \sigma_k u_k v_k^\mathsf{T} - \sum_{k=1}^{s} \sigma_k u_k v_k^\mathsf{T}\right\|_F$$

$$= \left\|\sum_{k=s+1}^{r} \sigma_k u_k v_k^\mathsf{T}\right\|_F$$

$$\le \sum_{k=s+1}^{r} \sigma_k \left\|u_k v_k^\mathsf{T}\right\|_F$$

$$= \sum_{k=s+1}^{r} \sigma_k \sqrt{\mathrm{tr}\left(v_k u_k^\mathsf{T} u_k v_k^\mathsf{T}\right)}$$

$$= \sum_{k=s+1}^{r} \sigma_k \sqrt{\mathrm{tr}\left(v_k v_k^\mathsf{T}\right)}$$

$$= \sum_{k=s+1}^{r} \sigma_k.$$

The inequality in the proof follows from repeated application of the triangle inequality.

(c) The range of the transformation $T(v) = M_s v$ is the span of the s linearly independent vectors $\{u_1, u_2, \cdots, u_s\}$. Thus, $\mathrm{Rank}(M_s) = \dim(\mathrm{Ran}(M_s)) = s$. $\qquad\square$

Theorem 7.5.11 shows us that a matrix transformation (or any matrix) can be written as a linear combination of **outer products** ($u_k v_k^\mathsf{T}$) of unit vectors. If M is $m \times n$ it is expressed as an array of

nm real values. However, if rank$(M) = r$, we can equivalently express the matrix with $(m + n + 1)r$ values (r singular values, r vectors $u_k \in \mathbb{R}^m$, and r vectors $v_k \in \mathbb{R}^n$). If $(m + n + 1)r < mn$, then the SVD provides a method of data compression.

Example 7.5.12 Consider the matrix

$$A = \begin{pmatrix} 2 & -1 & 0 & 0 \\ 4 & -2 & 2 & 1 \\ 5 & -3 & 3 & 0 \end{pmatrix}.$$

The singular value decomposition of A is given by the orthogonal matrices (expressed to three decimal places):

$$U = \begin{pmatrix} -0.199 & +0.376 & +0.905 \\ -0.591 & -0.783 & +0.195 \\ -0.782 & +0.496 & -0.378 \end{pmatrix}, \quad V = \begin{pmatrix} -0.774 & -0.294 & -0.562 & 0.000 \\ +0.445 & -0.322 & -0.445 & -0.707 \\ -0.445 & +0.322 & +0.445 & -0.707 \\ -0.071 & -0.841 & +0.537 & 0.000 \end{pmatrix},$$

and singular values $\sigma_1 = 8.367$, $\sigma_2 = 0.931$, $\sigma_3 = 0.363$. (The SVD is typically computed using mathematical software. The above were computed using the MATLAB/OCTAVE code [U,S,V]=svd(A);) We consider a rank-1 approximation of A given by $A_1 = \sigma_1 u_1 v_1^\mathsf{T}$. (In Exercise 7, the reader is asked to show that such an outer product is, indeed, rank 1.) We have $\|A\|_F = 8.544$ and from Theorem 7.5.11, the error in the approximation satisfies $\|A - A_1\|_F \le \sigma_2 + \sigma_3 = 1.294$. We find that

$$A_1 = \sigma_1 u_1 v_1^\mathsf{T} = \begin{pmatrix} +1.287 & -0.741 & +0.741 & +0.118 \\ +3.826 & -2.203 & +2.203 & +0.349 \\ +5.059 & -2.912 & +2.912 & +0.462 \end{pmatrix}.$$

And, in fact, $\|A - A_1\|_F = 0.999$. We can also provide a rank-2 approximation for A:

$$A_2 = A_1 + \sigma_2 u_2 v_2^\mathsf{T} = \begin{pmatrix} +1.185 & -0.854 & +0.854 & -0.176 \\ +4.040 & -1.968 & +1.968 & +0.962 \\ +4.923 & -3.061 & +3.061 & +0.074 \end{pmatrix},$$

where $\|A - A_2\|_F = 0.363$. Finally, we note that because A is a rank 3 matrix, the rank three approximation is exact, $A_3 = A$. □

In addition to approximating matrices, we can also determine approximations of linear transformations through their matrix representations. This is particularly useful if the transformation has a nullspace of large dimension compared to its rank.

The matrix approximations of Theorem 7.5.11 can be applied to any two-dimensional array of numerical values. Grayscale images, as an array of pixel values, can be treated as a matrix and represented in compressed form using the singular value decomposition. RGB images can be compressed by compressing the three channels separately.

Example 7.5.13 (Data Compression) Consider the bottom right sub-image of the seedling tree in Figure 7.22. This image is a pixel array using 8-bit integer values (values between 0 and 255) in red, blue, and green channels. Let $M^{(r)}$, $M^{(b)}$ and $M^{(g)}$ be the 772 by 772 matrices containing the red, blue, and green 8-bit integer values. The values in each channel can be approximated with a rank-s

approximation through the singular value decomposition. In particular, we compute

$$M_s^{(r)} = \sum_{k=1}^{s} \sigma_k^{(r)} u_k^{(r)} v_k^{(r)\mathsf{T}} = \widetilde{U}_s^{(r)} \widetilde{\Sigma}_s^{(r)} \widetilde{V}_s^{(r)\mathsf{T}},$$

where the singular value decomposition of $M^{(r)}$ is $\widetilde{U}^{(r)} \widetilde{\Sigma}^{(r)} \widetilde{V}^{(r)\mathsf{T}}$ (and similarly for the blue and green channels). The resulting RGB approximate image is defined by the values in $M_s^{(r)}$, $M_s^{(b)}$, and $M_s^{(g)}$. The rank-s SVD approximations are not necessarily 8-bit integer values.

Example rank-s compressed images are shown in Figure 7.22 for $s = 8, 16, 32, 64, 128$. The image arrays have rank-772 and rank-875, respectively. Notice that even the rank-8 approximation captures a good proportion of the larger shape and color features. The rank-16 approximations are recognizable scenes and require less than 5% of the memory storage as the full image. The rank-128 approximations, which are visually indistinguishable from the full image, require less than one-third of the memory storage. $\qquad\square$

7.5.2 Computing the Pseudo-Inverse

The singular value decomposition of a matrix T identifies an isomorphism between the row-space, $\mathrm{Null}(T)^\perp$, and column-space, $\mathrm{Ran}(T)$, of a matrix. In particular, we have the pseudo-inverse theorem.

> **Theorem 7.5.14 [Pseudo-Inverse]**
> Let M be an $m \times n$ real matrix of rank r with singular value decomposition $M = U\Sigma V^\mathsf{T}$. Then for all integers $1 \le s \le r$,
>
> $$P_s = \widetilde{V}_s \widetilde{\Sigma}_s^{-1} \widetilde{U}_s^\mathsf{T}$$
>
> is the unique $n \times m$ real matrix of rank s such that $P_s M x = x$ for all $x \in \mathrm{Span}\{v_1, v_2, \cdots, v_s\}$.

Proof. Suppose M is an $m \times n$ real matrix of rank r with singular value decomposition $M = U\Sigma V^\mathsf{T}$. Let $1 \le s \le r$. Then for $x \in \mathrm{Span}\{v_1, v_2, \cdots, v_s\}$, there are scalars, $\alpha_1, \alpha_2, \ldots, \alpha_s$ so that

$$x = \alpha_1 v_1 + \alpha_2 v_2 + \ldots + \alpha_s v_s.$$

Assume that $v_1, v_2, \ldots, v_n$ are the columns of V, $u_1, u_2, \ldots, u_m$ are the columns of U, and that $\sigma_1, \sigma_2, \ldots, \sigma_{\min\{m,n\}}$ are the nonzero diagonal entries of Σ. Then

$$v_k^\mathsf{T} x = \begin{cases} \alpha_k & \text{if } 1 \le k \le s \\ 0 & \text{if } k > s \end{cases}.$$

Therefore, we have

$$P_s M x = \left(\widetilde{V}_s \widetilde{\Sigma}_s^{-1} \widetilde{U}_s^\mathsf{T} \right) \left(U\Sigma V^\mathsf{T} x \right)$$

Fig. 7.22 Illustration of image compression using the singular value decomposition. The uncompressed images M are shown at the bottom right of each image sequence. The other five images in each sequence are compression examples M_s using s singular vectors/values. The s values are given in the corner of each image. The RGB color channels of these images have been compressed independently.

$$= \left(\widetilde{V}_s \widetilde{\Sigma}_s^{-1} \widetilde{U}_s^\mathsf{T} \right) U \begin{pmatrix} \sigma_1 v_1^\mathsf{T} x \\ \vdots \\ \sigma_s v_s^\mathsf{T} x \\ 0 \\ \vdots \\ 0 \end{pmatrix}$$

$$= \left(\widetilde{V}_s \widetilde{\Sigma}_s^{-1} \widetilde{U}_s^\mathsf{T} \right) \sum_{k=1}^{s} \sigma_k u_k v_k^\mathsf{T} x$$

$$= \left(\widetilde{V}_s \widetilde{\Sigma}_s^{-1} \widetilde{U}_s^\mathsf{T} \right) \left(\widetilde{U}_s \widetilde{\Sigma}_s \widetilde{V}_s^\mathsf{T} x \right)$$

$$= x.$$

Therefore, P_s satisfies $P_s M x = x$. The uniqueness of P_s follows from Corollary 4.2.27. $\square$

The matrix P_s is the pseudo-inverse that we seek. The pseudo-inverse matrix is written as a product of three matrices. Revisiting the ideas that led to Equation 7.9, for the transformation $M = U\Sigma V^T$ of rank r we can write

$$P_s = \widetilde{V}_s \widetilde{\Sigma}_s^{-1} \widetilde{U}_s^\mathsf{T} = \sum_{k=1}^{s} \frac{v_k u_k^\mathsf{T}}{\sigma_k}. \qquad (7.10)$$

for all $0 < s \le r$. Here, we give the formal definition of the pseudo-inverse of rank s.

Definition 7.5.15

The matrix P_s of Theorem 7.5.14 is the **rank s pseudo-inverse** of M. When $r = \text{Rank}(M)$, the rank r pseudo-inverse, P_r is called the **pseudo-inverse** of M and is denoted, simply, P. $\square$

Let us now consider a couple of examples of pseudo-inverse computations.

Example 7.5.16 Consider the matrix

$$A = \begin{pmatrix} 4 & 2 & 2 & 2 & 6 \\ 1 & 1 & 2 & 1 & 5 \\ 2 & 3 & 4 & 4 & 3 \\ 5 & 2 & 1 & 2 & 4 \end{pmatrix}.$$

This 4×5 matrix is non-invertible and has rank $r = 3$. The singular value decomposition $A = U\Sigma V^\mathsf{T}$ provides the pseudo-inverse

$$P = \widetilde{V}_r \widetilde{\Sigma}_r \widetilde{U}_r^\mathsf{T} \approx \begin{pmatrix} +0.0296 & -0.1584 & -0.0455 & +0.2027 \\ -0.0190 & -0.0543 & +0.0856 & +0.0258 \\ -0.0268 & +0.0498 & +0.1350 & -0.0899 \\ -0.0390 & -0.0724 & +0.1413 & +0.0139 \\ +0.0870 & +0.1900 & -0.0903 & -0.0596 \end{pmatrix}.$$

The pseudo-inverse matrix is 5×4. The reader should check to see that $PA \ne I_5$ and $AP \ne I_4$. Consider the three vectors

$$v_1 = \begin{pmatrix} 2 \\ 1 \\ 1 \\ 1 \\ 3 \end{pmatrix}, \quad v_2 = \begin{pmatrix} -2 \\ 3 \\ -4 \\ 2 \\ 1 \end{pmatrix}, \quad v_3 = \begin{pmatrix} 1 \\ 1 \\ 1 \\ 1 \\ 1 \end{pmatrix}.$$

The reader can also verify that $v_1 \in \text{Row}(A)$ and indeed, $P A v_1 = v_1$. Notice that $v_2 \in \text{Null}(A)$, that is, $A v_2 = 0$. So $P(A v_2) = 0$. Thus, v_2 is not recoverable from the pseudo-inverse transformation. For the third vector, we find

$$P A v_3 = P \begin{pmatrix} 18 \\ 10 \\ 16 \\ 14 \end{pmatrix} \approx \begin{pmatrix} 0.998 \\ 0.882 \\ 0.970 \\ 1.108 \\ 1.014 \end{pmatrix}.$$

We see that the result of $P A v_3$ is only an approximation to v_3 and v_3 is not exactly recovered from $A v_3$ by P. This means that $v_3 \notin \text{Row}(A)$. $\qquad \square$

Example 7.5.17 Consider the matrix representation, T, of the transformation $D : \mathcal{P}_3(\mathbb{R}) \to \mathcal{P}_2(\mathbb{R})$ defined by $D(f(x)) = f'(x)$. Using the standard basis for $\mathcal{P}_3(\mathbb{R})$, the reader should verify that

$$T = \begin{pmatrix} 0 & 1 & 0 & 0 \\ 0 & 0 & 2 & 0 \\ 0 & 0 & 0 & 3 \end{pmatrix}, \text{ and } P = \begin{pmatrix} 0 & 0 & 0 \\ 1 & 0 & 0 \\ 0 & 1/2 & 0 \\ 0 & 0 & 1/3 \end{pmatrix}.$$

From calculus, we expect that P should represent an antiderivative transformation. Let $f(x) = a + bx + cx^2 + dx^3$ be an arbitrary vector in $\mathcal{P}_3(\mathbb{R})$ where $a, b, c, d \in \mathbb{R}$. For notational simplicity, we will use $[v]$ to indicate a coordinate vector according to the standard basis. We have

$$[f(x)] = \begin{pmatrix} a \\ b \\ c \\ d \end{pmatrix}, \quad T[f(x)] = \begin{pmatrix} b \\ 2c \\ 3d \end{pmatrix}, \text{ and } PT[f(x)] = \begin{pmatrix} 0 \\ b \\ c \\ d \end{pmatrix}.$$

The pseudo-inverse integration transformation does not recover the constant of integration a because the derivative transformation is not invertible. However, if $f(x) \in \text{Span}\{x, x^2, x^3\}$, then $PT[f(x)] = [f(x)]$. $\qquad \square$

In general, we have found that if P is the pseudo-inverse of M, then $P M x \neq x$. If the pseudo-inverse is to be used to recover, or partially recover, transformed vectors, it is important to understand the relationship between $P M x$ and x for all x.

Corollary 7.5.18
Let M be an $m \times n$ real matrix of rank r with singular value decomposition $M = U \Sigma V^{\mathsf{T}}$. Let $X_s = \text{Span}\{v_1, v_2, \cdots, v_s\}$. Then $P_s M x = \text{proj}_{X_s} x = \widetilde{V}_s \widetilde{V}_s^{\mathsf{T}} x$ for all $x \in \mathbb{R}^n$ and all integers $1 \leq s \leq r$.

The proof is similar to the argument in the proof of Theorem 7.5.14 applied to an arbitrary vector $x \in \mathbb{R}^n$. The proof is the subject of Exercise 12.

Example 7.5.19 Recall Example 7.5.17. The derivative transformation matrix T has rank $r = 3$. Let $X = \text{Row}(T) = \text{Span}\{[x], [x^2], [x^3]\}$. We also have $\text{Null}(T) = \text{Span}\{[1]\}$. For arbitrary $f(x) = a + bx + cx^2 + dx^3 \in \mathcal{P}_3(\mathbb{R})$, we have

$$\text{proj}_X[f(x)] = [bx + cx^2 + dx^3] = \begin{pmatrix} 0 \\ b \\ c \\ d \end{pmatrix} = PT[f(x)].$$

$\square$

Example 7.5.20 Recall the example matrix A in Example 7.5.16. Let $X = \text{Row}(A)$. Because $v_1 \in X$, $\text{proj}_X v_1 = v_1$ and $P A v_1 = \text{proj}_X v_1 = v_1$ as previously demonstrated. Also, Corollary 7.5.18 shows us that

$$\text{proj}_X v_3 = P A v_3 \approx \begin{pmatrix} 0.998 \\ 0.882 \\ 0.970 \\ 1.108 \\ 1.014 \end{pmatrix}.$$

$\square$

Example 7.5.21 Recall the radiographic transformation T of Example 7.5.7. We found that $\text{Row}(T) = \text{Span}\{v_1, v_2, v_3\}$ and $\text{Ran}(T) = \text{Span}\{u_1, u_2, u_3\}$. The pseudo-inverse is the transformation $P : \text{Ran}(T) \to \text{Row}(T)$ defined as

$$P = \tilde{V}_r \tilde{\Sigma}_r \tilde{U}_r^{\mathsf{T}} = \frac{1}{12} \begin{pmatrix} 5 & -3 & 3 & -1 & -1 & 3 \\ -1 & 3 & 3 & -1 & 5 & -3 \\ 3 & -1 & -1 & 3 & -3 & 5 \\ -3 & 5 & -1 & 3 & 3 & -1 \end{pmatrix},$$

where $r = 3$, the rank of T. And,

$$PT = \frac{1}{4} \begin{pmatrix} 3 & 1 & 1 & -1 \\ 1 & 3 & -1 & 1 \\ 1 & -1 & 3 & 1 \\ -1 & 1 & 1 & 3 \end{pmatrix}.$$

Now, consider the object vector $x = \begin{pmatrix} 1 & 2 & 3 & 4 \end{pmatrix}^{\mathsf{T}}$, which transforms to radiograph vector $y = Tx = \begin{pmatrix} 7/2 & 13/2 & 3 & 7 & 9/2 & 11/2 \end{pmatrix}^{\mathsf{T}}$. The reconstructed object vector is $Py = \begin{pmatrix} 1 & 2 & 3 & 4 \end{pmatrix}^{\mathsf{T}} = x$. So, x must be a vector in the row space of T. In fact, $x = \frac{5}{2} v_1 + v_2 + \frac{1}{2} v_3$.

As a second example, consider the object vector $x = \begin{pmatrix} 1 & 2 & 2 & 2 \end{pmatrix}^{\mathsf{T}}$, which transforms to the radiograph vector $y = Tx = \begin{pmatrix} 3 & 4 & 3 & 4 & 7/2 & 7/2 \end{pmatrix}^{\mathsf{T}}$. The reconstructed object vector is $Py = \begin{pmatrix} 5/4 & 7/4 & 7/4 & 9/4 \end{pmatrix}^{\mathsf{T}} \neq x$. So, x must have a nullspace component and PTx is the projection of x onto the row space of T. In this case, the actual object desciption x is not exactly recoverable from the radiograph y. $\square$

☞ **Path to New Applications**

Data Analysts use principal component analysis (PCA) to find a small set of vectors that describe their data well, reducing the dimension of the problem. These *principal vectors* are merely translated singular vectors. The principal vectors then form an orthonormal basis for an *affine subspace* (a set formed by translating a subspace away from the origin) that describes the data. That is, the principal vectors form a basis for the subspace prior to translating. See Section 8.3.2 to learn more about PCA and other tools for data applications.

7.5.3   Exercises

Compute the singular value decomposition and pseudo-inverse of each of the matrices in Exercises 1–6 using the methods illustrated in the Examples.

1. $M = \begin{pmatrix} 1 & 2 \\ 0 & 2 \end{pmatrix}$.

2. $M = \begin{pmatrix} 3 & 1 \\ 2 & 1 \end{pmatrix}$.

3. $M = \begin{pmatrix} 1 & 1 & 1 \\ 2 & 2 & 2 \end{pmatrix}$.

4. $M = \begin{pmatrix} 0 & 1 \\ 1 & 2 \\ 2 & 3 \end{pmatrix}$.

5. $M = \begin{pmatrix} 1 & 0 \\ 0 & 1 \end{pmatrix}$.

6. $M = \begin{pmatrix} 1 & -1 & 0 & 1 \\ -2 & 2 & 0 & 1 \\ 0 & 0 & -1 & -1 \end{pmatrix}$.

Additional Exercises

7. Show that if $u \in \mathbb{R}^m$ and $v \in \mathbb{R}^n$ are both nonzero, then the corresponding $m \times n$ matrix given by the outer product, uv^T has rank 1.

8. Show that Equation 7.9 holds for the case $r \leq m < n$.

9. Argue that the singular value decomposition of a matrix is not unique.

10. Let M be a real-valued matrix. Prove that if M^{-1} exists, then M^{-1} is equal to the pseudo-inverse of M.

11. Suppose matrix M has singular value decomposition $M = U\Sigma V^T$. Find a singular value decomposition for M^T.

12. Prove Corollary 7.5.18.

13. Consider Corollary 7.5.18. Show that, for transformation T with singular value decomposition $T = U\Sigma V^T$, pseudo-inverse P, and rank r, $PT = \widetilde{V}_r\widetilde{V}_r^T$.

14. Find the rank s pseudo-inverses (for all possible s) of the transformation from Section 7.4.

15. Consider the radiography Examples 7.5.7 and 7.5.21. Suppose the (unknown) object is

$$x = \begin{array}{|c|c|} \hline 1 & 1 \\ \hline 0 & 1 \\ \hline \end{array}.$$

(a) Using the standard bases, determine $[x]$ and the radiograph $[y] = T[x]$.
(b) Compute $[\hat{x}]$, the object reconstructed from $[y]$ using the pseudo-inverse transformation.
(c) Show that $[x]$ and $[\hat{x}]$ differ by a null vector of T.
(d) Use vector norm concepts to measure the similarity between $[x]$ and $[\hat{x}]$.

16. Determine a singular value decomposition for the general heat diffusion transformation. (Recall that the transformation is symmetric.)
17. Consider Equation 7.10. Compute the rank-1 through rank-r pseudo-inverse matrices of the transformation matrix of Exercise 6.

7.6 Explorations: Pseudo-Inverse Tomographic Reconstruction

We now have the linear algebra tools for performing brain image reconstructions from radiographic data. We have seen that the radiographic transformation $T : \mathbb{R}^N \to \mathbb{R}^M$ may not have ideal properties such as invertibility. In fact, it need not be one-to-one, nor onto. Making use of the singular value decomposition of T, we can construct a pseudo-inverse transformation P, which is the isomorphism between $\mathrm{Ran}(T)$ and $\mathrm{Null}(T)^\perp$. We have seen that this means $PTx = x$ for all $x \in X = \mathrm{Null}(T)^\perp$. So, if we are given a radiograph y, the pseudo-inverse reconstruction produces $\hat{x} = Py$. Our goal is to determine a pseudo-inverse procedure so that $\hat{x}$ is a "good" approximation of x. This section brings together many linear algebra topics. Completion of this exploration requires some synthesis and creative thought. We encourage the reader to be prepared for both. We hope the reader will find enjoyment in this last activity. We expect that such an activity can be enjoyed as a project extending a linear algebra course using this text.

In order to complete this exploration, you will need the following OCTAVE/MATLAB code and data file:

- tomomap.m
- Lab6Radiographs.mat

Here, we explore a radiographic scenario that uses 108 by 108 voxel brain image layers and 30 radiographic views of 108 pixels each. (Recall your answer to Exercise 2 of Chapter 1.) The radiographic views are at equally spaced angles beginning with $0°$ and ending at $174°$. The object voxels and image pixels have the same width.

7.6.1 The First Pseudo-Inverse Brain Reconstructions

We begin by constructing the pseudo-inverse transformation corresponding to the radiographic scenario. Then, we apply this transformation to our radiographic data. We refer the reader to Section 4.1 to be reminded of the appropriate use of tomomap.m.

1. Construct the transformation T using the tomomap function. What is the size of T?
2. Next, we will carefully construct the pseudo-inverse P using the singular value decomposition $T = U\Sigma V^\mathsf{T}$. We do not need the full decomposition of T to perform the reconstruction (See Theorem 7.5.11). In this exercise, we obtain the decomposition $T = \widetilde{U}\widetilde{\Sigma}\widetilde{V}^\mathsf{T}$. The matrices $\widetilde{U}$, $\widetilde{\Sigma}$, and $\widetilde{V}^\mathsf{T}$ are obtained using the svd command.

```
>> [U,S,V]=svd(T,'econ');
```

This computation may take a few minutes for your computer to complete. The second input argument asks the svd command to only return the matrices $\widetilde{U}$, $\widetilde{\Sigma}$ and $\widetilde{V}$, which are called U, S, and V for clarity in the code. The inverse of $\widetilde{\Sigma}$ can be directly computed because it is diagonal:

```
>> SI=diag(1./diag(S));
```

The pseudo-inverse is then computed as

```
>> P=V*SI*U';
```

(a) What is the rank of T? (The numerically stable method for finding rank of a large matrix is by observing the singular values. What kind of observation should you make here?)

(b) What is the dimension of $\text{Null}(T)^{\perp}$?

(c) What is $\text{Nullity}(T)$?

(d) Is T one-to-one? Onto?

3. Now, load the data file **Lab6Radiographs.mat** that contains the radiograph data as a matrix B. The first 181 columns of B are noise-free radiographic data vectors for each brain slice. The next 181 columns of B are low-noise radiographic data vectors for each brain slice.

```
>> load Lab6Radiogaphs.mat;
```

Next, compute and view the pseudo-inverse reconstructions for both noise-free and low-noise radiographic data. All 362 reconstructions are obtained through a left multiplication by the pseudo-inverse operator.

```
>> recon=P*B;
```

Use the following lines of code to display the noise-free and low-noise reconstructions for a single brain slice.

```
>> slice=50;                          % choose a slice to view
>> brain1=recon(:,slice);             % noise-free recon
>> brain1=reshape(brain1,108,108);
>> brain2=recon(:,slice+181);         % low-noise recon
>> brain2=reshape(brain2,108,108);
>> figure('position',[0 0 1200 450]); % figure size
>> imagesc([brain1 brain2]);
>> set(gca,'visible','off');          % don't draw the axes
>> caxis([-32 287])                   % set color scale
>> colormap(gray);
>> colorbar('fontsize',20);
```

The figures that you get show the slice reconstruction from noise-free data (on the left) and from noisy data (on the right).

Observe and describe reconstructions of several slices of the brain. You need not choose slice 50, choose any number between 1 and 181. Note that choosing slices above 160 (slice 1 is near the chin and slice 181 is above the top of the skull) are not all that interesting. Recall that images in this vector space are expected to be 8-bit grayscale (values from 0 to 255).

You have now computed brain slice reconstructions! Some do not look that great. We see that the noisy radiographs did not give useful reconstructions. Our next step is to figure out how to attempt reconstructions from noisy data.

7.6.2   Understanding the Effects of Noise.

We can understand why the low-noise reconstructions have such poor quality by examining the details of the transformation P. Consider Equation 7.10 (with $s = r = \text{rank}(T)$):

$$P = \tilde{V}_r \tilde{\Sigma}_r^{-1} \tilde{U}_r^{\mathsf{T}} = \sum_{k=1}^{r} \frac{v_k u_k^{\mathsf{T}}}{\sigma_k}.$$

Reconstruction of a brain slice image is obtained by left multiplication by P. Let y be a noise-free radiograph and b a low-noise radiograph of the same object. These differ by an unknown noise vector η. That is, assume $b = y + \eta$.

4. Expand Pb to be of the form $\hat{x} + x_\eta$ using the representation for P above and our assumptions that $b = y + \eta$. Here, we use $\hat{x}$ to be the noise-free reconstruction and x_η is the contribution to the reconstruction due to the noise.

5. Explain the nature of your initial brain slice reconstructions in terms of how P interacts with η. (Consider how it might occur that the contribution of the reconstruction due to noise might create the effects you saw above. In particular, what properties of η and of T (seen in x_η) might be reasons for x_η to obscure $\hat{x}$?)

6. Test your ideas by examining the individual terms in your expansion for some particular radiograph slices. For example, let $b = \text{recon}(:, \text{slice} + 181)$.

7.6.3   A Better Pseudo-Inverse Reconstruction

The pseudo-inverse transformation suffers from noise amplification along some directions v_k. We can understand why this happens based on the size of the singular values σ_k of T. If a particular σ_k is very small, then any noise contribution along u_k results in an amplified reconstruction contribution along v_k—possibly dominating the noiseless part of the reconstruction. We can alleviate this problem by carefully crafting alternative or approximate pseudo-inverse transformations.

7. In order to understand a strategy for reconstruction from noisy data, we want to consider how the noise might have a contribution along some u_k. We begin with an understanding of noise. Create a list of noise-like properties that a vector might have. Do any of the basis vectors u_k have noise-like properties?

8. As we stated above, a larger contribution in a particular direction can occur based on the size of the singular values. Create a visualization of the singular values of T that will be useful in determining their relative size.

9. Of course, if we knew exactly what the data noise η is, we would just remove it and reconstruct as if there was no noise. But, η is unknown and cannot be removed before attempting a reconstruction of the brain slice. How might one utilize the *known* information contained in the singular value decomposition to reduce noise amplification? Devise one or more modified pseudo-inverse transformations that are not as sensitive to noise amplification. Recall the study of pseudo-inverse construction in Section 7.5.

10. Use your modified pseudo-inverse to perform brain reconstructions on the low-noise radiographs. Compare your new reconstructions to the pseudo-inverse reconstructions obtained from the noise-free radiographs. Some example reconstructions of slice 50 are shown in Figure 7.23. These reconstructions use modified pseudo-inverse transformations applied to the low-noise data.

11. Next, explore how one might choose a "best" modified pseudo-inverse.

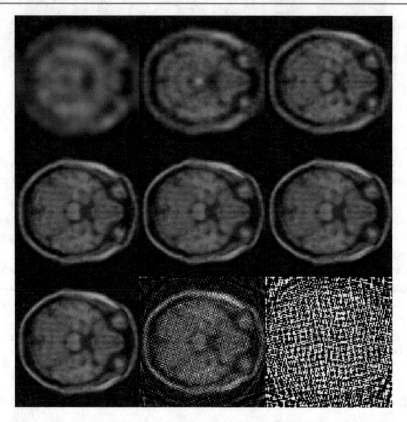

Fig. 7.23 Example modified pseudo-inverse reconstructions of brain slice 50. Reconstructions are inverted from low-noise radiographic data.

(a) Suppose, for the sake of argument, that the actual brain slice description x is known. How would you determine the best pseudo-inverse transformation for recovering x from noisy data b? Remember, the noise vector η is unknown. You might consider how you would measure how well the reconstruction approximates the actual brain slice.

(b) Now, suppose more realistically, that x is unknown, but the magnitude of the noise, $\beta = \|\eta\|$, is known or can be reasonably estimated. How would you determine the best pseudo-inverse transformation for recovering x from noisy data b?

12. Using some of the tools found in Questions 9 and 11, find "good" pseudo-inverse reconstructions for several brain slices. In this exercise, both the actual object, x, and the magnitude of the noise, β, are unknown. Comment on the quality of the reconstructions and estimate the magnitude of the noise vector. It is important to recognize here that you are exploring, using ideas above, to adjust the pseudo-inverse and find the corresponding reconstruction. Your goal is to gain more information about how a "good" reconstruction can be obtained.

7.6.4   Using Object-Prior Information

We have seen that the pseudo-inverse transformation $P : \text{Ran}(T) \to \text{Null}(T)^{\perp}$ is an isomorphism, where T has rank r. So, if $b \in \text{Ran}(T)$, then $x = Pb$ is the unique vector in the domain of T for which

$Tx = b$. Also, if $b \notin \text{Ran}(T)$, say $b = y + w$ where $y \in \text{Ran}(T)$ and $w \in \text{Ran}(T)^{\perp}$, then $x = Pb = Py$ is the unique vector in the domain of T for which $Tx = y$. That is, $w \in \text{Null}(P)$.

In a tomographic situation, the radiographic data b may be noisy, say $b = y + \eta$. If $\eta \in \text{Ran}(T)^{\perp} = \text{Null}(P)$ then it does not affect the reconstruction. In this case, we obtain the desired result $Pb = Py = x$. But if η has a component in $\text{Ran}(T)$, say $\eta = \eta_y + \eta_w$ where $\eta_y \in \text{Ran}(T)$ and $\eta_w \in \text{Ran}(T)^{\perp}$, then it is possible that $\|x\| \ll \|P\eta_w\|$ so that our reconstruction is dominated by noise effects. And, since we do not know η (or η_w), it is difficult to obtain meaningful brain images.

The singular value decomposition of T and the rank-s pseudo-inverse P_s provided a means of reducing the effects of noise amplification allowing us to obtain meaningful brain image reconstructions. However, deciding how to choose an appropriate P_s is unclear unless one has some knowledge of the noise vector, such as its magnitude.

In our next task, we consider additional methods for improving the tomographic process. We may also have some knowledge about the properties of the unknown object.

13. Choose two or three of the best reconstructions you have found in the previous exercises. Consider reconstructions from both noise-free and low-noise radiographs. List and justify reasons why you know that these reconstructions are *not* accurate.

14. The radiographic transformation T from which the pseudo-inverse transformation P was obtained is not one-to-one. List properties (and their radiographic interpretations) of transformations that are not one-to-one. Discuss how one or more of these properties may assist in finding a more realistic brain image from the pseudo-inverse reconstruction. It may be useful to review Sections 4.6 and 4.7.

15. Consider the specific goal of finding a brain slice image $x + w$ where $Pb = x$, b is the radiographic data, $w \in \text{Null}(T)$, and the entries of $x + w$ are nonnegative.

    (a) Would it make sense for entries of $x + w$ to be negative?
    (b) What does the size of $\text{Nullity}(T)$ tell us about the type/amount of object detail found in $\text{Null}(T)$?
    (c) How might a suitable null vector w be determined using linear algebra tools?
    (d) To obtain a reconstructed object that produces the radiographic data we are given, we want to be sure we are changing the reconstruction using "invisible" object data. How do we make sure that we are making adjustments close to our desired corrections while keeping the adjustments "invisible?"

16. Suppose we have a candidate (non-null) correction vector $z \notin \mathcal{N} = \text{Null}(T)$. We do not have a basis for $\mathcal{N}$. So, how can we compute the the closest null vector $w = \text{proj}_{\mathcal{N}} z$? See Theorem 7.2.19 and the subsequent Corollaries 7.2.20 and 7.2.21. Keep in mind that we have only the transformation T along with $\widetilde{U}$, $\widetilde{\Sigma}$ and $\widetilde{V}$.

17. Choose two brain reconstructions of the same slice—one from noise-free data (x_1), and another from low-noise data (x_2). Using the results of questions 15 and 16, and for both reconstructions:

    (a) Compute correction vectors w_1 and w_2,
    (b) Verify that w_2 and w_2 are null vectors of T,
    (c) Compute the corrected reconstructions $x_1 + w_1$ and $x_2 + w_2$.

Carefully examine the corrected reconstructions and draw some conclusions about the success of the procedure. In particular, how well did the corrections w_1 and w_2 "fix" the known problems in x_1 and x_2?

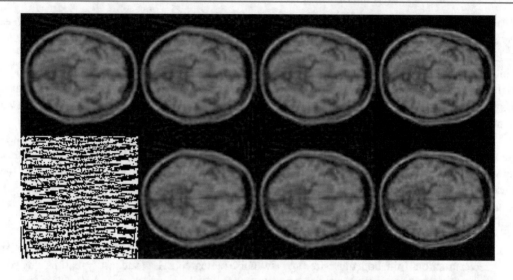

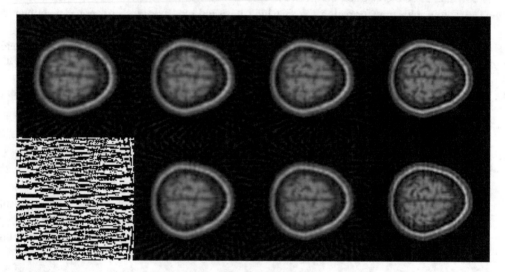

Fig. 7.24 Example brain image reconstructions of slice 62 (upper set of images) and slice 140 (lower set of images). The first row of each set are reconstructions from noise-free radiographs and the second row from low-noise radiographs. The columns correspond to the pseudo-inverse, the rank-2950 pseudo-inverse, the first null-space correction, and the iterative null-space correction, respectively.

18. How might you further improve your corrected reconstructions using an iterative procedure? Apply your procedure to the two reconstructions of question 17 and discuss the results in detail. Figure 7.24 shows example results for brain slices 62 and 140.

19. The current set of explorations has focused on the physical reality that brain densities cannot be negative. Describe several other prior knowledge conditions that could be applicable to the tomography problem. Describe several other types of linear inverse problems that could benefit from prior knowledge null-space enhanced corrections.

7.6.5   Additional Exercises

1. Plot several (left) singular vectors u_k of T some which correspond to small singular values and some to large singular values. Which k values appear more prone to noise amplification in a pseudo-inverse process? Justify your conclusions.

2. Compute several null-vector enhanced brain slice reconstructions using the prior knowledge that the reconstructed density image, ρ, must have values that lie in a given range: $[0, \rho_{max}]$.

3. Suppose object densities are known to come from a finite specified list $\{\rho_1, \rho_2, \ldots, \rho_m\}$. How can you implement this prior knowledge in a null-vector enhanced reconstruction scheme?

4. A fiducial object is an object of known geometry and density used in radiographic testing. Suppose one or more fiducial objects are placed in the object space during the radiographic process. Describe how the fiducial information can be used in a null-vector enhanced reconstruction scheme.

Conclusions

8

We began with two main application ideas: Radiography/tomography of brain scan images and heat diffusion for diffusion welding and image warping. In this chapter, we would like to recognize what we did and how we move forward.

8.1 Radiography and Tomography Example

In this text, we considered the example of brain scan tomography. Given a radiograph made from several views of an object, we want to find the object that produced such a radiograph. See Figure 8.1. In the case of brain scans, we recognize that the more views used, the more radiation the patient experiences. To limit such exposure, our goal was to find good reconstructed brain images from few views. We recognize that the radiographic transformation is, in general, not one-to-one. This means that every radiograph can be produced from many different objects. We saw that the two objects in Figure 8.2 produce the same radiograph, though there are subtle differences. This led us to ask which object, among infinitely many possibilities, produced the given radiograph.

Because no inverse exists for the radiographic transformation T, we sought a method for "inverting" the transformation. Really, we were just looking for one solution to $Tx = b$. Using orthogonal coordinates, we were able to simplify the pseudo-inverse computations. After finding a solution, we realized that just any solution was not good enough, especially in the event that the radiographic data included measurement noise (see Figure 8.3). In the end, we were able to create a pseudo-inverse algorithm that gave good approximate reconstructions, relying heavily on the linearity of the radiographic transformation and the idea that all objects that produce the same radiograph differ by nullspace objects. Determining a desired adjustment vector in the nullspace via orthogonal projection provided a new reconstruction that we found to be better. In the end, we produced good approximate reconstructions (see Figure 7.24).

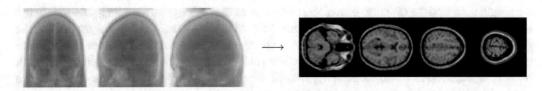

Fig. 8.1 How do we reproduce the brain slices (right) from the radiographs (left)?

© Springer Nature Switzerland AG 2022
H. A. Moon et al., *Application-Inspired Linear Algebra*, Springer Undergraduate Texts in Mathematics and Technology, https://doi.org/10.1007/978-3-030-86155-1_8

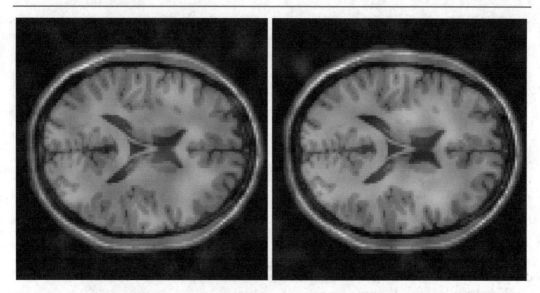

Fig. 8.2 Two different brain images that produce the same radiograph in a particular 30-view radiographic geometry.

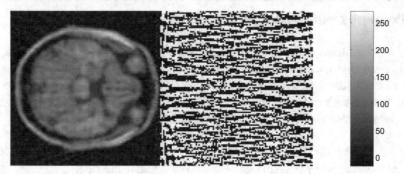

Fig. 8.3 Left: Reconstructed brain image slice, from noise-free data, using a pseudo-inverse. Right: Reconstructed brain image slice, from low-noise data, using a pseudo-inverse.

The problem of object reconstruction arises in many applications and is a current problem in image analysis. We have just scratched the surface in this field of study. There is still much to do to create good reconstructions. We encourage the interested reader to explore more.

8.2 Diffusion

In this text, we also considered the example of heat diffusion for diffusion welding. In particular, we sought a method for describing long-term behavior as heat diffuses along a long thin rod. In our exploration, we found particular heat states that diffused in a way that only scaled the vector (see Figure 8.4). More importantly, we found that these eigenvectors formed a basis for the diffusion transformation. This was an exciting find as we were able to describe the long-term behavior without repeated application of (multiplication by) the diffusion matrix. Instead, we used the powers of the eigenvalues to describe long-term behavior in two ways:

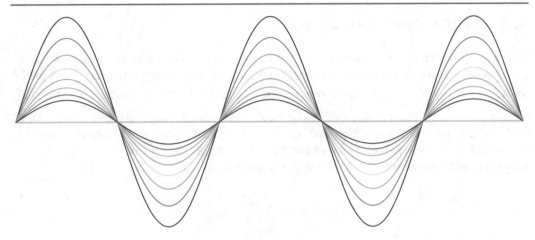

Fig. 8.4 An eigenvector of the diffusion matrix diffuses by scaling in amplitude only.

- First, we were able to create a diagonalization of the diffusion matrix, E using eigenvectors $v_1, v_2, \ldots, v_n$ as columns of a matrix V and the corresponding eigenvalues $\lambda_1, \lambda_2, \ldots, \lambda_n$ as diagonal entries of the diagonal matrix D, giving for any time step k

$$h(k) = E^k h(0) = V D^k V^{-1} h(0).$$

Because D is diagonal, D^k is computed by raising each diagonal entry to the kth power.
- We also viewed the heat state at the kth time step by writing it as a linear combination of the eigenvectors as follows:

$$h(0) = \alpha_1 v_1 + \alpha_2 v_2 + \ldots \alpha_n v_n.$$
$$h(k) = \alpha_1 \lambda_1^k v_1 + \alpha_2 \lambda_2^k v_2 + \ldots + \alpha_n \lambda_n^k v_n.$$

Again, this computation is simplified because we need only raise the eigenvalues to the kth power. But, more interesting is that this representation shows that the larger the value of λ_ℓ, the more contribution to the kth heat state comes from v_ℓ.

Simplifying the computations gives us a means for predicting the best time to remove a rod from the diffusion welder. In other Markov processes, we see a similar ability to predict the state of the process at a particular iteration.

8.3 Your Next Mathematical Steps

We promised in Chapter 1 that you would be prepared to use linear algebra tools in other areas of mathematics. Below, we provide brief discussions to connect linear algebra tools to a few (of many) areas of active research. We hope that it will encourage you to seek out more detailed resources and continue your investigations.

8.3.1 Modeling Dynamical Processes

Modeling dynamic processes means that we are finding functions based on how variables change with respect to others. From calculus, we know that this means we have derivatives like $\frac{d^n y}{dx^n}$ or partial derivatives like $\frac{\partial^n y}{\partial x_1^{k_1} \partial x_2^{k_2} ... \partial x_m^{k_m}}$. Observations and various scientific laws instruct scientists, economists, and data analysts about how to form differential equations that describe how a function changes based on the independent variables. A differential equation is an equation where the unknown is a function, and the structure of the equation involves terms that include the function and its derivatives. We saw a partial differential equation related to heat diffusion given by (in 3D)

$$c \nabla_x \cdot \nabla_x y = c \sum_{i=1}^{3} \frac{\partial^2 y}{\partial x_i^2} = \frac{\partial y}{\partial t}.$$

In solving this equation, the goal is to find a function $y : \mathbb{R}^3 \to \mathbb{R}$ that has first and second partial derivatives that satisfy the above equation. Notice that this equation says that the value of y changes over time in a way that is related to how y curves in the x_i directions.

Researchers interested in providing a tool for early detection of tsunamis create partial differential equations (and sometimes systems of partial differential equations) using information about how the wave speed and shape changes due to the impedance created by the topography of the ocean floor, currents, wind, gravitational forces, energy dissipation, and water temperature. Partial differential equations are also used to characterize and predict vegetation patterns. Such patterns arise as the result of resource competition and various climatic and terrain conditions. Similar partial differential equations can be used to describe skin coloration patterns in mammals such as zebras and leopards.

Based on the properties of differentiation given in calculus, we learned that the derivative operator is linear. One can then find eigenfunctions f_k for the differential operator D ($D f_k = \lambda f_k$) and construct solutions based on initial values and boundary conditions. Approximate solutions can be found by discretizing the function and its derivatives thereby creating matrix equations. One can then apply the various techniques given in the invertible matrix theorem (Theorem 7.3.16) to determine the existence of solutions.

8.3.2 Signals and Data Analysis

In this text, we have seen data in the form of images. But, we hear about data that are collected through financial and business, political, medical, social, and scientific research. Data exist on just about everything we can measure. As we mentioned in Section 1.1.2, researchers are interested in determining ways to glean information from this data, looking for patterns for classification or prediction purposes. We also discussed how machine learning is a tool to automate these goals. Here, we will connect the aforementioned machine learning tools with the necessary linear algebra tools.

Fourier analysis is used to recognize patterns in data. The Fourier transform uses an orthonormal basis (of sinusoidal functions) in an infinite dimensional function space to rewrite functions in Fourier space. Function inner products are used to determine which patterns are most prevalent (larger inner products suggest a larger level of importance in the data). Discrete Fourier analysis is similar, but with vector inner products.

Regression analysis is a tool for predicting a result based on prior data. Given a set of user supplied functions, $\{f_k : \mathbb{R}^n \to \mathbb{R} \mid k = 0, 1, ..., m\}$, a regression analysis seeks to find a vector $a^* \in \mathbb{R}^m$ so that for given data, $\{(d_i, u_i) \mid x_i \in \mathbb{R}^n, u_i \in \mathbb{R}, i = 1, 2, ..., N\}$, the vector $a^* = (a_0^*, a_1^*, \ldots, a_m^*)$

minimizes the total least squares deviation LSD,

$$\text{LSD} = \sum_{i=1}^{N} \left(u_i - \sum_{k=0}^{m} a_k f_k(d_i) \right)^2 .$$

The equation $u = \sum_{k=0}^{m} a_k^* f_k(x)$ is the regression equation used for prediction. Using multivariable calculus, LDS minimization can be reduced to solving a matrix equation. We note that if the set of user-supplied functions is not linearly independent, then there is no unique minimizing vector a^*. By choosing a linearly independent set of functions, we recognize that we are finding the coordinate vector from the respective coordinate space that is closest to the given data. On the other hand, if your data fit a function that is not in the span of the set of supplied functions, the least-squares deviation could be large, meaning that the regression may not be an effective predictor.

Data clustering is a tool that uses relatively few feature vectors $\{v_1, v_2, \ldots, v_k\}$ to classify vectors into k clusters. In general, a data vector u belongs to cluster j if u is more similar to vector v_j than any other v_i, where $i \neq j$. Similarity can be defined in many ways. One such way is using inner products. The challenges are finding the "best" feature vectors and an optimal number of these "best" feature vectors.

In the event that a researcher has collected data that can already be classified, yet no choice for feature vectors is possible, support vector machines use the clustered data to classify any new data points. For example, suppose scientists have measured eight key physical and biological attributes for a large sample (training set) of kittens and have regularly monitored their health for their first 5 years. Kittens that developed kidney problems during this time represent one cluster. Kittens did not represent the other cluster. Each kitten is represented as a data vector in $\mathbb{R}^8$. Veterinarians then use a support vector machine to predict the likelihood of a kitten developing kidney problems long before any issues arise. The way that support vector machines work is to transform (using a nonlinear transformation) the set of kitten vectors from $\mathbb{R}^8$ to $\mathbb{R}^n$ for some $n > 8$ in which data clustering is performed using inner products as described above. The challenges exist in finding the appropriate transformation and the appropriate n so that the classification tool is robust, minimizing both false positives and false negatives.

Principle component analysis (PCA) is used to find relevant features or structures in data. This technique is a simple extension from singular value decomposition (SVD). In fact, the only difference is that the data need not be centered around the origin for principal component analysis. The first and last step in PCA are translating the data so that it is centered at the origin and then back after finding the SVD. Now, since SVD finds the subspace that most closely represents the data, PCA finds the affine subspace (a subspace translated from the origin) that most closely represents the data. The vectors obtained after translating the singular vectors back are called the *principal components*. Notice that an affine subspace is not actually a subspace, rather it is a translation of a subspace.

8.3.3   Optimal Design and Decision Making

Optimal Design is the process of determining parameter values that describe a best product design among all possible valid designs. For example, in designing a rudder for a new sailboat, one might wish to maximize the responsiveness. The design is specified by several geometric and material choice parameters. The parameter choices are limited by weight, clearance, manufacturing capability, financial considerations, etc.

Optimal decision making is the process of determining the best strategy among all possible strategies. For example, in planning the employee work schedule for the upcoming week, a company would like a strategy that minimizes paid wages. Some considerations that limit the company's choices include ensuring adequate employee presence, allowing for time off, appropriate matching of employee tasks and skill sets, and assigning tasks that do not degrade morale. Other considerations might include limited employee availability or limited overtime.

These types of optimization problems, and myriad others, are modeled mathematically as an objective function $f : \mathbb{R}^n \to \mathbb{R}$, which quantifies the quality of a particular choice of variables, and constraint equalities $h(x) = 0$ ($h : \mathbb{R}^n \to \mathbb{R}^m$) and inequalities $g(x) \leq 0$ ($g : \mathbb{R}^n \to \mathbb{R}^p$), which limit the choices of possible variable values. The idea is to find the valid choice (satisfy all constraints), which maximizes (or minimizes) the objective function. Typically, there are infinitely many valid choices, but only one optimal choice.

If the model functions, f, g and h, are linear, then we expect that linear algebra will be a key tool. In general, there are infinitely many valid solutions defined by the solution set of a system of linear equations, and optimal solutions are obtained by examining the basic solutions of the system. For a system with m equations and n unknowns, the basic solutions are found by setting $n - m$ variables (free variables) to zero and solving for the remaining (basic) variables. Invertibility properties of submatrices of the coefficient matrix lead to the presence or absence of basic solutions. This is the study of *linear programming*.

Even when optimization problems involve *nonlinear* objective and constraint functions, linear algebra is a necessary tool. Efficient methods for solving these types of problems involve local quadratic function approximations. Suppose we have quantitative derivative information of the function at $x = a$, then the quadratic approximation of f near a is

$$f(x) \approx f(a) + \nabla f(a)^{\mathsf{T}}(x - a) + \frac{1}{2}(x - a)^{\mathsf{T}}\nabla^2 f(a)(x - a),$$

where $\nabla f(a)$ is the vector of first partial derivatives, the ith entry of which is $[\nabla f(a)]_i = \frac{\partial f(a)}{\partial x_i}$, and $\nabla^2 f(a)$ is the Hessian matrix of second partial derivatives, the i, j-entry of which is $\left[\nabla^2 f(a)\right]_{ij} = \frac{\partial^2 f(a)}{\partial x_i \partial x_j}$. Function and gradient evaluations involve linear algebra techniques and the eigenvalues of the Hessian matrix reveal information about the curvature of the function. Together, ∇f and $\nabla^2 f$, provide specific information on solutions. For example, if the gradient is the zero vector ($\nabla f(x^*) = 0$) and all eigenvalues of the Hessian are nonnegative ($\nabla^2 f(x^*)v = \lambda v$ and $v \neq 0$ implies $\lambda \geq 0$), then x^* is a local minimizer of f.

8.4 How to move forward

In this text, we explored particular applications for which Linear Algebra topics presented solution techniques. Solution paths for these applications followed a fairly typical procedure. That is, the path we took in this book is very similar to the path an applied mathematician takes to solve other real-world problems. We recap by describing our procedure very simply as

1. Explore the data, looking for characteristics that suggest the use of particular techniques that we can employ.
2. Create simplified and smaller data sets (called "toy data") with similar characteristics (starting with fewer characteristics than are actually present in the data).

3. Apply mathematical techniques that were considered in Step 1 on the toy data. Adjust the current technique so that it can be used on the actual data.
4. Determine what shortcomings we find in the current technique and what how the technique moves us toward our goals.
5. Explore creative and new techniques that might be used to overcome the shortfalls of the previous techniques.
6. Add more richness to the toy data and continue the same process with the new toy data and new technique ideas. At each iteration, increase the complexity of the data until we find a "good" solution to our original problem with the actual data.

These techniques are valid for many other applications such as those mentioned in Section 8.3, and even for applications that are yet to be discovered. Willingness to explore and play with data is key to finding novel solutions to real-life problems.

8.5   Final Words

There is still much to do with the applications that we focused on in this text. The interested reader is encouraged to explore newer techniques that create even better solutions. Further validation for these methods is also needed. The methods encouraged by the exploration in Section 7.6 have been used with many applications, but the validation and refinement of the method, at the time of publication, is still an open problem.

The interested reader is encouraged to seek out texts that delve deeper into applications that we discussed. We encourage you to talk to experts and become part of a community looking for solutions to many of these problems. We encourage you to find new applications and new uses for the tools of linear algebra.

Correction to: Application-Inspired Linear Algebra

Correction to:
H. A. Moon et al., *Application-Inspired Linear Algebra*,
Springer Undergraduate Texts in Mathematics and Technology,
https://doi.org/10.1007/978-3-030-86155-1

The original version of the book has received belated corrections from the author, which have been updated. The book and the chapters have been updated with the changes.

The updated version of this book can be found at https://doi.org/10.1007/978-3-030-86155-1

© Springer Nature Switzerland AG 2023

H. Moon et al., *Application-Inspired Linear Algebra*, Springer Undergraduate Texts
in Mathematics and Technology, https://doi.org/10.1007/978-3-030-86155-1_9

Appendix A
Transmission Radiography and Tomography: A Simplified Overview

This material provides a brief overview of radiographic principles prerequisite to Section 4.1. The goal is to develop the basic discrete radiographic operator for axial tomography of the human body. To accomplish this goal, it is **not** necessary to completely understand the details of the physics and engineering involved here. We wish to arrive at a mathematical formulation descriptive of the radiographic process and establish a standard scenario description with notation.

A.1 What is Radiography?

Transmission radiography and tomography are familiar and common processes in today's world, especially in medicine and non-destructive testing in industry. Some examples include

- Single-view X-ray radiography is used routinely to view inside the human body; for example, bone fracture assessment, mammography, and angiographic procedures.
- Multiple-view X-ray radiography is realized in computerized axial tomography (CAT) scans used to provide 3D images of body tissues.
- Neutron and X-ray imaging is used in industry to quantify manufactured part assemblies or defects which cannot be visually inspected.

Definition A.1.1

Transmission Radiography is the process of measuring and recording changes in a high-energy particle beam (X-rays, protons, neutrons, etc.) resulting from passage through an object of interest. □

Definition A.1.2

Tomography is the process of inferring properties of an unknown object by interpreting radiographs of the object. □

X-rays, just like visible light, are photons or electromagnetic radiation, but at much higher energies and outside of the range of our vision. Because of the wavelength of typical X-rays (on the order of a nanometer), they readily interact with objects of similar size such as individual molecules or

---

© Springer Nature Switzerland AG 2022

H. A. Moon et al., *Application-Inspired Linear Algebra*, Springer Undergraduate Texts in Mathematics and Technology, https://doi.org/10.1007/978-3-030-86155-1

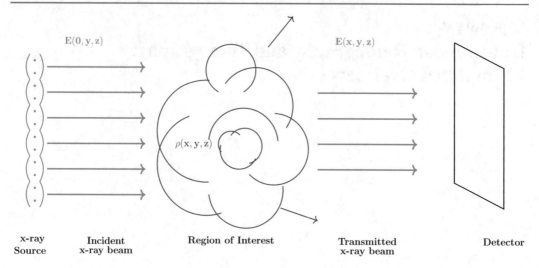

E(0, y, z)   E(x, y, z)

$\rho(x, y, z)$

x-ray   Incident   Region of Interest   Transmitted   Detector
Source   x-ray beam   x-ray beam

Fig. A.1 Typical radiographic experiment.

atoms. This property makes them particularly useful in transmission imaging. Figure A.1 is a cartoon of a typical X-ray radiographic experiment or procedure. An X-ray beam is produced with known energy and geometric characteristics. The beam is aimed at a region of interest. The photons interact with matter in the region of interest, changing the intensity, energy and geometry of the beam. A detector measures the pattern (and possibly the distribution) of incident energy. The detection data, when compared to the incident beam characteristics, contains the known signature of the region of interest. We consider the mathematics and some of the physics involved in each step with the goal of modeling a radiographic transformation appropriate for mixed soft and hard tissue axial tomography of the human body.

A.2   The Incident X-ray Beam

We begin with an X-ray beam in which the X-ray photons all travel parallel to each other in the beam direction, which we take to be the positive x-direction. Additionally we assume that the beam is of short time duration, the photons being clustered in a short pulse instead of being continuously produced. A beam with these geometric characteristics is usually not directly achievable through typical X-ray sources (see supplementary material in Section A.8 for some discussion on X-ray sources). However, such a beam can be approximated readily through so-called collimation techniques which physically limit the incident X-rays to a subset that compose a planar (neither convergent nor divergent) beam.

While not entirely necessary for the present formulation, we consider a monochromatic X-ray source. This means that every X-ray photon produced by the source has exactly the same "color." The term "monochromatic" comes from the visible light analog in which, for example, a laser pointer may produce photons of only one color, red. The energy of a single photon is proportional to its frequency ν, or inversely proportional to its wavelength λ. The frequency, or wavelength, determine the color, in exact analogy with visible light. In particular, the energy of a single photon is $h\nu$ where the constant of proportionality h is known as Planck's constant.

The intensity (or brightness), $E(x, y, z)$, of a beam is proportional to the photon density. The intensity of the beam just as it enters the region of interest at $x = 0$ is assumed to be the same as the intensity at the source. We write both as $E(0, y, z)$. It is assumed that this quantity is well known or independently measureable.

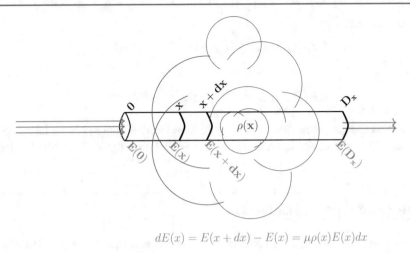

$$dE(x) = E(x + dx) - E(x) = \mu\rho(x)E(x)dx$$

Fig. A.2 X-ray beam attenuation computation for a beam path of fixed y and z.

A.3 X-Ray Beam Attenuation

As the beam traverses the region of interest, from $x = 0$ to $x = D_x$ (see Figure A.2), the intensity changes as a result of interactions with matter. In particular, a transmitted (resultant) intensity $E(D_x, y, z)$ exits the far side of the region of interest. We will see that under reasonable assumptions this transmitted intensity is also planar but is reduced in magnitude. This process of intensity reduction is called attenuation. It is our goal in this section to model the attenuation process.

X-ray attenuation is a complex process that is a combination of several physical mechanisms (see Section A.9) describing both scattering and absorption of X-rays. We will consider only Compton scattering which is the dominant process for typical medical X-ray radiography. In this case, attenuation is almost entirely a function of photon energy and material mass density. As we are considering monochromatic (monoenergetic) photons, attenuation is modeled as a function of mass density only.

Consider the beam of initial intensity $E(0, y, z)$ passing through the region of interest, at fixed y and z. The relative intensity change in an infinitesimal distance from x to $x + dx$ is proportional to the mass density $\rho(x, y, z)$ and is given by

$$dE(x, y, z) = E(x + dx, y, z) - E(x, y, z) = -\mu\rho(x, y, z)E(x, y, z)dx,$$

where μ is a factor that is nearly constant for many materials of interest. We also assume that any scattered photons exit the region of interest without further interaction. This is the so-called single-scatter approximation which dictates that the intensity remains planar for all x.

Integrating over the path from $x = 0$ where the initial beam intensity is $E(0, y, z)$ to $x = D_x$ where the beam intensity is $E(D_x, y, z)$ yields

$$\frac{dE(x, y, z)}{E(x, y, z)} = -\mu\rho(x, y, z)dx$$

$$\int_{E(0,y,z)}^{E(D_x,y,z)} \frac{dE(x, y, z)}{E(x, y, z)} = -\mu \int_0^{D_x} \rho(x, y, z)dx$$

$$\ln E(D_x, y, z) - \ln E(0, y, z) = -\mu \int_0^{D_x} \rho(x, y, z) dx$$

$$E(D_x, y, z) = E(0, y, z) e^{-\mu \int_0^{D_x} \rho(x,y,z)dx}.$$

This expression shows us how the initial intensity is reduced, because photons have been scattered out of the beam. The relative reduction depends on the density (or mass) distribution in the region of interest.

A.4 Radiographic Energy Detection

The transmitted intensity $E(D_x, y, z)$ continues to travel on to a detector (e.g., film) which records the total detected energy in each of m detector bins. The detected energy in any bin is the intensity integrated over the bin cross-sectional area. Let p_k be the number of X-ray photons collected at detector bin k. p_k is then the collected intensity integrated over the bin area and divided by the photon energy.

$$p_k = \frac{1}{h\nu} \iint_{(bin\ k)} E(0, y, z) \left(e^{-\mu \int_0^{D_x} \rho(x,y,z)dx} \right) dy dz.$$

Let the bin cross-sectional area, σ, be small enough so that both the contributions of the density and intensity to the bin area integration are approximately a function of x only. Then

$$p_k = \frac{\sigma E(0, y_k, z_k)}{h\nu} e^{-\mu \int_0^{D_x} \rho(x,y_k,z_k)dx},$$

where y_k and z_k locate the center of bin k. Let p_k^0 be the number of X-ray photons initially aimed at bin k, $p_k^0 = \sigma E(0, x, y)/h\nu$. Due to attenuation, $p_k \leq p_k^0$ for each bin.

$$p_k = p_k^0 e^{-\mu \int_0^{D_x} \rho(x,y_k,z_k)dx}.$$

Equivalently, we can write (multiply the exponent argument by σ/σ):

$$p_k = p_k^0 e^{-\frac{\mu}{\sigma} \int_0^{D_x} \sigma\rho(x,y_k,z_k)dx}.$$

The remaining integral is the total mass in the region of interest that the X-ray beam passes through to get to bin k. We will call this mass s_k. Now we have

$$p_k = p_k^0 e^{-s_k/\alpha},$$

where $\alpha = \sigma/\mu$. This expression tells us that the number of photons in the part of the beam directed at bin k is reduced by a factor that is exponential in the total mass encountered by the photons.

Finally, we note that the detector bins correspond precisely to pixels in a radiographic image.

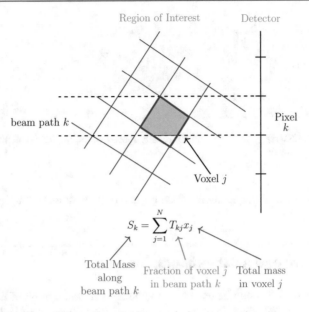

$$S_k = \sum_{j=1}^{N} T_{kj} x_j$$

Total Mass along beam path k Fraction of voxel j in beam path k Total mass in voxel j

Fig. A.3 Object space and radiograph space discretization.

A.5   The Radiographic Transformation Operator

We consider a region of interest subdivided into N cubic voxels (3-dimensional pixels). Let x_j be the mass in object voxel j and T_{kj} the fraction of voxel j in beam path k (see Figure A.3). Then the mass along beam path k is

$$s_k = \sum_{j=1}^{N} T_{kj} x_j,$$

and the expected photon count at radiograph pixel k, p_k, is given by

$$p_k = p_k^0 e^{-\frac{1}{\alpha} \sum_{j=1}^{N} T_{kj} x_j},$$

or equivalently,

$$b_k \equiv \left(-\alpha \ln \frac{p_k}{p_k^0} \right) = \sum_{j=1}^{N} T_{kj} x_j.$$

The new quantities b_k represent a variable change that allows us to formulate the matrix expression for the radiographic transformation

$$b = Tx.$$

This expression tells us that given a voxelized object mass distribution image $x \in \mathbb{R}^N$, the expected radiographic data (mass projection) is image $b \in \mathbb{R}^m$, with the two connected through radiographic transformation $T \in \mathcal{M}_{m \times N}(\mathbb{R})$. The mass projection b and actual photon counts p and p^0 are related as given above. It is important to note that b_k is defined only for $p_k > 0$. Thus, this formulation is only valid for radiographic scenarios in which every radiograph detector pixel records at least one hit. This

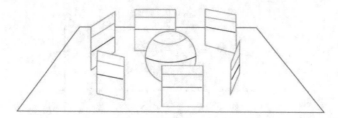

Fig. A.4 Example axial radiography scenario with six equally spaced views. Horizontal slices of the object project to horizontal rows in the radiographs.

is always the case for medical applications which require high contrast and high signal-to-noise ratio data.

A.6   Multiple Views and Axial Tomography

Thus far, we have a model that can be used to compute a single radiograph of the region of interest. In many applications, it is beneficial to obtain several or many different views of this region. In some industrial applications, the region of interest can be rotated within the radiographic apparatus, with radiographs obtained at various rotation angles. In medical applications the radiographic apparatus is rotated about the region of interest (including the subject!). In this latter case, the voxelization remains fixed and the coordinate system rotates. For each of a view angles the new m pixel locations require calculation of new mass projections T_{kj}. The full multiple-view operator contains mass projections onto all $M = a \cdot m$ pixel locations. Thus, for multiple-view radiography: x is still a vector in $\mathbb{R}^N$, but b is a vector in $\mathbb{R}^M$ and T is a matrix operator in $\mathcal{M}_{M \times N}(\mathbb{R})$.

Finally, we make the distinction between the general scenario and axial tomography (CAT scans). In principle, we could obtain radiographs of the region of interest from any direction (above, below, left, right, front, back, etc.). However, in axial tomography the physical limitations of the apparatus and subject placement dictate that views from some directions are not practical. The simplest scenario is to obtain multiple views by rotating the apparatus about a fixed direction perpendicular to the beam direction. This is why CAT machines have a donut or tube shaped appearance within which the apparatus is rotated. The central table allows the subject to rest along the rotation axis and the beam can pass through the subject along different trajectories.

This axial setup also simplifies the projection operator. If we consider the ℓth slice of the region of interest, described by an $n \times n$ arrray of N voxels, the mass projections of this slice will only occur in the ℓth row of pixels in each radiographic view see Figure A.4. As a result, 3D reconstructions can be obtained by a series of independent 2D reconstructed slices. For example, the brown slice of the spherical object (represented in $\mathbb{R}^N$) is related to the collection of brown rows of the radiographs (represented in $\mathbb{R}^M$) through the projection operator $T \in \mathcal{M}_{M \times N}(\mathbb{R})$. The black slice and black rows are related through the *same* projection operator.

A.7   Model Summary

The list below gathers the various mathematical quantities of interest.

- N is the number of object voxels.
- M is the number of radiograph pixels.

- $x \in \mathbb{R}^N$ is the material mass in each object voxel.
- $b \in \mathbb{R}^M$ is the mass projection onto each radiograph pixel.
- $p \in \mathbb{R}^M$ is the photon count recorded at each radiograph pixel.
- $T \in \mathcal{M}_{N \times M}(\mathbb{R})$ is voxel volume projection operator. T_{ij} is the fractional volume of voxel j which projects orthogonally onto pixel i.
- p_k^0 is the incident photon count per radiograph pixel.
- $b = -\alpha \ln \frac{p}{p^0}$.
- $b = Tx$ is the (mass projection) radiographic transformation.

The description of images (objects and radiographs) as vectors in $\mathbb{R}^N$ and $\mathbb{R}^M$ is computationally useful and more familiar than a vector spaces of images. One should keep in mind that this is a particular representation for images which is useful as a tool but is not geometrically descriptive. The price we pay for this convenience is that we no longer have the geometry of the radiographic setup (pixelization and voxelization) encoded in the representation.

A vector in $\mathbb{R}^3$, say $(1, 2, 5)$, is a point in a 3-dimensional space with coordinates described relative to three orthogonal axes. We can actually locate this point and plot it. An image represented in $\mathbb{R}^3$, say $(1, 2, 5)$, is *not* a point in this space. Without further information about the vector space of which it is a member, we cannot draw this image. The use of $\mathbb{R}^3$ allows us to perform scalar multiplication and vector addition on images *because these operations are equivalently defined on* $\mathbb{R}^3$.

A.8 Model Assumptions

The radiography model we have constructed is based on a number of approximations and assumptions which we list here. This list is not comprehensive, but it does gather the most important concepts.

- **Monochromaticity.** Laboratory sources generate X-rays with a wide range of energies as a continuous spectrum. We say that such X-ray beams are polychromatic. One method of approximating a monochromatic beam is to precondition the beam by having the X-rays pass through a uniform material prior to reaching the region of interest. This process preferentially attenuates the lower energy photons, leaving only the highest energy photons. This process is known as beam-hardening. The result is a polychromatic beam with a narrower range of energies. We can consider the beam to be approximately monochromatic, especially if the attenuation coefficient(s) of the material, μ, is not a strong function of photon energy.
- **Geometric Beam Characteristics.** Laboratory sources do not naturally generate planar X-ray beams. It is more characteristic to have an approximate point source with an intensity pattern that is strongly directionally dependent. Approximate planar beams with relatively uniform intensity $E(0, y, x)$ can be achieved by selective beam shielding and separation of source and region of interest. In practice, it is common to use the known point source or line source characteristics instead of assuming a planar beam. The model described here is unchanged except for the computation of T itself.
- **Secondary Radiation.** Our model uses a single-scatter approximation in which if a photon undergoes a Compton scatter, it is removed from the analysis. In fact, X-rays can experience multiple scatter events as they traverse the region of interest or other incidental matter (such as the supporting machinery). The problematic photons are those that scatter one or more times *and* reach the detector. This important secondary effect is often approximated by more advanced models.
- **Energy-Dependent Attenuation.** The attenuation coefficient μ, which we have taken to be constant, is not only somewhat material dependent but is also beam energy dependent. If the beam

is truly monochromatic this is not a problem. However, for a polychromatic beam the transmitted total energy will depend on the *distribution* of mass along a path, not just the total mass.

- **Other Attenuation Mechanisms.** We have included only Compton scattering in the model. Four other mechanisms (outlined in the supplementary material) contribute to the attenuation. While Compton scattering is the dominant contributor, photoelectric scattering will have some effect. It becomes important at the lower energies of interest and for materials of relatively high atomic number—such as calcium which is concentrated in bone. The major effect of ignoring photoelectric scattering is quantitative mass uncertainty.

- There are a number of **Detector-Related Effects** which affect radiograph accuracy. Energy detection efficiency can be a function of beam intensity, photon energy and even pixel location. Detectors are subject to point-spread effects in which even an infinitesimally narrow beam results in a finitely narrow detection spot. These types of effects are usually well-understood, documented and can be corrected for in the data. Detection is also prone to noise from secondary radiation, background radiation, or simply manufacturing variances.

A.9 Additional Resources

There is a variety of online source material that expands on any of the material presented here. Here, are a few starting points.

```
https://www.nde-ed.org/EducationResources/CommunityCollege/Radiography/c_rad_index.htm

http://web.stanford.edu/group/glam/xlab/MatSci162_172/LectureNotes/01_Properties%20&%20Safety.pdf

http://radiologymasterclass.co.uk/tutorials/physics/x-ray_physics_production.html
```

Several physical processes contribute to absorption and scattering of individual photons as they pass through matter. Collectively, these processes alter the geometry and intensity of a beam of such photons. What follows is a brief description of each.

- The **Photoelectric Effect** is the absorption of an X-ray photon by an atom accompanied by the ejection of an outer shell electron. This ionized atom then re-absorbs an electron and emits an X-ray of energy characteristic of the atom. The daughter X-ray is a low-energy photon which is quickly re-absorbed and is effectively removed from the X-ray beam. Photoelectric absorption is the dominant process for photon energies below about 100keV and when interacting with materials of high atomic number.

- **Rayleigh Scattering** is the process of a photon interacting with an atom without energy loss. The process is similar to the collision of two billiard balls. Rayleigh scattering is never the dominant mechanism, at any energy.

- **Compton Scattering** occurs when an X-ray photon interacts with an electron imparting some energy to the electron. Both electron and photon are emitted and the photon undergoes a directional change or scatter. Compton Scattering is the dominant process for soft tissue at photon energies between about 100keV through about 8MeV.

- **Pair Production** is the process in which a photon is absorbed producing a positron-electron pair. The positron quickly decays into two 510keV X-ray photons. Pair Production is only significant for photon energies of several MeV or more.

- **Photodisintegration** can occur for photons of very high energy. Photodisintegration is the process of absorption of the photon by an atomic nucleus and the subsequent ejection of a nuclear particle.

Exercises

1. Given $T_{kj} = 0.42$, what does this value mean?
2. What is accomplished in the matrix multiply when multiplying an object vector by the kth row of T?
3. Explain how you change the radiographic operator when changing from one view to many.
4. Why do we use pixel values b_k where b_k is defined as

$$b_k \equiv -\alpha \ln \left(\frac{p_k}{p_0} \right)$$

instead of the expected photon count p_k?

Appendix B
The Diffusion Equation

The 1-dimensional diffusion equation, or heat equation, is the partial differential equation

$$\frac{\partial f}{\partial t} = \frac{\partial}{\partial x}\left(\kappa(x,t)\frac{\partial f}{\partial x}\right),$$

where κ is the position- and time-dependent diffusion coefficient. The diffusing quantity, f, also depends of both position and time. For example, $f(x,t)$ could describe the distribution of heat on a rod, or the concentration of a contaminant in a tube of liquid. For uniform static materials, we can assume $\kappa = 1$ and use the homogeneous diffusion equation

$$\frac{\partial f}{\partial t} = \frac{\partial^2 f}{\partial x^2}.$$

Solutions to this simplified diffusion equation, $f(x,t)$ describe how the quantity f evolves in time for all positions x. The rate at which f changes in time is equal to the second derivative of f with respect to position x. When the second derivative is positive, the function $f(x)$ is concave up, and f is increasing in this region where f is locally relatively small. Similarly, when the second derivative is negative, the function $f(x)$ is concave down, and f is decreasing in this region where f is locally relatively large. We also notice that the degree of concavity dictates how fast $f(x)$ is changing. Sharp peaks or valleys in $f(x)$, change faster than gentle peaks or valleys. Notice also that where $f(x)$ has nearly constant slope, the concavity is nearly zero and we expect $f(x)$ to be constant in time. Finally, the sign of the rate of change guarantees that $\lim_{t\to\infty}\frac{\partial^2 f}{\partial x^2} = 0$.

Boundary Conditions

We consider the one-dimensional diffusion equation on the interval $[a, b]$ with $f(a, t) = f(b, t) = 0$. This particular boundary condition is appropriate for our discussion and solution of the heat state evolution problem used throughout this text.

© Springer Nature Switzerland AG 2022

H. A. Moon et al., *Application-Inspired Linear Algebra*, Springer Undergraduate Texts in Mathematics and Technology, https://doi.org/10.1007/978-3-030-86155-1

Discretization in Position

We will *discretize* $f(x, t)$ (in position) by sampling the function values at m regularly spaced locations in the interval $[a, b]$. For $L = b - a$, we define $\Delta x = \frac{L}{m+1} = \frac{b-a}{m+1}$. Then, the sampling locations are $a + \Delta x, a + 2\Delta x, \ldots, a + m\Delta x$. We then define the discretized sampling as a vector u in $\mathbb{R}^m$.

$$u = [u_1, u_2, \ldots, u_m] = [f(a + \Delta x), f(a + 2\Delta x), \ldots, f(a + m\Delta x)],$$

where we have suppressed the time dependence for notational clarity. Notice that if $u_j = f(x)$ for some $x \in [a, b]$ then $u_{j+1} = f(x + \Delta x)$ and $u_{j-1} = f(x - \Delta x)$.

Now, because derivatives are linear, intuitively we believe that it should be possible to get a good discrete approximation for $\partial f / \partial x$ and $\partial^2 f / \partial x^2$. We will use the definition of derivative

$$\frac{\partial f}{\partial x} = \lim_{h \to 0} \frac{f(x + h) - f(x)}{h}.$$

Notice that in a discrete setting, it is not possible for $h \to 0$. The smallest h can be is the sampling distance Δx. Let's use this to make the matrix representation for the second derivative operator, which we call D_2. That is, $D_2 u$ approximates $\partial^2 f / \partial x^2$.

$$\begin{aligned}
\frac{\partial^2 f}{\partial x^2} = \frac{\partial}{\partial x} \frac{\partial f}{\partial x} &= \frac{\partial}{\partial x} \left(\lim_{h \to 0} \frac{f(x + h) - f(x)}{h} \right) \\
&\approx \frac{\partial}{\partial x} \left(\frac{f(x + \Delta x) - f(x)}{\Delta x} \right) \\
&= \frac{1}{\Delta x} \lim_{h \to 0} \left[\frac{f(x + \Delta x) - f(x) - f(x + \Delta x - h) + f(x - h)}{h} \right] \\
&\approx \frac{1}{(\Delta x)^2} [f(x + \Delta x) - 2f(x) + f(x - \Delta x)].
\end{aligned}$$

Notice that we have used both a forward and backward difference definition of the derivative in order to make our approximation symmetric. This helps us keep the later linear algebra manageable. Applying this result to our discrete state u:

$$\frac{\partial^2 u_j}{\partial x^2} = \frac{1}{(\Delta x)^2} (u_{j+1} - 2u_j + u_{j-1}). \tag{B.1}$$

The matrix representation for this second derivative operator (in the case of $m = 6$) is

$$\frac{1}{(\Delta x)^2} D_2 = \frac{1}{(\Delta x)^2} \begin{pmatrix} -2 & 1 & 0 & 0 & 0 & 0 \\ 1 & -2 & 1 & 0 & 0 & 0 \\ 0 & 1 & -2 & 1 & 0 & 0 \\ 0 & 0 & 1 & -2 & 1 & 0 \\ 0 & 0 & 0 & 1 & -2 & 1 \\ 0 & 0 & 0 & 0 & 1 & -2 \end{pmatrix}.$$

For any discrete vector $u \in \mathbb{R}^m$, $\frac{1}{(\Delta x)^2} D_2 u \in \mathbb{R}^m$ is the discrete approximation to the second spatial derivative.

Discretization in time

The quantity f also changes in time. Again, using the definition of derivative, we have

$$\frac{\partial f}{\partial t} = \lim_{h \to 0} \frac{f(t + h) - f(t)}{h}.$$

In this case, we consider a finite time step Δt and approximate the derivative as

$$\frac{\partial f}{\partial t} \approx \frac{f(t + \Delta t) - f(t)}{\Delta t}.$$

Then, using our discrete representation,

$$\frac{\partial u_j}{\partial t} \approx \frac{u_j(t + \Delta t) - u_j(t)}{\Delta t}.$$

Discrete Diffusion Equation

Now, combining our discretized approximations in both time and position, we have the discrete diffusion equation

$$\frac{u(t + \Delta t) - u(t)}{\Delta t} = \frac{1}{(\Delta x)^2} D_2 u(t).$$

Or, more simply,

$$u(t + \Delta t) = u(t) + \frac{\Delta t}{(\Delta x)^2} D_2 u(t)$$

$$= \left(I + \frac{\Delta t}{(\Delta x)^2} D_2 \right) u(t).$$

Finally, we define $m \times m$ matrix E as $E = I + \frac{\Delta t}{(\Delta x)^2} D_2$. Then we have the discrete diffusion evolution matrix equation

$$u(t + \Delta t) = Eu(t).$$

The matrix E is given by

$$E = \begin{pmatrix} 1 - 2\delta & \delta & & & & \\ \delta & 1 - 2\delta & \delta & & \mathbf{0} & \\ & \delta & \ddots & \ddots & & \\ & & \ddots & \ddots & \delta & \\ \mathbf{0} & & & \delta & 1 - 2\delta & \delta \\ & & & & \delta & 1 - 2\delta \end{pmatrix}, \tag{B.2}$$

where $\delta \equiv \frac{\Delta t}{(\Delta x)^2}$. E is a symmetric matrix with nonzero entries on the main diagonal and on both adjacent diagonals. All other entries in E are zero.

Time Evolution

Now, we consider the evolution over more than one time step.

$$u(t + \Delta t) = Eu(t)$$
$$u(t + 2\Delta t) = Eu(t + \Delta t) \ = \ EEu(t) \ = \ E^2 u(t).$$

This means k time steps later,

$$u(t + k\Delta t) = E^k u(t).$$

Appendix C
Proof Techniques

In the course of reading and writing your own careful mathematical justifications (proofs), you will gain a deeper understanding of linear algebra. In addition, you will develop your skills in reading and communicating mathematics. To give the reader examples and practice, we present several proofs and request several proofs in exercises as well. In addition, we cover, in this appendix, techniques and etiquette commonly seen in proofs. First, we give rules of logic to help understand what follows. Then we will outline the various standard proof techniques that will be helpful in studying linear algebra. We wrap up this appendix with rules of etiquette for proof writing. It is important to note that this is not a complete discussion about proof writing. We encourage readers to consult other resources on mathematical communication and proof writing to gain more perspectives and a bigger picture of mathematical discourse.

C.1 Logic

In this section, we layout standard rules of mathematical logic and the interpretation of words such as 'or' and 'and.' We begin with the mathematical definition of a 'statement,' which is different from the colloquial definition as a declaration in speech or writing.

Definition C.1.1

A **statement** is a sentence with a truth value. □

Before giving examples, we need to understand 'truth value.'

A statement has a truth value if it is either true or it is false. There are many statements for which we do not know the truth value, but it is clear that there is a truth value. For example,

"There is life in another galaxy."

is a statement. We do not have the technology to check all galaxies for life and we haven't found life yet. (Who knows, maybe by the time this book is published, we do know the truth about this statement.) But, we do know that it must be either true or false.

Example C.1.2 Here, are a few more statements.

1. 5 is an odd number.
2. 6 is the smallest integer.
3. If $x = 0$ then $3x + 1 = 1$.

© Springer Nature Switzerland AG 2022
H. A. Moon et al., *Application-Inspired Linear Algebra*, Springer Undergraduate Texts in Mathematics and Technology, https://doi.org/10.1007/978-3-030-86155-1

4. Whenever $3x^2 + 2 = 5$ is true, $x = 6$ is also true.
5. Everyone who reads this book is taking a linear algebra course.
6. Some cats are pets.   □

As you can see, not all of the statements above are true. Indeed, statements 1, 3, and 6 are true, but statements 2, 4, and 5 are false.

Example C.1.3 A statement cannot be a phrase, command, exclamation, an incomplete idea, nor an opinion. For example:

A. Blue
B. Cats are the best pets.
C. Wow!
D. Prove that $x = 2$.   □

(A.) in Example C.1.3 is not even a sentence. The sentence C.1.3(B.) is not a statement. You may want to argue that it is true, but it is only an opinion. Sentence (C.) is not something about which we would discuss the truth. Finally, though $x = 2$ may or may not be true, (D.) has no truth value as it is commanding one to prove a statement.

We can combine statements to make new statements using the conjunctions *and*, *or*, *if...then*, and *if and only if*. There is a precise way to determine the truth of the resulting statements, using the following rules.

When 'and' is used to connect two statements, the new statement is true only when both original statements are true. Hence, the statement "5 is an odd number **and** 6 is the smallest integer" is false because "6 is the smallest integer" is false.

When 'or' is used to connect two statements, the new statement is false only when both are original statements are false. In other words, the new statement is true when at least one of the component statements is true. The statement "5 is an odd number **or** 6 is the smallest integer" is true because "5 is an odd number" is true.

When using "If...then..." to create a new statement from two others, we call the statement that directly follows "If" is called the **hypothesis**. The statement that follows "then" is called the **conclusion**. If-then statements are false whenever the hypothesis is true and the conclusion is false. They are true in every other situation. The statement "If unicorns exist, then I can ride them to the moon." is true because there the hypothesis, "Unicorns exist" is false. "If ...then..." statements are called **implications**. Another way to phrase an implication is as a "... implies ..." For example, the statement

"The number x is divisible by 4 implies that x is even."

is equivalent (i.e., they have the same truth value) to

"If x is divisible by 4 then x is even."

Switching the hypothesis and conclusion results in the statement

"If x is even then x is divisible by 4."

We know that x is even does not imply that x is divisible by 4. Therefore, the resulting statement is false.

When using "if and only if" to combine two statements, we are writing an equivalence between two statements. That is, an if-and-only-if statement is true when both statements are true or when both are false. The statement "The integer x is even if and only if 2 divides x." is a true statement. We know this because either x is even or x is odd. If x is even, then "x is even" and "2 divides x" are both true. If x is odd, then x is even" and "2 divides x" are both false.

In the truth table below, we have indicated the truth for the various cases that occur when connecting two statements. Given two statements, P and Q, the following notation will allow us to compactly summarize a variety of situations in Table C.1.

Notation	Meaning
T	true
F	false
$P \wedge Q$	P and Q,
$P \vee Q$	P / Q,
$P \Rightarrow Q$	P implies Q or If P then Q,
$\sim P$	not P

The truth values (T or F) for various logic cases are shown in Table C.1. Each row of the table shows the logical truth value for conjunctions and implications formed from P and Q based on the given truth values in the first two columns. Notice that the columns of Table C.1 corresponding to $P \wedge \sim Q$ and $P \Rightarrow Q$ have exactly opposite truth values. This will be very useful later when we discuss proof by contradiction. Notice, also, that $P \Rightarrow Q$ and $\sim Q \Rightarrow \sim P$ have the same truth value. This will be very important when discussing the contrapositive.

P	Q	$\sim P$	$P \wedge Q$	$P \vee Q$	$P \Rightarrow Q$	$P \wedge \sim Q$	$\sim Q \Rightarrow \sim P$	$P \Longleftrightarrow Q$
T	T	F	T	T	T	F	T	T
T	F	F	F	T	F	T	F	F
F	T	T	F	T	T	F	T	F
F	F	T	F	F	T	F	T	T

Table C.1 A truth table for conjunctions and implications formed using statements P and Q.

C.2 Proof structure

In this section, we describe what we consider a good general structure for a proof. This should only serve as a starting point for your proof writing. You can (and should!) add your own voice to your proofs without losing the necessary components for a good proof.

Here is the general structure with which you should write your proofs.

Statement
This is the statement that you will be proving.
Proof. ⟵Every proof should begin with this. In cases of particular types of proofs, you may indicate the technique you will be using. More on that later.
Hypotheses: Here, you list all the relevant assumptions. You use words like *Assume that...* or

Suppose... or *Let....*
Plan: For more complicated proofs, you give the reader an idea about how the proof will work.
We will show that...
Proof Body: Here is where you begin listing statements that link to definitions, theorems, algebra, arithmetic,... these will all lead to the result being proved. These statements should create a path that can be traversed to get from the hypotheses to the conclusion. It is important to know your audience. Leaving large gaps between statements can make it difficult or impossible for your reader to traverse the argument.
Conclusion: Here, you state what you just proved.
Every proof should end with some sort of symbol indicating your proof is done. One very common example is the symbol here.$\longrightarrow$ □

Now, we move to various proof techniques. In these examples, we use only one proof technique. In writing proofs, you may use multiple techniques in a single proof.

C.3   Direct Proof

In this section, we will show some examples of writing direct proofs. Whenever a direct proof is achievable it is more proper to give a direct proof of a statement than any other type of proof. This is because direct proofs tend to be more straightforward and easier to follow. Of course, this is not always the case and we discuss other proof techniques in upcoming sections.

Proofs use theorems and definitions to make a clear argument about the truth of another statement. To illustrate this, we will use the following definition.

Definition C.3.1

A number $n \in \mathbb{Z}$ is **even** if there exists $k \in \mathbb{Z}$ so that $n = 2k$.   □

In the following examples, we will add comments in blue to emphasize important features.

Example C.3.2 If $n \in \mathbb{Z}$ is even, then n^2 is even.

Proof. Suppose $n \in \mathbb{Z}$ is even.   We always start by telling the reader what we are assuming.

We will find $k \in \mathbb{Z}$ so that $n^2 = 2k$.   We follow up with our goal or our plan.
We know that $n = 2m$ for some $m \in \mathbb{Z}$.   Write what our assumptions mean. We can work with this information.

Notice $n^2 = 4m^2 = 2(2m^2)$.   We use algebra to reach our goal.[1]
Since $2m^2 \in \mathbb{Z}$, we can use the definition of   We always end a proof stating the result.
even to say that n^2 is even. □

Example C.3.3 If n and m are both even integers, then $n + m$ is an even integer.

Proof. Suppose n and m are even integers.

Write the assumptions.

We will show that there is a $k \in \mathbb{Z}$ so that $n + m = 2k$.

Indicate the plan.

We know that there are integers x and y so that $n = 2x$ and $m = 2y$.

Write what the assumptions imply.

Then $n + m = 2x + 2y = 2(x + y)$.

Use Algebra to write a statement implied by the last statement.

Since $x + y \in \mathbb{Z}$, $n + m$ is even. $\square$

The conclusion follows from the previous statement.

All the assumptions in the above proofs came from the hypothesis of the statement. And, the proof ended with the conclusion of the statement. In the next example, we will refrain from adding the blue comments. Try to find the same elements in this next proof.

Example C.3.4 Suppose $x, y \in \mathbb{R}$. If x and y are positive then $x + y \geq 2\sqrt{xy}$.

Proof. Assume $x, y \in \mathbb{R}$ are positive. We will show that $x + y - 2\sqrt{xy} \geq 0$. Notice that

$$x - 2\sqrt{xy} + y = (\sqrt{x})^2 - 2\sqrt{x}\sqrt{y} + (\sqrt{y})^2$$
$$= (\sqrt{x} - \sqrt{y})(\sqrt{x} - \sqrt{y})$$
$$= (\sqrt{x} - \sqrt{y})^2 \geq 0.$$

Thus, $x + y \geq 2\sqrt{xy}$. $\square$

In linear algebra, we prove many things about sets. Here, we show some examples of proofs about sets. In the beginning of this text, we seek to know whether certain objects lie in a given set. Let's see how such a proof might look.

Example C.3.5 Let $A = \left\{ \begin{pmatrix} a & b \\ c & d \end{pmatrix} \mid a, b, c, d \in \mathbb{R} \right\}$. If $u, v \in A$ then $u + v \in A$.

Proof. Suppose A is the set given above and suppose that $u, v \in A$. We will show that there are real numbers α, β, γ, and δ so that $u + v = \begin{pmatrix} \alpha & \beta \\ \gamma & \delta \end{pmatrix}$. We know that there are real numbers $a_1, b_1, c_1, d_1, a_2, b_2, c_2$, and d_2 so that

$$u = \begin{pmatrix} a_1 & b_1 \\ c_1 & d_1 \end{pmatrix} \text{ and } v = \begin{pmatrix} a_2 & b_2 \\ c_2 & d_2 \end{pmatrix}.$$

Notice that

$$u + v = \begin{pmatrix} a_1 & b_1 \\ c_1 & d_1 \end{pmatrix} + \begin{pmatrix} a_2 & b_2 \\ c_2 & d_2 \end{pmatrix}$$
$$= \begin{pmatrix} a_1 + a_2 & b_1 + b_2 \\ c_1 + c_2 & d_1 + d_2 \end{pmatrix}.$$

Since $a_1 + a_2, b_1 + b_2, c_1 + c_2$, and $d_1 + d_2$ are all real numbers, we know that $u + v \in A$. □

Let us consider one more proof of this form.

Example C.3.6 Suppose $P = \{ax^2 + bx + c \mid a, b, c \in \mathbb{R}$ and $a + b - 2c = 0\}$. If $v \in P$ then $\alpha v \in P$ for all $\alpha \in \mathbb{R}$.

Proof. Suppose P is the set given above. Let $\alpha \in \mathbb{R}$ and $v \in P$. We will show that $\alpha v = ax^2 + bx + c$ for some $a, b, c \in \mathbb{R}$ satisfying $a + b - 2c = 0$. We know that there are real numbers r, s, and t so that $r + s - 2t = 0$ and $v = rx^2 + sx + t$. Notice that

$$\alpha v = \alpha r x^2 + \alpha s x + \alpha t.$$

Notice also that $\alpha r, \alpha s$, and αt are real numbers satisfying

$$\alpha r + \alpha s - 2\alpha t = \alpha(r + s - 2t) = \alpha \cdot 0 = 0.$$

Thus, $\alpha v \in P$. Since α was an arbitrarily chosen real number, we conclude that $\alpha v \in P$ for all $\alpha \in \mathbb{R}$. □

To compare sets, we need the following definition.

Definition C.3.7

Let A and B be sets. We say that B is a subset of A and write $B \subseteq A$ if $x \in B$ implies $x \in A$. We say that $A = B$ if $A \subseteq B$ and $B \subseteq A$. □

We will use the definition to show the following statement.

Example C.3.8 Let

$$A = \{2x + 3y \mid x, y \in \mathbb{Z}\} \text{ and } B = \{6n + 12m \mid n, m \in \mathbb{N}\}.$$

Then $B \subseteq A$.

Proof. Suppose A and B are the sets given above and suppose $b \in B$. We will show that there are integers x, y so that $b = 2x + 3y$. We know that there are natural numbers n, m so that

$$b = 6n + 12m$$
$$= 2(3n) + 3(4m).$$

Now, since $3n$ and $4m$ are integers, we see that $b \in A$. Thus, $B \subseteq A$. □

In some instances, we are faced with proving a statement that is difficult to use a direct proof. An example is one that is of the form $\sim P \implies \sim Q$. An example follows.

Example C.3.9 Let $x \in \mathbb{R}$. Suppose $x^2 + 2x - 3 \neq 0$, then $x \neq 1$ and $x \neq -3$.

Proof. Let $x \in \mathbb{R}$. Suppose $x^2 + 2x - 3 \neq 0$. We will show that $x \in \{a \in \mathbb{R} \mid a \neq 1$ and $a \neq -3\}$. Notice that

$$x^2 + 2x - 3 = (x - 1)(x + 3).$$

We see that neither $x - 1$ nor $x + 3$ can be zero. So then $x \neq 1$ and $x \neq -3$. $\square$

The above proof feels a bit forced at the end. We see that our understanding of quadratic polynomials and lines helps us to see the ending statements must be true. But, it feels less like a proof and more like a list of statements we just know. In the next section we will prove this statement again, but with a smoother proof.

C.4   Contrapositive

Recall, from Table C.1, that the statement $P \implies Q$ is equivalent to $\sim Q \implies \sim P$. In this section, we discuss when it is more appropriate to prove $\sim Q \implies \sim P$ as a means to proving $P \implies Q$. We begin by defining the contrapositive of a statement.

Definition C.4.1

Given two statements P and Q. The **contrapositive** of the statement

If P is true then Q is true. ($P \implies Q$.)

is the statement

If Q is false then P is false. ($\sim Q \implies \sim P$.) $\square$

Now, we return to Example C.3.9.

Example C.4.2 Let $x \in \mathbb{R}$. Suppose $x^2 + 2x - 3 \neq 0$, then $x \neq 1$ and $x \neq -3$.

Proof. Let $x \in \mathbb{R}$.	We write our assumptions.
Instead, we will prove the contrapositive.	Here, we make clear the proof style we will use.
That is, we will show	
If $x = 1$ or $x = -3$ then $x^2 + 2x - 3 = 0$.	This is the contrapositive.
We will use cases on x to prove this statement.	Whenever there is an "or" in the hypothesis
	We consider cases assuming each of the smaller statements.
First, assume $x = 1$. Then $x - 1 = 0$ and	
$x^2 - 2x - 3 = (x - 1)(x + 3) = 0 \cdot 2 = 0.$	
Now, assume $x = -3$. Then $x + 3 = 0$ and	
$x^2 - 2x - 3 = (x - 1)(x + 3) = -4 \cdot 0 = 0.$	
In either case, we see that $x^2 + 2x - 3 = 0$. $\square$	We finish with the result.

Notice that proving the contrapositive flowed much more smoothly than trying a direct proof. We see that this statement sets itself up to be proved using the contrapositive because both the hypothesis and the conclusion have "not equal." Negating these statements change these to "equals," which is much easier to prove and much easier to use as an assumption. Let's look at another example where the contrapositive is clearly easier to prove.

Example C.4.3 Let X, Y be sets. If $x \notin X$ then $X \notin X \cap Y$.

Proof. Let X, Y be sets. We will, instead, show that the contrapositive is true. That is, we will prove

   If $x \in X \cap Y$ then $x \in X$.

Suppose $x \in X \cap Y$. By definition of the intersection, this means that $x \in X$ and $x \in Y$. Thus, if $x \notin X$ then $x \notin X \cap Y$. $\square$

In some instances, a proof of the contrapositive is useful even though the hypothesis and conclusion statements do not explicitly have "not" in them.

Example C.4.4 Suppose $m, n \in \mathbb{Z}$. If mn is even, then m or n is even.

Proof. Let $n, m \in \mathbb{Z}$. We will show, instead, the contrapositive. That is, we will show

   If m and n are odd, then mn is odd.

Now, suppose m and n are odd. Then there are integers k and ℓ so that $m = 2k + 1$ and $n = 2\ell + 1$. Notice that

$$\begin{aligned}
mn &= (2k + 1)(2\ell + 1) \\
&= 4k\ell + 2k + 2\ell + 1 \\
&= 2(2k\ell + k + \ell) + 1.
\end{aligned}$$

Since $2k\ell + k + \ell \in \mathbb{Z}$, mn is odd. Thus, if mn is even, m or n is even. $\square$

All statements in this section have a contrapositive because they area all implications. In subsequent sections, we will introduce other methods of proof that are not only useful for proving some implications but can also be used to prove statements that are not implications.

C.5   Proof by Contradiction

In this section, we discuss the method of proof by contradiction. We know that P and $\sim P$ have opposite truth values and we, also, see in Table C.1 that $P \implies Q$ and $P \wedge \sim Q$ have opposite truth values. That is, when one is true, the other is false. That is, $\sim (P \implies Q)$ and $P \wedge \sim Q$ have the same truth value. We will make use of this fact in this section. Here, we outline the method.

Statement: P

Proof. (By Contradiction) By way of contradiction, assume $\sim P$ is true. (That is, P is false.) We will search for an absurd statement.
$\vdots$

Logical path through statements starting with a statement that follows from the assumption.
$\vdots$

A clearly false statement.
$\rightarrow \leftarrow$ (the symbol to recognize a contradiction.)
Thus, the original assumption must be false and P must be true. $\square$

One of the most common first proofs by contradiction is seen in the following example. First, we define the set of rational numbers.

Definition C.5.1

Let x be a real number. We say that x is rational (or that $x \in \mathbb{Q}$) if there are integers a, b, with 1 being the only common divisor, so that $x = \frac{a}{b}$.   □

For a number to be irrational, it must be real, but not rational. Notice as you read the statement that it is not an implication and cannot be proved using a contrapositive. In the following example, we begin by assuming that $\sqrt{2}$ is rational. This, after some algebra, leads to an absurd statement.

Example C.5.2 The number $\sqrt{2}$ is irrational.

Proof. (By Contradiction)	Indicate to the reader the type of proof.
Assume the statement is false.	
That is, assume that $\sqrt{2} \in \mathbb{Q}$.	Write the negation of the statement.
We will search for a contradiction.	Give the reader your plan.
Then there are $a, b \in \mathbb{Z}$ so that	Here, the proof feels the same.
$\sqrt{2} = \frac{a}{b}$ and	
a, b share only 1 as a common divisor.	
So, $2b^2 = a^2$.	
This means that a is even.	
Therefore, there is an integer k	
so that $a = 2k$.	
This means that $2b^2 = 4k^2$.	
Thus, $b^2 = 2k^2$, meaning that b is even.	
So, 2 is a common factor of a and b.	
This means that $2 = 1$. $\rightarrow \leftarrow$	Clearly, indicate the absurd statement.
Therefore, our assumption that $\sqrt{2}$	State what is actually true
is rational was false and so $\sqrt{2}$ must	by recognizing the false assumption
be irrational. □	and writing the true statement.

In the next example, we prove an implication using a proof by contradiction. Recall that $\sim (P \implies Q)$ and $P \wedge \sim Q$ have the same truth value. So, in such a case, we begin by assuming the implication is false by assuming P is true and Q is false. (Typically, we negate the statement Q and assume the negation is true.)

Statement: $P \implies Q$

Proof. (By Contradiction) By way of contradiction, assume the statement is false. That is, assume P and $\sim Q$ are true. We will search for an absurd statement.

$\vdots$

Logical path through statements starting with a statement that follows from the assumption.

$\vdots$

A clearly false statement.

$\rightarrow \leftarrow$ (the symbol to recognize a contradiction.)

Thus, the original assumption must be false and $P \implies Q$ must be true.   □

Example C.5.3 If X and Y are sets, then $X \cap (Y \cup X)^c = \emptyset$.

If A is a set, then A^c contains all elements from some universal set, except those in A.

Proof. (By Contradiction) By way of contradiction, suppose the statement is false. That is suppose X and Y are sets and $X \cap (Y \cup X)^c \neq \emptyset$. We will search for a contradiction. We know that there is an element $x \in X \cap (Y \cup X)^c$. This means that $x \in X$ and $x \notin Y \cup X$. Since $x \notin Y \cup X$, x can be an element of neither X nor Y. That is, $x \notin X. \rightarrow \leftarrow$ therefore our original assumption that $X \cap (Y \cup X)^c \neq \emptyset$ is false. So $X \cap (Y \cup X)^c = \emptyset$.   □

Let us now spend a little time making sense of the logic behind a proof by contradiction. We do this with the following illustration.

Suppose we have a statement P for which we want to find a proof. Consider the following logical path of statements in a hypothetical proof by contradiction.

$$\text{Statement path: } \sim P \implies R \implies Q$$

Now, suppose that Q is an obviously false statement. Since every statement that we put into a proof must be true (except for possibly our assumptions), we can follow the following logical progression backwards find the truth of P. We do this using the fact that given two statements S_1 and S_2, if S_2 is false and $S_1 \implies S_2$ is true, then S_1 must also be false (see Table C.1 column 7).

Statement	Truth Value	Reason
Q	False	Q is the obviously absurd statement.
$R \implies Q$	True	Statement in the proof
R	False	Table C.1, column 7
$\sim P \implies R$	True	Statement in the proof
$\sim P$	False	Table C.1, column 7
P	True	Table C.1, column 3

C.6   Disproofs and Counterexamples

Up to this point, we have proved statements. In this section, we talk about how to show that a statement is false. We have two choices to indicate how we know a statement is false: using a counter example or writing a disproof. It is not the case that both choices work for any false statement. Let's look at some examples.

Example C.6.1 Let $x \in \mathbb{R}$ then $x^2 - 5x + 4 \geq 0$.

Note: In this example, we see that the statement says that no matter which real number we choose for x, we will get a nonnegative result for $x^2 - 5x + 4$. To show that the statement is false, we only need to choose one real number for x that makes $x^2 - 5x + 4$ negative.

Counter Example. Let $x = 2$. Then $x^2 - 5x + 4 = 4 - 10 + 4 = -2 < 0$.   □

Consider the following example for which a counter example is not attainable.

Example C.6.2 There exist prime numbers p and q for which $p + q = 107$.

Disproof. Assume the statement is true. We will show that this leads to a contradiction. Let p and q be prime numbers so that $p + q = 107$. Since 107 is odd, we know that p and q cannot both be odd (nor can they both be 2). Thus, either $p = 2$ or $q = 2$. Without loss of generality, let us assume $p = 2$. Then $q = 105 = 19 \cdot 6. \rightarrow \leftarrow$. Therefore, there are no two primes whose sum is 107.   □

Notice that we used the words "without loss of generality" in this disproof. We use this phrase to indicate that had we chosen $q = 2$, the proof would have not been different except to say the exact same things about p as we said about q.

In linear algebra, many times, we want to prove or disprove whether a list of properties is satisfied. In the next example, we show that a set V is not a vector space (see Definition 2.3.2), using a disproof.

Example C.6.3 Let $V = \{ax + b \mid a, b \in \mathbb{R} \text{ and } a + b = 1\}$. V is a vector space.

Counter Example. We notice that V is not closed under addition. Indeed, Let $u = 1x + 0$ and $v = 0x + 1$, then $u, v \in V$, but the sum $u + v = x + 1 \notin V$. Thus, V is not a vector space.   □

C.7   The Principle of Mathematical Induction

In this section, we discuss statements of the form $P(n)$, where, for each $n \in \mathbb{N}$, we have a different statement. If we know that for any $n \geq 1$, $P(n + 1)$ can be inferred from the statement $P(n)$, then we can apply the principle of mathematical induction. The idea is to think of this like a domino affect. If you push the first domino, then the next will fall, causing the toppling of the entire design as in Figure C.1.

Fig. C.1 Mathematical Induction is like toppling dominoes.

Theorem C.7.1 (The Principle of Mathematical Induction)
Suppose $P(n)$ is a statement defined for every $n \in \mathbb{N}$, then

(1) $P(1)$ is true and
(2) Whenever $P(k)$ is true for some $k \in \mathbb{N}$, then $P(k+1)$ is also true ($P(k) \implies P(k+1)$),

then $P(n)$ is true for all $n \in \mathbb{N}$.

Next, we prove a statement using mathematical induction to see how this works. We will use a statement we use in Calculus.

Example C.7.2 For any $n \in \mathbb{N}$,

$$\sum_{j=1}^{n} j = \frac{n(n+1)}{2}.$$

Proof. (By Induction) First, we define the statement $P(n)$ as follows
$\sum_{j=1}^{n} j = \frac{n(n+1)}{2}$.
(1) Notice that $P(1)$ is the trivial statement $1 = \frac{1(2)}{2}$.
(2) Let $k \in \mathbb{N}$. Now, we will show that $P(k) \implies P(k+1)$.
Suppose $P(k)$ is true. That is, suppose $\sum_{j=1}^{k} j = \frac{k(k+1)}{2}$.
We will show that $\sum_{j=1}^{k+1} j = \frac{(k+1)(k+2)}{2}$ is also true.
Notice that
$\sum_{j=1}^{k+1} j = k+1+\sum_{j=1}^{k} j$.
Using the Induction Hypothesis, we have
$\sum_{j=1}^{k+1} j = k+1+\frac{k(k+1)}{2} = \frac{2(k+1)+k(k+1)}{2}$
$= \frac{(k+1)(2+k)}{2}$.
Thus, $P(k) \implies P(k+1)$.
Therefore, by the Principle of Mathematical Induction, $\sum_{j=1}^{n} j = \frac{n(n+1)}{2}$ is true for all n. $\square$

We indicate the type of proof at the start. This step is not necessary, but does ease notation.
Base Case: $P(1)$ is true.

Here, we use a direct proof. This is called the "induction hypothesis."

In every proof by induction you will employ the induction hypothesis.

General case is true.
We call upon the principle of mathematical induction to use the truth of the base case and the general case to the statement being proved.

In reading through the above proof, you can see the base case is the same as hitting the first domino. Then, because the general case is true, the second domino is hit by the first. In turn, the third domino is hit by the second, and so on. We get the following sequence of statements:

$$P(1) \qquad\qquad\qquad \text{True}$$
$$P(1) \implies P(2) \qquad\qquad \text{True}$$
$$P(2) \qquad\qquad\qquad \text{True}$$
$$P(2) \implies P(3) \qquad\qquad \text{True}$$
$$P(3) \qquad\qquad\qquad \text{True}$$
$$\vdots$$

Here, we show another example.

Example C.7.3 Let $n \in \mathbb{N}$, then $(1+x)^n \geq 1 + nx$ for all $x \in \mathbb{R}$, $x > -1$.

Proof. (By Induction) Let $n \in \mathbb{N}$, $x \in \mathbb{R}$, and $x > -1$. Define $P(n)$ to be the statement

$$(1+x)^n \geq 1 + nx.$$

(1) Notice that $P(1)$ is the trivial statement $(1+x) = (1+x)$. Thus, $P(1)$ is true.

(2) Fix $k \in \mathbb{N}$. We will show that $P(k) \implies P(k+1)$. That is, if $(1+x)^k \geq 1 + kx$ then $(1+x)^{k+1} \geq 1 + (k+1)x$.
Suppose $(1+x)^k \geq 1 + kx$. Notice that

$$(1+x)^{k+1} = (1+x)^k(1+x) \text{ and } 1 + x > 0.$$

Thus, by the induction hypothesis,

$$\begin{aligned}
(1+x)^{k+1} &\geq (1+kx)(1+x) \\
&= 1 + kx + x + kx^2 \\
&= 1 + (k+1)x + kx^2.
\end{aligned}$$

Since, $kx^2 \geq 0$, we know that $(1+x)^{k+1} \geq 1 + (k+1)x$. Thus, $P(k) \implies P(k+1)$.
By the principal of mathematical induction, we know that $(1+x)^n \geq 1 + nx$ for all $n \in \mathbb{N}$. $\qquad\square$

In both examples, $P(1)$ was a trivial statement. This is not always the case. It should also be noted, that if we were to prove a statement that is only true for all $n \in \{k \in \mathbb{N} \mid k \geq N\}$ for some given number N, the only change to an induction proof is the base case. This is, in essence, the same as knocking over the 10th domino and then seeing the toppling effect happen from the 10th domino onward.

C.8   Etiquette

With some basics on proof-writing down, we now turn to the etiquette of proof-writing. It should be noted that the suggestions, rules, and guidelines in this section may be slightly different than those another mathematician will give. We try to give very basic guidelines here. You should know your audience when writing proofs, be somewhat flexible to adjust your proof writing to communicate clearly with those whom you are trying to communicate. To better understand the need for such etiquette, it is a good exercise to understand what makes mathematics easier for you to understand. The following is a compilation of questions that you can ask yourself to begin understanding some etiquette you would like to have in place. It may be that we do not agree on etiquette, but it is always good for you to form an opinion also. With all that said, read through and answer the following questions to determine your preference for each. Unfortunately, we will not give you a character or personality profile at the end. This is, only, for your own use.

1. Which of the following is most clear?

   (a) $\forall u, v \in V, v + u = u + v$.
   (b) For every pair of elements, $u, v \in V, u + v = v + u$.
   (c) The set V has the commutative property for addition.
   (d) The set V has the property so that if you take two elements and add them together in one order, you get the same thing as when you add them together in the other order.

2. Which of the following best defines m as a multiple of k.

   (a) Let $k \in \mathbb{Z}$, then m is a multiple of k if $m = kn$.
   (b) $m = kn$.
   (c) Let $k \in \mathbb{Z}$, then m is a multiple of k if $m = kn$ for some $n \in \mathbb{N}$.
   (d) m and k are integers and n is a natural number and $m = kn$ means that m is a multiple of k.

3. When reading mathematics, which of the following do you prefer?

   (a) Let $k \in \mathbb{Z}$. Suppose $x = 5k - 6$ and $y = 3k + 2$. Then $x + y = 5k - 6 + 3k + 2 = 5k + 3k - 6 + 2 = 8k - 4 = 4(2k - 1)$.
   (b) Let $k \in \mathbb{Z}$. Suppose $x = 5k - 6$ and $y = 3k + 2$. Then

$$
\begin{aligned}
x + y &= 5k - 6 + 3k + 2 \\
&= 8k - 4 \\
&= 4(2k - 1).
\end{aligned}
$$

   (c) Let $k \in \mathbb{Z}$. Suppose $x = 5k - 6$ and $y = 3k + 2$. Then $x + y = 4m$ where $m \in \mathbb{Z}$.
   (d)

$$
\begin{aligned}
x + y &= 5k - 6 + 3k + 2 \\
x + y &= 8k - 4 \\
x + y &= 4(2k - 1).
\end{aligned}
$$

4. Which of the following reads better?

   (a) Let $x \in \mathbb{Z}$ be even and $y \in \mathbb{Z}$ be odd. Then $x = 2k$ for some $k \in \mathbb{Z}$ and $y = 2m + 1$ for some $m \in \mathbb{Z}$. Then $x + y = 2k + 2m + 1$. Then $x + y$ is odd.
   (b) Let $x, y \in \mathbb{Z}$ and let x be even and y be odd. Then there are integers k and m so that $x = 2k$ and $y = 2m + 1$. Thus, $x + y = 2k + 2m + 1$. Therefore, $x + y$ is odd.

5. Which of the following flows better?

   (a) Let $x \in \mathbb{R}$ be positive and suppose $x^2 - 2x - 3 = 0$. We will first factor to get $(x - 3)(x + 1) = 0$. Setting each factor to zero gives $x - 3 = 0$ and $x + 1 = 0$. In the first, we add 3 to both sides to get $x = 3$. Subtracting 1 on both sides in the second gives $x = -1$. Since x is positive, $x = 3$.
   (b) Let $x \in \mathbb{R}$ be positive and suppose $x^2 - 2x - 3 = 0$. Notice that

$$0 = x^2 - 2x - 3$$
$$= (x - 3)(x + 1).$$

Thus, $x = 3$ or $x = -1$. Since x is positive, $x = 3$.

(c)

$$x^2 - 2x - 3 = 0$$
$$(x - 3)(x + 1) = 0$$
$$x - 3 = 0, \quad x + 1 = 0$$
$$x = 3 \text{ or } x = -1$$

(d) Let $x \in \mathbb{R}$ be positive and suppose $x^2 - 2x - 3 = 0$. We can solve the quadratic equation $x^2 - 2x - 3 = 0$ by factoring. This leads to $(x - 3)(x + 1) = 0$. To solve this equation, we set each factor to zero giving us two equations. These equations are $x - 3 = 0$ and $x + 1 = 0$. To solve $x - 3 = 0$, we add 3 to both sides. We get the solution $x = 3$. To solve the equation $x + 1 = 0$, we subtract 1 from both sides. We get the solution $x = -1$ which is not positive. So our only solution is $x = 3$.

6. Which of the following best shows that the sum rule for differentiation is true?

(a) *Proof.* Let $f(x) = x + 1$ and $g(x) = x^2 - 2x + 1$. Then $f'(x) = 1$ and $g'(x) = 2x - 2$. So $f'(x) + g'(x) = 1 + 2x - 2 = 2x - 1$. $f(x) + g(x) = x + 1 + x^2 - 2x + 1 = x^2 - 2x + 2$. So $(f(x) + g(x))' = 2x - 2$. $\square$

(b) *Proof.* Let f and g be differentiable functions. Then, by definition, both

$$f'(x) = \lim_{h \to 0} \frac{f(x + h) - f(x)}{h} \quad \text{and} \quad g'(x) = \lim_{h \to 0} \frac{g(x + h) - g(x)}{h}$$

exist.

$$\begin{aligned}
(f + g)'(x) &= \lim_{h \to 0} \frac{(f + g)(x + h) - (f + g)(x)}{h} \\
&= \lim_{h \to 0} \frac{f(x + h) - f(x) + g(x + h) - g(x)}{h} \\
&= \lim_{h \to 0} \frac{f(x + h) - f(x)}{h} + \frac{g(x + h) - g(x)}{h} \\
&= \lim_{h \to 0} \frac{f(x + h) - f(x)}{h} + \lim_{h \to 0} \frac{g(x + h) - g(x)}{h} \\
&= f'(x) + g'(x).
\end{aligned}$$
$\square$

(c) *Proof.*

$$\begin{aligned}
(f + g)'(x) &= \lim_{h \to 0} \frac{(f + g)(x + h) - (f + g)(x)}{h} \\
&= \lim_{h \to 0} \frac{f(x + h) - f(x) + g(x + h) - g(x)}{h}
\end{aligned}$$

$$f'(x) + g'(x) = \lim_{h \to 0} \frac{f(x+h) - f(x)}{h} + \frac{g(x+h) - g(x)}{h}.$$

So

$$\lim_{h \to 0} \frac{f(x+h) - f(x)}{h} + \frac{g(x+h) - g(x)}{h} = \lim_{h \to 0} \frac{f(x+h) - f(x) + g(x+h) - g(x)}{h}.$$

Therefore,

$$\lim_{h \to 0} \frac{f(x+h) - f(x)}{h} + \frac{g(x+h) - g(x)}{h} = \lim_{h \to 0} \frac{f(x+h) - f(x)}{h} + \frac{g(x+h) - g(x)}{h}.$$

$\square$

7. Which of the following recognizes that the results are part of a collective effort by mathematicians?

(a) Let $x, y \in \mathbb{N}$ be even. We will show that $x + y$ is also even. We know that there are integers m, n so that $x = 2m$ and $y = 2n$. Thus, $x + y = 2m + 2m = 2(m + n)$. Therefore, we have that $x + y$ is also even.

(b) Let $x, y \in \mathbb{N}$ be even. I will show that $x + y$ is also even. I know that there are integers m, n so that $x = 2m$ and $y = 2n$. Thus, $x + y = 2m + 2m = 2(m + n)$. Therefore, I have that $x + y$ is also even.

(c) Let $x, y \in \mathbb{N}$ be even. They will show that $x + y$ is also even. They know that there are integers m, n so that $x = 2m$ and $y = 2n$. Thus, $x + y = 2m + 2m = 2(m + n)$. Therefore, they have that $x + y$ is also even.

(d) Let $x, y \in \mathbb{N}$ be even. It will be shown that $x + y$ is also even. It is known that there are integers m, n so that $x = 2m$ and $y = 2n$. Thus, $x + y = 2m + 2m = 2(m + n)$. Therefore, it has been shown that $x + y$ is also even.

Now that you have completed the questionnaire, we will give some guidelines on the etiquette and rules used in proof writing.

Etiquette Rule #1: Do not overly use words or symbols. Be concise, yet clear by mixing both. Only use terminology if it is clear the audience knows its meaning.

If you notice, in question 1. above, (a) is full of symbols and less clear, at a glance, to parse, (c) uses terminology that is useful to anyone intimately familiar with such, (d) is overly verbose and one can easily get lost in all the language. We choose (b) because it simply gives the statement with some mix of words and symbols and terminology is at a minimum. Our second choice is (c).

Etiquette Rule #2: Never begin a statement with symbols, but instead start with words.

In question 1 notice, also that neither (b) nor (c) begin with symbols. This gives our next rule.

Etiquette Rule #3: Always define your variables clearly being sure, when appropriate, to add quantifiers such as "there exist," "for some," and/or "for all." This is less about etiquette as the mathematics is improper if not done this way.

Notice in question 2. above, (a) does not define n so we cannot be sure that we are still talking about integers when discussing m and n. In (b), we see an equation and nothing more. In (d), we know the sets in which m, n, and k lie, but we are not sure which of these we are given and which we find. Our choice is (c) because it clearly fixes k and then m is defined as n multiplied by k and we are told it doesn't matter which natural number n is as long as such an n exists. Really the idea is that we don't have questions about which of k, m, and n are known and which are not nor do we question in which sets each are found.

Etiquette Rule #4: In longer computations, determine which steps are key and which are obvious. Be sure to leave in the key steps and leave out obvious steps that are not key to illustrating the goal.

In question 3, response (a) has a long string of mathematical steps which flow from one line to the next. This is not visually appealing and can be difficult (depnding on length) to follow.

Etiquette Rule #5: In longer computations, arrange your steps so that they are visually appealing, using vertical alignment to make the steps easier to follow.

In question 3, response (c) leaves out all details, giving the reader work to do. Response (d) looks like a basic algebra student's homework response. It is missing words and a clear path through the mathematics. Really, it is just a list of equations. We choose response (b). It is clear in (b) what we are assuming and how one gets from $x + y$ to $4(2k - 1)$. Basic steps like using the commutative law of addition are left out.

Etiquette Rule #6: Use transition words to make your proofs flow, but be sure to change words so that you are not transitioning with the same word every time.

Question 4 has a clear winner (to us). We choose (b) because in (a), the repetition of "then" gets monotonous. It is nicer to read when the transition words change.

Etiquette Rule #7: Do not treat your reader as if they are learning high school algebra. Rather, get to the point using clear and concise statements, focusing on the level of mathematics for which the proof is being written. In question 5, (a), (c), and (d) all have a flavor resembling the work of a basic algebra student. Although, (a) and (d) add some discussion and this takes these two above (c) which is just low level scratch work. Notice in (b), the assumptions are clear and the transitive property of equality clearly gets us from 0 to $(x - 3)(x + 1)$ without an excess of words. We choose (b).

Etiquette Rule #8: Examples, in general, do not make proofs. Question 6 shows two very common errors in writing proofs. In (a), an example is given to show a general statement. There are many statements that are false, but can be stated as a true statement for one example. For example, it is not true that an even number times an even number is zero, but $0 \cdot 2 = 0$.

Etiquette Rule #9: When proving an equality or inequality, it is best to start with one side and pave a clear path to the other. Never write a string of statements that leads to an obvious statement. Continuing with question 6, we see in (c), the end result is a trivial statement and tells us no new information. Notice that the second to last line is also a bad stopping place as it tells us something obvious, but does not make the connection. Notice that in (b) (our choice), we obtain the clear result without any breaks from the beginning, $(f + g)'(x)$ to the end $f'(x) + g'(x)$.

Etiquette Rule #10: In mathematical writing, the pronoun we use is "we."

We as mathematicians have been building new maths by standing on the shoulders of those before us. If it is not clear with the very pointed question 7 or the way we have been writing this chapter, we hope to make it clear now that mathematicians use the pronoun "we" to indicate that we do all this together, using past mathematical work and the minds of the reader to verify new mathematical ideas. Looking at response (b), you should notice how very egocentric it sounds. In response (c), we still ask the age old question, who are "they?" Response (d) leads one to wonder who will do and has done this work?

Appendix D
Fields

In this appendix, we give a very brief overview of what we need to know about fields. In this book, many examples required that you understand the two fields $\mathbb{R}$ and Z_2. Here, we discuss other examples as well. A more complete discussion of fields is the subject of abstract algebra courses. It should be noted that we are only scratching the surface of fields to glean what we need for linear algebra. We begin with the definition of a field.

Definition D.0.1

A field $\mathbb{F}$ is a set with operations $+$ and $\cdot$ with the following properties:

- $\mathbb{F}$ is closed under $+$ and $\cdot$: If $a, b \in \mathbb{F}$ then so are $a + b$ and $a \cdot b$.
- The operations $+$ and $\cdot$ are associative: If $a, b, c \in \mathbb{F}$ then $a + (b + c) = (a + b) + c$ and $a \cdot (b \cdot c) = (a \cdot b) \cdot c$.
- There is an additive identity, that is, there is an element $z \in \mathbb{F}$ so that $z + a = a$ for all $a \in \mathbb{F}$.
- There is a multiplicative identity, that is, there is an element $i \in \mathbb{F}$ so that $i \cdot a = a$ for every $a \in \mathbb{F}$.
- There are multiplicative inverses: For all $a \in \mathbb{F}, a \neq z$, there is an element $b \in \mathbb{F}$ so that $a \cdot b = i$.
- The operations $+$ and $\cdot$ are commutative: If $a, b \in \mathbb{F}$, then $a + b = b + a$ and $a \cdot b = b \cdot a$.
- Every element has an additive inverse: If $a \in \mathbb{F}$ then there is an $\tilde{a} \in \mathbb{F}$ so that $a + \tilde{a} = z$.
- Multiplication distributes over addition: If $a, b, c \in \mathbb{F}$, then $a \cdot (b + c) = a \cdot b + a \cdot c$. $\square$

For the rest of this chapter, we will look at several examples of fields. We begin with the two fields we use throughout the text.

Example D.0.2 In this example, we recognize that $\mathbb{R}$, the set of real numbers, with the usual definition of $+$ and $\cdot$ satisfies the field conditions. Many of these conditions are built into the operations. That is, we define both operations to satisfy the commutative, associative, and distributive properties.

We know that 1 is the real number that acts as the multiplicative identity and 0 acts as the additive identity. We know that we use the negative real numbers to act as additive inverses for the positive numbers and vice versa ($a \in \mathbb{R}$ has additive inverse $-a$) with 0 being its own additive inverse. We use reciprocals as multiplicative inverses ($a \in \mathbb{R}$ and $a \neq 0$ then $1/a \in \mathbb{R}$ is its multiplicative inverse). $\square$

© Springer Nature Switzerland AG 2022

H. A. Moon et al., *Application-Inspired Linear Algebra*, Springer Undergraduate Texts in Mathematics and Technology, https://doi.org/10.1007/978-3-030-86155-1

Example D.0.3 Let p be a prime number and define the set $Z_p = \{0, 1, 2, \ldots, p - 1\}$. We define $\oplus_p$ (addition modulo p) and $\odot_p$ (multiplication modulo p) as follows:
For $a, b, c \in \mathbb{Z}$

$$a \oplus_p b = c \text{ means } \frac{a + b - c}{p} \in \mathbb{Z} \text{ and } 0 \leq c < p$$

$$a \odot_p b = c \text{ means } \frac{c - ab}{p} \in \mathbb{Z} \text{ and } 0 \leq c < p.$$

Notice that the above says that $a \oplus_p b$ is the remainder of $(a + b) \div p$ and $a \odot_p b$ is the remainder of $(ab) \div p$.   □

Theorem D.0.4

$(Z_2, \oplus_2, \odot_2)$ is a field.

Proof. Because the associative, distributive, and commutative properties are inherited from $\mathbb{Z}$, we need only show that Z_2 is closed under $\oplus_2$ and $\odot_2$ and that there are additive and multiplicative identities and inverses in Z_2. Closure properties for both $\oplus_p$ and $\odot_p$ are evident by the fact that the remainder of division by p is always a positive integer less than p.

Since $0 \oplus_2 1 = 1$ and $0 \oplus_2 0 = 0$, we see that 0 is the additive identity for Z_2. The multiplicative identity is $1 \in Z_2$ since $1 \odot_2 0 = 0$ and $1 \odot_2 1 = 1$.

We can see that the additive inverse of every element of Z_2 is itself. So, Z_2 has additive inverses. The multiplicative inverse of 0 is 0 and the multiplicative inverse of 1 is also itself. Thus, Z_2 has multiplicative inverses.   □

Theorem D.0.5

$(Z_3, \oplus_3, \odot_3)$ is a field

Proof. We see from the proof that Z_2 is a field, that Z_3 is also closed under $\oplus_3$ and $\odot_3$. Also, similar to the proof for Z_2, we know that 0 and 1 are the additive and multiplicative identities, respectively. So, we need only show the existence of inverses.

To show that additive inverses exist, we show a more general result. In Z_p, $a \oplus_p (p - a)$ is the remainder of $p \div p$. Thus, $a \oplus_p (p - a) = 0$ and the additive inverse of a is $p - a$. Thus, Z_p has additive inverses.

To show that multiplicative inverses exist in Z_3, we will find each one. Since $1 \odot_p 1 = 1$ the multiplicative inverse of 1 is 1 in Z_p. We can also see that $(p - 1) \odot_p (p - 1)$ is the remainder of $(p^2 - 2p + 1) \div p$. Thus, $(p - 1) \odot_p (p - 1) = 1$ and $p - 1$ is its own multiplicative inverse in Z_p. Thus, $1, 2 \in Z_3$ both have multiplicative inverses.   □

We defined Z_p for prime numbers, p. If p is not prime, we run into problems finding multiplicative inverses.

Example D.0.6 Consider $Z_4 = \{0, 1, 2, 3\}$. The element 2 has no multiplicative inverse in Z_4. Indeed, $0 \odot_4 2 = 0$, $1 \odot_4 2 = 2$, $2 \odot_4 2 = 0$, and $2 \odot_4 3 = 2$. □

In considering Z_p for a general prime, p, we only mention the fact that it is a field. This fact is one that students learn in an Abstract Algebra class and is beyond the scope of this text.

Fact D.0.7
For any prime number p, $(Z_p, \oplus_p, \odot_p)$ is a field.

Example D.0.8 Let $\mathbb{C} = \{a + bi \mid a, b \in \mathbb{R}\}$. We call $\mathbb{C}$ the set of complex numbers. We define $\oplus$ and $\odot$ as follows.

$$(a + bi) \oplus (c + di) = (a + c) + (b + d)i \text{ and } (a + bi) \odot (c + di) = (ac - bd) + (ad + bc)i$$

□

Theorem D.0.9
The set of complex numbers $\mathbb{C}$ with $\oplus$ and $\odot$ defined as above is a field.

Proof. We will show each of the field properties. Let $a, b, c, d, g, f \in \mathbb{R}$. Notice that

$$[(a + bi) \oplus (c + di)] \oplus (f + gi) = [(a + c) + (b + d)i] \oplus (f + gi) = ((a + c) + f) + ((b + d) + g)i.$$

Therefore, by the associative property of addition on $\mathbb{R}$, we see that the associative property for addition on $\mathbb{C}$ also holds.

Similarly, we see that the commutative property for addition is also inherited from the real numbers.

To see that the associative property of multiplication holds, we observe the following:

$$
\begin{aligned}
[(a + bi) \odot (c + di)] \odot (g + fi) &= [(ac - bd) + (ad + bc)i] \odot (g + fi) \\
&= [g(ac - bd) - f(ad + bc)] + [g(ad + bc) + f(ac - bd)]i \\
&= (acg - bdg - adf - bcf) + (adg + bcg + acf - bdf)i \\
&= [(a(cg - df) - b(cf + dg)] + [(a(cf + dg) + b(cg - df)]i \\
&= (a + bi) \odot [(cg - df) + (cf + dg)i] \\
&= (a + bi) \odot [(c + di) \odot (g + fi)].
\end{aligned}
$$

To see that the distributive property of multiplication holds, we observe the following:

$$(a + bi) \odot [(c + di) \oplus (f + gi)] = (a + bi) \odot [(c + f) + (d + g)i]$$
$$= [a(c + f) - b(d + g)] + [a(d + g) + b(c + f)]i$$
$$= (ac + af - bd - bg) + (ad + ag + bc + bf)i$$
$$= (ac - bd) + (ad + bc)i + (af - bg) + (ag + bf)i$$
$$= (a + bi) \odot (c + di) \oplus (a + bi) \odot (f + gi).$$

Notice that $0 + 0i \in \mathbb{C}$ and $(a + bi) \oplus (0 + 0i) = a + bi$. Thus, $0 + 0i$ is the additive identity. Notice that $-a, -b \in \mathbb{R}$. Thus, $-a - bi \in \mathbb{C}$ and $(-a - bi) \oplus (a + bi) = 0 + 0i$. So there exist additive inverses in $\mathbb{C}$. Notice also that $1 + 0i \in \mathbb{C}$ and

$$(1 + 0i) \odot (a + bi) = a + bi.$$

Thus, $1 + 0i$ is the multiplicative identity in $\mathbb{C}$. Finally, notice that if a, b are not both zero and $\alpha = a^2 + b^2$ then $\frac{a}{\alpha} - \frac{b}{\alpha}i \in \mathbb{C}$. Also, notice that

$$(a + bi) \odot \left(\frac{a}{\alpha} - \frac{b}{\alpha}i \right) = \left(\frac{a^2}{\alpha} + \frac{b^2}{\alpha} \right) + \left(\frac{-ab}{\alpha} + \frac{ab}{\alpha} \right)i$$
$$= \frac{a^2 + b^2}{\alpha} = 1$$

Thus, $\frac{a}{\alpha} - \frac{b}{\alpha}i$ is the multiplicative inverse of $a + bi$.

Therefore, $(\mathbb{C}, \oplus, \odot)$ is a field. It is common practice to write $(\mathbb{C}, +, \cdot)$. $\qquad\square$

In no way have we exhausted the study of fields. A more in depth study of fields happens in Abstract Algebra courses.

D.1 Exercises

1. Show that $a \in Z_p, a \neq 0$ has a multiplicative inverse, thereby proving that $(Z_p, \oplus_p, \odot_p)$ is a field.
2. The set of complex numbers, $\mathbb{C}$, is a vector space over the field of complex numbers, with the standard definitions of addition and multiplication.
3. Determine whether or not the set of quaternions, Q defined below is a field.

$$Q = \{a + bi + cj + dk \mid a, b, c, d \in \mathbb{R}\}.$$

Here, i, j, k are defined as the numbers so that $i^2 = j^2 = k^2 = ijk = -1$ and their products are defined by the following multiplication table.

$\cdot$	i	j	k
i	-1	k	$-j$
j	$-k$	-1	i
k	j	$-i$	-1

Addition is defined in the usual sense and multiplication is defined as usual taking into consideration the table above.

Index

© Springer Nature Switzerland AG 2022
H. A. Moon et al., *Application-Inspired Linear Algebra*, Springer Undergraduate Texts
in Mathematics and Technology, https://doi.org/10.1007/978-3-030-86155-1

Printed in the United States
by Baker & Taylor Publisher Services